T0140172

Lecture Notes in Networks and Systems 625

The series "Lecture Notes in Networks and Systems" publishes the latest developments in Networks and Systems—quickly, informally and with high quality. Original research reported in proceedings and post-proceedings represents the core of LNNS.

Volumes published in LNNS embrace all aspects and subfields of, as well as new challenges in, Networks and Systems.

The series contains proceedings and edited volumes in systems and networks, spanning the areas of Cyber-Physical Systems, Autonomous Systems, Sensor Networks, Control Systems, Energy Systems, Automotive Systems, Biological Systems, Vehicular Networking and Connected Vehicles, Aerospace Systems, Automation, Manufacturing, Smart Grids, Nonlinear Systems, Power Systems, Robotics, Social Systems, Economic Systems and other. Of particular value to both the contributors and the readership are the short publication timeframe and the world-wide distribution and exposure which enable both a wide and rapid dissemination of research output.

The series covers the theory, applications, and perspectives on the state of the art and future developments relevant to systems and networks, decision making, control, complex processes and related areas, as embedded in the fields of interdisciplinary and applied sciences, engineering, computer science, physics, economics, social, and life sciences, as well as the paradigms and methodologies behind them.

Indexed by SCOPUS, INSPEC, WTI Frankfurt eG, zbMATH, SCImago.

All books published in the series are submitted for consideration in Web of Science.

For proposals from Asia please contact Aninda Bose (aninda.bose@springer.com).

Mohamed Lazaar · El Mokhtar En-Naimi ·
Abdelhamid Zouhair · Mohammed Al Achhab ·
Oussama Mahboub
Editors

Proceedings of the 6th International Conference on Big Data and Internet of Things

Editors
Mohamed Lazaar
ENSIAS
Mohammed V University
Rabat, Morocco

El Mokhtar En-Naimi
FST
Abdelmalek Essaâdi University
Tangier, Morocco

Abdelhamid Zouhair
FST
Abdelmalek Essaâdi University
Tangier, Morocco

Mohammed Al Achhab
ENSA
Abdelmalek Essaâdi University
Tetuan, Morocco

Oussama Mahboub
ENSA
Abdelmalek Essaadi University
Tetouan, Morocco

ISSN 2367-3370 ISSN 2367-3389 (electronic)
Lecture Notes in Networks and Systems
ISBN 978-3-031-28386-4 ISBN 978-3-031-28387-1 (eBook)
https://doi.org/10.1007/978-3-031-28387-1

This Springer imprint is published by the registered company Springer Nature Switzerland AG
The registered company address is: Gewerbestrasse 11, 6330 Cham, Switzerland

Preface

We are happy to present you this book, Big Data and Internet of Things, which is a collection of papers that were presented at the 6th International Conference on Big Data Cloud and Internet of Things, BDIoT 2022. The conference took place on October 25–27, 2022, Tangier—Morocco.

The book consisted of 49 chapters, which correspond to the four major areas that are covered during the conference, namely Big Data and Cloud Computing, CyberSecurity, Machine Learning, Deep Learning, E-Learning, Internet of Things, Information System and Natural Language Processing.

Every year BDIoT attracted researchers from all over the world, and this year was not an exception—we received 98 submissions from seven countries. More importantly, there were participants from many countries, which indicates that the conference is truly gaining more and more international recognition as it brought together a vast number of specialists who represented the aforementioned fields and share information about their newest projects. Since we strived to make the conference presentations and proceedings of the highest quality possible, we only accepted papers that presented the results of various investigations directed to the discovery of new scientific knowledge in the area of Big Data, IoT and their applications. All the papers were reviewed and selected by the Program Committee, which comprised 96 reviewers from over 58 academic institutions. As usual, each submission was reviewed following a double process by at least two reviewers. When necessary, some of the papers were reviewed by three or four reviewers. Our deepest thanks and appreciation go to all the reviewers for devoting their precious time to produce truly through reviews and feedback to the authors.

Mohamed Lazaar
El Mokhtar En-Naimi
Abdelhamid Zouhair
Mohammed Al Achhab
Mahboub Oussama

Contents

Computational Intelligence & Big Data

Computational Intelligence & Big Data

Construction of a Training Device in Pedagogical Engineering from the "My Scenari" Model

Driss Elomari[1](✉), Younès Daife[1], Najemeddin Soughati[2], Elassad Elharbaoui[3], Youssef Elyaacoubi[1], Dalal Doukha[2], and Ouidad Elmaamri[2]

[1] Sidi Mohamed Ben Abdellah University, Fez, Morocco
driss.elomari@usmba.ac.ma
[2] University Ibn Tofail, Kenitra, Morocco
[3] University of Carthage, Tunis, Tunisia

Abstract. This work intends to present a design work of a research design which aims to experiment with a training device in pedagogical engineering. Thanks to the "Scenari" suite, the training on the implementation of a pedagogical project made it possible to test the validity of the hypothesis that the instrumental design of a training in pedagogical engineering develops the professional skills of teachers. During the duration of the workshops, participatory observation raised several factors of performance and failure.

Keywords: Pedagogical Engineering · Training Workshop · Design Research · Training Device · Participant Observation

1 Theoretical and Conceptual Framework

1.1 Theoretical Frame

This work presents the result of an online device experiment. It aims to test the validity of the question relating to the professional development of teachers through training in pedagogical engineering [1]. However, the question has been if a general and pedagogical competence such as resource design can become an issue for teacher training in terms of initial training and in-service training [2]. It should be recalled that the initial and in-service training of teachers deploys a training module that aims to learn about the use of technology rather than its use in teaching. Based on this, our work has been to verify whether the field of pedagogical engineering cannot become a starting point to explore the prospects of a possible digital transformation. In practice, it was necessary to set up training objectives on a documentary level which specify the pedagogical issues [3]. Our specifications have been developed to describe this training path, while indicating the types of activity to be carried out in the mastery of the different knowhow. This description is based on the notion of project as a general framework of our approach [4], and on the mobilization of the different concepts relating to a digital project before moving on to the content of the training but in general the design of a training based on criteria that require the fact of adapting to various audiences and contexts [5]. Therefore, it

M. Lazaar et al. (Eds.): BDIoT 2022, LNNS 625, pp. 3–9, 2023.
https://doi.org/10.1007/978-3-031-28387-1_1

was necessary to reflect on a techno-pedagogical model that respects both the principle of digital transformation and the professional development of teachers [6]. The position of the researcher or participant determines the construction of an inverted model according to the objectives of the actors (Fig. 1):

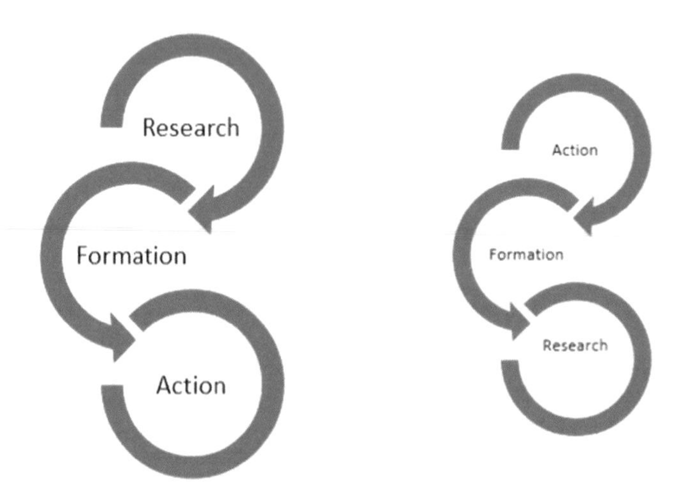

Fig. 1. Construction process of the experimental instrument

From these two models, we can identify a process that falls within the field of intervention research. This is a course that interests a trainer vis-à-vis the participants, but also the other way around. The involvement of these actors has favored the development of this research based on the data collected.

1.1.1 The Framework of Training

In order to define translatable objectives, the training activity is called upon to choose the pedagogical methods relating to the production of practical aspects of training. In this context, a design model had to be followed. The use of the ADDIE model seemed to us to be the appropriate substantive structure for the project. This choice is justified by its technological and pedagogical purpose, in contrast to other models which sometimes focus on technical aspects. Teachers are also found in this type of model since they manage to compare it with the didactic models of their disciplines. Indeed, depressor ADDIE overlaps with the various didactic purposes of languages to consider the training needs at the heart of the system. The development of training resources has been the subject of collaborative and cooperative work. The objective of this modality of cooperation has been to divide the work by mobilizing group interaction. The exchanges between the different teams made it possible to identify the skills built during the training course. The scenario adopted favored the involvement of the actors of this project, since the pedagogical teams shared and evaluated the part of their peer. This methodology of the project facilitated the design of the different phases of the implementation while providing us with qualitative data on the nature of the exchanges and relationships maintained between teachers and their peers [7].

The transition to the digital environment required an adaptation of these elements to set up a training strategy [8]. This approach must respect a specific scenario that responds to the challenge of research, namely, to verify whether a training in engineering training is a factor in the development of a teacher's professional skills [9]. This summary table presents the different correlations between the "Opal" items and the pedagogical engineering terminology:

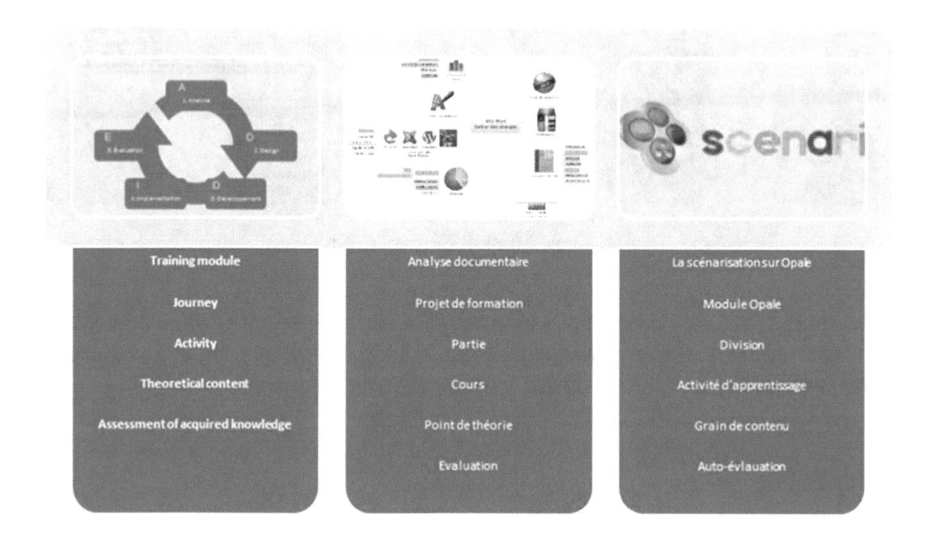

Table 1. Comparison between "Opale" terminology and didactic language [10]

Items	Literature review	Scriptwriting on Opale
Training module	Training project	Opal Module
Route	Part	Division
Activity	Course	Learning activity
Theoretical content	Point of theory	Grain of content
Assessment of prior learning	Evaluation	Auto-évlauation

The transition to the scripting of the educational content was dictated according to our specifications [10]. The media coverage of training through videos required the use of "Webmedia". However, the practicality caught the attention of the participants by its practical functionality in terms of synchronized scripting between audio and video.

1.1.2 Implementation of the Project on "Opale"

The "scenarios" suite offers different documentary solutions from a pedagogical scripting perspective [11]. We adopted it to associate techno-pedagogical with our research needs. The migration from the documentary phase to the script phase required basic training on "Opal" tools [11]. It should be noted that the development of the mastery of this tool requires a cognitive load insofar as each activity calls for notions of pedagogical engineering. However, a technical challenge looms as to the user experience between participants. Foreach sequence, a group label was detected from the "ADDIE" process (Fig. 2):

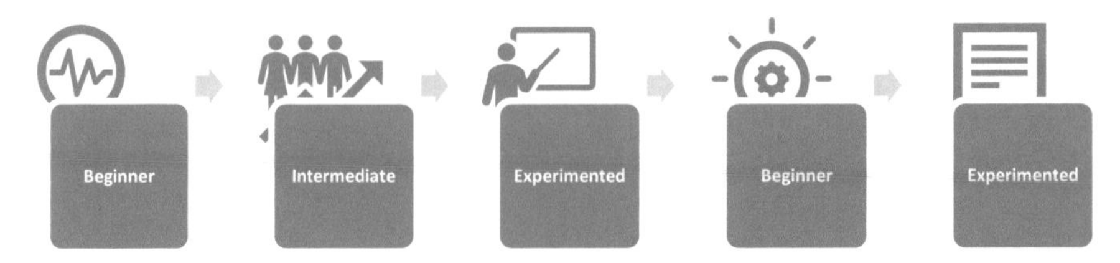

Fig. 2. A different Nivel of training

According to a peer review, it was necessary for the group to have evaluated a work by assigning a profile. In this way, the parts of the training have been categorized according to three levels: Beginner, Intermediate, and Experienced. The particularity of our use of "Opal 3.9" lies in its multiplicity of uses both documentary and technical, during our drafting of specifications, during the pedagogical scripting, and finally after the online implementation:

Table 2. The cycle of implementation the training

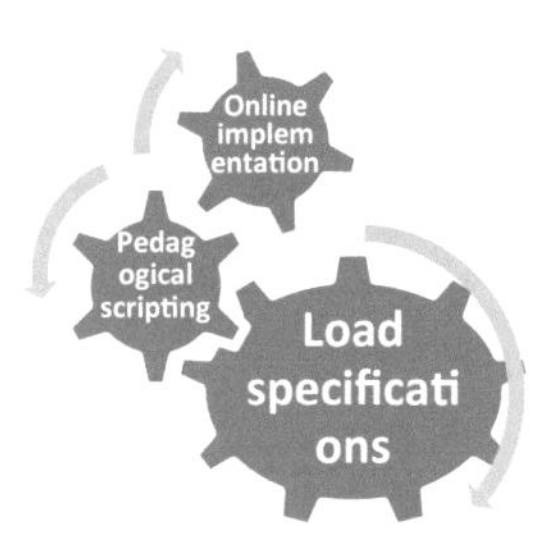

Our documentary model is based on an engineering approach, which has led us to structure the training around three components, namely a first part aimed at understanding and analyzing the operational needs of the project. In the same way, it will be a question of carrying out the design of the research to develop, implement, and evaluatethe content. Moreover, the very principle of documentary production at "Opale" offersa diversity of choices in the generation of output formats (Fig. 3):

This freedom of format has favored the transition from a remote mode to a face-toface mode. This action strategy was the basis for the construction of training references, which is only one version of training engineering training. For the teachers, the presence of an object to be reproduced was the contribution of the work of collaboration and cooperation

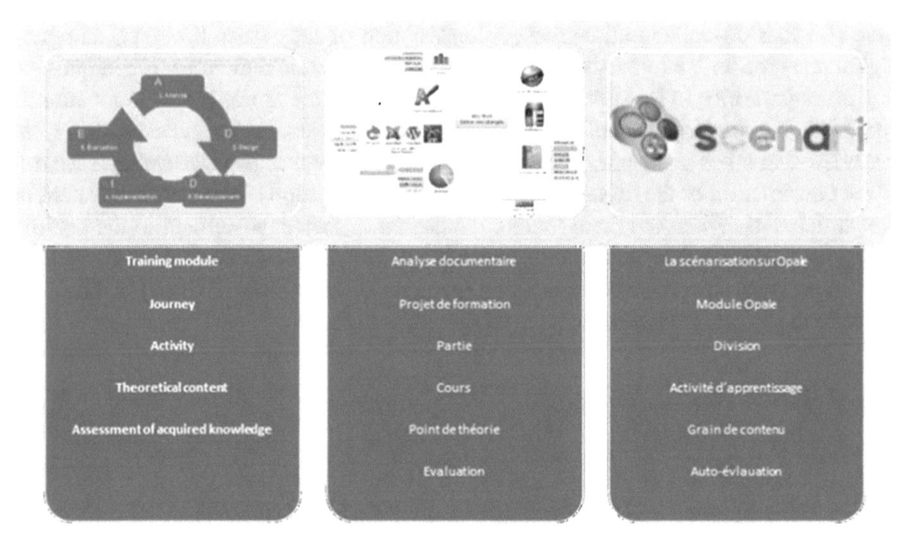

Fig. 3. The design phases of the training system

which undoubtedly made it possible to extend another phase of refinement of the whole system. In addition to its socio-constructivist aspect, the mobilization around a resource has given rise to a dynamic of pooling knowledge and know-how on raw data, which have been refined using scripting on Opal (Fig. 4):

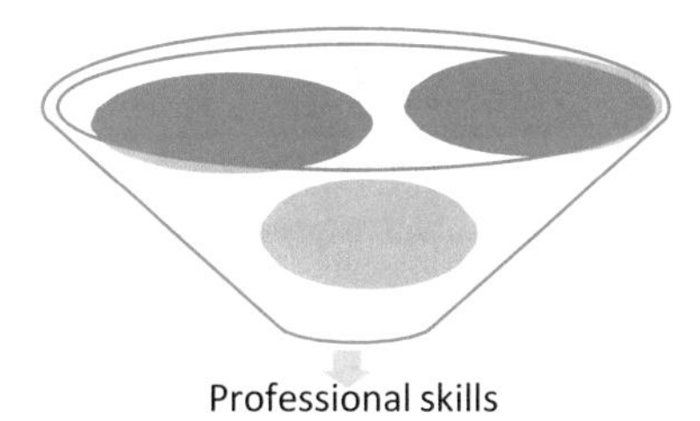

Fig. 4. The interaction of training cycle

2 Results

The finalization of the project was a moment to gather data on the valorization of project-based learning initiated through a program around pedagogical engineering. On the documentary level, it must be understood that the online launch of the module was a starting point for the experimentation of the project. Many comments were made on the format and design of the content. As designers, the teachers involved in this project did not express their opinion on the effectiveness of the course in pedagogical engineering, but on their experience in the design and use of the "Opal" tool. Their involvement gave rise to a co-collaboration and co-construction of the research object, namely a digital resource with a formative aim. The mobilization around the resource in question was an opportunity to test the degree of acceptance of technology, and innovation in general in the development of teachers' professional skills. At the research level, the interview that takes place after the implementation phase was used to evaluate the role of the construction of the digital resource in the appropriation of the training system by the participants. We asked them to note the degree of their involvement in the positive situation. The interview with the teachers is guided by questions that try to identify their impression of collaborative work and requires a rating (from 1 to 10) at the end to measure the effort in the development of the project (Fig. 5):

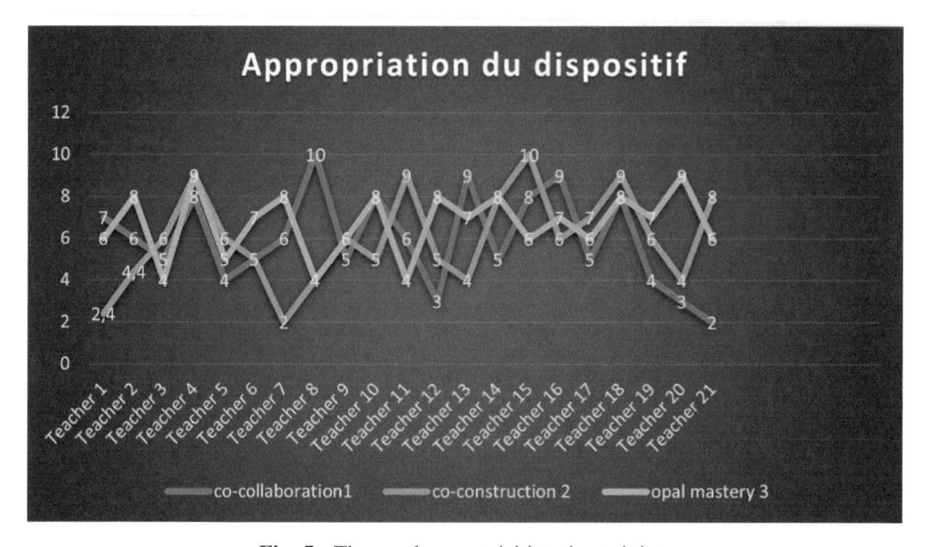

Fig. 5. The result at acquisition the training

This final survey was able to record a balance in terms of the contribution of the designers. Despite the informal nature of the s ratings, nevertheless it remains a plausible thermometer to the final product. Indeed, the participation of the teaching group in the peer evaluation remains sufficient, since it is difficult to quantify the personal contribution.

3 Conclusion

In conclusion, the development of a digital resource required a choice of design instrument for the implementation of the training. The choice of the "Opal" track made it possible to generate the different phases of the project according to an "ADDIE" Model [12]. The advantage of this type of structure is that it is possible to work on the training project from an inverted process. Indeed, through our development of specifications, we obtained a preliminary version of our training module. After the construction phase of a design, the generation of different formats favored the online release of the final version of the training. According to our research objectives, this design was able to reveal a questioning of the modalities of collaboration and cooperation between teachers. This goes without saying that the notion of digital resource is likely to be the subject of other debates, on the process of designing the design of this resource among teachers, and how this process is able to create synergy around teacher training.

References

1. Class, B., Schneider, D.: La Recherche Design en Education : vers une nouvelle approche? Frantice.net. **7**, 5–16 (2013)
2. Adler, P., Adler, P.: Membership Roles in Field Research. SAGE Publications, Inc., 2455 Teller Road, Thousand Oaks California 91320 United States of America (1987). https://doi.org/10.4135/9781412984973
3. Mishra, P., Koehler, M.J.: Technological pedagogical content knowledge: a framework for teacher knowledge. Teach. Coll. Rec. **108**, 1017–1054 (2006). https://doi.org/10.1111/j.1467-9620.2006.00684.x
4. Sorden, D.S.: A cognitive approach to instructional design for multimedia learning. Informing Sci. Int. J. Emerg. Transdiscipl. **8**, 263–279 (2005). https://doi.org/10.28945/498
5. Skaf-Molli, H., Ignat, C., Rahhal, C., Molli, P.: New Work Modes For Collaborative Writing. Presented at the July (2007)
6. Baran, E., Canbazoglu Bilici, S., Albayrak Sari, A., Tondeur, J.: Investigating the impact of teacher education strategies on preservice teachers' TPACK: The impact of teacher educaion strategies on TPACK. Br. J. Educ. Technol. **50**, 357–370 (2019). https://doi.org/10.1111/bjet.12565
7. Soughati, N., Maamri, E.O.: La e-formation par les pairs ou l'accompagnement des adultes par les jeunes; Retour d'expérience. Presented at the March (2018)
8. Pogent, F., Albero, B., Guérin, J.: Professional and personal transformations in a hybrid training situation: the case of a schoolteacher with the mixed learning system Mgistère. Distances Médiations Savoirs (2019)
9. Gauvreau, S.A., Hurst, D., Cleveland-Innes, M., Hawranik, P.: Online professional skills workshops: perspectives from distance education graduate students. Int. Rev. Res. Open Distrib. Learn. **17** (2016). https://doi.org/10.19173/irrodl.v17i5.2024
10. Koponen, I.T.: Systemic view of learning scientific concepts: A description in terms of directed graph model. Complexity **19**, 27–37 (2014). https://doi.org/10.1002/cplx.21474
11. Antonenko, P.D.: The instrumental value of conceptual frameworks in educational technology research. Educ. Technol. Res. Dev. **63**, 53–71 (2015). https://doi.org/10.1007/s11423-014-9363-4
12. Burton, R., et al.: Vers une typologie des dispositifs hybrides de formation en enseignement supérieur. Distances Savoirs. **9**, 69–96 (2011). https://doi.org/10.3166/DS.9.69-96

Machine Learning-Based Intrusion Detection System: Review and Taxonomy

Omar Chaieb[1](✉), Nabil Kannouf[2], Rachida Amjoun[3], and Mohammed Benabdellah[1]

[1] ACSA Laboratory, FSO, UMP Oujda, Oujda, Morocco
omar.chaieb.men@gmail.com
[2] LSA laboratory, SDIC team, ENSAH, Hoceima, Morocco
[3] Sheridan Institute of Technology and Advanced Learning, Oakville, ON, Canada

Abstract. The widespread use of smart devices (smartphones, smart locks, etc.), and the rush by companies to digitize their resources, as well as the lack of awareness among users about the dangers around us while surfing the web, have left the world full of vulnerabilities that have created a fertile ground for hackers to try all kinds of hard-to-detect attack techniques. On the other side, cybersecurity researchers have redoubled their efforts to develop effective intrusion detection systems, capable of detecting not only well-known attacks but also ones that have never been seen before. This paper will highlight the different stages of designing Network Intrusion Detection Systems (NIDS) using Machine Learning (ML) techniques, including, benchmark datasets, feature reduction, hyperparameter optimization, detection methods, and evaluation metrics. In addition, we will conduct a detailed study of recent articles (2018–2022) in which we will discuss each work's strengths and shortcomings. Finally, we will use the drawbacks of proposed approaches to list the difficulties that researchers can face while developing ML-based NIDS.

Keywords: Intrusion Detection System · Machine Learning · Anomaly Detection · Feature Reduction

1 Introduction

As technologies advance and spread, an increasing number of devices are linked to the internet. Thus, preventing malicious behavior became the primary concern of network security experts, since failure to detect malicious behavior may compromise the confidentiality, integrity, or availability of data [1]. This failure may cause enormous damages to any living organism, including system paralysis, privacy breaches, exposure to blackmail (ransomware), and reputation damage.

Intrusion Detection System (IDS) is a security solution that permanently scans traffic to identify unusual behavior that breaches security policies [2]. IDS can be classified into several classes. On the one hand, according to the data source; we divide IDS into network IDS, that use network traffic as data source, and Host IDS that use logs of the operating system or applications. On the other hand, depending on the detection method,

M. Lazaar et al. (Eds.): BDIoT 2022, LNNS 625, pp. 10–21, 2023.
https://doi.org/10.1007/978-3-031-28387-1_2

we divide IDS into knowledge-based and anomaly-based IDS. The first approach looks for signatures of known attacks in a signature database, which is very effective against known attacks. The second approach describes the user's ordinary behavior and treats any deviation from this behavior as an intrusion, which is extremely efficient against new attacks.

Over the past few years, several researchers have been working on improving the effectiveness of IDSs, by using Machine Learning (ML) algorithms, which has successfully solved complex problems in various fields such as image recognition, traffic prediction, and healthcare. Furthermore, ML algorithms need to learn from historical data, using benchmark datasets which are good basis to train ML algorithms, and evaluate the effectiveness of proposed NIDSs [3]. However, because of the massive volume of data contained datasets, ML-based IDS does not perform well when predicting if an event is a normal traffic or not, that is why feature-reduction (FR) techniques are employed to minimize data dimensionality, by eliminating unnecessary features. Moreover, we can classify ML techniques into supervised learning, which tries to make predictions using pre-classified data samples, unsupervised learning, which utilizes unlabeled data to train the ML algorithm, and Semi- Supervised learning that uses a non-totally labeled data. Once the machine-learning algorithm is trained and tested, evaluation measurements are applied to define its efficiency.

The remainder of the paper is organized as follows: In the second section we discuss the general concept of IDS, as well as a classification based on several criteria will be given. In Sect. 3, we will introduce the different datasets used in recent research papers to validate IDSs, while discussing characteristics of each. In addition, we will discuss various preprocessing and feature reduction operations that the dataset goes through before ML algorithms can use it. Section 4 will be devoted to present the latest research based on ML methods to improve NIDS. We will also discuss the benchmark datasets, Machine Learning/optimization methods, and performance metrics used in each paper, while analyzing each work's strengths and shortcomings.

2 General Concept of IDS

In this section, we outline the major components of IDS before classifying it, based first on data source, then on the detection method.

2.1 Intrusion Detection System (IDS)

Intrusion is considered as any illegal behavior which may harm the information system [4]. Thus, any behavior which is intended to exploit a system vulnerability to compromise the confidentiality, integrity, and availability is regarded as intrusion. For example, an action that gathers data from the system to use it after for malicious purposes is considered as an intrusion.

An IDS is a software application or a device, that monitors network traffic to detect anomalous behaviors, then informs administrators when such behavior occurs, by generating reports or triggering alarm. The primary propose of IDS is to identify harmful activities that are invisible to a traditional firewall or an antivirus [4]. Figure 1 presents an illustration of NDIS implementation.

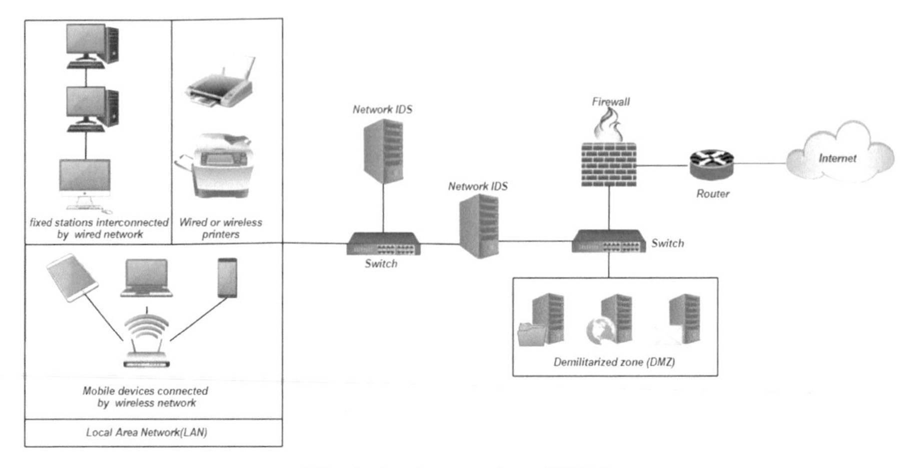

Fig. 1. implementation of NIDS

2.2 Classification of Intrusion Detection System

Different methods of IDS classification have been proposed [2, 4]. Here, we choose to classify them into two categories: the first is based on the input data source and the second is based on detection methods. The Fig. 2 illustrates the suggested categorization of IDS.

Based on Data Sources. Based on source of data, IDS are divided into two types: Host IDS (HIDS) and Network IDS (NIDS). The first uses data from logs of operating system or application and is effective against insider attacks but consumes host resources; the second uses network traffic, and is simple to use, but requires a specialized equipment.

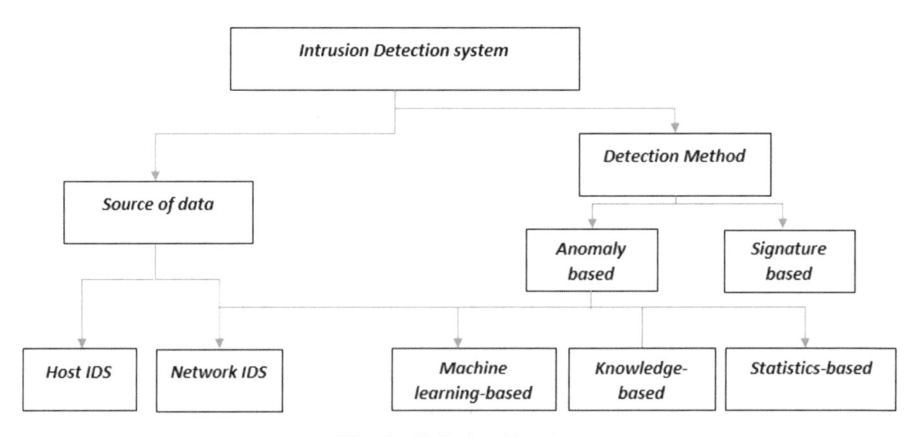

Fig. 2. IDS classification

Based on Detection Methods. We can categorize IDS depending on the techniques used to design it: Anomaly-based IDS, knowledge-based IDS, and Hybrid-based IDS.

The knowledge-based IDS or Signature-based detection relies on a database of signatures of known attacks, if an event corresponds to one of the signatures that exist in the database, network administrators will be notified.

The anomaly-based IDS can be categorized in three main groups: first, statistics-based IDS that creates a distribution model for normal activity profiles, then detects low frequency occurrences, and marks them as probable intrusion. Second, knowledge-based IDS that relies on creating a knowledge base representing the legal traffic pattern, then traffic that deviates from this standard profile is regarded as an intrusion [4]. Third, a ML based IDS, which consists of modeling during a learning period the normal behavior, then considers as suspect in detection phase, any deviation from this behavior.

The hybrid-based IDS that uses benefits of the two methodologies to accomplish better outcomes.

3 Benchmark Datasets, Features Reduction, and Evaluation Metrics

This section will first introduce different benchmark datasets that were used on recent research, to evaluate the proposed methods. It then details the different steps a dataset goes through before being submitted to the ML algorithm. Furthermore, we will discuss the various metrics for the model performance evaluation. Figure 3 summarizes the various steps of ML based IDS design from the preprocessing phase to the performance evaluation phase.

3.1 Intrusion Detection Datasets

The quantity and quality of the data shows how powerful the model is, however, the paid IDS datasets are expensive and not easy to get due to confidentiality concerns. That's why, to evaluate the efficiency of the ML algorithm, we have to choose between the few publicly available IDS datasets [3].

KDD 99: Dataset contains 41 different features, with 5 million data samples and more than 20 different attacks which can be classified into four categories of attacks: Denial of Service Attacks (DOS), Privilege Escalation, probing [3]. In order test the effectiveness of their models, several research papers used these datasets [5–7].

NSL_KDD: Dataset was made by removing duplicates from KDD 99 dataset. It contains 150 000 data samples with 41 attributes and the same classes of attack. [8–11] used this dataset to evaluate the effectiveness of their algorithms.

AWID: Dataset focuses on 802.11 networks with 156 features and 16 different attacks against the 802.11 network. This can be grouped into three attack categories namely flooding, injection, and impersonation. It also includes 1795574 data samples for training, and 575642 for testing. This dataset was used in [11] as evaluation dataset.

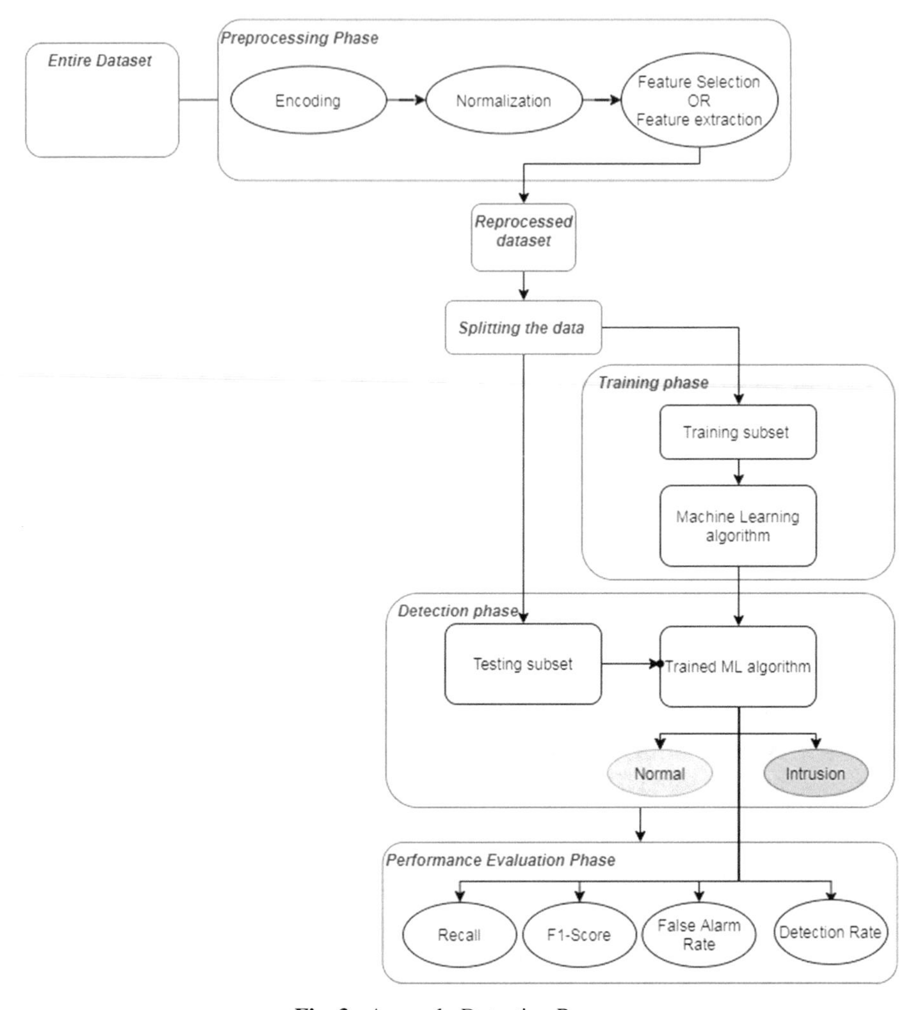

Fig. 3. Anomaly Detection Process

CIC-IDS-2017: Dataset contains 15 different classes, 80 features, and 7 categories of attacks. [9, 11, 12] used this dataset to validate their model.

CSE-CIC-IDS-2018: Dataset contains 18 classes, 80 features and has the same seven attack types as the previous dataset. This dataset was used in [12], to evaluate four unsupervised algorithms.

UKM-IDS20: This dataset was created by [13] and contains 46 features with four types of attack, namely ARP poisoning, Dos attacks, Probing attacks, and exploits. The quality of this dataset was compared against well-known datasets, which are KDD99 and UNSW-NB15 then the results show a higher complexity and relevancy of UKM-IDS 2018. This dataset is publicly available.

3.2 Feature Selection

As previously stated, ML algorithms are known for learning from the historical data given to them, so in case of the data is of poor quality, the resulting algorithm is inherently bad. For this reason, it is crucial to prepare the data before using it, so it has to be cleaned, filtered, and normalized. These procedures are called preprocessing.

In addition, choosing the best features plays an important role to ensure the performance of ML algorithms. Besides it is known that large datasets are known to require more memory and computational costs, and in the worst case it leads to overfitting [14]. Therefore, before moving to the training phase, it is important to keep only useful features for model development, and this is where the use of feature selection techniques, comes in.

Feature selection (FS) is the technique that involves finding the most important features for ML algorithm, while removing the irrelevant features. We can divide this technique into filter and wrapper methods. The first chooses features based on performance assessment regardless of the data modeling technique employed, and it is known to be faster in term of time processing. The second subtracts the best features by measuring their performance using the learning algorithm, and it is better at finding the most efficient features. In the literature, there are several research papers using FS techniques to minimize the dimensionality of datasets, hence enhancing the performance of their proposed NIDS model.

In [15], a wrapper FS method based on SVM classifier, and a filter FS ranker-based algorithm, were used, to reduce the number of features in the NSL-KDD dataset. The wrapper FS method was successful in selecting 17 features from 41, while the filter FS method achieved only 35 features from 41.

In [16], Information gain based on attribute evaluation was used on NSL-KDD dataset, to select 11 features from 41 features.

In [17] a Differential Evolution (DE) Wrapper FS technique was proposed to minimize the number of the 41 NSL-KDD dataset features, and evaluated the selected features using Extreme Learning Machine(ELM) algorithm.

All results showed that, using FS techniques improves the efficiency of the proposed models.

3.3 Evaluation Metrics

This part of paper will be devoted to the evaluation metrics that have been most widely employed in recent research to quantify the performance of the NIDS, in other words, how well the model has generalized to unseen data.

Confusion matrix: Most evaluation metrics are based on several attributes included in two dimensional matrix namely confusion matrix, which includes information about known truth and what NIDS predicted [2]. These metrics are defined as follows:

True Positive Rate (TPR): The data samples correctly identified as an intrusion by NIDS.

False Positive Rate (FPR): The normal event identified as intrusion by NIDS.

True Negative Rate (FNR): The data samples correctly identified as normal event by NIDS.

False Negative Rate (FNR): The data samples that are intrusions and identified as normal events by NIDS.

In addition to the confusion matrix method, which is described in Table 1, other methods are used to better evaluate the performance of ML models, such as F1-score, Accuracy, and precision.

Table 1. Confusion matrix

		Actual	
		Intrusion	Normal
Predicted	**Intrusion**	True Positive	False Positive
	Normal	False Negative	True Negative

4 The ML-Based NIDS

Using ML to improve the performance of NIDS has lately received a lot of attention. We may describe machine learning as the process of developing a prediction system for future data using historical data.

In this part, we review the recent research that used ML methods to increase NIDS efficiency. We will also discuss optimization functions, benchmark datasets, and metrics used to evaluate each algorithm. The ML detection methods that will be discuss in this section are divided into three different sub-sections: supervised, unsupervised, and ensemble detection methods.

4.1 Supervised Learning Methods

This type of learning technique relies on pre-classified data samples. Several supervised ML algorithms have been proposed in the literature, to achieve effective NIDS, such as Naïve Bayes, Decision Tree (DT), Artificial Neural Network (ANN), Support vector machine (SVM). However, here we choose to review the most used ones.

Support Vector Machine (SVM). SVM is a supervised learning algorithm that can handle regression and classification problems. The idea behind this method is to build an optimal hyperplane that divides all training data samples into two classes, while maximizing the margin between this hyperplane and the support vectors. This hyperplane is afterward used to categorize unlabeled data samples during the testing phase.

Moreover, SVM classifier can classify linear and non-linear data. In case of linear data, and given data input xi (i = 1 … M) where, M is the number of data samples, the hyperplane that separate these data is defined as:

$$f(x) = w^T x + b \sum_{j=1}^{M} w_j x_j + b = 0 \quad (1)$$

where w is M-dimensional vector and b a scalar.

In case of non-linear data, this classifier employs kernel functions meant to move data samples to a higher dimension to better categorize them [14], such as Polynomial, Gaussian, Sigmoid, Laplace Radial Basis Function(RBF).

In addition, one of the key tasks of SVM modeling is to pick the best values for hyper-parameters, to increase the performance of this classifier [18]. To do so, several optimization techniques are used such as, Genetic Algorithms [5, 19], Particle Swarm Optimization (PSO) [20], and Binary Gravitational Search Algorithm (BGSA) [8].

Furthermore, several publications have appeared in recent years, using SVM classifiers to implement an efficient NIDS. In [5], an optimization technique basing on genetic algorithm was proposed, to minimize the dimensionality of the KDD CUP 99 dataset and to optimize SVM's hyper-parameters. The results showed that this method surpassed other similar methods by improving the detection rate, while decreasing false positive rate and processing time.

In [8], the SVM classifier was combined with Crossover based Binary Gravitational Search Algorithm (CBGSA) to optimize SVM's hyper-parameters and to select the best Features, and has been evaluated using NSL-KDD dataset. CBGSA outperformed other similar methods.

In [20] the authors proposed a framework which used the MULTI-PSO optimization technique to optimize the SVM's hyper-parameters. To validate their model, the authors used KDD 99 dataset, without feature reduction, and evaluated the framework using precision-recall rate as performance metrics. The results showed that MULTI-PSO technique outperformed similar optimization methods namely PSO, Grid Search, and Gradient descent.

Although the results of the above methods have shown superiority over other similar methods. Their drawbacks are that they are based on SVM classifier which is known for not giving exact reading when dealing with high volume of data. In addition, the authors had to use various datasets to evaluate the true effectiveness of their approaches.

Artificial Neural Network (ANN). ANN is a supervised ML algorithm made of nodes (called neurons), connections between them, and coefficients (weights) connected with the connection. Given parameters $\beta \in \mathbb{R}$ and $\theta \in \mathbb{R}^d$ and a real function $f : \mathbb{R} \to \mathbb{R}$, the artificial neuron $n_{\beta,\theta,f}$ is the function:

$n\beta, \theta, f \colon \mathbb{R}^d \to \mathbb{R}$;

$$x \to f\left[\beta + \theta^T x\right] = f\left[\beta + \sum_{j=1}^{d} \theta_j x_j\right]. \quad (2)$$

The weight $\theta_1, \ldots, \theta_d$ are the neuron's sensitivities to the different inputs, the bias β is the neuron's overall sensitivity, and the function f is known as the activation function.

The artificial neurons can be added and concatenated into new functions $\mathbb{R}^{d} \rightarrow \mathbb{R}$: given parameters $\zeta \in \mathbb{R}$, $\gamma \in \mathbb{R}w$ *and* $\beta 1 \ldots \beta w \in \mathbb{R}$, $\theta 1 \ldots \theta w \in \mathbb{R}d$ and activation functions $g, f1 \ldots fw$: $\mathbb{R} \rightarrow \mathbb{R}$,

$$\mathbb{R}^{d} \rightarrow \mathbb{R}$$

$$x \rightarrow n_{\zeta,\gamma},_{g}\begin{bmatrix} n_{\beta^{1},\theta^{1}},_{f^{1}}[x] \\ \cdot \\ \cdot \\ \cdot \\ n_{\beta^{w},\theta^{w}},_{f^{w}}[x] \end{bmatrix} = g\left[\zeta + \sum_{k=1}^{w} \gamma k f^{k}\left[\beta^{k} + \sum_{j=1}^{d}\left(\theta^{k}\right)_{j} x_{j}\right]\right] \quad (3)$$

These functions are combined, resulting in the complex process called artificial neural networks.

This approach has been employed in various research articles in recent years to solve classification problems such as intrusion detection, because of its self-learning capability. However, the major drawback of artificial neural network is the long processing time needed for big neural network. The following discussion focuses on recent research, using ANNs to improve NIDSs.

In [21], Multi-Layer Perceptron (MLP) with two hidden layers, each with 30 hidden neurons was used to implement the NIDS. This optimal ANN structure was found using grid search technique, then a repeated 10-fold Cross-Validation was applied to ensure that the classifier generalizes well to unseen data. In addition, a typical set of normal network traffic data, and malicious shell code files, was used as a validation dataset. Results showed that this proposed method improved accuracy, sensitivity, and precision rate.

In order to overcame the ANN's time consumption issue, [6] proposed a novel NIDS called PSO-FLN, which is an enhanced Fast Learning Network (FLN) based on PSO. The authors showed that using PSO to optimize FLN parameters outperforms other FLN optimization techniques, such as, Genetic Algorithm (GA), and Harmony Search Optimization (HSO). Furthermore, they found that increasing the number of hidden neurons improves the accuracy of the algorithm. However, the only disadvantage of this model is that it has encountered difficulties in detecting certain attacks classes due to the limited amount of available training data.

Furthermore, with the intention to evaluate how hyper-parameters affect effectiveness of ANN-based NIDS, [9] tested various ANN topologies using NSL-KDD and CICIDS2017 datasets. The authors have proved that little changes in hyper-parameters setting could have a considerable impact on the accuracy of the ANN topology of both benchmark datasets.

Unsupervised Learning Methods. Unsupervised learning is a ML technique that learns from unlabeled data, which can become extremely useful in cases when annotated data are difficult to find, especially in cybersecurity field. This type of learning uses a grouping technique called clustering which involves dividing data into groups based on their similarity, and considers the groups with small size as intrusion, since the occurrence of normal traffic is known to be quite superior compared to malicious traffic.

To build an efficient multi-class NIDS classifier, [7] proposed a method combining improved K-Means with SVM and Extreme Learning Machine(ELM). The purpose of using updated K-Means was to minimize the dimensionality of the dataset, which helps the SVM and ELM classifiers to achieve better accuracy while reducing the consuming time. To validate their model, they used the KDD 99 dataset, and grid search technique to optimize the SVM parameters. Although the results indicated that the suggested model surpassed previous similar methods, the false positive rate remained high, and it was difficult to detect certain attack classes, due to their similarity to normal instances.

In [10], they proposed an unsupervised IDS method, combining Sub-Space Clustering (SCC) and One-Class SVM (OCSVM) using NSL-KDD dataset, then they applied the F-test as a feature selection technique. Results showed that the proposed solution outperformed other solutions such as k-means, DBSCAN, and SCC-EA in terms of DR and FAR. However, the proposed solution made more processing time than other approaches.

Four unsupervised ML algorithms are proposed in [12], namely autoencoder, one-class SVM, isolation forest, and principal component analysis. To validate these models, they used CIC-IDS-2017 and CSE-CIC-IDS-2018 datasets. The authors evaluated the performance of each method by calculating recall, precision, f1-score, accuracy. Furthermore, to evaluate the generalization strength of the earlier algorithms, they used an evaluation strategy named inter-dataset. This strategy uses two different, but related datasets previously cited, and rather than defining the final model's performance on the test set of the same dataset, they used the test set of the second related dataset. The results show that all four algorithms achieved high classification scores, but the scores dropped when using the inter-dataset evaluation strategy.

Ensemble Learning. This strategy entails training numerous ML models, and then considers their entire predictions. In other words, the ensemble learning technique consists of building a strong classifier by combining several weak classifiers while selecting the best predictions using a voting algorithm [2].

In [16], an ensemble learning method used multiple classifiers namely IBK(KNN), Random Tree, REP Tree, j48graft, and Random Forest classifier on NSL_KDD dataset. Experimental results showed that using multiple classifiers outperforms using a single classifier by improving accuracy, while reducing processing time, and false positive rate.

In [11], they proposed a heuristic CFS-BA based method to reduce the dimensionality of several datasets. Then they combined the decisions of several classifiers namely C4.5, Random Forest, and Forest PA, by using a voting classifier. All results showed that the proposed ensemble learning techniques performs better compared to individual classification approaches, or similar ensemble learning techniques.

5 Discussion and Conclusion

This paper provides a clear investigation of various aspects involving the design of ML-based NIDS, starting with a description of the validation dataset, and ending with a presentation of the various NIDS evaluation metrics. Despite recent research efforts to improve the effectiveness of NIDS, by using ML techniques, there is no way to ensure that the proposed solutions are able to identify new attacks in real-world environment, given the lack of up-to-date datasets with a significant amount of recent attacks. Furthermore, the detection rate still needs to be improved, especially for attack types that are difficult to detect due to their resemblance to normal traffic; Adding to that, false alarm rate and processing time still need to be reduced. Finally, in this paper, we focused on ML algorithms, while excluding deep learning methods is due to the belief that there is still work to be done in enhancing NIDS basing on ML techniques. A detailed review on the use of Deep-Learning techniques in NIDS improvement will be considered soon. As a future work, we intend to design a NIDS based on ML able to overcome the above challenges.

References

1. Kannouf, N., Labbi, M., Benabdellah, M., Azizi, A.: Security of information exchange between readers and tags. In: Security and Privacy in Smart Sensor Networks, pp. 368–396. IGI Global (2018)
2. Ahmad, Z., Shahid Khan, A., Wai Shiang, C., Abdullah, J., Ahmad, F.: Network intrusion detection system: a systematic study of machine learning and deep learning approaches. Trans. Emerg. Telecommun. Technol. **32**(1), e4150 (2021)
3. Ring, M., Wunderlich, S., Scheuring, D., Landes, D., Hotho, A.: A survey of network-based intrusion detection data sets. Comput. Secur. **86**, p. 147–167 (2019)
4. Khraisat, A., Gondal, I., Vamplew, P., Kamruzzaman, J.: Survey of intrusion detection systems: techniques, datasets and challenges. Cybersecurity **2**(1), 1–22 (2019). https://doi.org/10.1186/s42400-019-0038-7
5. Tao, P., Sun, Z., Sun, Z.: An improved intrusion detection algorithm based on GA and SVM. IEEE Access **6**, 13624–13631 (2018)
6. Ali, M.H., Al Mohammed, B.A.D., Ismail, A., Zolkipli, M.F.: A new intrusion detection system based on fast learning network and particle swarm optimization. IEEE Access **6**, 20255–20261 (2018)
7. Al-Yaseen, W.L., Othman, Z.A., Nazri, M.Z.A.: Multi-level hybrid support vector machine and extreme learning machine based on modified K-means for intrusion detection system. Expert Syst. Appl. **67**, 296–303 (2017)
8. Manghnani, T., Thirumaran, T.: Computational CBGSA – SVM model for network based intrusion detection system. In: Applications and Techniques in Information Security, pp. 185–191. Singapore (2019)
9. Choraś, M., Pawlicki, M.: Intrusion detection approach based on optimised artificial neural network. Neurocomputing **452**, 705–715 (2021)
10. Pu, G., Wang, L., Shen, J., Dong, F.: A hybrid unsupervised clustering-based anomaly detection method. Tsinghua Sci. Technol. **26**(2), 146–153 (2021)
11. Zhou, Y., Cheng, G., Jiang, S., Dai, M.: Building an efficient intrusion detection system based on feature selection and ensemble classifier. Comput. Netw. **174**, 107247 (2020)
12. Verkerken, M., D'hooge, L., Wauters, T., Volckaert, B., De Turck, F.: Towards model generalization for intrusion detection: unsupervised machine learning techniques. J. Netw. Syst. Manag. **30**(1), 12 (2021)
13. Al-Daweri, M.S., Abdullah, S., Zainol Ariffin, K.A.: An adaptive method and a new dataset, UKM-IDS20, for the network intrusion detection system. Comput. Commun. **180**, 57–76 (2021)
14. Mohammadi, M., et al.: A comprehensive survey and taxonomy of the SVM-based intrusion detection systems. J. Netw. Comput. Appl. **178**, 102983 (2021)

15. Taher, K.A., Mohammed Yasin Jisan, B., Rahman, M.: Network intrusion detection using supervised machine learning technique with feature selection. In: 2019 International Conference on Robotics, Electrical and Signal Processing Techniques (ICREST), pp. 643–646 (2019)
16. Kunal, Dua, M.: Attribute selection and ensemble classifier based novel approach to intrusion detection system.Procedia Comput. Sci. **167**, 2191–2199 (2020)
17. Almasoudy, F.H., Al-Yaseen, W.L., Idrees, A.K.: Differential evolution wrapper feature selection for intrusion detection system. Procedia Comput. Sci. **167**, 1230–1239 (2020)
18. Kalita, D.J., Singh, V.P., Kumar, V.: A Survey on SVM Hyper-Parameters Optimization Techniques. In: Shukla, R.K., Agrawal, J., Sharma, S., Chaudhari, N.S., Shukla, K.K. (eds.) Social Networking and Computational Intelligence. LNNS, vol. 100, pp. 243–256. Springer, Singapore (2020). https://doi.org/10.1007/978-981-15-2071-6_20
19. Kannouf, N., Labbi, M., Chahid, Y., Benabdellah, M., Azizi, A.: A key establishment attempt based on genetic algorithms applied to RFID technologies. Int. J. Inf. Secur. Priv. IJISP **15**(3), 33–47 (2021)
20. Kalita, D.J., Singh, V.P., Kumar, V.: SVM hyper-parameters optimization using Multi-PSO for intrusion detection. In: Social Networking and Computational Intelligence, pp. 227–241. Singapore (2020)
21. Shenfield, A., Day, D., Ayesh, A.: Intelligent intrusion detection systems using artificial neural networks. ICT Express **4**(2), 95–99 (2018)

Retrospective Study and Evaluation of School Failure (Junior High School) in Scientific Subjects

Youssra El Janous[1(✉)], Mohamed Laafou[1], and El Hassan El-Hassouny[2]

[1] University Abdelmalik Essaadi, ENS Tetouan, Tetouan, Morocco
youssra.eljanous@etu.uae.ac.ma
[2] Higher Institute of Nursing Professions and Health Technique, Tetouan, Morocco

Abstract. This article aims, on the one hand, to theoretically analyze the fact of school failure by identifying and describing its extent, specifically in the province of Ouezzane in northern Morocco, on the other hand, it aims to describe the effect of hybrid education caused by the COVID-19 health crisis on student results in the 2020/2021 school year as well as to make a comparative analysis of school failure rates following an exploratory approach for previous school years; namely, the years 2015/2016, 2016/2017, 2017/2018 and 2018/2019. In order to carry out this study, we proceeded with an in-depth analysis of the marks of the students relating to the scientific subjects, in particular: mathematical sciences, sciences of life and earth and physics, resulting from the school curriculum obtained at the regional examination for the third year of college. Finally, we have suggested some recommendations regarding the technology plan that aim to reduce the rate of this failure in this province in particular and can be generalized in the other parts of the kingdom.

Keywords: School Failure · COVID-19 · Mathematics · Science of Life and Earth · Physics Sciences

1 Introduction

Please Educational achievement is the foundation of a country's human development. However, school failure is a crucial indicator of the performance and quality of education systems, and of the value of a society's human capital.

The reforms of this education system would allow Moroccan students to better succeed in their studies, regardless of the intrinsic and/or extrinsic factors. Certainly, significant progress has been made in reducing the school failure rate in Moroccan schools, but there is still a lot of room for improvement, to prevent many students, who do not have basic knowledge to drop out of school; an inevitable consequence of school failure.

However, the galloping socio-economic change, the unbridled urbanization and the appearance of more technical jobs led, from 1959 onwards, to make schooling compulsory for the majority of population [1, 2]. Thus, the discourse of the public authorities

M. Lazaar et al. (Eds.): BDIoT 2022, LNNS 625, pp. 22–32, 2023.
https://doi.org/10.1007/978-3-031-28387-1_3

focused more and more on the role of the school in correlation with economic development. As a result, a growing number of students found themselves on the school (college secondary cycle) benches, giving rise to school failure, since a good number of them, according to Perrenoud [3], "found themselves brought together and compared with a view to selection for entry to secondary school". But, according to the pedagogy and the differentiated school system, most of these students are not able to succeed in their transition from primary to secondary school, which leads to a significant increase in school failure.

2 Context

At the end of the colonial era, the schooling rate in Morocco was 11% in 1953 [4, 5]. Since independence, Morocco has thus lost at least half a century in the global race for schooling [6]. However, school education (primary and secondary) has seen a significant increase in terms of student numbers. In fact, the increase in secondary school enrolment rose in thirty years (1955–1985) from 0.3% to 28%, and was multiplied almost 16 times between 1955 and 2004 [7]. By 2050, the population of Morocco will increase by 30%, which represents a major challenge for the Moroccan education system [8].

Thus, significant efforts have been made to ensure access to school in very remote areas of the country. However, the distance between children's homes and schools remains significant, which has critical repercussions on the levels of schooling and school failure in Morocco [7].

In Morocco, the issue of school failure has recently reached a worrying level. The latest reforms of the education system have enabled pupils to succeed without acquiring knowledge and skills. Although significant progress has been made in reducing the failure rate in Moroccan schools, there are many disappointments.

Due to the COVID-19 health crisis, and the epidemiological situation in Morocco during the 2019/2020 school year, the National Education Ministry saw the need to consider a new study system. This can be summed up in the fact that 50% of the courses were organized by the presence of students in schools while the other half were carried out online using the communication and collaboration application launched by Microsoft, Called "Microsoft Teams".

3 The School Body in the Province of Ouezzane in Morocco

The province of Ouezzane has one urban commune and 16 rural communes, with a population exceeding 298,000 inhabitants, and a very remarkable youth rate of 60%, according to the general population and housing census carried out by the High Commission for Planning in 2014 [9]. The school body in the province of Ouezzane in the 2021/2022 school year includes 98 primary schools and 39 secondary schools, including 22 middle schools and 17 qualifying secondary schools. A total of 2648 teachers are present in these schools. According to the school year 2021/2022, is divided as follows: 1543 teachers of primary education, 505 teachers of secondary college education and 600 teachers of secondary qualifying education.

Concerning the distribution of these college secondary establishments between rural and urban areas, we found that 16 establishments among these college secondary schools are located in rural areas and only 6 colleges in the urban commune, two of which are private schools.

4 Theoretical Framework

4.1 School Failure

Literally, and according to the Larousse dictionary, failure is the negative result of an attempt, an undertaking, a lack of success, a defeat, a non-success, or a setback. We can add, according to Diedra Andenas and Kassandre Lapin [10], that the school failure is not just a difficulty affecting the student at the specific grade level only, but it also influences the student's psyche. In other words, the student is unable to meet the demands of the school and this is how this unease between the student and the school occurs.). For Best [11], "school failure can be detected through indicators such as the number of repeaters, the marks obtained in the CE2/sixth grade exams, and the dropout of pupils without a diploma or qualification". As a result, and considering that school failure is an unstable concept over time, since it is always correlated with a constantly changing socio-economic context [12]. Similarly, pedagogical notions must have a meaning in reality and have a relationship with the practices in the field of their implementation [13]. So, for the school to be able to reduce academic failure and provide responses adapted to each individual, it must integrate partnership projects or workshops carried out within the classes in order to provide an alternative in the remediation of difficulties in the schools [14].

There are three explanatory theories of school failure, the first of which is psycho-medical and considers that school failure may be caused by mental deficiencies. Pinell and Zafiropoulos [15] consider that:

> *"The abnormality of the retarded, which psycho-pedagogues at the beginning of the 20^{thème} century characterized as an intellectual deficiency that prevents the subject from acquiring school knowledge, goes unnoticed outside of school (in this, the retarded differs from the idiot and the imbecile). It is the emergence of this new figure of "mental deficiency" that makes it possible to construct a new theory of idiocy, based on the hypothesis that there is a continuum of intelligence going from the most profound states of mental deficiency to normal intelligence, and to develop a test (the intelligence metric scale) making it possible to establish the diagnosis of debility".*

The second theory is based on social pathology; school failure is only the inevitable result of the social environment from which the pupil comes, in fact if the child comes from a disadvantaged social environment. He or she is more likely to fail at school. According to Bourdieu and Passeron [16], the school institution, language and cultural practices are the means by which the dominant categories maintain social differences and ensure that their powers are inherited intellectually from generation to generation.

The third theory is based on the habitus which, according to Le Baron [17], designates a coherent set of acquired dispositions that guide an agent's practices, independently of any explicitly stated conscious end. An individual's primary habitus results from his or her family upbringing, the secondary habitus from the set of acquisitions linked to the school system, etc. The habitus is constantly changing as a result of social experience (family, professional, etc.), but it has a certain stability and consistency. Differences in habitus can be identified, in surveys or from qualitative observations, through systematic and coherent differences in behavior, attitudes, representations and life- styles that organize and structure social space. The biological basis of the habitus lies in cerebral plasticity. Thus, society gives men the reasons to exist, and society also generates situations in which it controls our actions and reactions. For this theory, school failure is a staging of the society in which we live.

5 Methods

5.1 Traget Audience

This work is to assess the failure rate of the province of Ouezzane based on the data of the grades of the scientific subjects: Mathematics, physical sciences and life and earth sciences, of the students of the third year of college of the regional examination to the academic year 2020/2021. The total number of students who took the regional exam for the college cycle certificate in the direction of Ouazzane is 4759, of which 3286 (69%) of students belong to rural communes.

5.2 Origin of the Data

The data of the school year 2020/2021 used in this work is from a database of the school management system MASSAR and the examination management system GEXAPLUS, adopted in June 2015, by the Moroccan Ministry of Education.

Massar is an online system that allows the entry and management of continuous assessment notes and exams; it is also used for the computerization of all teacher information, the management of timetables, the follow-up lateness and absences, and student sanctions. While GEXAPLUS is an online system for managing certified exams such as the 6th year of primary school, 3rd year of college, and 1st and 2nd years of the baccalaureate, it allows the entry and management of exam scores, the calculation of the averages, and ranks the candidates in order of merit. This application is meant only for the head of the provincial examination centers.

5.3 Statistical Analysis

The statistical analysis was performed with Excel, and the pass average (MR) is calculated as follows (with MCC: the average of the continuous assessment; MENL1S: the average of the first semester local standardized examination and MENN: the average of the regional standardized examination):

$$MR = \frac{\mathrm{MCC}*30 + \mathrm{MENL1S}*30 + \mathrm{MENN}*40}{100}$$

6 Results and Interpretation: Exploratory Approach "Comparative Analysis"

6.1 Learning Environment (Rural/urban) and Success Rate

Our comparative study of the success rate of the 17 schools showed that the success rate is directly related to two main factors; the first one consists in the number of students per class, and the quality of infrastructure to facilitate the integration of ICT in the learning process (library, multimedia room,...), while the second one is summarized in the fact that girls remain more motivated to succeed than boys. Indeed, the success rate reaches 100% in private schools, and 81 to 91% in a public school in rural areas (Secondary school 11 January) with a good infrastructure and a maximum number of students that does not exceed 28 per class. Paradoxically, this rate does not exceed 37% in public schools located in rural areas, without a library or multimedia room and with an excess number of students per class (>38). Thus, we note that the infrastructure and the number of students per class have a close relationship with the success or failure of students throughout the province.

6.2 The Effect of Learning Certain Science Subjects on Success Rates

To answer this question, we will determine for each (science) subject the number of students with a mark above 10/20 and those who did not achieve the average (Fig. 1).

6.3 Life and Earth Sciences

The evaluation of the students' marks according to the mark out of twenty of the final examination of the subject of life and earth sciences, showed that 76% of the students' marks are between 0/20 and 08,25/20, thus these students could not reach the average mark. The peak of the graph (Fig. 1) which appears first, reaches almost 225 students and corresponds to the mark 03/20, which is undoubtedly, a 'bad mark'. We also notice that the curve increases very rapidly over the interval [0; 03], then it regresses rapidly over the interval [03; 08.25], and at the last interval [08.25; 20] it becomes very regressive (Fig. 2).

6.4 Physics Sciences

The evaluation of the students' marks according to the mark out of twenty of the final examination in physics sciences showed that half of the marks are between 0/20 and 09.5/20. The graph contains two peaks: the first corresponds to 225 students whose abscissa is the mark 04.50/20; and the second to 200 students whose abscissa is the mark 05.50/20.

According to Bloom's taxonomy, the exam contains six categories of the cognitive domain which are as follows: knowledge, understanding, application, analysis, evaluation and creation.

When we apply this taxonomy on the curve, we deduced that the first part is the interval [0; 05], the curve in this interval is increasing which means that the students

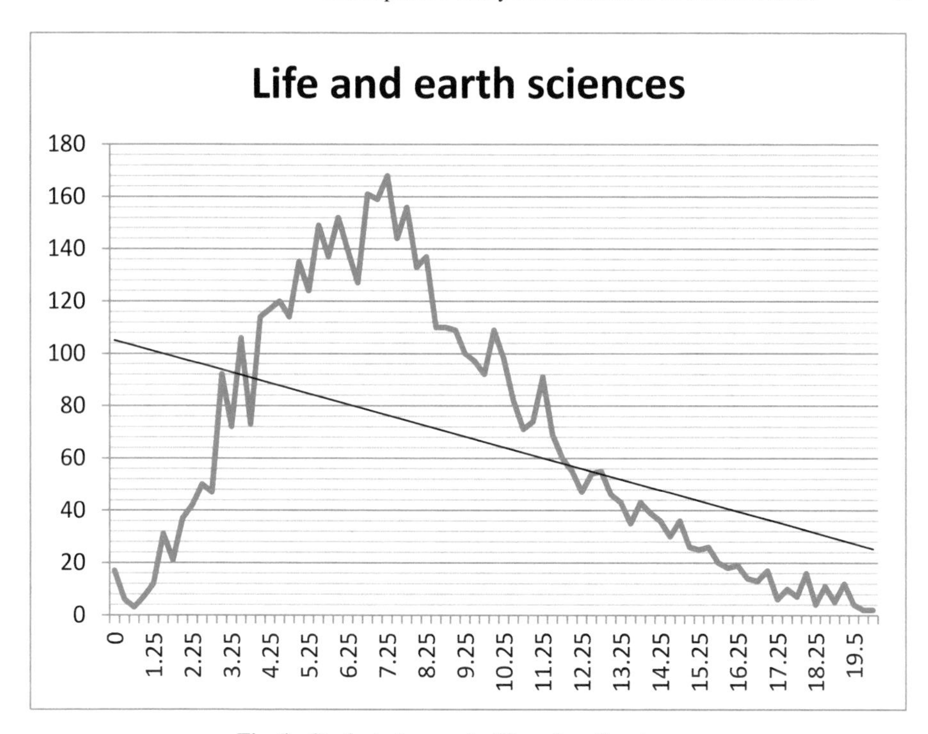

Fig. 1. Students 'scores in life and earth sciences.

answer the questions of knowledge and comprehension, the second part of the curve is the interval [05; 12.5], the curve in this part is decreasing i.e. 63% of the students face major difficulties when the questions of the examination relate to the application and the analysis to the subject of the physical sciences The third part of the curve is the interval [12.5; 20] with a number of students always lower than 20 students, i.e. less than 7% of the students pass the cognitive domains "evaluation" and "creation";

6.5 Mathematics

From Fig. 3, we see that more than 75% of the students' marks in mathematics are between 0/20 and 7.25/20. The peak of the graph corresponds to 400 students having a mark of 0/20. The linear trend line has a very strong decrease. It is generally decreasing on the interval [0,20], which shows the inversely proportional tendency of the number of pupils in relation to the marks. Moreover, from 10/20 to 20/20, the number of students does not exceed 20 out of 4759 students on the ordinate axis. We also notice that the curve decreases at an irregular rate with great speed. The decrease is very fast between the marks 0/20 and 2.75/20 compared to the decrease on the interval [2.75; 7.75]. Thus, there is a significant decline in the numbers of students in relation to the grade in mathematical sciences.

The drop in marks in mathematical sciences shows that students have serious gaps in their understanding of mathematical concepts. Most of the students received a mark of

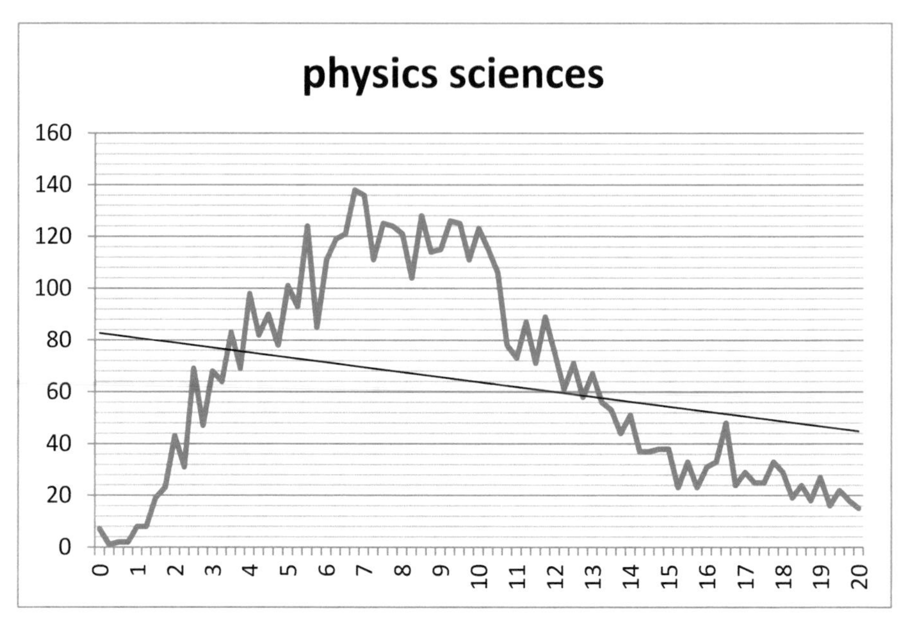

Fig. 2. Students 'scores in physics sciences.

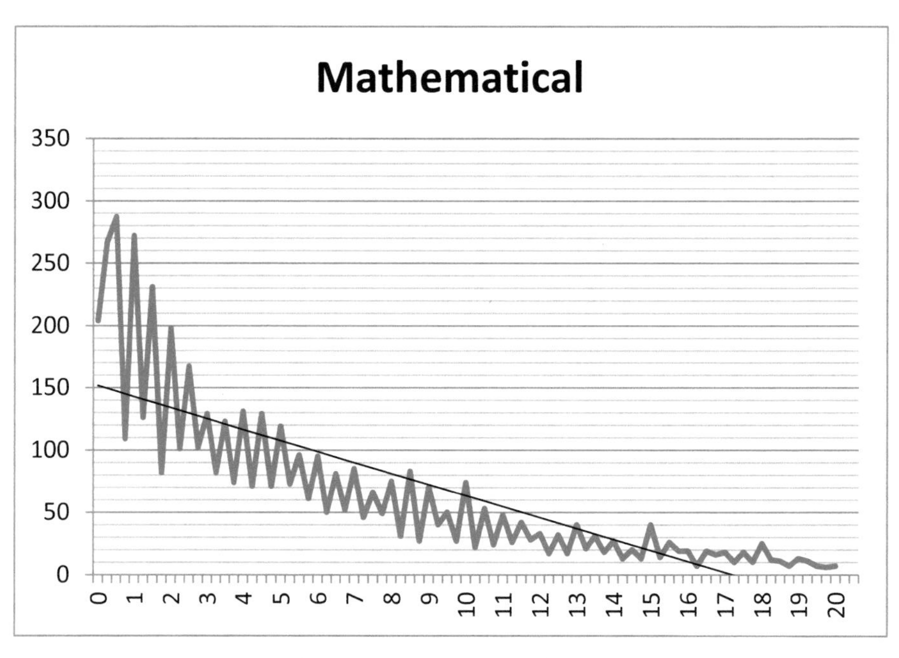

Fig. 3. Students 'scores in mathematical.

0/20, which shows that these students are unable to answer at least one question correctly, even though they have had lessons on these concepts. Among the causes of these results is the threshold of success, which is always and in all the regions of Morocco, very lower than the average (10/20), arriving sometimes until 06/20, which can generate problems of mastery of the prerequisites to pass to the class of a higher level. We have also noticed that, on the one hand, the mathematics program is too heavy in relation to the amount of time devoted to the subject. On the other hand, students need support in mathematics classes, as the number of students who do not manage to get a grade higher than 05/20 is very high.

Fallowing these results, we can affirm that the learning of scientific subjects (Math, Life and Physical Science) represents a major obstacle to the success of students in this year.

To reinforce these results, we will compare the success averages of students in the promotion of the year 2021, with those of the promotions of the school years 2019, 2018, 2017 and 2016 in order to know if the academic failure exists in all promotions, or the promotion of 2021 presents an exception, since the kingdom in this year, adopted hybrid education (in presence and online learning). It should be noted that the year 2019/2020 is not included in this comparison since the realization of a unified regional examination was not possible due to the unfavorable epidemic situation in most of the country's provinces (Fig. 4).

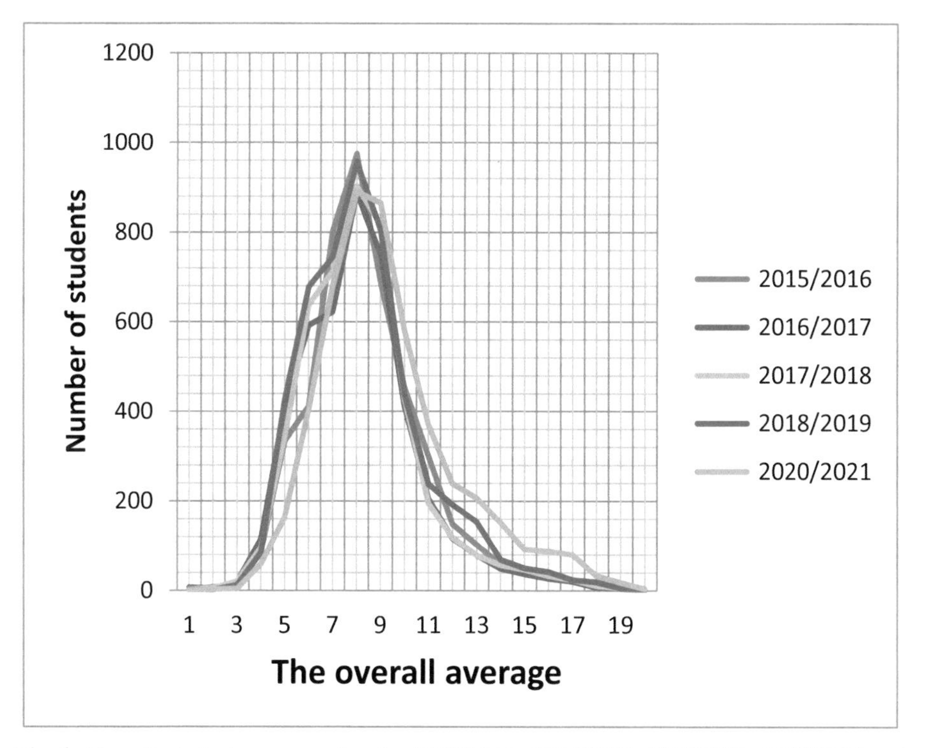

Fig. 4. Overall student averages for the 2015/2016, 2016/2017, 2017/2018, 2018/2019 and 2020/2021 school years.

From these four graphs, we can see that the four curves are almost identical, and that the peaks of all these curves are included in the interval [07,8: 08,8].

We will divide and analyze the four curves into four parts:

- The curves are almost constant on the interval [0; 03[, and the number of pupils on this interval does not exceed 20 pupils for the 4 curves.
- The curves are increasing at the interval [03; 07[, and 56% of the students are included on the latter, which shows that the results of the students in all the subjects taught are based only on the restitution of information, and on comprehension.
- The curves are decreasing on the interval [07; 11[, then, the average success is inversely proportional to the number of students, which comes to affirm the ranking of Morocco in the PISA result for the year 2019 where Morocco was located in 75th place on a list of 79 countries.
- The curves are very regressive on the interval [11; 20[which shows that the students did not have good general averages. Also, the number of students on this interval does not exceed 75 for the 4 promotions.

The four graphs visually illustrate the extent of academic failure in the province, since more than 7% of students obtained an overall average between [0;05[, and 75% of averages are included in the interval [05; 09], in other words, the majority do not reach the threshold of success, and only 18% of students obtained an average higher than 09/20.

We believe that this academic failure is mainly related to problems such as.

- Problems adjusting to the school environment.
- Cognitive difficulties and weaknesses in learning.
- Problems of prerequisites for moving from one school year to another.

School failure is a very complex phenomenon since it requires a precise redefinition of its concept and a multidimensional critique in order to be able to identify the most influential causes of this phenomenon.

Finally, when we compare the present data with those of the school year 2021, which saw a reduction of half of the hours of students attending schools in presence and an adoption of distance education to compensate for the remaining hours, it becomes clear that the educational model itself does not affect the profitability of students in this region.

7 Recommendations and Proposals

After the analysis of the databases, we will propose the integration of technology as a solution to improve school results, as well as to fight against school failure in the province of Ouezzane. Indeed, the ministry has already taken the initiative to create the taalimtice.ma portal for this purpose, and it has distributed ready-to-use digital resources too, but the main problem it faced was that teachers did not have a prior knowledge of these elements. To overcome such adversity, we propose that the ministry create announcements and publications to inform these teachers about them. Also, we must create an infrastructure that facilitates and encourages the use of technology, such as equipping all the rooms with at least one computer, a video projector and a modem to access the Internet, we must also change blackboards with interactive whiteboards (IWB). In the same direction, the Ministry of National Education must create mobile applications for cell phones to facilitate the accessibility to courses and lessons anywhere and at any time. In addition, we recommend encouraging the teachers to use the inquiry approach when learning science subjects with the integration of ICT.

8 Conclusion

The present study and analysis of data concerning school failure in the province of Ouazzane have enabled us to reach the following conclusions. The rate of grades of students with an average of 10/20 in science subjects in the third year of college does not exceed 40% in 2019. In other words, 60% of the students' marks are in the interval [0,10[. Indeed, the analysis of the databases of school results over the five years confirms the existence of a drop in level in these years. Clearly, in the province of Ouezzane, school failure is quite clear, especially, in 2016 in the grades of the third college year of the school year, and it persists until the 2020/2021 school year. We also conclude that academic failure does not only affect the class of 2021 (under the circumstances of a global pandemic) in particular, but also other classes with the same academic problems. The causes of school failure are multiple, and many factors contribute to this loss of momentum. Among these factors are those internal and external to the school, such as the infrastructure, the location of the school, and problems with the evaluation method itself [18]. The phenomenon of school failure has an important impact on the development of the country, hence the importance of urgently implementing appropriate measures to rectify it.

References

1. Carle, J.C.: Obligation de scolarité et contrôle de l'obligation scolaire. París, Sénat (504), 1997–1998 (1998). https://www.senat.fr/rap/l97-504/l97-504_mono.html
2. Legrand, A., Solaux, G.: Du CAP et de ses usages...(1959–1992). Revue française de pédagogie, 47–58 (1992). https://www.jstor.org/stable/41200289
3. Perrenoud, Ph.: La pédagogie à l'école des differences, 2nd ed. Éditions ESF, Paris (1995). http://www.unige.ch/fapse/SSE/teachers/perrenoud/php_main/php_livres/php_ecole.html
4. Zouggari, A.: Le système d'enseignement sous le protectorat français et espagnol. In: Lamrani (ed.) Systèmes éducatifs, savoir, technologies et innovation, pp. 451–469. Cinquantenaire de l'Indépendance du Royaume du Maroc, Rabat (2005) http://www.rdh50.ma/fr/pdf/contributions/GT4-12.pdf
5. Kateb, K., Diguet, D.: Bibliographie critique. Population **1**(62), 179–207 (2007). https://doi.org/10.3917/popu.701.0179
6. Akkari, A.: La scolarisation au Maghreb : de la construction à la consolidation des systèmes éducatifs. Carrefours de l'éducation **1**(27), 227–244 (2009). https://doi.org/10.3917/cdle.027.0227
7. Lamrini, A.: Systèmes Éducatifs, Savoir, Technologies et Innovation. Report thématique. 50 ans de développement humain (2006). http://www.ondh.ma/Pdf_doc%5CRap_SESTI.pdf

8. Ouadah-Bedidi, Z., Vallin, J.: Maghreb: la chute irrésistible de la fécondité. Médecine/Sciences **17**(2), 247 (2012). https://doi.org/10.4267/10608/1904
9. HCP: Annuaires statistique régional 2014. Haut Commissariat au plan, Direction régionale Tanger_Tétouan_Al Hoceima (2015). http://www.apdn.ma/apdn/images/stories/file/statistique/statistique_en_chiffres/ANNUAIRE_STATISTIQUE_REGIONAL_2014.pdf
10. Diedra, A., Kassandre, L.: Échec scolaire et difficultés scolaires : la pédagogie différenciée, une réponse?. Education (2013). https://dumas.ccsd.cnrs.fr/dumas-01017113
11. Best, F.: L'échec scolaire. Presses Universitaires de France. « Que sais-je? » (1996). https://doi.org/10.3917/puf.best.1996.01
12. Chauveau, G., Rogovas-Chauveau, E.: À l'école des banlieues. Éditions Esf Editeur, France (1995). ISBN-10 2710111179
13. Anne, A.: Conceptualisation et dissemination des "bonnes pratiques" en éducation: Essai d'une approche internationale à partir d'enseignements tirés d'un projet. Developpement curriculaire et "Bonne practique" en Éducation, pp. 1–17. https://unesdoc.unesco.org/ark:/48223/pf0000173642.locale=fr
14. Deborde, J.: L'art pour sortir de l'échec scolaire. Nectart **10**, 64–69 (2020). https://doi.org/10.3917/nect.010.0064
15. Pinell, P., Zafiropoulos, M.: La medicalisation de l'échec scolaire. Actes de la Recherche en Sciences Sociales (24), 23–49 (1978). https://www.persee.fr/doc/arss_0335-5322_1978_num_24_1_2614
16. Bourdieu, P., Passeron, Jc.: LES HERITIERS. Les étudiants et la culture. Les Éditions de Minuit, Paris (1964)
17. Le Baron, F.: La sociologie de A à Z. Éditions Dunod, Paris (2008) http://livre21.com/LIVREF/F38/F038017.pdf
18. Youssra, E.J., Mohamed, L., Rachid, J.-I., Mourad, M.: La perception des enseignants et des élèves sur l'examen normalisé et le contrôle continu pour la matière de science de la vie et de la terre. European Scientific Journal ESJ **17**(20), 110 (2021). https://doi.org/10.19044/esj.2021.v17n20p110

The Use of "Mathematical Modeling" in Physics: Case of Representations and Teaching–Learning Practices

Fouad El-hars(✉), Rachid Souidi, Issam Benqassou, Abdelhamid Lechhab, and Moulay Mustapha Hafid

Higher School of Education and Training, Ibn Tofail University, Kenitra, Morocco
fouad.elhars@uit.ac.ma

Abstract. Professors and students rarely use mathematical modeling as a means for understanding physical phenomena, which negatively impacts on the quality of learning. The current study probes the theoretical landscape surrounding mathematical modeling in the context of teaching physics in Moroccan high schools. In this study, a survey, based on two questionnaires intended for the two actors of the teaching–learning operation will be implemented. The analysis of the surveys allows us to characterize the modeling approach that is supposed to be taught at this school level. We test the ability of teachers and students to detect modeling in the physics classroom, the criteria that define it and its limitations. The forms of modeling that interest us are used in connection with the concepts of force and differential equations. These govern the dynamics of kinetic variables and electrical phenomena in circuits. These important results show the difficulties encountered in the use of mathematical modeling in physics. We propose resolution models to restore the balance between these two axes.

Keywords: Mathematical modeling · Concept · Representation · Investigation · Differential equation · Force

1 Introduction

A great deal of learning research points to the difficulties students face in acquiring knowledge, particularly in the science disciplines [1]. These are due, in physics for example, to the limitation of the implementation of experimentation activities and the lack of scientific equipment in school laboratories [2]. This state is also the subject of international reports that place Morocco in critical positions. The sources of this problem are multiple. Misconceptions, mathematical modeling, teachers' professional practices, lack of laboratory work and the low rate of ICT integration are just a few examples [3, 4]. Indeed, the use of mathematical symbolism in physics is justified by the consideration of mathematics as a powerful "toolbox" [5]. This is because modeling is a major scientific practice that concerns a growing number of fields and is one of the main modalities of interaction between mathematics and other sciences. The model is an approach to theory that allows for the interpretation of objects and events. However,

M. Lazaar et al. (Eds.): BDIoT 2022, LNNS 625, pp. 33–46, 2023.
https://doi.org/10.1007/978-3-031-28387-1_4

the use of mathematical symbolism can become a real barrier to learning. This study therefore proposes a reflection on mathematical modeling in physics and on the impact, it can have on the representations of physics acquisition in learning situations.

2 Theoretical Framework

In physics, theory is not always the basis of observation and experimentation activities, just as the activity of modeling the material world at the heart of physics makes it possible to link theory and practice, it makes theory an experience and experience an abstraction [6]. Two dimensions are then to be considered: the domain of theoretical constructions of knowledge and specific intermediate models on the one hand, and the space of experiences formed by objects, events with their own lexicon and representations on the other hand [7]. The idea is that the student, in order to learn physics, is inspired by his knowledge of the everyday world divided into objects and events; the teacher, on the other hand, constructs the knowledge of physics with reference to the everyday knowledge of the physical world and that of the theory and the model. Tiberghien and Gaidioz [7] propose the example of a falling ball. According to the student, this event is due to the material cause "Land", by which he/she puts a concept and an object-event on the same footing, and to the explanatory cause, by which he/she refers the movement of the ball to the force; the teacher, for his/her part, directs this superficial interpretation of the student towards Newton's laws, towards the theory, the model [8]. To bridge the gap between theory and practice, TJAD (Theory of Joint Action in Didactics) presents itself as a didactic solution to be experimented in the teaching of physics. The TJAD model generic use to which [he] refers refers to the use of the notion of play as a model, which makes people see human activity as a game. "Indeed, the game is the space where theory-model and praxis meet because it" has defining rules (which correspond roughly, in "conventional" games, to the rules of the game), which can often be reduced to the constitutive rules put forward by Searle [9]. It requires strategic rules, which, as describes [10], make explicit how to play the game well (they can, for example, be transmitted by a connoisseur of the game to a lesser connoisseur), and (effective) strategies, which constitute for the player the concrete way of acting in a determined praxis, revealing (more or less) a certain sense of the game" [11].

3 Research Methodology

This study therefore proposes a reflection on mathematical modeling in physics and on the impact, it can have on the representations of physics acquisition in learning situations. The objective was to lead the student to make a problem explicit using their own conceptions [12]. To carry out our research, we therefore opted for the quantitative methodology. It allows, on the one hand, to exploit the differences between the representations of the students and the teachers, and those which exist between the students who succeeded in solving the proposed didactic situation and those who did not [13, 14]. It clarifies, on the other hand, the studied phenomena and raises new questions [15]. Within the framework of the present research, two questionnaires are constructed, one targeting a population of 68 teachers practicing in secondary schools of the Regional Academy of Education and Training (RAET) of Beni-Mellal, Provincial Direction of Khénifra-

Morocco and the other targeting their learners. 64.7% (44) of the 68 teachers are in the qualifying cycle, the 35.3% (24/68) are teachers in the secondary college cycle. The majority of the faculty have less than 6 years of tenure, 23.5% have more than 6 years and less than 12 years of tenure and 7% have more than 12 years of tenure.

4 Results and Discussions

4.1 Questionnaire 1 Addressed to Teachers

The first questionnaire includes 10 items whose objective is to identify the "teachers'" representations of modeling in physics and of mathematical modeling in physics in particular.

Item 1.

Q1- The physico-mathematical relation is a relation

✓ Complementarity (mathematics complements physics and vice versa);
✓Of inclusion (physics implicitly includes mathematics);
✓Implication (understanding physics necessarily implies mastering math).

It defines the physics-mathematics relationship: the answer "complementary" is equivalent to say that mathematics and physics are inseparable, hence its importance in the phase of mathematization of physical concepts in particular. Mathematics serves, in this case, to give fluidity to the imaginations of a physicist. The choice of the answer "inclusion" suggests that the construction of concepts is impossible without mathematics. This implies the necessity of detecting real causes generating representations from modeling, because during teaching–learning situations in physics, the majority of teachers use mathematical tools (vectors, derivative, point, line, ...) to explain the physical concepts to be institutionalized without insisting on the physical content of the concept. The "implication" answer shows that the learner can understand physical phenomena and use experiments without using mathematical theorems and propositions.

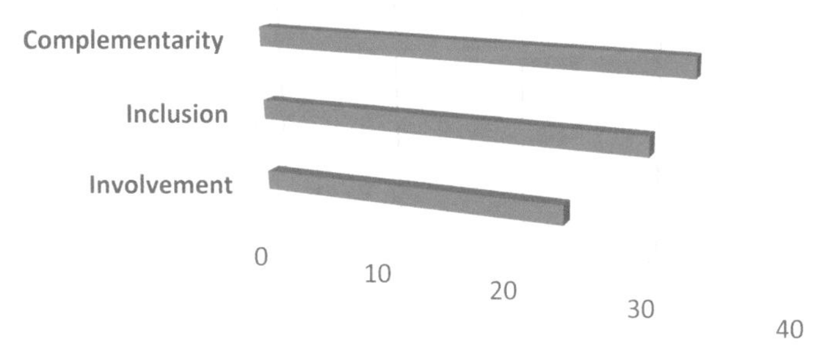

Fig. 1. The physical–mathematical relationship.

The responses to item 1 reveal, as explained in Fig. 1, that the majority of teachers find physics and mathematics to be in a complementary relationship. The use of the second is a normality in the exercise of the profession because they allow the construction of physical concepts, which will generate more or less undesirable effects in the learners.

Items 2 et 3.

Q2- The importance of mathematics in physics in the Moroccan secondary school curriculum is nearly 100%, 90%, 70%, 30%, 10% or 0%)

Q3- The units in this program that require more mathematical modeling are

✓Unit 1: Nuclear;
✓Unit 2: Waves;
✓Unit 3: Mechanics.
✓Unit 4: Electricity.

These two items devoted to mathematical modeling in physics are used to determine the degree of importance of mathematics in physics and to evaluate its use in classroom practices (Q2) among physics teachers on the one hand; on the other hand, to probe this level through the identification of units where mathematical modeling is in use (Q3). Figure 2a and b below show the responses obtained:

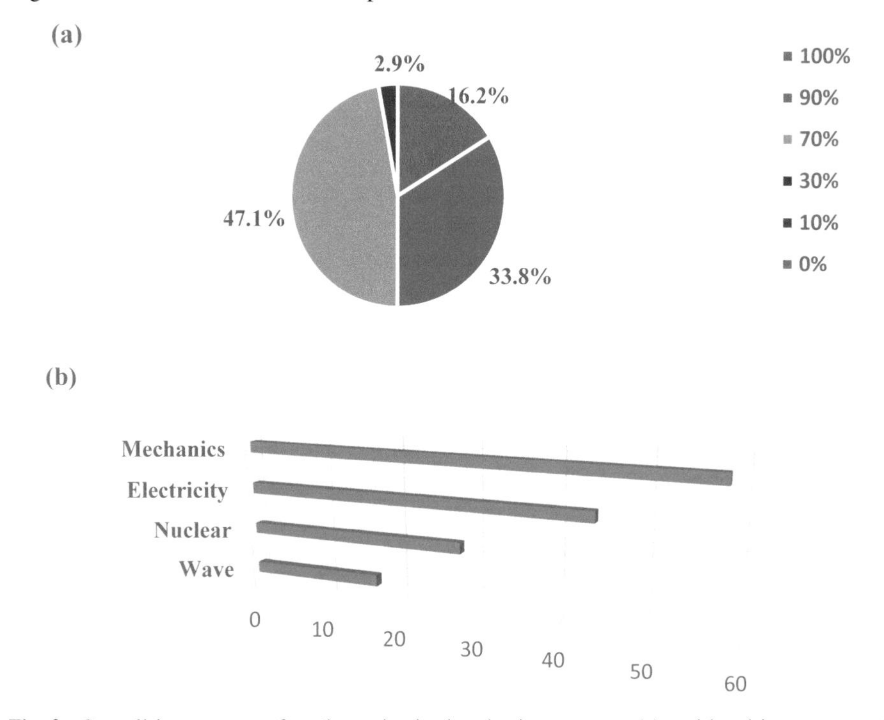

Fig. 2. Overall importance of mathematics in the physics program (a) and local importance of mathematics in the physics program (b).

The results obtained show that 47.1% of the teachers surveyed are convinced that the physics curriculum includes up to 70% of mathematics (Fig. 2a), while 16.2% of them think that physics is purely mathematical (100%), 33.8% see that 90% of the physical concepts are mathematized while a small percentage think that mathematics occupies only 30% of the space in physical science. Figure 2b shows the units that require a greater use of mathematics, namely electricity and mechanics. It is thus revealed that the majority of teachers need mathematics to build the physics course.

Items 4 and 5

Q4- In your opinion, mathematical modeling is

✓ Necessary and sufficient;

✓ Necessary, but not sufficient;

✓ Not necessary and not sufficient.

Q5- The modeling of a force by a vector in teaching is

✓ Sufficient;
✓ Satisfactory;
✓ Insufficient;
✓ Fatal.

Item 4 addresses the debate of the sufficiency and necessity of mathematical modeling when building physical concepts (Fig. 3).

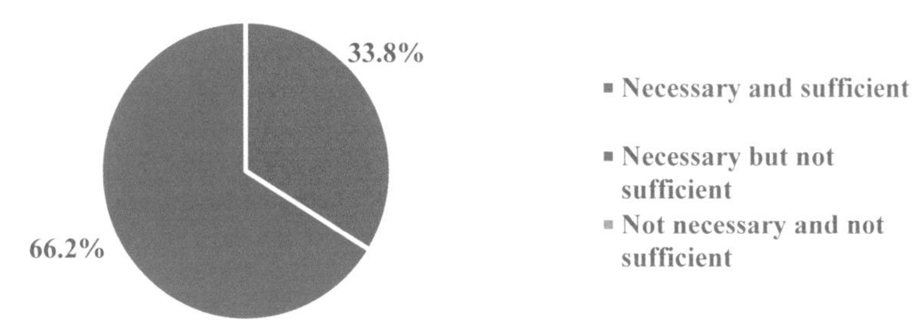

Fig. 3. Necessity and sufficiency of mathematics modeling from the teacher's point of view

While 66.2% of the teachers surveyed believe that mathematical modeling is necessary but not sufficient, 33.8% of them believe that modeling is necessary and sufficient. The third choice had no voters. It is clear that mathematical modeling in the construction of physical concepts is both necessary and insufficient (Q4).

Item 5, while repeating item 4, limits the question to an example from a physics course (Modeling a force by a vector).

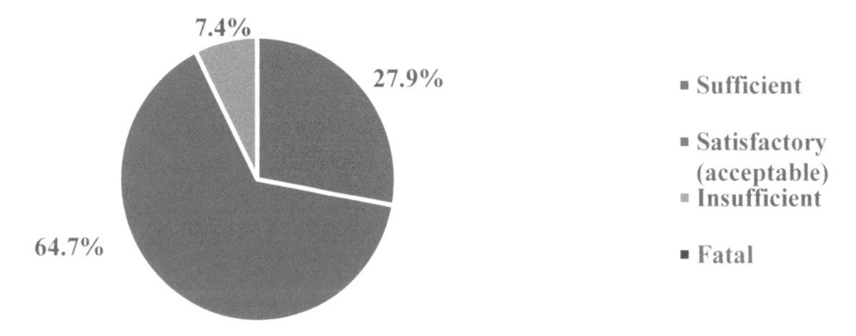

Fig. 4. Satisfaction of the modeling of a force by a vector

In Fig. 4, the majority of teachers 64.7% say that this modeling is satisfactory while 27.9% see it as sufficient, only 7.4% of teachers recognize its insufficiency.

Items 6 and 7.

Q6- Do you know the limit of modeling a force by a vector?

✓ Yes;

✓ No.

Q7- If yes, give a precise example

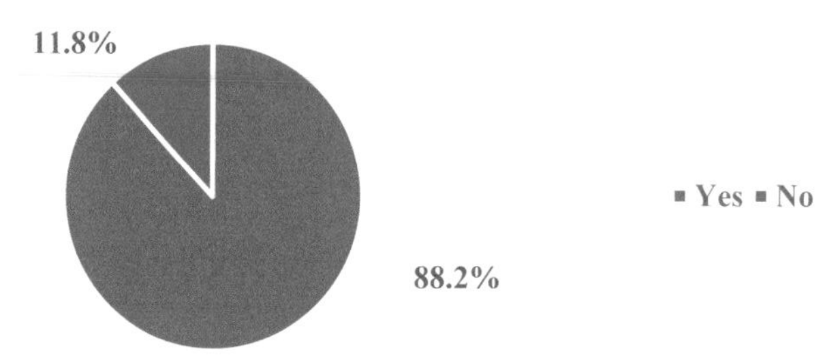

Fig. 5. Limit of the modeling of a force by a vector

Figure 5 shows that all 88.2% of the teachers were unaware of the modeling limitation. The remaining 11.8% answered yes to question Q7, but none of them offered a correct example representing this limitation. This leads to a reconsideration of the concept of mathematical modeling in physics among the teachers of physical sciences in the region.

Item 8.

Q8- Does modeling influence students' conceptions (representations)?

✓ Yes
✓ No

The majority of these concepts were indeed understood in a mathematical way, and the physical concept is not important to the learners, because the majority of the evaluations even in the national examinations, do not take into account the physical meanings of the concepts and the experimental side of the phenomenon, which makes the mathematization of the phenomena and the optimal solution either in the cognitive or didactic side.

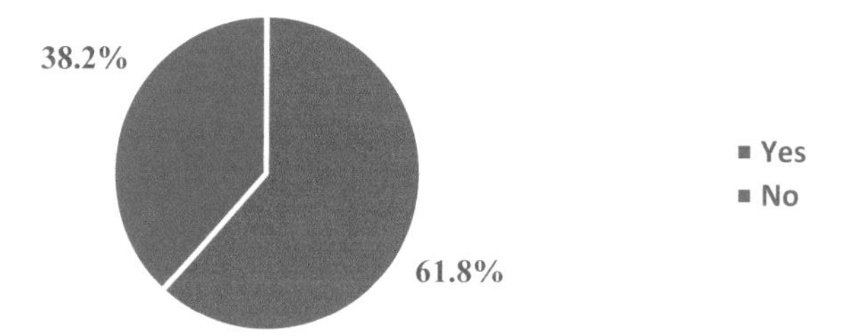

Fig. 6. Influence of modeling on learners' conceptions

We notice (Fig. 6) that for 61.8% of the teachers, mathematical modelling has an important impact on the students' representations of physics learning. A large percentage of teachers do not detect these representations, because this identification requires a good mastery of both physical and mathematical concepts, in addition to modeling, its criteria and its limits.

Items 9 and 10.

Q9- Do you consider your students' conceptions during a teaching session?

✓ Yes, often;
✓ Yes, sometimes;
✓ No.

Q10- Do you think it can be important to consider students' conceptions in a teaching session?

✓ Yes;
✓ No;
✓ I don't know.

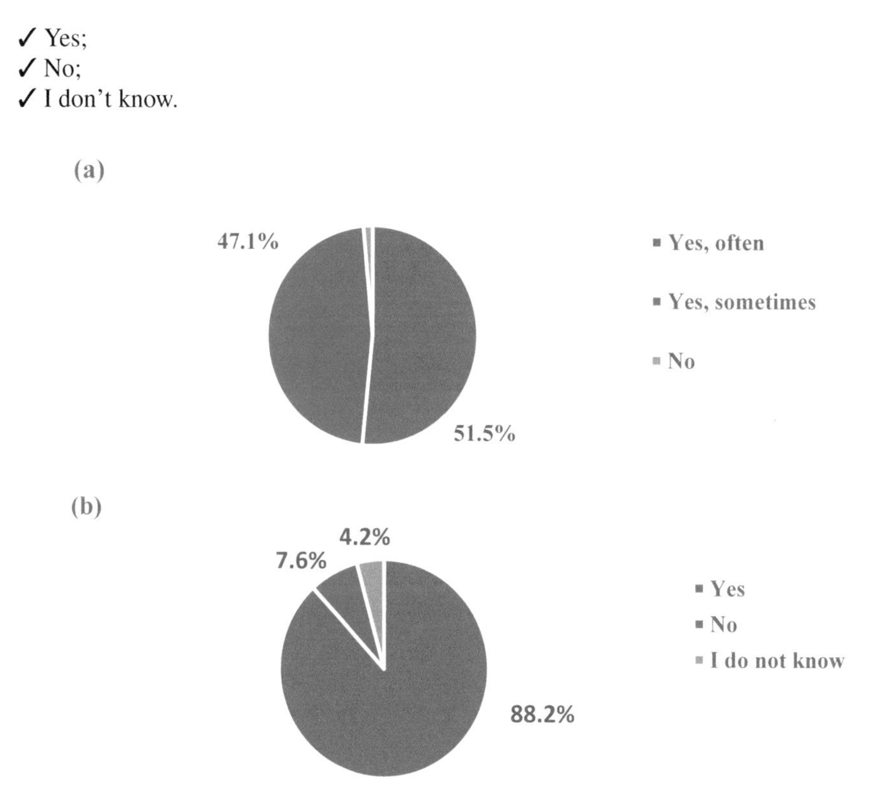

Fig. 7. Consideration of students' representations by teachers (a) and importance of taking into account the students' representations from the teacher's point of view (b).

For 51.5%, i.e., half of the respondents, the teachers take into account the representations of their learners of the use of mathematical modelling in physics (Fig. 7a). Of course, this does not deny the awareness (88.2%) that they have to consider these representations impacting on the problems encountered by the students in the mastery of physics and what the different evaluations reveal (Fig. 7b).

Synthesis (questionnaire 1)

.

The epistemic and didactic foundations that underlie this reflection make sense of the theory and support the models used to analyze the teaching situations; they also help to give consistency to the interpretations of the data. This is why their explanation is important in class, especially in physical sciences. However, it turns out that the survey conducted among the teachers of the provincial directorate of Khénifra revealed that the majority of these teachers:

✓ Fails to locate the importance of the model;
✓ Ignores the criteria for modeling and its limitations;
✓ Does not give particular importance to the practical treatment of representations;
✓ Suffers from a lack of mastery of the epistemological and didactic factors involved in the concepts and laws of physics.

The importance of Epistemical-didactic foundations lies in their ability to account for the different types of knowledge that are woven together in a classroom.

4.2 Questionnaire 2 (Addressed to Learners)

Items 1 and 2.
Q1- After the red car is launched and before it collides with the black car, indicate where the force on the red car comes from (check the right answer):

✓ The land
✓ The red car
✓ The experimenter
✓ The rails
✓ The table

✓ I don't know

Q2- After the red car is launched and before it collides with the black car, indicate where the force on the rails comes from (check the right answer):

✓ The red car
✓ The black car
✓ The experimenter
✓ The table
✓ The land

These two questions are designed to test the students' ability to identify the systems involved in the case of a car moving in a straight line on parallel rails.

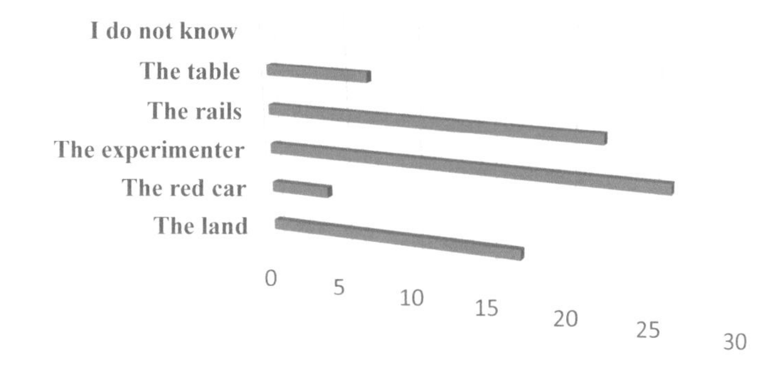

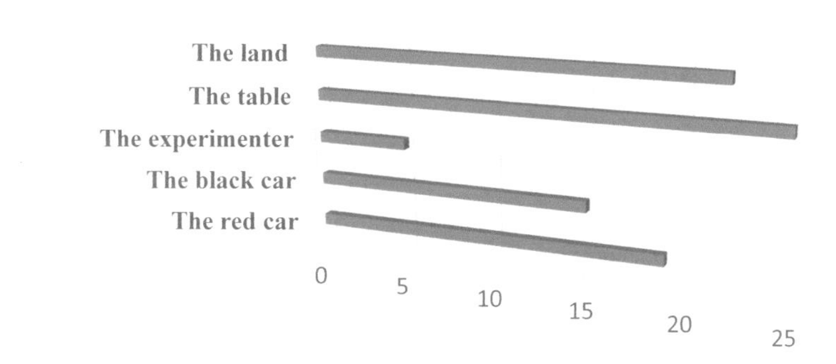

Fig. 8. The actors of forces on the red car (a) and Force players on track (b)

Learners (Fig. 8a) are mostly unable to identify the origin of the force on the red car; they point to the experimenter as part of the actor system, which is not the case. This one is indeed not in contact with the studied system, it does not exert a direct force on the red car and is satisfied to give an initial quantity of movement, it is the variation of this quantity of movement which generates the concept of the force (second law of Newton). Similarly, the action of the experimenter is unclassifiable in the known classes of force (remote and contact). In Fig. 8b, learners seem to understand the theoretical principle behind the force, and for good reason they point to the "earth" and the "table"; the experimenter is present in 11.6% of learners. The fact remains that students, for the most part, remain unable to accurately identify force actors; they have not fully grasped the concept of force.

Item 3.

Q3- Can we represent the interaction by a force?

✓ Yes, always
✓ Not necessarily
✓ I do not know

The third question is designed to detect students' ability to distinguish between force and interaction, where force is, in this case, a preliminary model of interaction. Figure 9, which follows, reveals that learners were unable to distinguish interaction from force and vice versa, indicating that they did not master the triplet of interaction, force and force vector.

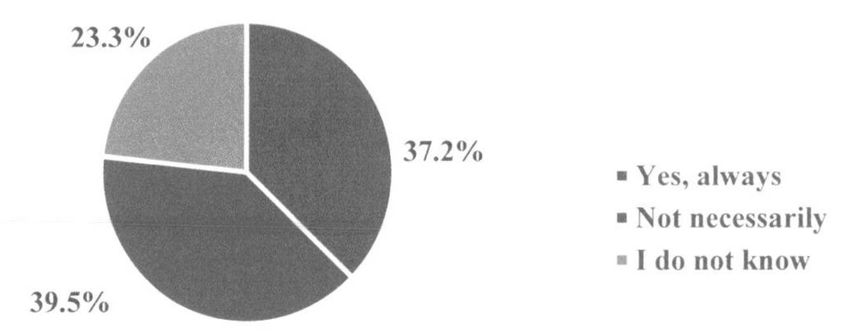

Fig. 9. Answers to the question concerning the representation of the interaction by a force

Item 4.

Q4- Why does the red car slow down after it is launched by the experimenter and before it collides with the black car?

✓ "1" The experimenter has given the red car a force whose value decreases over time.
✓ "2" Friction is braking the red car.
✓ "3" The force given to the car by the experimenter is eventually offset by friction.
✓ "4" I don't know.

With this question (Q4), we try to detect and define the representations of the learners regarding the friction force. These forces are difficult to understand because they are non-conservative; no ultimate model has been posed, especially since the elements of knowledge at this level present in the qualifying secondary school curriculum in physics concern the effects of friction. Most students do not understand the characteristics of this force and are unable to characterize the force of friction, this is mainly due to (Fig. 10).

✓ The lack of practical models that describe friction forces
✓ And to learners' focus on mathematizing physical concepts; friction forces are problematic for learners because they do not have models that describe them.

The results obtained in (Q4) show that the force of friction is an important example to illustrate the impact of the mathematization of force, as the majority of learners are unable to detect the characteristics of this force.

Item 5.

Q5- This differential equation $RC\frac{di}{dt} + i = 0$ corresponds to

✓ The charge of the capacitor;
✓ The discharge of the capacitor;

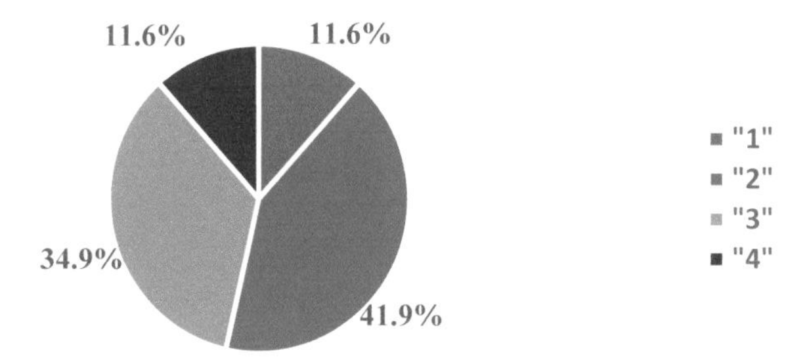

Fig. 10. Answers to the question about slowing down the car

✓ Other

Question Q5 is used to measure the insufficiency of mathematical modeling to explain the phenomenon. A differential equation without a second member was asked, two answers were proposed. 68.3% of the learners surveyed consider that the second answer, i.e. "the capacitor discharge" is the right one (Fig. 11), why?

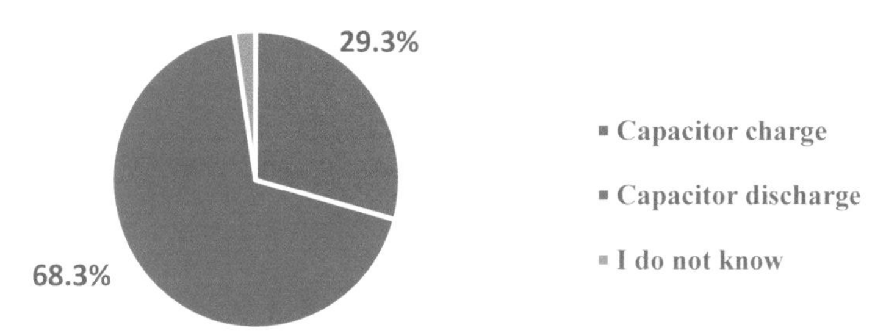

Fig. 11. Answers to the differential equation $RC\frac{di}{dt} + i$

If the students could not answer correctly, it is because the absence of the second member and their inability to assume, makes it impossible for them to understand the equation properly. In any case, they do not know that in all cases, the discharge and the load have the same differential equation; what will differentiate them is the initial conditions of these equations, these conditions are mostly not taken into consideration by the learners.

Item 6.

Q6- How many solutions correspond to this differential equation:

$$RC\frac{dUc}{dt} + Uc = E$$

✓ Single solution,

✓ Two solutions or

✓ Infinity of solutions?

Knowing that learners had to answer the equation proposed in Q5 and Q6, related to the differential equation:

$$RC\frac{dUc}{dt} + Uc = E$$

was asked to test the ability of these learners to use the same proposed model to solve this new equation. It turns out that the majority of them (80%) could not do it because they do not know that this differential equation has an infinite number of solutions and that the initial conditions serve to fix the unique solution. It is important that the mathematical modeling itself be mastered because, legitimately, it does not influence the concept (Fig. 12).

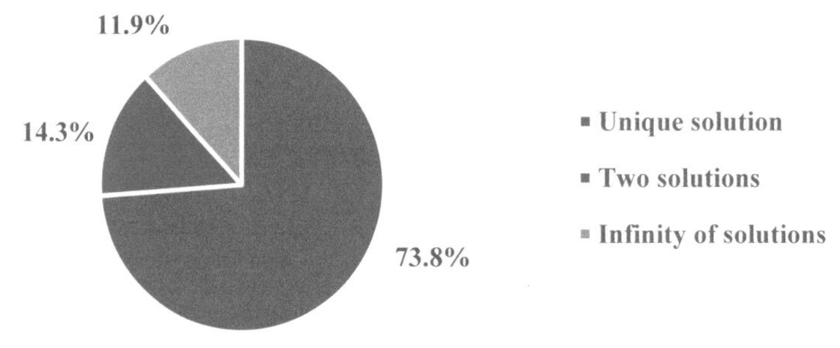

Fig. 12. Answers to the question of the differential equation $RC\frac{dUc}{dt} + Uc = E$

Item 7.
Q7- Can we consider this writing a differential equation: $I_0 = \frac{q}{\Delta t}$

✓ Yes?
✓ No?

The purpose of question Q7 is to test the learners' mastery of the definition of the differential equation. To do so, item 7 proposes a differential equation verified by the electric charge during a linear charge of the capacitor. This question is included in the National Examination for Mathematical Sciences Option A and B in Morocco.

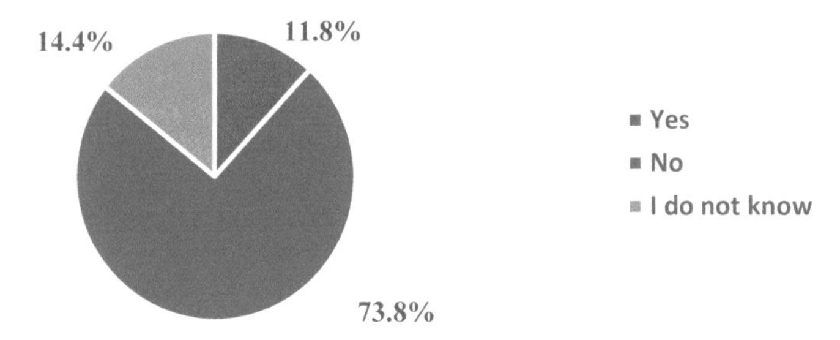

Fig. 13. Answers to the question of writing $I_0 = \frac{q}{\Delta t}$ as a differential equation

From Fig. 13, it appears that almost 80% of learners are unable to answer this question. This is mainly due to their ignorance of the definition of the differential equation and the lack of mastery of the mathematical aspect and the phenomenon. We consider in this case that teaching physics requires a significant knowledge of mathematics and elementary concepts of physics.

5 Conclusion

Mathematical modeling is an essential tool for understanding physics, but it remains insufficient. Indeed, in the course of this study, we have noticed that teachers are not effectively trained for the treatment of the Concept-Modeling duality; they are, as a result, unable to locate the concept in the appropriate cognitive field due to their lack of mastery of physical and mathematical concepts. In the initial training of physics teachers, it is necessary geta training in the epistemological study of science, especially to allow them to improve themselves through the epistemological mastery of the concept. This analysis showed that a significant number of teachers are not aware of the limitations and modeling criteria for fairly important and widely used concepts such as force and differential equation. On the learners' side, the research reveals the inability of the majority to identify the elementary characteristics of the force, hence the inadequacy of mathematical modeling in understanding the phenomenon; they do not know how to classify a force (at a distance or contact) and almost 80% of them cannot link the model (the differential equation) to the appropriate phenomenon (load/discharge) or determine the criteria of a differential equation. In summary, rigorous mathematical modeling does not affect the physical concept, but it enhances it. However, incomplete or without limits, it impacts negatively on the physical content of the concept, and then on the students' learning.

References

1. Mazouze, B., Lounis, A.: Résolution de problèmes et apprentissage des ondes: quels types de difficultés rencontrent les élèves? Review of Science, Mathematics and ICT Education **9**(2), 25–40 (2015). https://doi.org/10.26220/rev.2216
2. Chekour, M., Laafou, M., Janati-Idrissi, R.: What are the adequate pedagogical approaches for teaching scientific disciplines? Physics as a case study. J. Educ. Soc. Res. **8**(2), 141–148 (2018). https://doi.org/10.2478/jesr-2018-0025

3. Chekour, M., Laafou, M., Janati-Idrissi, R.: Distance training for physics teachers in Pspice simulator. Mediterr. J. Soc. Sci. (2015). https://doi.org/10.5901/mjss.2015.v6n3s1p232
4. Chekour, M.: The impact perception of the resonance phenomenon simulation on the learning of physics concepts. Phys. Educ. **53**(5), 055004 (2018). https://doi.org/10.1088/1361-6552/aac984
5. Atteia, M.: Physique et Mathématiques. Éditions Universitaires Européennes. https://perso.math.univ-toulouse.fr/yak/files/2011/02/Atteia_physMath.pdf (2017)
6. Tiberghien, A.: Modélisation des savoirs dans la classe en didactique de la physique. Recherches en éducation, (29) (2017). https://doi.org/10.4000/ree.2957
7. Gaidioz, P., Tiberghien, A.: Un outil d'enseignement privilégiant la modélisation. Bull. Un. Phys. **97**(850), 71–83 (2003). http://pegase.ens-lyon.fr/sites/default/files/2019-05/BUP_Modelisation_SESAMES_1.pdf
8. Tiberghien, A.: Causalité dans l'apprentissage des sciences. Intellectica **38**(1), 69–102 (2004). https://www.persee.fr/doc/intel_0769-4113_2004_num_38_1_1709
9. Searle, J.R.: What is an institution? J. Inst. Econ. **1**(1), 1–22 (2005). https://doi.org/10.1017/S1744137405000020
10. Hadji, H.: Développement de la compétence métasyntaxique dans un contexte d'apprentissage plurilingue. Cas d'élèves de 6e AEP. ACTES 5, 28. https://tacd-2021.sciencesconf.org/data/pages/TACD_2021_Actes_volume_5_final.pdf#page=30
11. Sensevy, G.: Le jeu comme modèle de l'activité humaine et comme modèle en théorie de l'action conjointe en didactique. Quelques remarques. Nouvelles perspectives en sciences sociales: Revue internationale de systémique complexe et d'études relationnelles **7**(2), 105–132 (2012). https://www.erudit.org/en/journals/npss/2012-v7-n2-npss0355/1013056ar.pdf
12. Besson, L., Borel, J.: Conceptions des élèves en physique: quels effets sur la pratique des enseignants vaudois? (2012). https://doc.rero.ch/record/233220/files/md_ms2_p13179_p23585_2012.pdf
13. Robert, A., Rogalski, J.: Le système complexe et cohérent des pratiques des enseignants de mathématiques: une double approche. Can. J. Sci. Math. Technol. Educ. **2**(4), 505–528 (2002). https://doi.org/10.1080/14926150209556538
14. Larsen-Freeman, D.: Teaching grammar. Teaching English as a second or foreign language 3, 251–266 (2001). https://www.uibk.ac.at/anglistik/staff/freeman/course-documents/tesfl_-_teaching_grammar.pdf
15. Ushioda, E.: Ema ushioda trinity college dublin, ireland language learning at university: exploring the role of motivational thinking. Motivation and Second Language Acquisition **23** (2001)

Towards Face-to-Face Smart Classroom that Adheres to Covid'19 Restrictions

Amimi Rajae(✉), Radgui Amina, and Ibn el haj el hassane

National Institute of Posts and Telecommunications, Rabat, Morocco
{amimi.rajae,radgui,ibnelhaj}@inpt.ac.ma

Abstract. Smart classrooms have improved dramatically in recent years, especially during the period of the pandemic of Covid-19 when the lockdowns were imposed in reaction to the virus's spread. Governments endeavor to maintain learning stability and seek technological solutions to establish teaching continuity. Therefore, distance learning was a viable interim solution; nevertheless, not all students worldwide have access to digital learning resources, and because of a variety of other issues, online education may be a poor choice. In this paper, we propose a model for an intelligent face-to-face classroom that adheres to Covid-19's constraints. Our proposed system transmits a real-time warning to the classroom supervisor (teacher) if students are not wearing their masks properly or are not respecting the physical distancing during the session. We base our suggested system components on existing research in artificial intelligence and deep learning-based facial recognition systems.

Keywords: Smart classrooms · Covid 19 · face mask detection · intelligent tutoring

1 Introduction

The concept of a modern classroom has attracted many researchers a long time ago. Since the 1980s with the advancement of information technology, such as networking, multimedia, and computer science; classrooms have become increasingly information-based at different levels. Thus, digital classrooms become popular as a new intelligent teaching methodology. Literature incorporates various technologies into Smart classrooms, ranging from students' basic usage of laptops to transforming the environment of the classroom into an emotionally aware area that analyzes students' engagement and improves the efficiency of teaching and learning activities.

Actually, responses to Covid-19 have taken the adoption of digital technologies in education to the next level. Online teaching, for instance, has been adopted by most countries around the world, and many tools have been used for this purpose, such as, learning management systems (LMS), Google classrooms, Zoom, Google Hangouts, Skype meet up, and YouTube [1]. Apart from its many benefits, including more flexibility, automated processes, and time savings; online

M. Lazaar et al. (Eds.): BDIoT 2022, LNNS 625, pp. 47–58, 2023.
https://doi.org/10.1007/978-3-031-28387-1_5

teaching has been recommended as the most effective way to decrease Covid-19 transmission in school areas and safeguard students and their families from infection.

However, in such a crisis, a pertinent question to ask is, whether online teaching is as effective as in-person teaching?

Certainly, in the face of the Coronavirus pandemic, online teaching presented a viable interim solution, but, not all students worldwide have access to digital learning resources, besides other challenges mentioned by Mukhtar and al. [1] such as follows:

- Limited interaction between students and lecturers.
- Lack of the necessary digital skills to teach online.
- Lack of students feedback.
- Difficulties in teaching practical and clinical materials.

In this article we propose a model for an intelligent face-to-face classroom that respects the restrictions of Covid-19. Our system automatically detects whether students in the classroom are wearing their masks or not, and calculates the distances between them, then compares them with the allowed distance. If the students do not comply with the rules, the system sends a warning to their supervisor in real-time.

We organize this paper as follows: In Sect. 2, we discuss technological advancement in the educational field during Covid-19, and deep learning approaches for face mask detection. In Sect. 3, we present our proposed system. In Sect. 4, we elaborate our experimental results, followed by a brief conclusion in Sect. 5.

2 Related Works

The pandemic has had a negative and positive impact on people's lives. On the positive side, virus spread and lockdowns have given rise to technological progress in many fields, notably e-commerce, healthcare, and the educational field [2,3]. In education, for example, many institutions began conducting online classes through internet platforms such as Zoom or Google classrooms to ensure that the lockdowns would not disrupt teaching, along with the usage of artificial intelligence-enabled robotic instructors as a part of these new online technologies [2].

Furthermore, while many researchers propose systems for detecting face masks and estimating distances to curtail virus spread in public spaces, this is the first article to our knowledge that implement such a system in a classroom environment. We suggest an in-person classroom that adheres to Covid-19 restrictions, enhanced by machine intelligence algorithms based on related works such as the following: In his study [4], Shashi Yadav describes a method for preventing the spread of the virus by monitoring in real-time if the person is using safe social distance and wearing face masks in public areas. The model was designed to operate on the Raspberry Pi 4 and the accuracy achieved was between 85% and 95%. It uses a combination of lightweight neural network MobileNetV2 and Single Shot Detector(SSD) using the transfer learning technique.

In another study, Tao Feng and al. [5] create their dataset by adopting a mask-to-face image blending approach to generate masked faces, then they present a novel loss function called Balanced Curricular Loss. After running their model, their method was efficient in eliminating noise and achieved accuracy up to 88%. Additionally, using hybrid machine learning approaches, Wang Bingshu and al. suggest a two-stage technique to detect mask wearers; the first stage uses a transfer learning model based on Faster RCNN and InceptionV2 to detect candidates wearing mask regions, and the second stage uses a broad learning system to verify facial masks. The suggested method reaches an overall accuracy of 97.32 percent for the basic scenario and 91.13 percent for the complicated scene.

3 The Proposed System

Our proposed method to build the system in Fig. 1 is divided into two steps:

Step 1: We detect automatically whether the students in the classroom are wearing their masks or not.

Step 2: We measure the distances between students and compare them to the maximum allowed.

If the students do not comply with the restrictions, the system sends a real-time warning to the teacher.

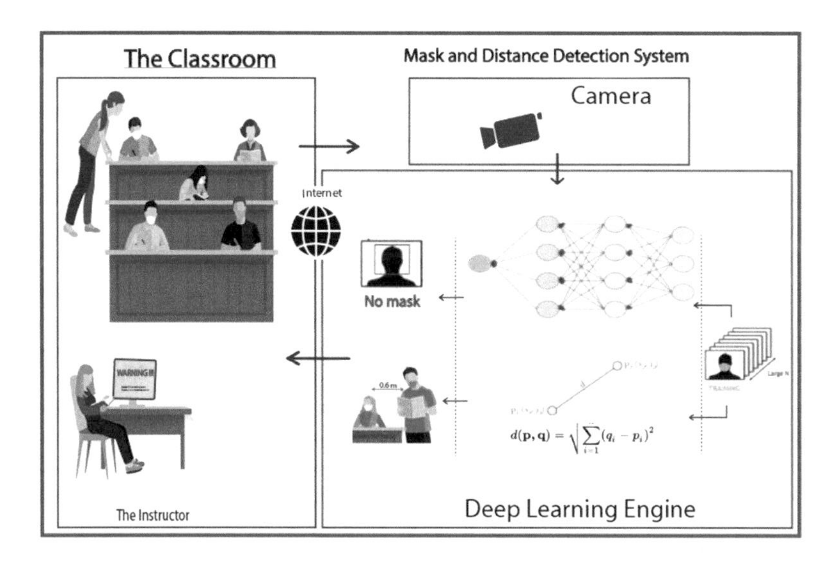

Fig. 1. Face mask recognition system

3.1 Dataset Description

Proposed Dataset: We have created our dataset with the help of 26 participants, it contains 284 images divided into two classes: 'with mask' and 'without a mask'. We have developed a tool with the Tkinter library in python to help us annotate easily and fast our data set as shown in Fig. 2. After the labeling, we got 171 images in the category without masks and 113 in the second category, and we use 80% of the images as a training set and 20% as a test set in each category. The dataset is small compared to the dataset that a neural network requires to be built efficiently from scratch, so we will use image augmentation in the Keras library to expand the size of the dataset before feeding it to the network using the following operations: rotation with a range of 20, zooming with a range of 0.15, shearing with a scale of 0.15, horizontal flipping, width and height shifting with a value of 0.2.

The use of the data augmentation technique is insufficient to obtain a large data set that can be fed to a neural network and produce accurate results, therefore, we recommend using the transfer learning technique described in Sect. 3.2, which only needs the use of a small size dataset and can still produce accurate results as it has been demonstrated in many classification tasks, such as in reference [6].

In Fig. 3, we provide a few examples from our dataset; as we can see, the data includes the faces of Moroccan/African students, which is the aspect that we couldn't find in online available datasets to train our model. Additionally, when we trained the model using a general set of data that almost exclusively included European, American, and Asian faces, the model didn't perform well, as we will demonstrate in the following section. The distinctive aspect of this data is that it includes faces with occlusion, such as scarves, beards, darker skin tones, and hands on the face which raises the performance of our proposed model.

Fig. 2. Dataset labeling

Fig. 3. Dataset samples

Face Mask Detection Dataset: This dataset is available online on Kaggle [7], and we use it to compare the results with the dataset we produced. It contains approximately 12000 images belonging to two classes: the first class labeled 'With mask', contains 5000 images in the training set and 483 in the test set, and the second class labeled 'Without mask', contains 5000 images in the training set and 509 in the test set. The samples of the used dataset are presented in Figs. 4 and 5.

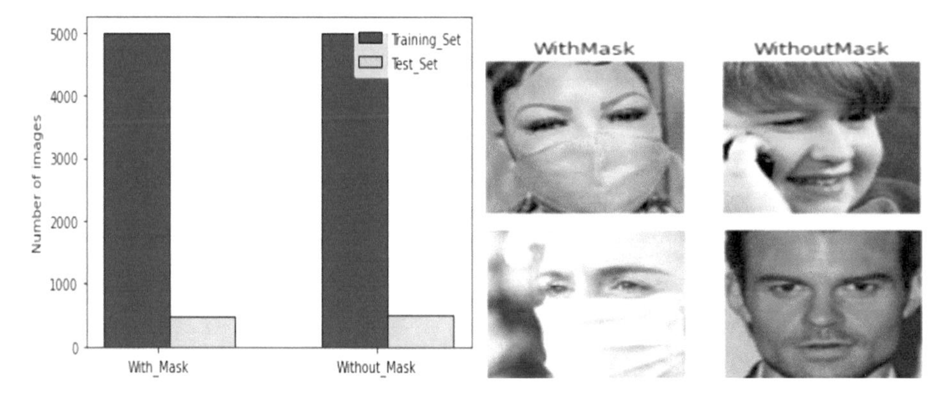

Fig. 4. Dataset visualisation

Fig. 5. Dataset samples

3.2 Face Mask Detection Approach

Image classification is one of the tasks in which deep neural networks achieve unrivaled success. However, their increased performance comes at a high cost in terms of computational complexity and time-consuming. Therefore, deep learning-based transfer learning has evolved to make the network train faster and more cost-effective [8]. Transfer learning is a machine learning research approach in which we reuse a pre-trained model as the basis for a new task for faster modeling progress, and improved performance, even when the new model is trained on a short dataset. For example, MobileNetV2, InceptionV3, VGG16, VGG19, AlexNet, etc..

In our proposed system, we start with a pre-trained MobileNetV2 [9] as a base model, we freeze all layers except the output layer, and we use fine-tuning that consists on adjusting the parameters to our classification problem, then we add five layers: an average-pooling layer with a 5×5 pool size, a flattening layer, a dense layer of 256 neurons with the rectified linear activation function (ReLU), and we regularize with a dropout of 0.5, then finally we add a dense layer with 2 neurons and the activation function (softmax). We build the model with adam optimizer and binary cross-entropy as parameters and a learning rate of 0.001. The model trains in 40 epochs and a batch size of 32. To choose the model hyper-parameters, we employ the Keras callbacks. These callbacks give insight into the internal states and data of the model during training. For instance, the EarlyStopping class stops training when a measured metric hits a threshold beyond which the training makes no progress.

Figure 6 represents the model architecture:

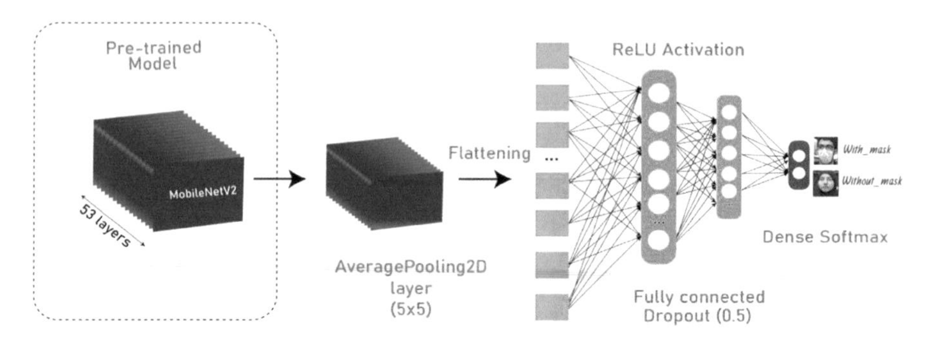

Fig. 6. Model architecture

3.3 Euclidean Distance

The Euclidean distance is the minimum distance between two points in any dimension [10]. It's the square root of the difference between two points' sum of squares as shown in the next equation:

$$d(p,q) = \sqrt{\sum_{i=1}^{n} (q_i - p_i)^2} \tag{1}$$

The numpy and scipy Python modules have many functions to calculate the euclidean distance. In our paper, we use the method "distance.euclidean()" function from scipy.spatial.

4 Experimental Results

Our system is based on a deep learning solution that uses TensorFlow and Keras libraries to train the model. For faster and more effective results, we use a pre-trained MobileNetv2 network as a transfer learning model. We calculate the distances between students and look for those that are greater than 130 cm.

We train the model on a PC with Windows 11 Operating System, Intel(R) Core(TM) i5-6300U with a 2.4 GHz processor, Tensorflow 2.3, 8 GB of RAM.

Besides, we evaluate our model results through the following parameters [11]:

$$Accuracy = \frac{(TP + TN)}{TP + TN + FP + FN} \tag{2}$$

$$Precision = \frac{(TP)}{TP + FP} \tag{3}$$

$$Recall = \frac{(TP)}{TP + FN} \tag{4}$$

$$F1 = \frac{(2 \times Precision \times Recall)}{Precision + Recall} \tag{5}$$

TP stands for true positive which is the element of the positive class correctly predicted.

TN stands for true negative which is the element of the negative class correctly predicted.

FP stands for false positive which is the element of the positive class incorrectly predicted.

FN stands for false negative which is the element of the negative class incorrectly predicted.

4.1 Results of the Proposed Dataset

In the test dataset, our approach can detect whether a person is wearing a mask or not with an accuracy score up to 95% and high as shown in Figs. 7 and 8.

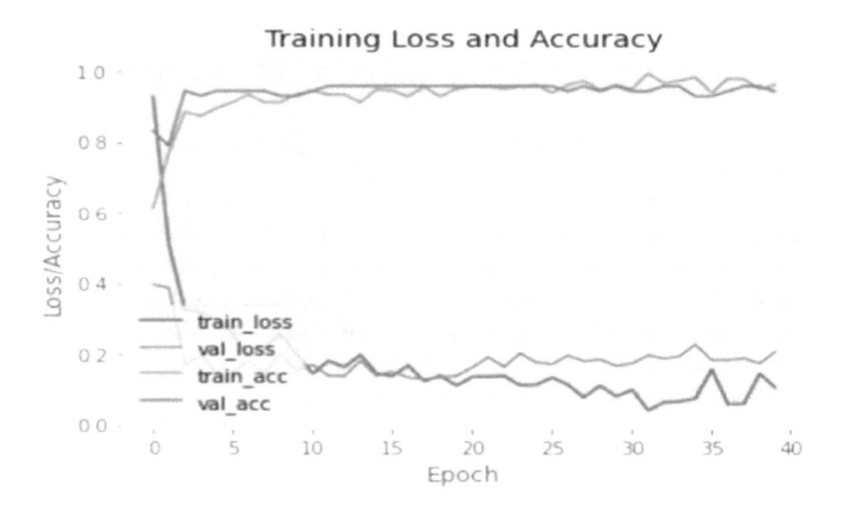

Fig. 7. Accuracy and loss of training and testing

	precision	recall	f1-score
Masked	0.96	0.89	0.93
NonMasked	0.93	0.98	0.95
accuracy			0.94
macro avg	0.95	0.93	0.94
weighted avg	0.94	0.94	0.94

Fig. 8. Evaluation parameters

In order to test the model's reliability in a real-world setting, we conducted an experiment in a university classroom and applied the model's prediction to pictures of students during the class session; the results are shown in Fig. 9.

Fig. 9. Students in classroom during the class session

The label "masked" (green color) or "no masked" (red color) is displayed on the detected face of the student as a result of the model's prediction, and the red color of the box surrounding the detected face indicates that the student is sitting too close to another student to be considered as safe (less than 130 cm) as shown in Figs. 10 and 11.

Fig. 10. Students in classroom during the class session

Fig. 11. Students in classroom during the class session

4.2 Results of Face Mask Detection Dataset

As shown in Figs. 12 and 13, the model achieved high accuracy when using the online dataset; it has an accuracy of up to 98%. However, when we test this model in a real-world setting, it exhibits significant limitations as opposed to when we use our dataset, and these limitations are evident in Fig. 14. We might highlight these limitations as follows: The presence of the girl wearing a scarf, the student covering his face with his hands, as well as the student with a beard, has impacted the model's performance. Therefore, it is evident that this dataset is not adequate for our environment.

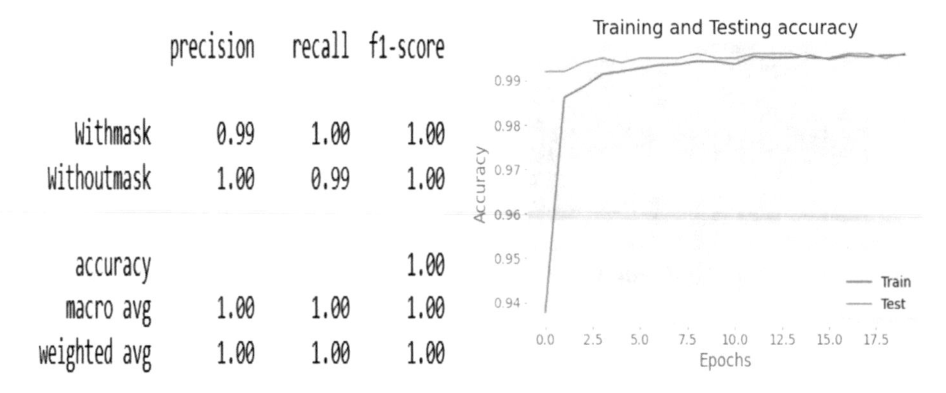

	precision	recall	f1-score
Withmask	0.99	1.00	1.00
Withoutmask	1.00	0.99	1.00
accuracy			1.00
macro avg	1.00	1.00	1.00
weighted avg	1.00	1.00	1.00

Fig. 12. Evaluation parameters

Fig. 13. Accuracy of training and testing

Fig. 14. Mask detection using general dataset

4.3 Comparison with Other Models

Table 1 compares our model trained on the database we created with other face mask detection models based on VGG19, InceptionV3, and EfficientV2L, in terms of: size, validation accuracy, and execution time.

Table 1. Comparison of models performance

Base model	Size (MB)	Validation accuracy	Time(s)
EfficientV2L	479	100%	53
VGG19	549	94%	32
InceptionV3	92	91%	8
MobileNetV2	**14**	**95%**	**4**

Notes and Comments. The size of the base models under comparison varies, and as we can see, EfficientV2L and VGG19 are the largest models. The larger size is an indication that the model has too many parameters and requires a longer time to train. So, the performance of the used machine will be essential for this feature; if the machine has great computational power, this won't be a problem. In terms of accuracy, EfficientV2L comes out on top with a score of 100%, while InceptionV3 shows the lowest score. While making predictions on an actual scene, the models produce relatively similar results, therefore the difference between their validation accuracy is not very significant. However, since our system is used in real-time, the time it takes to predict the test set is crucial. As a result, we choose the MobileNetV2 model as a base model, which performs efficiently with a value of 4 s compared to other models that exceed double this value.

5 Conclusion

The Smart classroom has altered the traditional classroom from a simple physical space that brings students and teachers together to study a subject, into a technologically enhanced environment. In this work, we propose a smart solution for limiting virus propagation inside schools. It is a system that detects the usage of masks by students and calculates the distances between them in order to provide real-time warnings to their supervisors if someone does not follow the rules. We use a transfer learning approach based on the MobileNetV2 network. We train our model on a dataset, that we have created from scratch, split into two categories 'with mask' and 'without a mask'. The proposed model achieves an accuracy of 95%. In our future work, we envision building a facial emotion recognition system that can detect the level of students' engagement using simply their gaze so it can work even when masks are worn.

Acknowledgement. The authors appreciate the support and funding provided by the National Center for Scientific and Technical Research (CNRST), as well as the participation of first-year students from the National Institute of Posts and Telecommunications in the experiment.

Acceptance of Participation and Ethical Approval. Before the study, the subjects gave their consent after being fully informed.

References

1. Mukhtar, K., Javed, K., Arooj, M., Sethi, A.: Advantages, limitations and recommendations for online learning during COVID-19 pandemic era. Pak. J. Med. Sci. **36**(COVID19-S4), S27 (2020)
2. Renu, N.: Technological advancement in the era of COVID-19. SAGE Open Med. **9**, 20503121211000910 (2021)
3. Keesara, S., Jonas, A., Schulman, K.: COVID-19 and health care's digital revolution. N. Engl. J. Med. **382**(23), e82 (2020)
4. Yadav, S.: Deep learning based safe social distancing and face mask detection in public areas for COVID-19 safety guidelines adherence. Int. J. Res. Appl. Sci. Eng. Technol. **8**(7), 1368–1375 (2020)
5. Feng, T., Xu, L., Yuan, H., Zhao, Y., Tang, M., Wang, M.: Towards mask-robust face recognition. In: Proceedings of the IEEE/CVF International Conference on Computer Vision, pp. 1492–1496 (2021)
6. Zanotti, M.: Transfer learning in image classification: how much training data do we really need? https://towardsdatascience.com
7. Jangra, A.: Face mask detection 12k images dataset. https://www.kaggle.com/datasets/ashishjangra27/face-mask-12k-images-dataset
8. Jignesh Chowdary, G., Punn, N.S., Sonbhadra, S.K., Agarwal, S.: Face mask detection using transfer learning of inceptionV3. In: Bellatreche, L., Goyal, V., Fujita, H., Mondal, A., Reddy, P.K. (eds.) BDA 2020. LNCS, vol. 12581, pp. 81–90. Springer, Cham (2020). https://doi.org/10.1007/978-3-030-66665-1_6
9. Sandler, M., Howard, A., Zhu, M., Zhmoginov, A., Chen, L.C.: Mobilenetv2: inverted residuals and linear bottlenecks. In: Proceedings of the IEEE Conference on Computer Vision and Pattern Recognition, pp. 4510–4520 (2018)
10. Wang, L., Zhang, Y., Feng, J.: On the Euclidean distance of images. IEEE Trans. Pattern Anal. Mach. Intell. **27**(8), 1334–1339 (2005)
11. Yacouby, R., Axman, D.: Probabilistic extension of precision, recall, and f1 score for more thorough evaluation of classification models. In: Proceedings of the First Workshop on Evaluation and Comparison of NLP Systems, pp. 79–91 (2020)

Supervised Machine Learning for Breast Cancer Risk Factors Analysis and Survival Prediction

Khaoula Chtouki[1(✉)], Maryem Rhanoui[1], Mounia Mikram[1], Siham Yousfi[1], and Kamelia Amazian[2]

[1] Meridian Team, LYRICA Laboratory, School of Information Sciences, Rabat, Morocco
{khaoula.chtouki,mrhanoui,mmikram,syousfi}@esi.ac.ma

[2] Faculty of Medicine and Pharmacy, Laboratoire Pathologie Humaine, Biomédecine et Environnement, Fez, Morocco
k.amazian@ispitsfes.ac.ma

Abstract. The choice of the most effective treatment may eventually be influenced by the breast cancer survival prediction. For the purpose of predicting the chances of a patient surviving, a variety of techniques were employed, such as statistical, machine learning, and deep learning models. In the current study, 1904 patient records from the METABRIC dataset were utilized to predict a 5-year breast cancer survival using a machine learning approach. In this study, we compare the outcomes of seven classification model to evaluate how well they perform using the following metrics: recall, AUC, confusion matrix, accuracy, precision, false positive rate, and true positive rate. The findings demonstrate that the classifiers for Logistic Regression (LR), Support Vector Machines (SVM), Decision Tree (DT), Random Forest (RD), Extremely Randomized Trees (ET), K-Nearest Neighbor (KNN), and Adaptive Boosting (AdaBoost) can accurately predict the survival rate of the tested samples, which is 75,4%, 74,7%, 71,5%, 75,5%, 70,3%, and 78%.

Keywords: Breast Cancer · Survival Prediction · Machine Learning · Classification · Supervised

1 Introduction

For more than four decades, the collection, intersection and analysis of massive data have been fundamental issues in the health sciences, requiring the use of artificial intelligence concepts, techniques, and tools to process, optimize and improve care by helping health professionals improve their efficiency, productivity, and consistency in the quality of care provided to patients [1,2]. Although it is mostly used in fields such as endocrinology-nutrition and hepato-gastroenterology, it now touches all medical specialties, including epidemiology, which is defined as the study of the relationship between diseases and the factors that may cause or influence their frequency, spatial distribution and evolution.

M. Lazaar et al. (Eds.): BDIoT 2022, LNNS 625, pp. 59–71, 2023.
https://doi.org/10.1007/978-3-031-28387-1_6

Many epidemiologists are looking to incorporate artificial intelligence (AI) tools and techniques into their research, as it can have many impacts in analytical research. AI can be defined as the ability of a machine to learn and recognize patterns and relationships from a sufficient number of representative examples and to use this information effectively to make decisions on unseen data [3]. For this reason, the use of AI methods has become essential in the field of epidemiology. It can find correlations between certain behaviors or socio-demographic characteristics and the presence of diseases, such as cancer, it can also predict the prevalence of infectious diseases and the survival time of a patient by considering many interdependent factors. And as in other areas of medicine, AI plays a role in detecting and mapping diseases [4,5], especially those closely related to environment and behavior, such as cancer.

In this paper, we concentrate on breast cancer, Many of us know someone who is struggling with breast cancer, or at least hear about the challenges faced by patients battling this cancer. Breast cancer (BC) is a cancerous tumor that originates in the breast tissue. It can spread straight to surrounding areas or to distant parts of the body. The cancer occurs almost exclusively in females, but males can also develop this type of cancer [6]. The malignant tumor is a group of cancerous cells that can spread to nearby tissues and destroy them. It can also extend to other parts of the body. It is the most common cancer diagnosed and one of the main causes of death in the female population. More than 2.26 million new breast cancer cases have been reported in women [7]. In 2020, 2.3 million women in the world were diagnosed with breast cancer and 685,000 died from it. By the end of 2020, 7.8 million women alive had been diagnosed with breast cancer in the preceding five years, which makes it the most common cancer in the world [8]. Based on the most recent GLOBOCAN [World Health Organization 2020] statistics, it is the number one commonly diagnosed cancer and the number five reason for cancer deaths in the world, accounting for 6.84% of all mortalities [9].

Many analytical investigations have predicted breast cancer survival using machine learning techniques in order to reduce the negative effects of BC on human health. In order to forecast the survival of patients with a specific disease over time, a variety of methodologies are used to data that are recorded in health datasets. This process is known as survival analytics. These techniques include, for instance, machine learning (ML) models [10]. A branch of artificial intelligence known as "machine learning" employs methods that let computers draw on past performance to get better at what they do. Malignant and benign cancers may be distinguished using machine learning algorithms, and breast cancer survival can also be predicted [11]. In general, these methods enable the creation of adaptive systems from different data sets, the discovery of latent links between data components, and the prediction of events [12].

The topic of this article is clinical survival prediction. We used METABRIC (Molecular Taxonomy of Breast Cancer International Consortium), which analyzes the patterns of molecules within tumors of close to two thousand women, for whom information about tumor characteristics had been meticulously recorded, to test the relationship between clinical features and outcomes. An exploratory data analysis must initially be performed on the METABRIC dataset. Risk factors are found using various techniques in the second stage. The precision, false positive rate, true positive rate, accuracy, recall, and confusion matrix are the metrics used to compare the performance of each classification method in the third stage.

The rest of the text is structured as follows: The contributions of artificial intelligence to epidemiology are discussed in Sect. 2, along with some examples of how it has been used in this discipline. The many approaches employed by various authors to forecast breast cancer survival are covered in Sect. 3, with particular emphasis on the datasets, factors included in the research under review, and machine learning algorithms. More information about the procedures employed in our investigation is provided in Sect. 4, and the findings are presented in Sect. 5.

2 Artificial Intelligence in Epidemiology

The study of diseases' relationships to potential causes or factors that may affect their incidence, spatial distribution, and evolutionary trends is known as epidemiology. It has a crucial role in public health strategies, particularly in recognizing or avoiding the emergence of new diseases or the recurrence of old ones. Only recently, in 1854, was the topic of epidemiology acknowledged as a legitimate academic discipline. Yet it is a cornerstone of both general care and preventive medicine in particular. Probability theories and statistical techniques have traditionally been used in epidemiology. In this way, it is one of the fields that has long utilized what is now known as big data. Artificial intelligence can be used in a variety of ways in this area of research to define the environmental causes of diseases through analytical studies. For instance, it can be used to determine links between particular habits or sociodemographic details and the emergence of ailments like diabetes or cancer [13].

Cancer is just one application of artificial intelligence in epidemiology. A similar connection between the environment and obesity rates was discovered in 2018 by a team of Americans from the University of Washington lead by Adyasha Maharana [14]. They conducted their study in two steps: first, they retrieved built-environment features from satellite data using a Convolutional Neural Network (CNN), and then they extracted and analyzed Point of Interest (POI) Data. They next developed a compact model utilizing net elasticity regression to evaluate the relationship between the built environment and the prevalence of obesity. As a result, various diseases that are closely tied to the environment can be found using this technique. The term "infectious disorders" immediately brings to mind COVID-19 (Coronavirus Disease 2019). The same method can be used by Deep Learning to comprehend the pandemic's evolution rates, forecast them, and modify health precautions. A team from the University of Tokyo predicted the spatiotemporal distribution of dengue in Taiwan in 2019 using data on sea temperature and rainfall [15]. We can battle infectious diseases like the Zika virus, dengue, or chikungunya based on the hypothesis that artificial intelligence can be used to model the distribution of mosquito nests [16].

3 Related Works

In order to identify the patterns associated with breast cancer patient survival, a number of machine learning techniques have already been employed to predict breast cancer survival, including logistic regression, KNNs, Bayesian networks, support vector machines, and decision trees. Many prognostic factors, including age, race, marital status, primary site, laterality, behavioral code, histology, tumor size, lymph node, extension, surgery, radiation, and TNM stage, were chosen in each of these studies. However, the handling of unknown or missing values is another issue that has been handled differently by various studies.

By using machine learning and deep learning approaches, Kalafi et al. [11] show a modest improvement in the accuracy of breast cancer survivability prediction. Their research finds the most crucial survival indicators (tumor size, age, total number of axillary lymph nodes removed, stage, and number of positive nodes). Then, they assessed the sensitivity, specificity, accuracy, F1 score, negative predictive value, false positive rate, false discovery rate, false negative rate, and Matthew correlation coefficient of the prediction models of Random Forest, Decision Trees, Support Vector Machine, and Multilayer Perceptron.

For the purpose of predicting survival rates for various subtypes of breast cancer, Montazeri et al. [17] describe a rule-based classification strategy that makes use of machine learning techniques. They used a dataset with eight attributes that contained the records of 900 patients, 876 (97.3‘%) of whom were female, and 24 (2.7%) of whom were male. They employed Naive Bayes, AdaBoost, Tree Random Forest, 1-Nearest Neighbor, Support Vector Machine, RBF Network, and Multilayer Perceptron in their research (MLP). Precision, sensitivity, accuracy, specificity, and AUC were used to gauge how well each ML approach performed (Area Under the ROC Curve).

Machine learning techniques were used by Ganggayah et al. [18] to examine breast cancer survival predictive variables. The random forest algorithm's model evaluation accuracy was a little bit better than that of the other algorithms. However, the level of accuracy demonstrated by each method appeared to be manageable. Their study found six factors to be particularly important: the cancer stage, the size of the tumor, the total number of axillary lymph nodes excised, the presence of positive lymph nodes, the technique of diagnosis, and the primary therapy type. The process of selecting variables in healthcare research, particularly when using ML approaches, may produce diverse findings based on the dataset, geography, and patient lifestyle. This study identifies the model's performance as well as the crucial factors influencing breast cancer patients' survival that may be applied in clinical practice. This study demonstrates how visualization of outcomes can be utilized to create predictive survival applications by building decision trees and survival graphs to enable validation of the major variables determining breast cancer survival.

4 Methods

On the METABRIC (Molecular Taxonomy of Breast Cancer International Consortium) dataset, the models were developed and evaluated. This dataset for 1904 breast cancer patients comprises 175 gene mutations, 31 clinical characteristics, and the mRNA z-scores for 331 genes. In the dataset, we solely kept the clinical variables. In order to determine whether the cancer patients would survive, we conducted analysis and created a model. There are a number of algorithms that can be applied to this binary classification problem. We evaluated these algorithms with the default settings to obtain a general feel of how well they performed before doing 10-fold cross-validation for each test because we are unsure yet whether one will perform better [19].

4.1 K-Fold Cross Validation Strategy

The 10-fold cross-validation approach, which separates the data into k subgroups, was used in this paper. Each of these subgroups is used for validation and k-1 is used for training in every k iterations. All data will be used exactly k times for training and once for testing as this technique is performed k times. The last estimated value is the average of the k-fold validation results [20].

4.2 Logistic Regression Classification

One technique that could be utilized to address binary classification issues is logistic regression (LR). This approach is based on the sigmoid function, an S-shaped curve that can be used to give a real number a value between 0 and 1, but never precisely within these bounds. When it is desired to be able to forecast whether cancer patients will survive based on the values of a number of predictor factors, LR is utilized. Despite being comparable to a linear regression model, it is suitable for models where the binary variable of interest [21,22].

The LR model for F independent variables can be expressed as follows:

$$F(X=1) = \frac{1}{1+e^{\beta_0+\beta_1 y_1+\beta_2 y_2+\ldots+\beta_p y_p}}$$

with F(X = 1) being the chance of survival of the cancer patient, and the regression coefficients are $\beta_0, \beta_1, \ldots, \beta_p$.

There exists a hidden linear model in the LR model. The natural logarithm of the ratio of F(X = 1) = (1− F(X = 1) leads to a linear model in y_i:

$p(y) = \ln \frac{F(X=1)}{1-F(X=1)} = \beta_0 + \beta_1 y_1 + \beta_2 y_2 + \ldots + \beta_p y_p$

Many of the necessary characteristics of a linear regression model are present in the p(y). The independent variables may include mixtures of categorical and continuous variables [23,24].

4.3 Support Vector Machines (SVM) Classification

In recent years, support vector machines (SVMs) have advanced quickly. The SVM learning problem is the term used to describe an uncertain and nonlinear dependence between a high-dimensional input vector x and a scalar output y. It is significant to note that no knowledge of the joint probability functions is accessible, necessitating the use of free distribution learning. The only information that is provided is a set of training data T = (xi,yi) ∈ X Y, i = 1, l, where l denotes the number of training data pairs and is thus equal to the size of the training data set D. Additionally, yi is marked as di, where d denotes a desired (target) value. Consequently, SVMs are a component of supervised learning approaches [25,26].

4.4 Decision Tree Classification

A decision tree (DT) is a form of tree structure resembling a diagram in which each leaf node denotes a result, each branch denotes a decision rule, and each internode denotes a feature. A DT is used to examine a piece of data in order to create a set of guidelines or inquiries that are used to forecast a class. By learning basic decision rules from the features in the data, a DT aims to create a model that can predict the value of a target variable. In this way, a DT selects the best attribute to divide the records, turns that attribute into a decision node, divides the data set into smaller subsets, and repeats this process iteratively until it starts to construct a tree [27].

4.5 Random Forest Classification

The supervised learning method known as random forests (RF) enables us to create multiple predictors before combining their various predictions to produce categorization predictive models. A group of decision trees that were typically trained using the bagging method make up these forests. The fundamental goal of bagging is to build an average of numerous noisy but roughly unbiased models in order to reduce variation [28]. The following algorithm is used to construct each tree:

- Consider X as the number of testing cases, Y as the number of variables in the classifier.
- Consider z as being the number of input variables to be applied to identify the decision in a particular node, z $\prec$ Y.
- Pick a training set for the tree and use the rest of the test cases to approximate the error.
- At each node of the tree, pick randomly z variables on which the decision should be based. Calculate the best training set score from the z variables.

A new case gets lowered in the prediction tree. The terminating node is then shown on the label. The process is really repeated for each tree in the forest, and the prediction is given for the tag with the greatest number of occurrences. We state that there are 100 trees in the forest as a whole [28,29].

4.6 Extremely Randomized Trees Classification

A machine learning approach called Extremely Randomized Trees or Extra Trees combines the predictions of various decision trees. It is comparable to the common RF algorithm. Despite using a simpler approach to create the decision trees that make up the ensemble, it frequently achieves performance that is comparable to or better than the RF algorithm [30].

4.7 K-Nearest Neighbor Classification

K-Nearest Neighbors (KNN) is a technique for keeping track of all examples that are accessible and evaluating new cases based on a similarity metric [31]. This approach is non-parametric because it doesn't assume anything about the distribution of the data at its core, and it's also less labor-intensive because it doesn't call for building a model from training data. The test stage makes use of all the training data. As a result, testing is more expensive and takes longer than training. In this strategy, the number of neighbors k is typically uneven if the total number of classes is 2. Distances like the Euclidean, Hamming, Manhattan, and Minkowski must be calculated in order to identify the closest spots [32].

4.8 Adaptive Boosting Classification

One of the boosting ensemble classifiers that Yoav Freund and Robert Schapire first presented in 1996 is adaptive boosting, also known as Ada-Boost. In order to create a strong and accurate classifier, it combines a number of weak classifiers [33]. It operates in accordance with the following steps:

Adaboost first selects a subset of training at random, then iteratively trains the model by selecting the training set based on the recent training's accurate predictions. Next, it assigns the highest weight to the misclassified observations so that they will have the highest probability of being classified in the upcoming iteration. It also assigns the highest weight to the trained classifier in each iteration based on the classifier's precision. The classifier with the highest accuracy will be given more weight. Finally, this is iterated until all training data matches accurately throughout or until the maximum number of specified estimators is reached, whichever comes first [34] The process is depicted in Fig. 1.

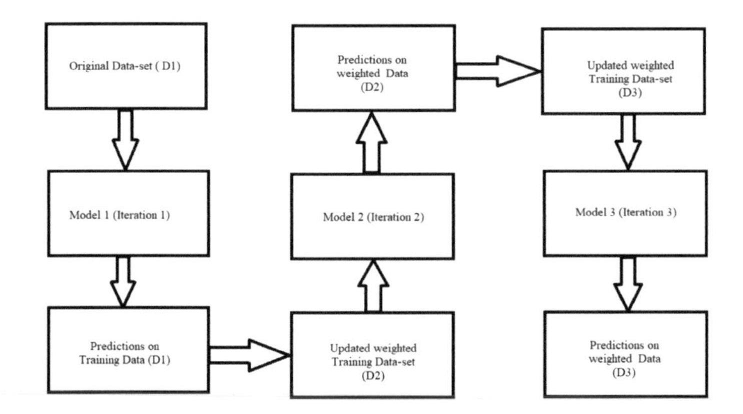

Fig. 1. The Ada Boost algorithm's steps

5 Results

5.1 Data Description

A Canadian-British collaboration produced the METABRIC (Molecular Taxonomy of Breast Cancer International Consortium) dataset, which includes chosen sequencing information from 1,980 primary breast cancer samples. The cBioPortal was used to upload the clinical and genetic data. Professor Sam Aparicio of the British Columbia Cancer Centre and Professor Carlos Caldas of the Cambridge Research Institute in Canada put together the dataset, which was then published in Nature Communications (Pereira et al., 2016). 1904 patients made up the data set for this investigation, of which 57.9% were still alive and 42.1% had passed away [19].

5.2 Performance of the Proposed Methods

Many variables are involved in some predictive modeling issues, which can slow down model development and learning and demand a lot of system memory. Additionally, some models' performance may worsen when input variables that are unrelated to the target variable are used. Because of this, many data analysts employ techniques that enable them to find the most crucial features. In our analysis, we employed SelectKbest, which calculates the relative relevance of each variable based on the highest correlation between the variables and survival time. According to Fig. 2, which displays the importance scores for each variable included in this analysis, TNM, age at diagnosis, cohort, and tumor stage were, in that order, the variables most important for predicting breast cancer survival. The recommended ML models are applied to forecast the prognosis of

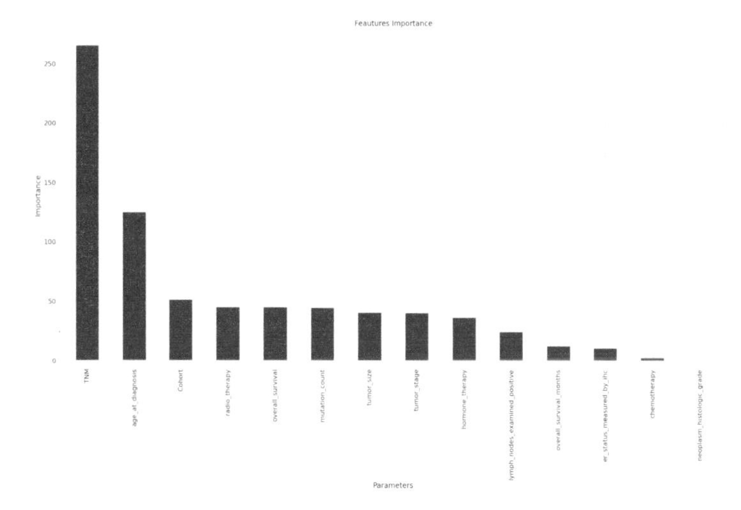

Fig. 2. The significance of many factors in determining breast cancer survival

Table 1. A comparison of all machine learning algorithms

MLA used	Train Accuracy	Test Accuracy	Precision	Recall	AUC
Logistic Regression	0.7754	0.7731	0.743169	0.727273	0.767718
AdaBoost	0.8244	0.7870	0.771429	0.0721925	0.779330
Extra Trees	1.0000	0.7153	0.716216	0.566845	0.697708
Random Forest	1.0000	0.7778	0.775758	0.684492	0.766736
SVM	0.7127	0.7361	0.721212	0.636634	0.724304
Decision Tree	1.0000	0.6875	0.644444	0.620321	0.6724304
K-Neighbors	0.8107	0.6875	0.639785	0.636364	0.681447

breast cancer patients. Seven distinct models: LR, SVM, DT, RF, ET, KNN, and Adaboost, were used in this study. Precision, false positive rate, true positive rate, recall, accuracy, AUC, and confusion matrix were just a few of the metrics used to gauge their effectiveness. When applying 10-fold cross validation on each model, the results provided in Table 1 and the box-plots in Fig. 3 demonstrate the superior performance of Adaboost. By contrasting the rates of true positives and the rates of false positives for various threshold levels, the ROC curve (Receiver Operating Characteristic) shown in Fig. 4 enables one to assess how well the algorithms we used in our study performed. It demonstrates that the AdaBoost algorithm is the most predictable algorithm.

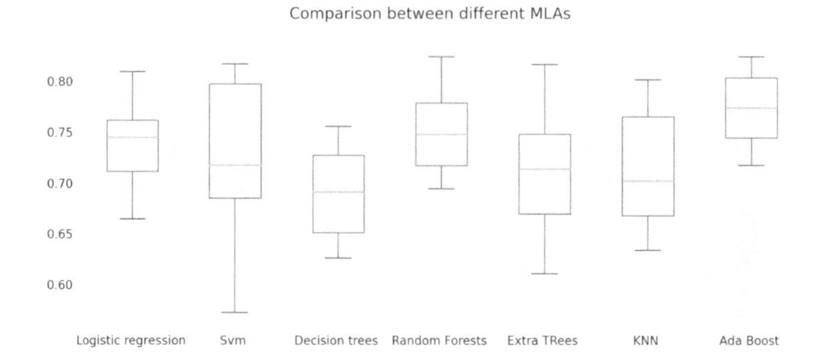

Fig. 3. Performance of LR, SVM, DT, RF, ET, KNN, and Adaboost based on accuracy with 10-fold cross validation

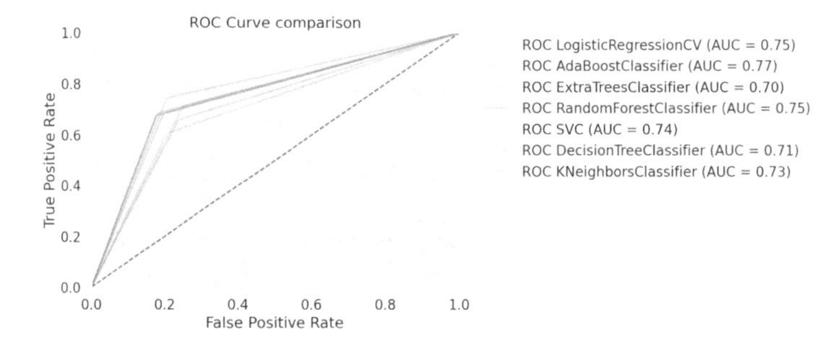

Fig. 4. Roc Curve Comparaison

The prediction level and significance test results for the survival prediction in LR, SVM, DT, RF, ET, KNN, and Adaboost are shown in Fig. 5. The confusion matrix reveals that the accuracy of LR, SVM, DT, RF, ET, KNN, and Adaboost is 75.4%, 74.1%, 71.1%, 75.1%, 70.1%, 73.1%, and 78.1%, respectively. Adaboost had the highest accuracy for survival prediction in breast cancer using the METABRIC dataset, and Extra trees had the lowest prediction accuracy of all classifiers. Adaboost achieved the best performance, dominating all evaluation metrics as shown in Table 1, Fig. 3, and Fig. 4.

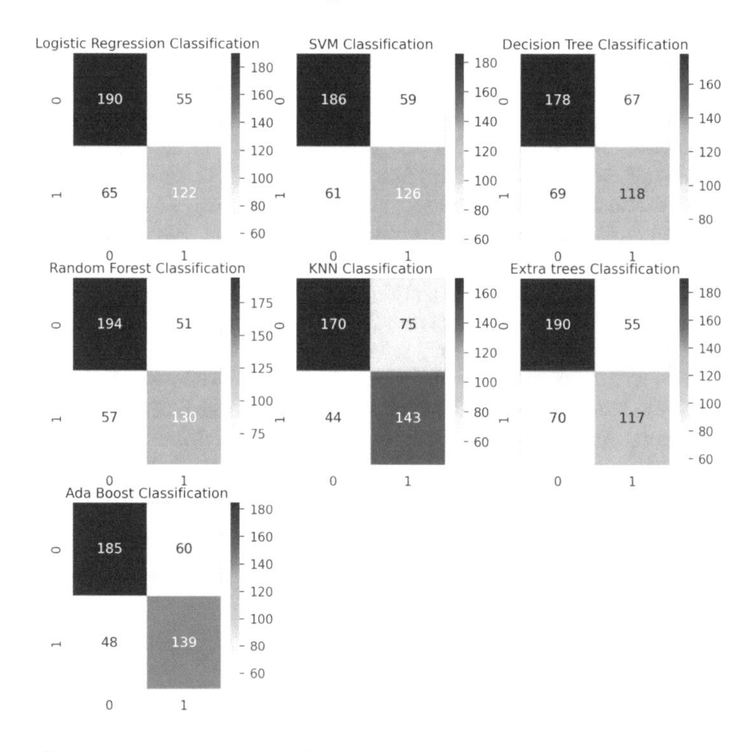

Fig. 5. Confusion matrix of LR, SVM, DT, RF, ET, KNN and Adaboost results

6 Conclusion and Recommendation

The proper diagnosis of breast cancer is crucial to save the lives of numerous people. Researchers are currently very interested in using machine learning classification algorithms to predict whether a cancer patient will survive, regardless of the usage of current diagnostic tools. With regard to machine learning classification approaches, such as LR, SVM, DT, RF, ET, KNN, and Ada Boost techniques, this study was conducted to assess their performance. These methods work well as survival prediction tools. With an accuracy of 78%, we discovered that the Ada Boost algorithm produced the most accurate results.

By gathering more data or creating new data that are statistically "near" to the existing data, we can enhance the performance of the models employed to better support the study's findings.

References

1. Yu, K.-H., Beam, A.L., Kohane, I.S.: Artificial intelligence in healthcare. Nat. Biomed. Eng. **2**(10), 719–731 (2018)
2. Harnoune, A., Rhanoui, M., Mikram, M., Yousfi, S., Elkaimbillah, Z., El Asri, B.: Bert based clinical knowledge extraction for biomedical knowledge graph construction and analysis. Comput. Methods Programs Biomed. Update **1**, 100042 (2021)

3. Thiébaut, R., Thiessard, F., et al.: Artificial intelligence in public health and epidemiology. Yearb. Med. Inform. **27**(01), 207–210 (2018)
4. Mikram, M., Moujahdi, C., Rhanoui, M., Meddad, M., Khallout, A.: Hybrid deep learning models for diabetic retinopathy classification. In: Lazaar, M., Duvallet, C., Touhafi, A., Al Achhab, M. (eds.) Proceedings of the 5th International Conference on Big Data and Internet of Things. BDIoT 2021. Lecture Notes in Networks and Systems, vol. 489, pp. 167–178. Springer, Cham (2021). https://doi.org/10.1007/978-3-031-07969-6_13
5. Abdoul-Razak, A.B., Mikram, M., Rhanoui, M., Ghouzali, S.: Hybrid machine and deep transfer learning based classification models for COVID 19 and Pneumonia diagnosis using X-ray images. In: Maleh, Y., Alazab, M., Gherabi, N., Tawalbeh, L., Abd El-Latif, A.A. (eds.) ICI2C 2021. LNNS, vol. 357, pp. 403–413. Springer, Cham (2022). https://doi.org/10.1007/978-3-030-91738-8_37
6. Al-shamasneh, A.R.M., Obaidellah, U.H.B.: Artificial intelligence techniques for cancer detection and classification: review study. Eur. Sci. J. **13**(3), 342–370 (2017)
7. Ferlay, J., et al.: Cancer statistics for the year 2020: an overview. Int. J. Cancer **149**(4), 778–789 (2021)
8. Sung, H., et al.: Global cancer statistics 2020: GLOBOCAN estimates of incidence and mortality worldwide for 36 cancers in 185 countries. CA: a Cancer J. Clin. **71**(3), 209-249 (2021)
9. Organization, W.H., et al.: Who report on cancer: setting priorities, investing wisely and providing care for all (2020)
10. Mostavi, M., Chiu, Y.-C., Huang, Y., Chen, Y.: Convolutional neural network models for cancer type prediction based on gene expression. BMC Med. Genomics **13**(5), 1–13 (2020)
11. Kalafi, E., Nor, N., Taib, N., Ganggayah, M., Town, C., Dhillon, S.: Machine learning and deep learning approaches in breast cancer survival prediction using clinical data. Folia Biol. **65**(5/6), 212–220 (2019)
12. Shinde, P.P., Shah, S.: A review of machine learning and deep learning applications. In: 2018 Fourth International Conference on Computing Communication Control and Automation (ICCUBEA), pp. 1-6 (2018). IEEE
13. Bellinger, C., Mohomed Jabbar, M.S., Zaïane, O., Osornio-Vargas, A.: A systematic review of data mining and machine learning for air pollution epidemiology. BMC Public Health **17**(1), 1–19 (2017). https://doi.org/10.1186/s12889-017-4914-3
14. Maharana, A., Nsoesie, E.O.: Use of deep learning to examine the association of the built environment with prevalence of neighborhood adult obesity. JAMA Netw. Open **1**(4), 181535 (2018)
15. Anno, S., et al.: Spatiotemporal dengue fever hotspots associated with climatic factors in Taiwan including outbreak predictions based on machine-learning. Geospatial Health **14**(2) (2019)
16. Jain, V.K., Kumar, S.: Effective surveillance and predictive mapping of mosquito-borne diseases using social media. J. Comput. Sci. **25**, 406–415 (2018)
17. Montazeri, M., Montazeri, M., Montazeri, M., Beigzadeh, A.: Machine learning models in breast cancer survival prediction. Technol. Health Care **24**(1), 31–42 (2016)
18. Ganggayah, M.D., Taib, N.A., Har, Y.C., Lio, P., Dhillon, S.K.: Predicting factors for survival of breast cancer patients using machine learning techniques. BMC Med. Inform. Decis. Mak. **19**(1), 1–17 (2019)
19. Curtis, C., et al.: The genomic and transcriptomic architecture of 2,000 breast tumours reveals novel subgroups. Nature **486**(7403), 346–352 (2012)

20. Refaeilzadeh, P., Tang, L., Liu, H.: Cross-validation. Encycl. Database Syst. **5**, 532–538 (2009)
21. Dreiseitl, S., Ohno-Machado, L.: Logistic regression and artificial neural network classification models: a methodology review. J. Biomed. Inform. **35**(5–6), 352–359 (2002)
22. Kurt, I., Ture, M., Kurum, A.T.: Comparing performances of logistic regression, classification and regression tree, and neural networks for predicting coronary artery disease. Expert Syst. Appl. **34**(1), 366–374 (2008)
23. Hosmer, D.W., Jr., Lemeshow, S., Sturdivant, R.X.: Applied Logistic Regression, vol. 398. John Wiley & Sons, New York (2013)
24. Menard, S.: Applied Logistic Regression Analysis, vol. 106. Sage, Newcastle upon Tyne (2002)
25. Noble, W.S.: What is a support vector machine? Nat. Biotechnol. **24**(12), 1565–1567 (2006)
26. Thissen, U., Van Brakel, R., De Weijer, A., Melssen, W., Buydens, L.: Using support vector machines for time series prediction. Chemom. Intell. Lab. Syst. **69**(1–2), 35–49 (2003)
27. Song, Y.-Y., Ying, L.: Decision tree methods: applications for classification and prediction. Shanghai Arch. Psychiatry **27**(2), 130 (2015)
28. Svetnik, V., Liaw, A., Tong, C., Culberson, J.C., Sheridan, R.P., Feuston, B.P.: Random forest: a classification and regression tool for compound classification and QSAR modeling. J. Chem. Inf. Comput. Sci. **43**(6), 1947–1958 (2003)
29. Speiser, J.L., Miller, M.E., Tooze, J., Ip, E.: A comparison of random forest variable selection methods for classification prediction modeling. Expert Syst. Appl. **134**, 93–101 (2019)
30. Dhananjay, B., Venkatesh, N.P., Bhardwaj, A., Sivaraman, J.: Cardiac signals classification based on extra trees model, pp. 402-406. IEEE (2021)
31. Peterson, L.E.: K-nearest neighbor. Scholarpedia **4**(2), 1883 (2009)
32. Mucherino, A., Papajorgji, P.J., Pardalos, P.M.: K-nearest neighbor classification, pp. 83-106 (2009)
33. Schapire, R.E.: Explaining AdaBoost. In: Schölkopf, B., Luo, Z., Vovk, V. (eds.) Empirical Inference, pp. 37–52. Springer, Heidelberg (2013). https://doi.org/10.1007/978-3-642-41136-6_5
34. Ying, C., Qi-Guang, M., Jia-Chen, L., Lin, G.: Advance and prospects of AdaBoost algorithm. Acta Automatica Sinica **39**(6), 745–758 (2013)

Applying Process Mining in Recommender System: A Comparative Study

Imane El Alama(✉) and Hanae Sbai

Mathematics, Computer Science, and Application Laboratory (LMCSA), University Hassan II of Casablanca, Mohammedia, Morocco
elalama.imane@gmail.com, hanae.sbai@fstm.ac.ma

Abstract. Process Mining is a combination of business process management and machine learning, which automatically allows one to discover the process model, compare it with an existing process to verify its conformity, and improve it. With the frequent use of the web and social media, recommender systems are increasingly used to build customer fidelity and smartly simplify access to services. In this paper, we conduct a comparative study between the latest works focusing on how to improve recommender systems using process mining. This study is considered the first step toward developing a new framework based on configurable process mining.

Keywords: Process Mining · Recommender System · Customer Journey · Weblog

1 Introduction

Today, companies aim to gain the fidelity of their customers and to offer them products and services that match their needs. That is the reason why the recommender system exists. It offers a better understanding of customer behavior to personalize their experience and improve their journey. In this context, the concept of the customer journey is introduced. Customer Journey is the set of interactions between customers and the company [1]. These interactions are simply an event or a collection of events where a customer looks to achieve a goal [1], for example, a customer looks to purchase a specific product or service, so the customer journey starts from the activity of research to purchase. However, the main problem is how to represent this journey and analyze it. Many web analytic tools allow companies to understand customer behavior. For example, google analytics provides a statistic on the number of people who visit the website and the most popular page visited, and other analyses. But they are very limited, and they don't offer a wide variety of information or a very fine level of granularity, so the type of data that can be analyzed is quantitative. However, the most important thing is to capture the customer's behavior. In this context, different approaches use process mining as a mature domain, a high-level and end-to-end solution to discover the process model of customer behavior [2]. Process mining techniques allow us to discover and analyze

M. Lazaar et al. (Eds.): BDIoT 2022, LNNS 625, pp. 72–83, 2023.
https://doi.org/10.1007/978-3-031-28387-1_7

customer journeys using a weblog, to make better and more successful recommendations. Different approaches use a weblog as a dataset and transform it into an event log, most transformation methods are manual (regular expression functions) [2–4] or use automatic clustering algorithms [4, 5], but to our knowledge, there is no standardization of this transformation.

After this step, they apply standard discovery algorithms on the event log obtained [1–5], the difficulty is how to obtain the representative Customer journey, to be able to assign a user to his representative journey in real-time, or to propose recommendations based on this Journey, or to identify on which step an action should be recommended.

On the other side, some researchers modify the recommendation algorithms by introducing features related to process mining. But most of them focus only on the discovery activity using process mining discovery algorithms.

The rest of this paper consists of the following sections: initially, some definition of the main element is presented in Sect. 2, in Sect. 3 a general paragraph of the related works is introduced, Sect. 4 contains the comparison criteria and the comparative tables between several existing approaches are given, in Sect. 5 a new framework is proposed, and finally the conclusion is presented in Sect. 6.

2 Background and Foundation

In this section, we will present the definition of the main elements of our study: Recommender system, weblog, Customer Journey, Process Mining, and Configurable Process Mining.

2.1 Recommender System

Nowadays, recommender systems exist everywhere. YouTube recommends videos, Spotify suggests music, Amazon lists products, and Netflix proposes movies. There are real-life examples of recommender strategies' power. So, a recommender system aims to provide a meaningful recommendation to a collection of users for items or products that might interest them [6].

A recommender system "is a subclass of information filtering system that seeks to predict the 'rating' or 'preference' a user would give to an item" [5]. So, the recommender strategy can be defined as follow:

Step 1: Collect customer Feedback. There are two types:

- Explicit feedback where the customer evaluates the contents, for example, using a rating on a scale of 1 (disliked) to 5 (liked) [6].
- Implicit feedback where the recommendation algorithms seek from data the elements that indicate whether the customer liked the content or not; for example, if a customer watches the entire video, we consider that he liked this video.

Step 2: predict the rating of items (by calculating similarity).
Step 3: recommend the top n item predicted.
The recommender systems are categorized as below:

- Collaborative Filtering: this system recommends to users the items based on the past rating of other users.
- Content-based recommending: this method recommends items similar to those liked by a user in the past.
- Hybrid approaches: These approaches combine the two previous types, Collaborative Filtering, and Content-based recommending. This type is based on textual information.

2.2 Weblog

Weblog files contain different data that depend on using a specific website [7], we can find multiple attributes like date and time of user access, IP Address of user that browses the website, HTTP Status [3], Browser and device [5]. Weblogs data are used to extract useful information and map it on an event log, it is considered as input in all research studies (Table 1).

Table 1. Weblog example.

Timestamp	Id	URL	IP Address	Device
2022-04-15 00:00	Id_1	.../Home	192.168.1.5	PC

This table shows an example of a weblog, a user identified by Id_1 visiting the home page of a website via a computer. Using these data, we can extract customer journey information from the weblog.

As we have seen in the introduction, a customer journey is a collection of events, so we can extract these events from the URLs of the weblog to create a customer journey map.

2.3 Customer Journey Map

To present the customer journey [8], CJMs are used to better understand customer's behavior and interest. This technique is too important for companies because it allows professionals to understand how users interact with different touchpoints (Steps of services) of a journey. There are two types of CJM:

- Expected Journey: the journey generated previously by the domain expert.
- Actual Journey: the journey followed by customers in real-life situations.

The customer journey map contains multiple components that are presented by [9] from their literature review, we find:

- Customer: the person who experiences the services.
- Journey: "A CJM contains at least one journey, which is a typical path followed by a customer"
- Goal: each customer journey is mapped to attempt a specific objective.
- Touchpoint: is the different interaction with products or services of companies experienced by users.

And other components like timeline, channel, stage, experience ... (Fig. 1).

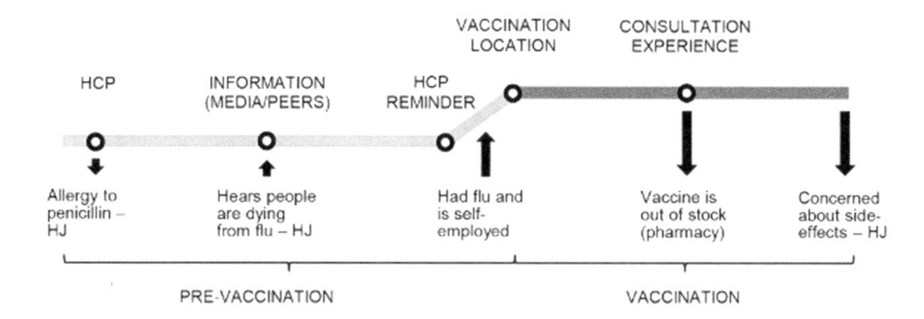

Fig. 1. Example of an analysis of the reasons for not vaccinating against the flu [10].

This example shows the reasons why people prefer not to get a vaccine against the flu, among the factors that influence their decisions, the allergy, the media, the unstoppable different rumors about side effects of the vaccine which can cause death, and the lack of resources in vaccination stocks.

The presentation of the customer journey map differs, there is no unification or standard of this map [9], has proposed a model inspired by the XES standard of the event log, to use process mining technologies.

In this context, process mining is used to discover an actual CJM, compare it with the expected one, and use recommender systems to improve it by recommending actions to attempt a certain goal.

2.4 Process Mining

Process mining is the combination of business process management and data analysis techniques. It aims to discover, monitor, and improve real processes by extracting knowledge from event logs, which are available in today's information systems. The important element in process Mining is the event log, each event in such a log refers to an activity (i.e., a well-defined step in some process) and is related to a particular case (i.e., a process instance) [11].

The event log contains extra information like resources (person or device) they executed the activity, timestamp of event, or data elements recorded with the event (e.g., the size of order) [11].

There are 3 types of process mining [11]:

- Discovery: using event log as input to discover a process model without any prior information, different discovery algorithms are used, fuzzy miner, heuristic miner, inductive miner, and other algorithms.
- Conformance Checking: compare the discovered process model with an event log from the same process to verify if the reality present in the log, conforms to the model.
- Enhancement: this type allows for improving an existing process model using information from the real process executed.

Process mining is used to discover a customer journey map or process model from a weblog, in our framework, we will use the configurable process mining, to include variability in the customer journey map, because several customer journeys share the same events or part of an event (touchpoint), and using conformance and checking in our framework, we can synchronize and compare expected customer journey map with the actual one.

2.5 Configurable Process Mining

In [12] the definition of the configurable process model is: "a Configurable Process is a process that merges multiple process variants into a single process". It Contains a different component [12], non-configurable elements represent the parts that are common between process variants, Configurable elements constitute the parts that allow for different choices, and Configuration constraints define the derivation rules for configurable process variants.

Each element of a configurable process support one of these types of variability: Functional variability concern variability of activities (atomic or composite), Behavioral variability represent sequence and control flow variability, Organizational variability considers resource variability, Informational variability refers to the variability of data and event.

As mentioned in the previous section process discovery allows to automatically discover a process model from an event log, [12] defines the variability discovery approach, to discover a configurable process instead of a simple process model.

In our framework, we will use this approach of configurable process discovery to avoid the redundancy of multiple parts in the process model result.

3 Related Works

The relationship between process mining and recommender systems is developed in several different ways, using process mining discovery is one of the ways that many approaches developed. They use machine learning algorithms to discover the process of navigation of users in web log data, and the result is analyzed by experts, to improve the recommendation strategies of a company, such as identifying key performance indicators to maximize or minimize.

The main works use weblog data as input, and after applying many transformations such as classification to obtain an event log, [2, 3] applied process mining in the news media domain. The authors of [2] use the article as the process instance and the source as the activity. The second method is to use the session as a process instance. The authors of [3] use the session as the process instance and the activities are the main page, article page, or section page. The activities represent the different sections of the pages, generally whatever the type of website, the URL is mapped to the activity in most of the works [2–5, 13]. They use a manual transformation using regular expression functions, or automatic clustering algorithms like Simple KMeans.

In the recommendation stage, only a few approaches integrate process mining, such as the ordering of activities in the OARA approach [1], or the integration of the relationship between activities [13].

The main works focus on the discovery of the process behavior or the customer journey map.

4 Comparative Study

In this part, we present our criteria of comparison and discussion of results.

4.1 Comparison Criteria

In this paragraph, we present the criteria used to compare the different approaches.

- **Data source**: There are two types of studies, the first type use weblog as input and applies process mining to discover the process of customer behavior of navigation on a website, and the second type extract from weblog customer journey information to discover a customer journey map, this criterion allows to distinguish between approaches that use the customer journey concept and approaches that are only interested in the study of the behavioral process model of a weblog.
- **Dataset type**: This criterion allows to know if the approach directly uses the normal dataset or applied modification in this dataset like filtering data, slice and dice method, saturating method, and data cleaning.
- **Process Mining tool**: There are two popular process mining tools used in most studies, Disco and ProM, ProM contain the largest variety of algorithms and Disco allows the processing of many events logs [5].
- **Discovery algorithm**: In the discovery step, they use a machine-learning algorithm to discover a process model of user behavior, then many different algorithms exist, fuzzy miner, heuristic miner, and inductive miner.

- **Transformation approach:** The input is a weblog, they must obtain the same form of an event log to use the discovery algorithm, so they applied many transformations, this criterion permits to identify if the transformations are automatic or manual.
- **Mapping:** When they transform a weblog to an event log, most approaches use a mapping function to define process instance, activity, and trace on an event log, the values are defined depending on each study, generally the URL is mapped on an activity, session on a Process Instance.
- **Recommendation method:** After obtaining a process model using the discovery algorithm, this criterion identifies the type of recommender system or algorithm used to provide the recommendations.
- **Navigation View:** To analyze user behavior, we can use three types of information [3]:
  - Syntactic information: identify the different links followed by users and how much time the user spent on each link.
  - Semantic information: trying to understand the meaning of links that are visited by users.
  - Pragmatic information: discover the goal and task of the user.

4.2 Result

In this section, the result of the comparison is summarized in the tables with the different criteria presented in the previous section.

Comparison Table: Part 1. This part of the comparison table show result of identifying the dataset and process mining approach used in the different studies.

Most of the studies use a weblog to apply the process mining approach [1–3, 5, 13], they use the information of a weblog to discover the process model of user behavior, [1, 5, 14] use the customer journey theory, they extract customer journey information from a weblog, and they apply process mining to discover customer journey map.

The dataset used in all studies is transformed in different ways. [4] uses saturating the dataset method to produce a new dataset by removing sessions that did not purchase a product or using clustering sessions. [2] applies the Slice and dice method, for example, creating a different classification for low, medium, and high popularity for different categories of articles. The compliance-based filtering is applied to remove anonymous sources or referrers under 1000 clicks [1–3, 5] directly uses the normal dataset, [5, 14] filters the dataset to remove web crawlers, a case under n event, duration of the event under n second and [15] clean their dataset.

The popular tool used to discover the process model is Disco, it is user-friendly, and too fast for big logs, in most studies, the algorithm used is Fuzzy miner [2, 4, 5, 13, 14].

In this part of the comparison, we deduce that in each approach a method of building the dataset is used to clean, complete and filter data, and the popular algorithm is Fuzzy miner using the Disco tool (Table 2).

Table 2. Comparative study results: part 1.

Article	Working element	Dataset type	Algorithm	Mining tool
[1]	Customer Journey	Real-life dataset (Stream data mining) filter depending on RFM (recency, frequency, monetary)	Algorithm not specified	Disco
[2]	Weblog	Slice and dice Compliance based filtering	Fuzzy miner	Disco
[3]	Weblog	Normal dataset	Alpha algorithm	Disco
[4]	Weblog	Saturating the dataset with customers Clustering sessions Normal dataset	Knowledge-Based miner Heuristic miner Fuzzy Miner	ProM
[5, 14]	Customer journey	Normal dataset Data filtering: web crawlers, a case under n event, duration of event < n sec trace clustering	Fuzzy miner Heuristic miner Inductive miner	Disco
[13]	Weblog	Normal dataset	Fuzzy miner	Disco
[9]	Weblog	Data cleaning	Algorithm not specified	ProM

Comparison Table: Part 2. This second part of the comparison table show result of identifying the mapping and transformation method to convert a weblog to an event log and the recommender system used in each study.

The transformation approach used in most of the research is the classification of weblog content, using sessions or categories of pages [2, 3, 5, 13, 14] or clusters [4, 5, 14]. Generally, the mapping used is transforming URL to activity, using a manual way [2–5, 13, 14], like regular expression functions, or using an automatic clustering algorithm [4, 5, 14], like SimpleKMeans, each study uses the mapping that is appropriate for the goal of the analysis to define the process instance and trace.

Many studies use process mining discovery to find the best recommender strategies or improve the existing ones [1, 5, 13, 14]. They analyze the process model of user behavior and find the decision point to achieve a specific goal. So, the rule of process mining is to capture behavior and interest to find KPI to minimize or maximize [5, 14]. On the other hand, [1] presents the OARA approach, a new recommender system that introduces the order of the activities. And the work of [5] introduces the relationship between activities in the recommender algorithm.

The navigation view of the weblog used in most studies is syntactic, and the process studies are based on activity.

Based on this comparison, the first problem we can find is the transformation method, some works use manual transformation, and there is not a unified and standard transformation. As a result of the comparison, process mining is mostly present in the discovery activity. The recommender systems are used in different ways using different algorithms modified by integrating characteristics of process mining as the order of the activity or the dependency between activities.

In the next section, we will present our proposed framework (Table 3).

Table 3. Comparative study: part 2.

Article	Transformation approach	Mapping	Recommender method
[3]	Clustering/Customer Journey segmentation	OARA	Ocular
[1]	Classification approach based on session or article (Manual)	Mapping 1: Activity = Source Trace = Set of same articles consulted Process Instance = Article Mapping 2: Activity = Category Trace = Set of sessions that visited the same article Process Instance = Session	No recommendation
[4]	Classification approach based on the content category of the web page (Manual)	Mapping 1: Activity = Type of page (main page, article page…) Trace = Set of sessions that visited the same types of pages Process Instance = Session Mapping 2: Activity = Sections page. Trace = set of sessions that have accessed the same types of sections page (sport, world, politic…)	No recommendation
[5]	A regular expression (Manual) SimpleKMeans (Automatic)	Activity = URL	No recommendation

(*continued*)

Table 3. (*continued*)

Article	Transformation approach	Mapping	Recommender method
[6, 7]	URL clustering based on page category (Manual) Unsupervised clustering (Automatic)	Mapping 1: Activity = URL (Home page, search, product) Mapping 2: Activity = (Cluster 1, Cluster 2…)	Collaborative filtering with sequence Memory-based sequence method Model-based sequence method
[8]	Classification of categories (Manual)	Trace = Session	Process-driven Strategy
[9]	Classification approach	Process Instance = Visit	No recommendation

5 Proposed Framework

This section details our proposed framework for the usage of process mining in a recommender system.

Figure 2 illustrates the three main phases of our framework:

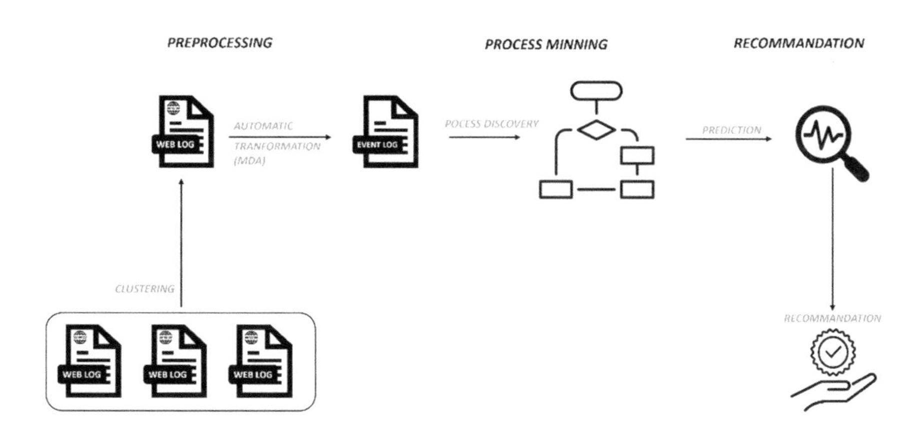

Fig. 2. Proposed framework.

Pre-processing: the input consists of multiple weblogs, after data cleaning, we apply the clustering method to have only one weblog to optimize the analysis, and the automatic transformation is applied to transform a weblog into an event log using the model-driven architecture standard to unify this transformation using the meta-model of the weblog and the event log by following the transformation rules.

Process Mining: in this step, most studies use a discovery algorithm to find a process model or customer journey map. In our framework, we have two scenarios: the first one is that company doesn't have a customer journey map, so we will use configurable process mining to discover the customer journey map. The second scenario is that the company has the expected customer journey map, and we will use configurable process verification to synchronize the expected journey with the actual one.

Recommendation: Based on the discovered process model, we will use the recommender algorithm to predict the rating of content and provide the best predicted one.

6 Conclusion

Recommender systems are an important element for companies, many types of recommender systems exist, but the challenge is always to choose the strategy to apply and to provide a better recommendation, existing algorithms use quantitative data such as the number of visits to a web page. In this context, process mining is applied to better understand the customer's behavior and interests by presenting a new approach based on the process model, it allows for analyzing the path followed by the customer to visit a page.

This comparative study discusses the different existing approaches that use process mining, especially the discovery part.

Based on this comparison, we would like to investigate the possibility of automatically using the process model result of process discovery as input in recommender algorithms.

References

1. Goossens, J., Demewez, T., Hassani. M.: Effective steering of customer journey via order-aware recommendation. In: International Conference on Data Mining Workshops 2018 (ICDMW), pp. 828–837. IEEE, Singapore (2018)
2. Epure, E.V., Ingvaldsen, J.E., Deneckere, R., Salinesi, C.: Process mining for recommender strategies support in news media. In: Tenth International Conference on Research Challenges in Information Science 2016 (RCIS), pp. 1–12. IEEE, France (2016)
3. Husin, H.S., Ismail, S.: Process mining approach to analyze user navigation behavior of a news website. In: The 4th International Conference on Information Science and Systems 2021, pp. 7–12. Association for Computing Machinery, Edinburgh United Kingdom (2021)
4. Poggi, N., Muthusamy, V., Carrera, D., Khalaf, R.: Business process mining from E-commerce web logs. In: Daniel, F., Wang, J., Weber, B. (eds.) BPM 2013. LNCS, vol. 8094, pp. 65–80. Springer, Heidelberg (2013). https://doi.org/10.1007/978-3-642-40176-3_7
5. Terragni, A., Hassani, M.: Analyzing customer journey with process mining: from discovery to recommendations. In: 6th International Conference on Future Internet of Things and Cloud 2018 (FiCloud), pp. 224–229. IEEE, Barcelona, Spain (2018)
6. Melville, P., Sindhwani, V.: Recommender systems. Encyclopedia of Machine Learning, pp. 829–838 (2011)
7. Hernandez, P., Garrigos, I., Mazon, J.-N.: Modeling Web logs to enhance the analysis of web usage data. In: Workshops on Database and Expert Systems Applications 2010, pp. 297-301. IEEE, Bilbao, Spain (2010)
8. Bernard, G., Andritsos, P.: CJM-ab: abstracting customer journey maps using process mining. In: Mendling, J., Mouratidis, H. (eds.) CAiSE 2018. LNBIP, vol. 317, pp. 49–56. Springer, Cham (2018). https://doi.org/10.1007/978-3-319-92901-9_5

9. Bernard, G., Andritsos, P.: A Process mining based model for customer journey mapping. In: Proceedings of the forum and doctoral consortium papers. In: 29th international conference on advanced information systems engineering 2017 vol. 1848. pp. 46–56. CEUR-WS (2017)
10. Wheelock, A., Miraldo, M., Parand, A., Vincent, C., Sevdalis, N.: Journey to vaccination: a protocol for a multinational qualitative study. BMJ Open **4**(1), e004279 (2014)
11. Van Der Aalst, W.: Process mining: overview and opportunities. ACM Trans. Manag. Inf. Syst. (TMIS) **3**(2), 1–17 (2012)
12. Sikal, R., Sbai, H., Kjiri, L.: Configurable process mining: variability discovery approach. In: 5th International Congress on Information Science and Technology 2018 (CiSt), pp. 137–142. IEEE, Marrakech, Morocco (2018)
13. Epure, E.V., Deneckere, R., Salinesi, C., Kille, B., Ingvaldsen, J.: Devising news recommendation strategies with process mining support. In: Interdisciplinary Workshop on Recommender Systems, pp. 1–7 (2017)
14. Terragni, A., Hassani, M.: Optimizing customer journey using process mining and sequence-aware recommendation. In: Proceedings of the 34th ACM/SIGAPP Symposium on Applied Computing, pp. 57–65. ACM, Limassol Cyprus (2019)
15. Filipowska, A., Kaluzny, P., Skrzypek, M.: Improving user experience in e-commerce by application of process mining techniques. Zeszyty Naukowe Politechniki Częstochowskiej Zarządzanie **33**(1), 30–40 (2019). https://doi.org/10.17512/znpcz.2019.1.03

An Improved Chaotic Sine Cosine Firefly Algorithm for Arabic Feature Selection

Meryeme Hadni[1(✉)] and Hassane Hjiaj[2]

[1] LAMIGEP, EMSI, Marrakech, Morocco
meryemehadni@gmail.com
[2] UAE, Faculty of Science, Department of Mathematics, Tetouan, Morocco

Abstract. Arabic documents are massively rising due to numerous contents utilized in electronic media and the web. The classification of such documents in labelled categories is a significant and vital task that deserves more attention. The main problem in Arabic text classification is how to reduce the size of the documents by selecting the most relevant features, improving the accuracy of classification. In this paper, a new variation such as chaotic sine cosine based Firefly Algorithm is proposed with hybridation of chaos theory and FA. The model is validated using kalimat Dataset, and SVM classifiers. Results reveal that the chaos-FA improve the performance of FA with an increase of 2%.

Keywords: Chaotic Firefly Algorithm · Sine Cosine · SVM Classifier

1 Introduction

The enormous growth of accessible Arabic text documents on the Web have engender a major problem on researchers to find better ways to treat with such large amount of information in order to give relevant information accurately. Which has become a complex task to satisfy the require of users.

Text Categorization (TC) is one of the supervised learning techniques we use to classify text into predefined categories. This technique has been used in topic modeling: spam detection [20], topic labeling [10], sentiment analysis [16] and mail filtering applications [19]. Automated Text classification is highly required to reduce time and space consumption, but the TC now face the problem of high dimensionality.

Feature selection [13] was a fundamental problem in classification because most existing learning algorithms were not designed to deal with a large feature space dimension. In the text classification problem, the selection of features aims to improve the computational precision and efficiency of models by removing irrelevant and redundant terms from the corpus. It was also used to select features with enough information about the text dataset. Few works have studied the effect of feature selection on Arabic text classification. For example, [8] has studied the impact of the maximum entropy method in classifying Arabic texts, and its accuracy was 0.80. [4] has showed that the SVM classifier, combined with the Chi-square-based feature selection, is an appropriate method to classify Arabic texts. [14] has compared TF.IDF, DF, Latent Semantic Indexing (LSI),

M. Lazaar et al. (Eds.): BDIoT 2022, LNNS 625, pp. 84–94, 2023.
https://doi.org/10.1007/978-3-031-28387-1_8

Stemming, and Light Stemming. Their work has shown that the former three methods outperformed the latter two stemming methods; some of them have successfully been applied in the Arabic language.

Recently, several Meta heuristic optimization algorithms were used to improve the attribute selection process. Among these approaches, the Firefly algorithm introduced by Yang in 2009 [26]. This algorithm is based on the principle of attraction between fireflies and simulates the behavior of a firefly swarm in nature. Although the Firefly algorithm has generally good efficiency for some optimization problems [15] and requires only a small number of iterations, it still has many shortcomings. However, some research have shown that the main drawbacks of FA are the high probability of falling into the local optimum, the premature convergence, as well as the step sizes moved in the firefly are highly random, which can lead to skipping the optimal solution.

In addition, Sine-Cosine (SC) [25] has introduced a new intelligent optimization algorithm based on the population; this algorithm guaranteed the diversity of the optimal solution, and including some adaptive random variables. Therefore, inspired by the search mechanism of this algorithm to overcome the problems related to FA, we hybridize FA and SC under the name: SCFA.

The rest of the paper is structured as follows: we present the related work in Sect. 2. The Sect. 3 provides the proposed chaotic Sine Cosine Firefly Algorithm. We evaluate our model and discuss the results of experiments in Sect. 4. Finally, in the last section, we present the conclusion and future work.

2 Related Works

Applying the selection feature method before TC is significantly required because of the size of large documents. Feature Selection (FS) involves detecting a subset of relevant features and significant words for use in the classification process.

In [5], the authors have proposed a new Feature Selection method based on Ant Colony Optimization (ACO) for Sentiment Analysis. Using customer review datasets, they used the k-NN technique to evaluate this proposed method's performance. The experiment results were showed a precision of 0.892. According to [17], Firefly Algorithm was developed to improve the accuracy of Web documents classification. The developed method was tested using the J48 classifier and Web KB dataset. The experiment results were showed that FA reduced the time of classifying Web documents and improved the accuracy.

The authors in [1] have proposed a new Particle Swarm Optimization (PSO) method for Feature Selection. They have investigated a new swarm intelligent technique such as an evolutionary strategy to handle Arabic Text Summarization. The proposed method reported precision of 0.67.

In [3], the authors have developed a new hybrid Feature Selection method combining Ant Colony Optimization (ACO) and Trace Oriented Feature Analysis (TOFA) for Text Classification. The results are satisfactory, and the ACO-TOFA yielded better results than TOFA. In [13] the authors have proposed a new heuristic method for feature selection for Arabic text categorization. This algorithm has been successfully applied to different combinatorial problems. The results were improved accuracy. In the paper [24], the authors have suggested two approaches: relation weighting and grouping schemes. The Naive Bayes was used as a classifier, and the results were show that the proposed methods outperform the Chi-Square and Information Gain (IG) statistical feature selection methods. The grouping methods enhance classification accuracy and reduce feature dimensionality.

A hybrid Feature selection method was proposed by [2]. This combined FS method comprises CHI squared, term frequency-inverse document frequency and mutual information. The authors have used the K-means classifier on well-known Arabic online newspapers. The hybrid approach achieved that the purity is improved by 28%.

The authors of [15] have developed a variant of the FA based Feature Selection by combining it with Simulated Annealing enhanced local and global promising solutions, chaotic-accelerated attractiveness parameters and diversion mechanisms to improved the obtained results. They used 11 regression benchmark data and 29 classification datasets to evaluate this proposed technique and compared it to different existing methods. The proposed FA variants enhance the results.

The authors in [28] have proposed an ant colony optimization approach for selecting relevant attributes. Sequential forward selection of attributes were used. Thus, they have selected the most relevant attributes in a sequential manner until the accuracy was improved. The accuracy was measured by the K-nearest neighbor classifier. The ant colony algorithm was applied to attribute selection because ants discover good paths of attributes in the search space.

In [27], the authors have proposed an optimization algorithm based on the genetic algorithm, which optimizes the parameter values for SVM. The GA-SVM method was applied to remove insignificant attributes and to efficiently find the best parameter values in the mammography (accuracy 0,781% and F-measure 0,78%) and Wisconsin breast cancer (with accuracy 0, 97% and F-measure 0,97%) in the medical datasets.

3 The SCFA Architecture

An enormous dimensionality problem is a kind challenging issue for most classifiers in which datasets have balanced number of samples and features.

In-text categorization, the major problem is the high number of features. Feature selection (FS) helps select the most informative words that discriminate between different categories in the dataset [24]. Therefore, FS is needed to reduce the high feature dimensionality without affecting the categorization accuracy. To solve this problem of growth data dimensionality, we propose a Sine Cosine Firefly Algorithm SCFA for feature selection to reduce the features of Arabic text categorization.

The proposed architecture (Fig. 1) composed on four parts which are as follow: (a) preprocessing is an important step for dataset, it consists to extract, normalize words by recovering their root, and finally transform those inputted texts into a vector. (b) Text representation: We used the BoW model to represent the documents in n-dimensional vector space where n is the number of unique words, and its weight is calculated by term frequency-inverse document frequency (TF-IDF) weighting scheme. (c) Feature selection is statistical analysis over the feature space to select a discriminative subset of features. The FA integrating the Sine-Cosine method for determining firefly position values for different fireflies. And finally (d) classification step is used to identify the category of new text on the basis of training data.

3.1 Pre-processing Phase

Preprocessing is a significant stage to accomplish meaningful information and accurate classification. It aims at prepare textual input for analysis and interpretation by a computational method. These works explore the segmentation text [7, 14], normalization, stemming [11], and the part of speech tagging of words [12]. The first step includes a few linguistic tools such as:

- Tokenization process use the spaces between words and punctuations to fragment text into words. The Arabic language words are written without short vowels. Text documents often contain white spaces, punctuation marks, and several mark-ups that indicate font changes, text subdivisions, special characters, digits, and numbers.
- Part of Speech Tagging required [12] to label words according to grammatical categories. We classify the words into verbs, adjectives, etc.
- Stop-words are repeated words in every document, so they are considered weak to be distinguished such as conjunctions, propositions, adverbs etc., and often referred to as function words. Example: (مثل ,فإن ,غالبا ,إلى ,إذا)
- Stemming is the process of substitutes words with their stem by eliminating the suffixes (هم، هن، ها، كن) and prefixes (و، ف، ب، ك). The stemmer we are using is developed based on the stemming approaches described in [11].

After extracting the distinct words for Arabic text documents in the preprocessing stage, each document is represented as a weighted vector of the terms. Term weighting is one of the important step in text classification based on the statistical approach. In this work, we used the TF-IDF weighting schemes to reflect the importance of each term in a documents or category.

3.2 Text Representation

The TF-IDF (Term Frequency – Inverse Document Frequency) reflect the relative importance of each term in a document (and category). The mathematical formula for this measure is defined as follows:

$$tfidf(t, d) = tf(t, d) * idf(t, n) \quad (1)$$

The frequency TF of a term t in a text document, measures the number of occurrences of the term t appeared in each document (number of all words).

$$tf(t, d) = \frac{count(t)}{|d|} \quad (2)$$

The IDF is used to determine whether a term is rare or common across a dataset. Common words have less value as opposed to ones that occur rarely. The IDF is defined as follows:

$$idf(t, N) = \log \frac{|N|}{n} \quad (3)$$

where N is the total number of documents in the dataset and n is the number of documents that contain the concerned term.

3.3 Feature Selection Phase

The firefly algorithm [13] is a metaheuristics algorithm inspired from the behavior of the real firefly's insect. The basic idea of this approach is that each firefly moves to the direction towards the brighter firefly in the search space to obtain the best solution. So, the goal of a firefly is to attract other fireflies regardless of their gender; the less bright fireflies will move towards, the brighter one and more brightness mean less distance between two fireflies.

In this paper, The FA algorithm has been combined with Sine Cosine method to improve feature selection process. The first step of SCFA is to generate an initial population such as the size of each document, one document represents one firefly, Initializing the number of fireflies, and Generating the position of the firefly.

After the initialization parameter, the discretization step modifies the position of the firefly to a discrete position (a binary number between [0,1] using Eq. (4).

$$S(x_{ij}) = \frac{1}{2\left(1 + x_{ij}\right)} S_0 \text{ with } 0.25 < S_0 < 1 \tag{4}$$

$$\textit{if } S(x_{ij}) < \textit{rand then the word j is selected in the document i}$$

where: indicate the probability of firefly's position (the jth word of document j) taking 1. a rand is a random number between 0 and 1.

The intensity step of the current firefly's position is used to compare between fireflies in order to decide which have the lower intensity to move with a controlled movement.

The intensity is defined as follows:

$$Intensity = \sum\nolimits_{i=0}^{N} Freq(x)_{ij} \tag{5}$$

where $Freq(x)_{ij}$ is the frequency of the firefly's position x_{ij}.

And finally, the proposed attractiveness of a firefly is proportional to the brightness: The attractiveness function is used to move towards the less firefly I_1 to the brighter firefly I_2, using Eq. (5).

$$\beta(r) = \frac{\beta_0}{(1+r)^m} \text{ with m} > 1 \tag{6}$$

and

$$\beta_0 = 1, \ r = \sum_{j=0}^{N} \left| x_{ij} - x_{ik} \right| \tag{7}$$

where β_0 is the attractiveness of a firefly at a distance r = 0, r. The Hamming distance between any two fireflies i and k, where N is the dimension of the firefly.

In this step, we adaptive sine-cosine [25] mathematics parameters were calculating the new position of the firefly, using the following formula:

$$X_i^{t+1} = \begin{cases} X_i^t + \beta_r * (x_j - x_i) + r_1 * \sin(r_2) * \varepsilon_i, \ r_4 < 0.5 \\ X_i^t + \beta_r * (x_j - x_i) + r_1 * \cos(r_2) * \varepsilon_i, \ r_4 \geq 0.5 \end{cases} \tag{8}$$

with $\varepsilon_i = \left| rand - \frac{1}{2} \right|$

where parameters r_1, r_2 and r_4 are the three parameters introduced by Sin Cosine Algorithm.

Here, $r_2 \in [0; 2\pi]$, is a random variable. r_4 are determines different search path, if the firefly moves sinusoidal or cosine.

Parameter r_1 is the convergence factor, it has a better performance in the optimization algorithm. The expression for the parameter r_1 is:

$$r_1 = r_{max} - (r_{max} - r_{min})\frac{t}{T} \tag{9}$$

where the maximum iteration number and the current iterations are represented by t and T respectively. And r_{max} and r_{min}, are constants, which can be set according to specific problem, and ε_i a vector of random values drawn from a uniform distribution or Gaussian distribution, are selected randomly in the interval [0,1].

Finally, the intensity must be updated using Eq. (5), and the best position is discretized to 0 and 1.

3.4 SVM Classifier

The resulting documents from the previous phase are treated as an input for the SVM classifier. Support Vector Machines (SVM) is a relatively new class of machine learning techniques first introduced by Vapnik [18] and has been introduced in TC by Joachims [23]. Based on the structural risk minimization principle from the computational learning theory, the SVM is a generic classifier and can be applied to solve classification problems in different domains. SVM search a decision surface to separate the training data points into two classes and make decisions based on the support vectors selected as the only practical elements in the training set.

Given a set of N linearly separable points $S = \{x_i \text{ c } R^n | i = 1, 2 \ldots L\}$, each point xi belongs to one of the two classes, labeled as y_{ic} $\{-1, +1\}$. Figure 3 is an example of an optimal hyperplane for separating two classes. SVM builds the classification model on the training data using a linear separating function to classify unseen instances. For linearly separable vectors, the kernel function is simple. The optimal separating hyperplane is the one that has the largest margin. The distance between the nearest vectors to the hyperplane is maximal.

SVM decides based on the OSH classification during classification instead of the whole training set. It simply finds out which side of the OSH the test pattern is located. Compared with other traditional pattern recognition methods, this property makes SVM highly competitive in computational efficiency and predictive accuracy [26].

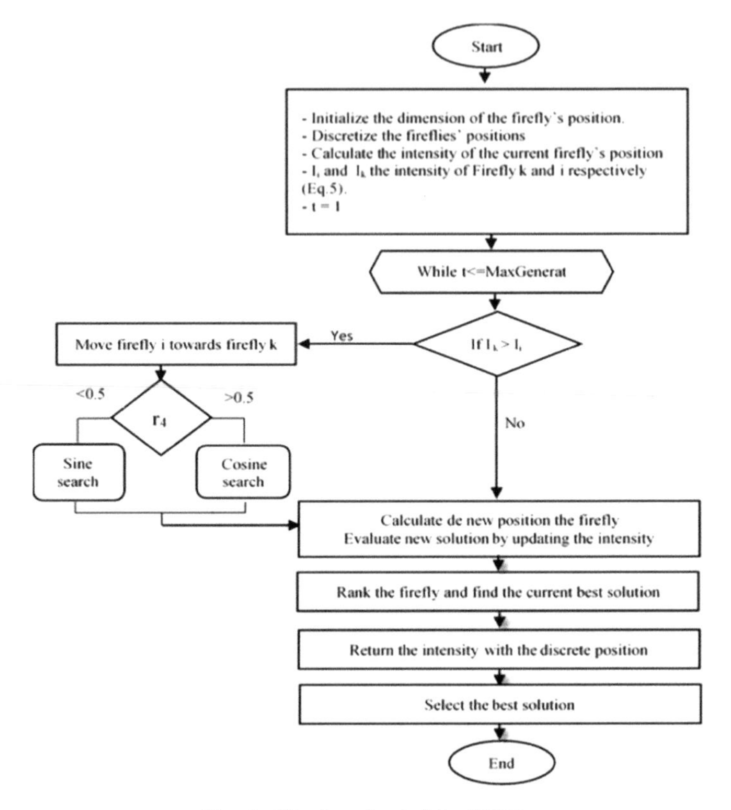

Fig. 1. The flowchart of the SCFA.

4 Evaluation and Discussion

4.1 Dataset

KALIMAT consists of: 1) 20,291 Arabic articles collected from the Omani newspaper Alwatan2 by (Abbas et al. 2011). 2) 20,291 extractive single-document system summaries. 3) 2,057 multi-document system summaries. 4) 20,291 Named Entity Recognized articles. 5) 20,291 part of speech tagged articles. 6) 20,291 morphologically analyze articles. The data collection articles fall into six categories: culture, economy, local-news, international-news, religion, and sports. Table 1 shows the collection statistics. The corpus is divided into 80% for training of the documents in each category and 20% for testing.

In our system four widely used performance metrics were utilized to evaluate the proposed method performance: precision(P), recall(R) and F-measure(F). To calculate the metrics specified by Eqs. (9)–(11).

$$P = \frac{TP}{(TP + FP)} \tag{10}$$

$$R = \frac{TP}{(TP + TN)} \tag{11}$$

Table 1. Kalimat Arabic Text Statistics.

Categories	Number of documents	Number of words
Culture	2.782	1.359.210
economy	3.468	3.122.565
local	3.596	1.460.462
international	2.035	855.945
religion	3.860	1.555.635
sports	4.550	9.813.366
Total	**20.291**	**18.167.183**

$$F = \frac{2RP}{(R+P)} \tag{12}$$

4.2 Experiment and Result

In this research, we analyze the experimental result. Text categorization system has been tested on kalimat corpus, FA and SCFA for feature selection, SVM classifier, and we adopted the different format of terms using Stem and No-Stem. The architecture (Fig. 2) is run in incremental mode on 64-bit machine with 4 GB RAM.

For classifying the document, the preprocessing phase is essential to filter the noisy and insignificant data by applying some transformation for each terms, such as: a) Normalizing the document by replacing the letter ("أآإ") with ("ا"), and replacing the letter ("ؤء") with ("ا"). b) Removing punctuation, numbers, words written in other languages, and any Arabic word containing special characters. c) Removing the diacritics of the words, if it exists.

After preprocessing, different statistical approach has been used to vector representation. In this study, TF-IDF technique has been used to extract the informative features. The result of this phase are exposed to feature selection step. In order to select the features, we used SCFA algorithm to improve feature selection process which achieve high classification accuracy compared to FA. And finally, we evaluate the ability of selected feature using SVM Classifier.

We compared the accuracy of the SCFA and FA methods for feature selection techniques of Kalimat Dataset. Table 2, present a comparison of SCFA and FA methods with TF-IDF weighting terms and without stemmer using SVM Classifier. The classifier precision of SCFA was 0.96, while the result of FA was 0.94, which indicates that SCFA be more efficient in feature selection.

To measure the impact of preprocessing (stemming) on the classification quality, we applied the ISRI-Stemmer to our database. Table 3 shows the obtained results.

Table 3 shows the results obtained when running the SCFA Feature selection on the Kalimat Dataset and ISRI-Stemmer using SVM classifier. As shown in Table 3 the classifier has achieved the highest classification accuracy using the ISRI-Stemmer with 98.7%.

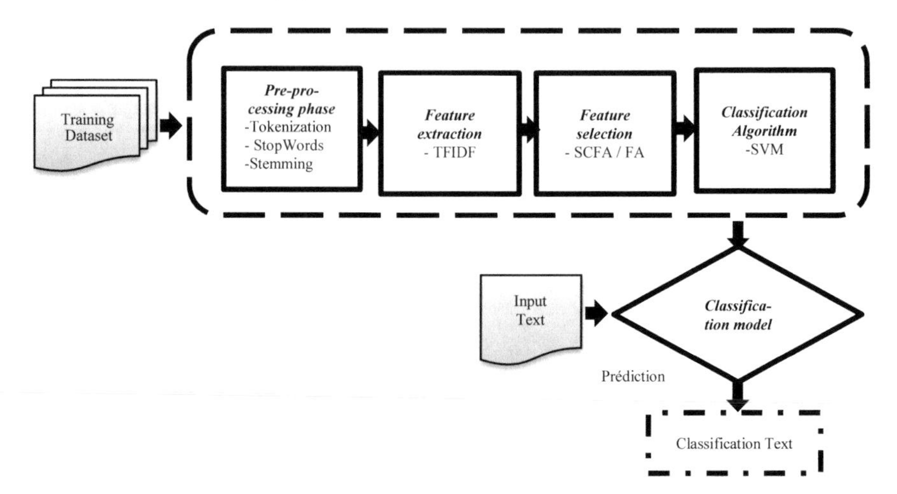

Fig. 2. Architecture of text classification using SCFA method

Table 2. Comparison results: SCF compared to FA

Class	FA			SCFA		
	P	R	F1	P	R	F1
Culture	0,981	0,957	0,97	0,994	0,99	0,99
Economy	0,988	0,98	0,98	0,854	0,967	0,91
Local	0,877	0,915	0,90	0,979	0,982	0,98
International	0,988	0,897	0,94	1	0,998	1,00
Religion	0,811	0,899	0,85	0,983	0,982	0,98
Sports	1	0,899	0,95	0,991	0,876	0,93

Table 3. Comparison results SCFA using ISRI-Stemmer for SVM classifier

Class	SCFA-No-Stem			SCFA-ISRI-Stemmer		
	P	R	F1	P	R	F1
Culture	0,994	0,990	0,992	0,990	0,988	0,989
Economy	0,854	0,967	0,907	0,966	0,956	0,961
Local	0,979	0,982	0,980	0,976	0,978	0,977
International	1,000	0,998	0,999	0,998	0,998	0,998
Religion	0,983	0,982	0,982	0,996	0,991	0,993
Sports	0,991	0,876	0,930	1,000	0,979	0,989

5 Conclusion and Future Work

In this paper, we proposed a new Arabic feature selection algorithm that combine Firefly Algorithm and Sine-Cosine method to improve classification accuracy. The algorithm was tested on Kalimat database, SVM classifier, with/without Stemmer in preprocessing. The experimental findings show that the suggested method allows choosing a more pertinent subset of characteristics. Still, the algorithm can be adopted by hybridizing it with other meta-heuristics algorithms and comparing it to other datasets in the future.

References

1. Abdalla, H., Al-Zahrani, A.M., Mathkour, H.: PSO-based feature selection for Arabic text summarization. J. Universal Comput. Sci. **21**(11), 1454–1469 (2015)
2. Alghamdi, H., Selamat, A.: The hybrid feature selection k-means method for Arabic webpage classification. Jurnal Teknologi **70**(5), 73–79 (2014)
3. Alshomrani, S., Alghamdi, H.S., Lilian Tang, H.: Hybrid ACO and TOFA feature selection approach for text classification. In: IEEE World Congress on Computational Intelligence, Brisbane, Australia, pp. 10–15 (2012)
4. Al-Harbi, S., Al-Muhareb, A., Al-Thubaity, M., Khorsheed, S., Al-Rajeh, A.: Automatic Arabic Text Classification. JADT: 9es, Journées internationales d'Analyse statistique des Données Textuelles, 77–87 (2008)
5. Azuraliza, A.B., SitiRohaidah, A., NurhafizahMoziyana, M.Y., Yaakub, M.R.: Statistical analysis for validating ACO-KNN algorithm as feature selection in sentiment analysis. In: International Conference on Electronics and Communication System (2017)
6. Ja'afaru, B., Ado SabonGari, N., Zubairu, B.: An analytical review on the recent performances of firefly algorithm (Fa). J. Eng. Res. Appl. **10** (2020)
7. Bessou, S., Saadi, A., Touahria, M.: Un système d'indexation et de recherche des textes en arabe SITRA. In: 1er séminaire national sur le langage naturel et l'intelligence artificielle LANIA, Université HAssiba ben Bouali, Département d'Informatique, Chlef DZ, 20–21 (2007)
8. El-Halees: Arabic text classification using maximum entropy. Islamic University J. 157–167 (2007)
9. El-Kourdi, M., Bensaid, A., Rachidi, T.: Automatic Arabic document categorization based on the Naïve Bayes algorithm. In: 20th International Conference on Computational Linguistics, Geneva (2004)
10. Greene, D., Cross, J.P.: Exploring the political agenda of the European parliament using a dynamic topic modeling approach. Polit. Anal. **251**, 77–94
11. Hadni, M., Ouatik, S., Lachkar, A.: Effective Arabic stemmer based hybrid approach for Arabic text categorization. IJDKP **3**(4), 1–14 (2013)
12. Hadni, M., Ouatik, S., Lachkar, A.: Hybrid part-of-speech tagger for non-vocalized Arabic text. Int. J. Nat. Lang. Comput. **2**(6), 1–15 (2013)
13. Larabi Marie-Sainte, S., Alalyani, N.: Firefly algorithm based feature selection for Arabic text classification. J. King Saud Univ. Comput. Inf. Sci. (2018)
14. Mesleh, A.: Chi-square feature extraction based SVMs Arabic language text categorization system. J. Comput. Sci. **2**, 430–435 (2007)
15. Peng, C., Limc, S., Chin Neoh, L., Zhang, K., Mistry, K.: Feature selection using firefly optimization for classification and regression models. Decis. Support Syst. **106**, 64–85 (2018)
16. Mohammad, S.: Sentiment Analysis: Automatically Detecting Valence, Emotions, and Other Affectual States from Text. https://doi.org/10.1016/B978-0-12-821124-3.00011-9.2021

17. Sarac, E., Aye Ozel, S.: Web page classification using firefly optimization. In: 2013 IEEE International Symposium on Innovations in Intelligent Systems and Applications INISTA (2013)
18. Suchanek, F., Kasneci, G., Weikum, G.: YAGO: a large ontology from Wikipedia and WN. J. Web Semant. **6**(3), 203–217 (2008)
19. Tandra, V.G., Yowen, Y., Tanjaya, R., Santoso, W.L., Nurul Qomariyah, N.: Short message service filtering with natural language processing in Indonesian language. In: 2021 International Conference on ICT for Smart Society ICISS, pp. 1–7 (2021)
20. Thirumagal Dhivya, S., Nithya, S., Sangavi Priya, G., Pugazhendi, E.: Email spam detection and data optimization using NLP techniques. Int. J. Eng. Res. Technol. **10** (2021)
21. Wang, L., Zhao, X.: Improved KNN classification algorithm research in text categorization. In: Proceedings of the 2nd International Conference on Communications and Networks CECNet, pp.1848–1852 (2012)
22. Yang, Y., Liu, X.: A re-examination of text categorization methods. In: 22nd Annual International ACM SIGIR Conference on Research and Development in Information Retrieval SIGIR 1999, pp. 42–49 (1999)
23. Yoshida, M., Ikeda, M., Ono, S., Sato, I., Nakagawa, H.: Person name disambiguation by bootstrapping. In: Proceedings of the 33rd International ACM SIGIR Conference on Research and Development in Information Retrieval, New York, USA, pp. 10–17 (2010)
24. Yousif, S.A.: Utilizing Arabic wordnet relations in Arabic text classification: new feature selection methods. IAENG Int. J. Comput. Sci. **64**, 750–761 (2019)
25. Mirjalili, S.: SCA: a sine cosine algorithm for solving optimization problems. Knowl. Based Syst. **96**, 120–133 (2016)
26. Yang, Y.: Firefly algorithms for multimodal optimization. In: Proceedings of the Stochastic Algorithms: Foundations and Applications in Computing Sciences, vol. 5792, pp. 178–178. Springer, Sapporo, Japan (2009)
27. Kumar, G., Ramachandra, G.-A., Nagamani, K.: An efficient feature selection system to integrating SVM with genetic algorithm for large medical datasets. Int. J. Adv. Res. Comput. Sci. Softw. Eng. **4**(2), 272–277 (2014)
28. El-Houby, E.-M.-F., Yassin, N., Omran, S.: Hybrid approach from ant colony optimization and K-nearest neighbor for classifying datasets using selected features. Informatica **41**(4), 495–506 (2017)

Improved Performance of Photovoltaic Panels by Image Processing

Zaidan Didi(✉) and Ikram El Azami

Computer Science Research Laboratory (LaRI), Faculty of Sciences, Ibn Tofail Kénitra University, Kénitra, Morocco
{Zaidan.didi,ikram.elazami}@uit.ac.ma

Abstract. In this paper, we have based on image processing with the internet of things technology to detect the position of the sun, seek its contour, and determine its center, we will propose with the smallest details an innovative method that allows us to better benefit and make the most of solar energy, this study focuses on improving the accuracy of solar spectrum tracking with respect to cloud cover problems. Our study is based on a Binary image processing algorithm to analyze the images acquired by a camera. We have integrated into this realization a webcam to capture the images of the sky at each time interval, and two servo motors to ensure the vertical and horizontal rotation of a joint of two axes, finally a Raspberry Pi with an open-source operating system, We exploited raspberry pi with the use of the Python programming language and the OpenCV library to perform image processing.

Keywords: Photovoltaic panel · Raspberry pi · servomotor · solar tracker

1 Introduction

The sun is a renewable and most promising and fast-growing energy that can be used continuously and regularly, especially in the sunniest countries of the world. This essential technology aims to replace the finite and costly consumption of fossil fuels. Studies show that the energy cost could be significantly minimal in regions that use this technology [1–3].

The use of cameras as the primary sensing element for solar tracking systems has recently been investigated [4–6], the high technology of the cameras these observation systems have shown significant efficiency in these studies to determine the shape and location of the sun, however, the problem of finding the threshold value to generate the binary image still arises.

In this paper, we present a design of a system that determines the accuracy of solar spectrum tracking based on a Binary image processing algorithm while integrating a webcam and a Raspberry Pi menu of a Linux operating system, as well as two servomotors.

M. Lazaar et al. (Eds.): BDIoT 2022, LNNS 625, pp. 95–104, 2023.
https://doi.org/10.1007/978-3-031-28387-1_9

2 Related Work

In recent years, most studies that focus on monitoring the sun incorporate components such as sensors as photodiodes and phototransistors [7, 8]. These studies are based on the detection of solar radiation and its transformation into an electrical signal. Disadvantages of this type of monitoring include high sensitivity to climatic conditions such as temperature, humidity, and rapid deterioration under extreme conditions [9, 10].

Other solar trackers incorporate a complex circuit-based control system [11, 12], although they overcome the difficulty of high sensitivity, but with relatively high maintenance costs. Note that other studies have used the ASC712 sensor and voltage sensor to calculate the parameters of a photovoltaic panel, such as current, voltage, and power [13].

To solve the problems related to the threshold to generate the binary image, a solar tracker based on image processing was developed in recent years using a webcam, this project was developed to locate the sun and its position [14–17].

3 Materials and Methods

3.1 Materials

In this study, the hardware design includes a Raspberry Pi 4 menu of a Raspbian distribution, two servomotors, and a rotary joint on two axes to ensure rotation according to both azimuth parameters and sun elevation, see Fig. 1.

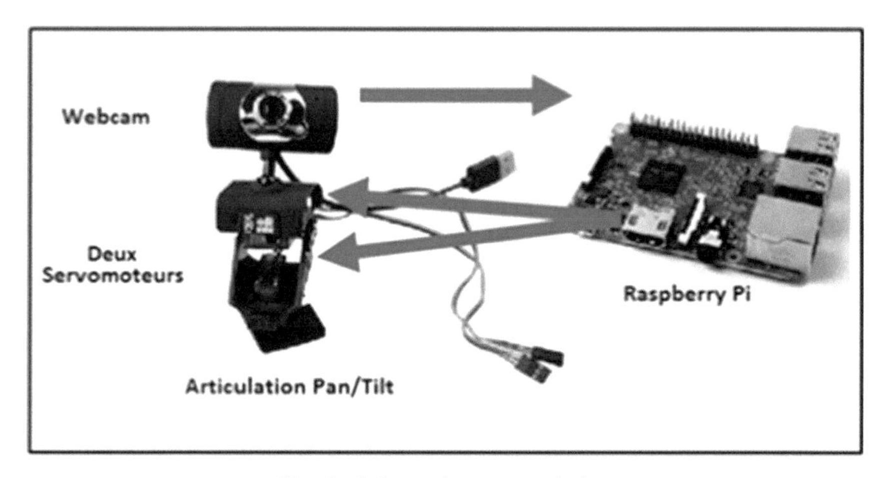

Fig. 1. Solar tracker system design

We used a Raspberry pi4 Model B with 4 GB of RAM, see Fig. 2, this 4th generation Raspberry pi has a very powerful Cortex A72 processor [18, 19], and it offers exceptional capabilities and performance. With a Debian-based free Raspbian operating system specifically designed for this hardware [20, 21].

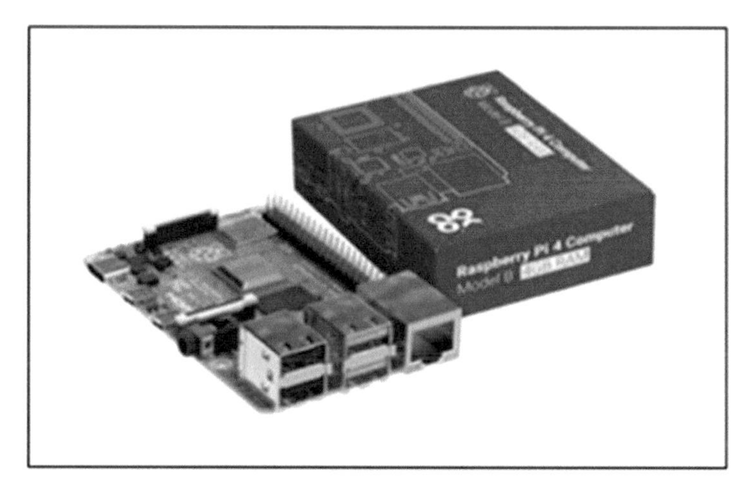

Fig. 2. Model B, Raspberry pi4 4GB

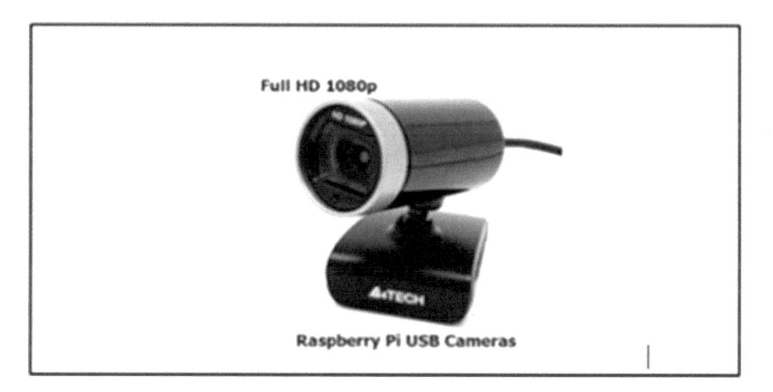

Fig. 3. Full HD 1080p USB camera

We used a USB Full HD 1080p camera to capture the image of the sky at each 30 min interval, see Fig. 3.

To find the best position of the sun, we used a rotating joint, this mechanism is based on a precise and effective joint on two axes Pan & Tilt, see Fig. 4.

3.2 Methods

The Raspberry Pi as the mainboard plays the role of an image processing center, it will control the servomotors. The webcam will capture the image of the sky each at intervals of 30 min», the captured images are sent to the Raspberry pi via a Universal Serial Bus for processing, the algorithm used to position the sun in the center of the image, then search for its contour and center. Note here that the image captured with the webcam is an RGB image encoded on 24 bits, so we have a color code of 3 digits between 0 and 255, Each digit represents the required dosage of each of the primary colors to obtain the desired color. The three codes represent the dose of red, green, and blue respectively.

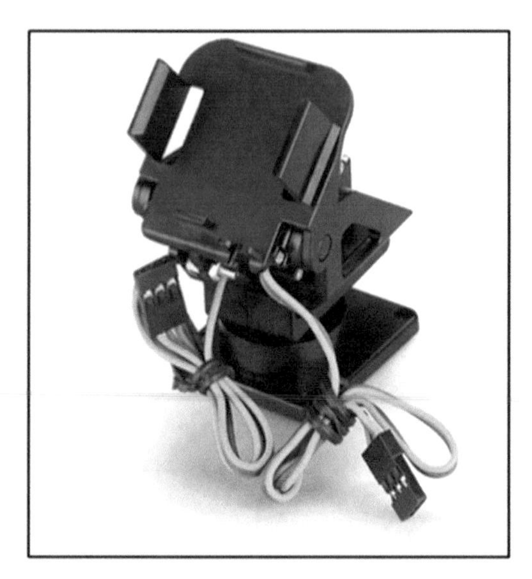

Fig. 4. Joint on two axes Pan & Tilt

To allow efficient processing, the previously captured RGB image will be converted to an 8-bit GrayScale-encoded image only. This conversion technique is based on the use of the OpenCV library of the Python programming language according to the formula (1).

$$\text{Gray} = \text{cv2.cvtColor}(\text{Image\_ Src}, \text{cv2.color\_ BGR2GRAY}) \tag{1}$$

- Image_Src: the source image of Webcam.
- cv2.COLOR_BGR2GRAY is the color code provided by OpenCV.
- BGR is blue, green, and red. (In OpenCV the image bytes are inverted it starts from blue, green, then red).

After this transformation, we obtain a gray image encoded on 8 bits instead of 24 bits. The next step is to convert the GrayScale image into a binary image to make it easier to detect the circular shape of the sun, this conversion is done by the formula (2).

$$\text{Ret, gray} = \text{cv2.threshold}(\text{Image\_ Gray}, 127, 256, \text{cv2.THRESH\_ BINARY}) \tag{2}$$

In this transformation formula the function used is cv2.threshold.

- Image_Gray is the first argument; it represents the GrayScale image.
- The second argument is 127, which is a threshold value for identifying pixel values.
- The third argument is 256, it's maxVal, it's a value we gave if the pixel value is higher than the threshold value.
- The fourth parameter is cv2.THRESH_BINARY, to present style thresholding.

This formula makes it possible to give a simple Thresholding Threshold. If the pixel value is greater than a threshold value, a value is assigned to it (maybe white), otherwise, another value (maybe black) is assigned to it.

We will then apply Gaussian convolution filters «smoothing filter» to reduce noise and the bypass filter to simplify the detection of contours with precision, these filters are widely integrated into the internal applications of digital cameras as well as in image processing software such as Photoshop on Windows or Gimp on Linux.

This part is dedicated to the detection of the sun, looking for its contour, and determining its center. To achieve our goal, we used The Hough transform, which is a form recognition technique used in binary image processing, we will integrate the OpenCV Python HoughCircles function which works very similarly to HoughLines.

To start detecting the sun's contour, the image previously captured with the Webcam must be converted to GrayScale and then to a binary image. Next, we will determine the contour of the sun with the formula (3) under Python

$$\mathrm{cv.Circle}(\mathrm{img},\ \mathrm{center},\ \mathrm{radius},\ \mathrm{color},\ \mathrm{thickness} = 1,\ \mathrm{lineType} = 8,\ \mathrm{shift} = 0) \quad (3)$$

- img: Image where the circle is drawn.
- Center: Center of the circle.
- radius: Radius of the circle.
- color: Circle color.
- thickness: Thickness of the circle outline, if positive. Negative thickness means that a filled circle is to be drawn.
- lineType: Type of the circle boundary. See the line() description.
- shift: Number of fractional bits in the coordinates of the center and in the radius value.

After the detection of the contour with the Hough transformation that targets the sunniest areas of the image, it remains to designate the contour and the center of the sun, we used the following python code:

```
if circles is not None:
circles = np.uint16(np.around(circles))
for i in circles[0, :]:center = (i[0], i[1])
# circle center
cv.circle(src, center, 1, (0, 255, 100), 2)
# circle outline
radius = i[2]
cv.circle(src, center, radius, (0, 255, 100),
```

The main flow chart is shown in Fig. 5.

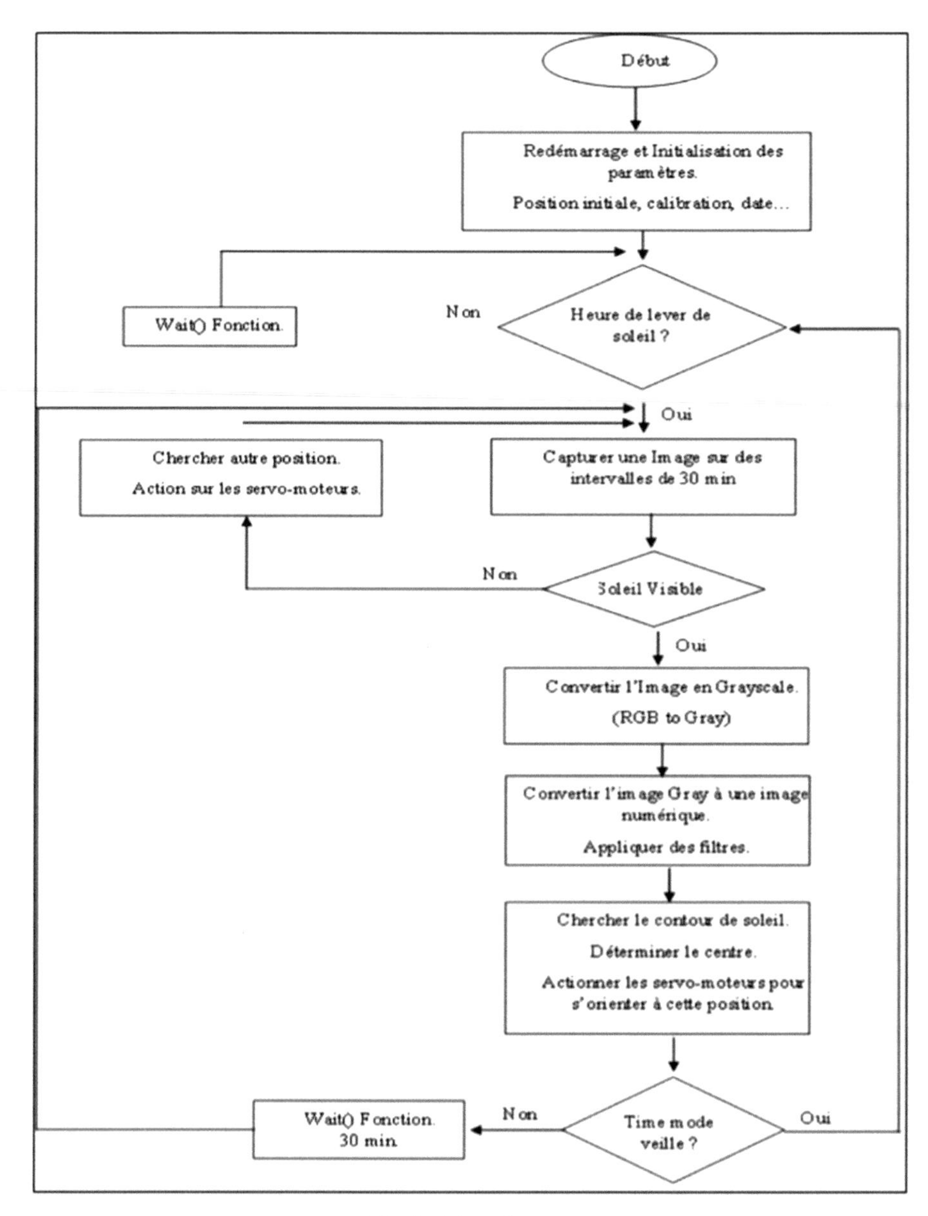

Fig. 5. Main diagram

4 Result and Discussion

4.1 Result

Our study has been implemented and tested, and the results obtained present a perfect witness of our realization, see Fig. 6(a) for an RGB image, Fig. 6(b) represents a GrayScale image, Fig. 6(c) represents a Binary image and finally Fig. 7(a) represents the contour detection and the center of the sun, Finally, the line tracing (Pan & Tilt), see Fig. 7(b).

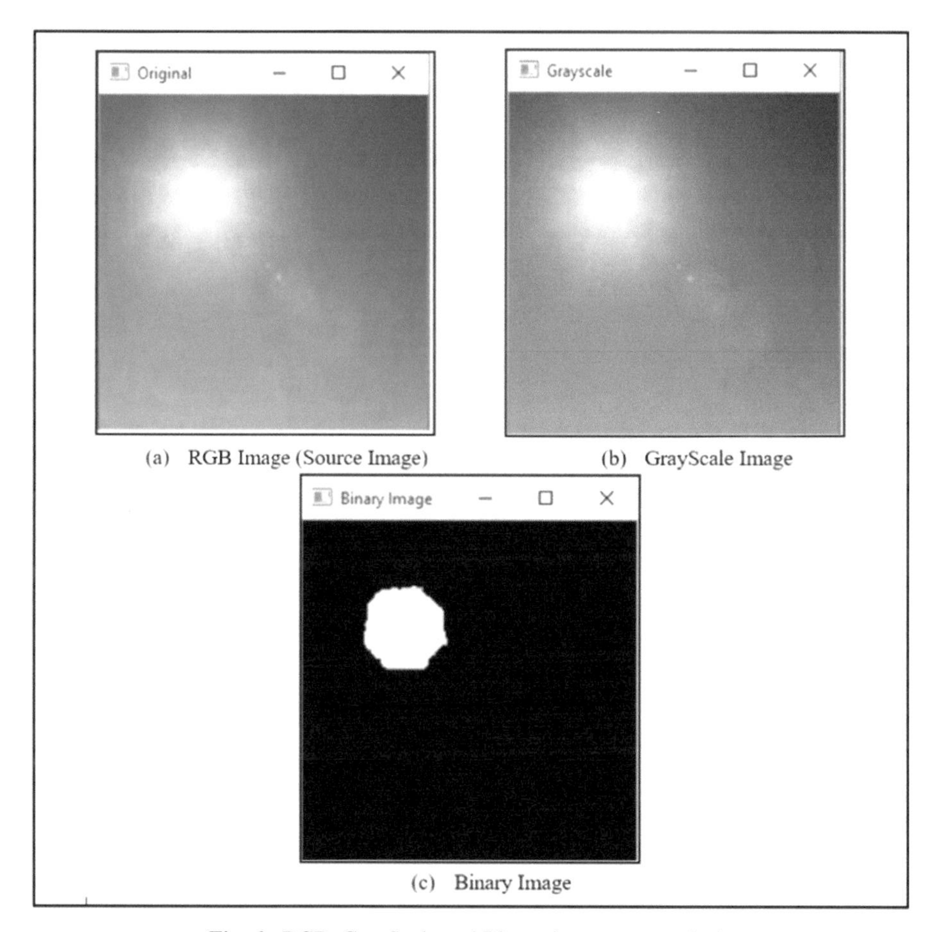

Fig. 6. RGB, GrayScale and Binary images respectively

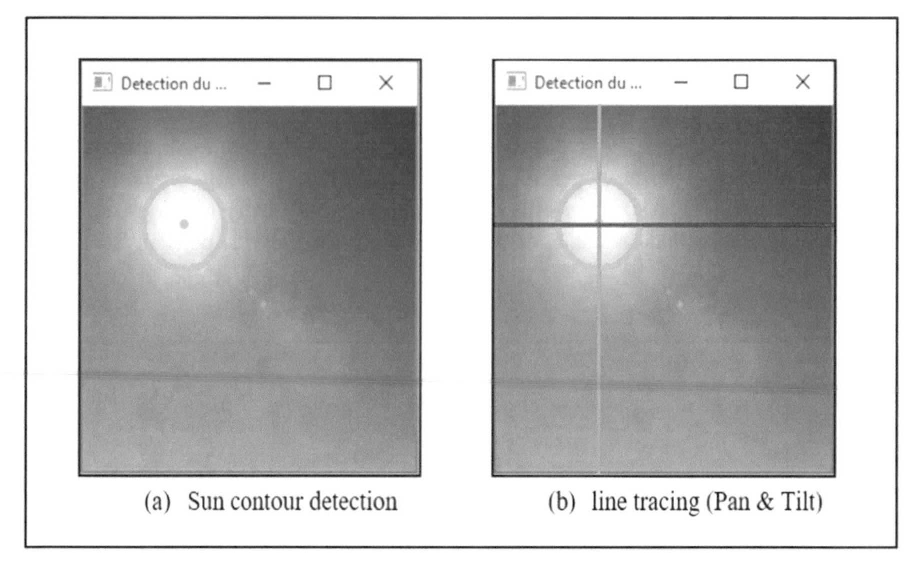

(a) Sun contour detection (b) line tracing (Pan & Tilt)

Fig. 7. Shape Detection

4.2 Discussion

Comparison with Similar Studies
In comparison with similar studies, this study has quite a lot of advantages since it is based mainly on Binary image processing, also noting that the cost can be reduced by using a webcam as an alternative to a high-cost camera.

This used Hough transform incorporates several advantages such as: simple concept, speed (by stochastic sampling), robustness to noise, and easily extensible to other domains than imaging. Unfortunately, in this study, the use of Hough transform includes the following disadvantages.

Inconvenient
We have noted here the problem of the homogeneity of space and its quantification complicates the problem of finding the threshold value to generate the binary image, in order to detect the shape and contour of the sun.

In purely specific cases, we have noticed a serious problem with the proof of convergence of the algorithm that makes the mechanism unstable, this anomaly results in a movement coming and going from the servomotors.

5 Conclusion

In this study, we extended a fundamental and active method based on the new Internet of Things technology, our study was implemented, and we noted great reliability. This shows that our study presents an acceptable solution to detect the position of the sun, the precision of the tracking of the solar spectrum gives a better efficiency for photovoltaic panels. Compared to other studies that target the same skills, our solution has strengths because it is based on digital image processing.

References

1. Renata, M., Fernando, O.: Comparative cost-benefit analysis of the energy efficiency measures and photovoltaic generation in houses of social interest in Brazil. Energy Build. **243**, 111013 (2021). https://doi.org/10.1016/j.enbuild.2021.111013
2. Muhammad, U., Naveed, R., Muhammad, Y.: Sensitivity analysis of capital and energy production cost for off-grid building integrated photovoltaic systems. Renewable Energy **186**, 195–206 (2022). https://doi.org/10.1016/j.renene.2022.01.003
3. Leonardo, M.: Energy and economic assessment of floating photovoltaics in Spanish reservoirs: cost competitiveness and the role of temperature. Renewable Energy **227**(2021), 625–634 (2021). https://doi.org/10.1016/j.solener.2021.08.058
4. Gerardo, G., Juan, R.: Fish-eye camera and image processing for commanding a solar tracker. Heliyon **5**(3), e01398 (2019). https://doi.org/10.1016/j.heliyon.2019.e01398
5. Carballo, J.A., Bonilla, J., Berenguel, M., et al.: New approach for solar tracking systems based on computer vision, low cost hardware and deep learning. Renewable Energy **133**, 1158–1166 (2019). https://doi.org/10.1016/j.renene.2018.08.101
6. Carballo, J.A., Bonilla, J., Berenguel, M., Fernández-Reche, J., García, G.: New approach for solar tracking systems based on computer vision, low cost hardware and deep learning. Renewable Energy **133**, 1158–1166 (2019). https://doi.org/10.1016/j.renene.2018.08.101
7. Nadia, A., Nor, I., Mohd, K.: Advances in solar photovoltaic tracking systems: a review. Renewable Sustainable Energy Rev. **82**(3), 2548–2569 (2018). https://doi.org/10.1016/j.rser.2017.09.077
8. Jiang, Y., Wang, R., Li, X., et al.: Photovoltaic field-effect photodiodes based on double van der Waals heterojunctions. ACS Nano **15**(9), 14295–14304 (2021). https://doi.org/10.1021/acsnano.1c02830
9. Roth, P., Georgiev, A., Boudinov, H.: Design and construction of a system for sun-tracking. Renewable Energy **29**(3), 393–402 (2004). https://doi.org/10.1016/S0960-1481(03)00196-4
10. Rizwin, M.Y., Mayadevi, N., Mini, V.P., Hari Kumar, R., Deepthi, S.: IOT-enabled economic planning of a solar PV system for long-term horizon under demand and climatic scenario. In: Komanapalli, V.L.N., Sivakumaran, N., Hampannavar, S. (eds.) Advances in Automation, Signal Processing, Instrumentation, and Control. LNEE, vol. 700, pp. 1749–1763. Springer, Singapore (2021). https://doi.org/10.1007/978-981-15-8221-9_163
11. Sharaf Eldin, S.A., Abd-Elhady, M.S., Kandil, H.A.: Feasibility of solar tracking systems for PV panels in hot and cold regions. Renewable Energy **85**(2016), 228–233 (2016). https://doi.org/10.1016/j.renene.2015.06.051
12. Ersan, K.: Review on novel single-phase grid-connected solar inverters: circuits and control methods. Solar Energy **198**, 247–274 (2020). https://doi.org/10.1016/j.solener.2020.01.063
13. Didi, Z., El Azami, I.: IoT design and realization of a supervision device for photovoltaic panels using an approach based on radiofrequency technology. In: Motahhir, S., Bossoufi, B. (eds.) ICDTA 2021. LNNS, vol. 211, pp. 365–375. Springer, Cham (2021). https://doi.org/10.1007/978-3-030-73882-2_34
14. Stinia, H., et al.: The light-on project: design and construction of a sun-tracking system using image processing. IOP Conf. Ser. Earth Environ. Sci. **642**, 012009 (2021). https://doi.org/10.1088/1755-1315/642/1/012009

15. Mahamudul Hasan, M., Hossain, A., Akhi, N.A., Abdur Razzak, M.: Design and implementation of a low-cost power diagnosis node for monitoring the stand-alone PV system in the mockery of IoT. In: Komanapalli, V.L.N., Sivakumaran, N., Hampannavar, S. (eds.) Advances in Automation, Signal Processing, Instrumentation, and Control. LNEE, vol. 700, pp. 1107–1116. Springer, Singapore (2021). https://doi.org/10.1007/978-981-15-8221-9_104
16. Pilakkat, D., Srinivasan, K.: Performance analysis of PV system under partial shading condition using predictive current control. In: Komanapalli, V.L.N., Sivakumaran, N., Hampannavar, S. (eds.) Advances in Automation, Signal Processing, Instrumentation, and Control. LNEE, vol. 700, pp. 1177–1186. Springer, Singapore (2021). https://doi.org/10.1007/978-981-15-8221-9_111
17. Pavithra, K., Kavitha, D., Krishnavenishri, R.: Optimal planning of PV-diesel hybrid systems. In: Komanapalli, V.L.N., Sivakumaran, N., Hampannavar, S. (eds.) Advances in Automation, Signal Processing, Instrumentation, and Control. LNEE, vol. 700, pp. 1499–1507. Springer, Singapore (2021). https://doi.org/10.1007/978-981-15-8221-9_140
18. Ibrahim, D.: Architecture of ARM microcontrollers. In: Arm-Based Microcontroller Multitasking Projects, pp. 13–32. Elsevier (2021). https://doi.org/10.1016/B978-0-12-821227-1.00002-5
19. Rathour, N., Singh, R., Gehlot, A.: Image and video capturing for proper hand sanitation surveillance in hospitals using euphony – a raspberry Pi and Arduino-based device. In: Singh Tomar, G., Chaudhari, N.S., Barbosa, J.L.V., Aghwariya, M.K. (eds.) International Conference on Intelligent Computing and Smart Communication 2019. AIS, pp. 1475–1486. Springer, Singapore (2020). https://doi.org/10.1007/978-981-15-0633-8_145
20. Appendix, A.: Adding PMU support to Raspbian for the Generation 1 Raspberry Pi. In: Bakos, J.D. (ed.) Embedded Systems, Morgan Kaufmann, pp. 233–235 (2016). https://doi.org/10.1016/B978-0-12-800342-8.09987-9
21. Karmakar, K., Varadharajan, V., Nepal, S., Tupakula, U.: SDN-enabled secure IoT architecture. IEEE Internet Things J. **8**(8), 6549–6564 (2021). https://doi.org/10.1109/JIOT.2020.3043740

Deep Graph Embeddings for Content Based-Book Recommendations

Amina Samih(✉), Abderrahim Ghadi, and Abdelhadi Fennan

Department of Computer Sciences, Data and Intelligent Systems (DIS) Team, FSTT, Abdelmalek Essaadi University, Tetouan, Morocco
amina.samih@etu.uae.ac.ma

Abstract. The artificial intelligence community has done much theorizing about how AI can help with the problem of successful information search in the vast reservoir of knowledge available on the Internet. Deep learning advances in speech recognition, image processing, and natural language processing have gotten much attention recently. Meanwhile, in several recent studies, deep learning is helpful in recommendation systems and information retrieval.

Recommendation systems provide individualized recommendations as a solution to this problem. Among the most common method for predicting these recommendations is content-based. This study employs this method to develop a system that provides more precise book recommendations.

Keywords: Book Recommendation · Keras Embeddings · Content filtering · Visualization

1 Introduction

The rapid growth of the Internet has ushered in a new era of data classification. The World Wide Web offers a new mode of communication that far outweighs traditional modes of communication (radio, telephone, and television). It has a significant impact not only on academic research but also on everyday life. We have arrived at a watershed moment in the evolution of how data is collected, stored, processed, presented, shared, and used. Data in the form of text, image, and video files are plentiful and easy to access. However, easy accessibility does not imply easy discovery.

Users are frequently confronted with situations in which they have too many options.

Because of the vast amount of information available on the Internet, one of the most important aspects may be overlooked if it remains unorganized and difficult to locate and use by a layperson. As a result, individuals require expert assistance in order to effectively and quickly narrow their preferences from the vast array of options available [1–3].

Many companies have implemented recommendation systems to assist users in navigating the avalanche of information. Although research in recommendation systems has been ongoing for nearly half a century, interest remains high due to the breadth of practical applications and the problem-rich domain. The Amazon.com book recommendation system is one example of an online recommendation system implemented and used [4].

M. Lazaar et al. (Eds.): BDIoT 2022, LNNS 625, pp. 105–115, 2023.
https://doi.org/10.1007/978-3-031-28387-1_10

Recommendation systems generate suggestions that users can accept or reject based on their preferences. The user can also provide implicitly (not picking) or explicit (disliking) feedback at any time during the process. Users' actions and feedback can be saved in the recommender database, which can then be used to generate new recommendations in subsequent user-system interactions. The user experience is enhanced by high-quality personalized recommendations [5, 6]. Web personalized recommendation systems have recently been used to provide users with various customized information. When it comes to book recommendations, the challenge of finding a good book will become increasingly difficult as time goes on [7].

Books are excellent sources of entertainment. Book features can frequently bring book ideas to an audience directly and immediately. The main elements that create the hype about any book are its descriptions, both after and before its release.

Many people know if they want to read a book or not by just looking at its description. The decision process is very straightforward and does not require any vote reading. Therefore, in addition to some standard book recommendation algorithms, A Neural Network Embedding is used to process book features and find similar books recommended to users. The goal is to mimic a human's ability and build an intuitive book recommender by looking at books' information based on deep learning. Decide whether or not to read the book based on the book descriptions. We can even predict the mood of a book just by looking at its textual features.

This study examines how data science methods are used to engineer information such as preferred book recommendations. The deep embedding model is used to discover the most effective and efficient method. This paper is organized as follows. First, we introduce related works in this area. Third, we present our proposed approach. then we present our results and discussion. Finally, we present our conclusions and directions for future work.

2 Related Work

The properties of items are the focus of content-based systems. The idea is to identify easily identifiable characteristics of the items, which will then be used to create a "profile." In this type of system, determining the similarity between items is equivalent to determining the similarity between their profiles. User profiles are another essential component of a content-based system. Vectors describing the user's preferences in the same dimensionality as the item profiles must be created. A utility matrix depicts the relationship between users and items once represented in the exact dimensions. The items that users like are an amalgamation of those items' profiles.

Profile vectors [1] for both the user and the items can be used to estimate the degree to which; When a user prefers an item, the cosine distance between the user's and the item's vectors is calculated.

Content-based filtering techniques [4] are used in traditional techniques for dealing with information overload. They compare items' contents to see how similar they are and then recommend similar items based on the users' previous preferences. [8] Typically, content-based filtering techniques use classifier-based approaches or nearest-neighbors methods to match items to users. Each user is related to a classifier as a profile in classifier-based approaches. The classifier takes an item as input and determines whether or not the item is preferred by associated users based on the contents of the item [9, 10]. Content-based filtering techniques are used in traditional techniques for dealing with information overload.

The content-based recommender system filters books using book descriptions [11]. The separation is done by the content recommendation system using content-based filtering. Content-based Recommender systems suffer from an overload of information, which poses numerous problems, including high cost, slow data processing, and low time complexity. So inspired by these limitations, we proposed to use neural network embeddings to represent data in low dimensional space before starting the recommendation and visualization process.

3 Proposed approach

This study employs both data science and computer science techniques. The fundamental methodologies used in this study are described in this section. Figure 1 depicts the overall of our proposition.

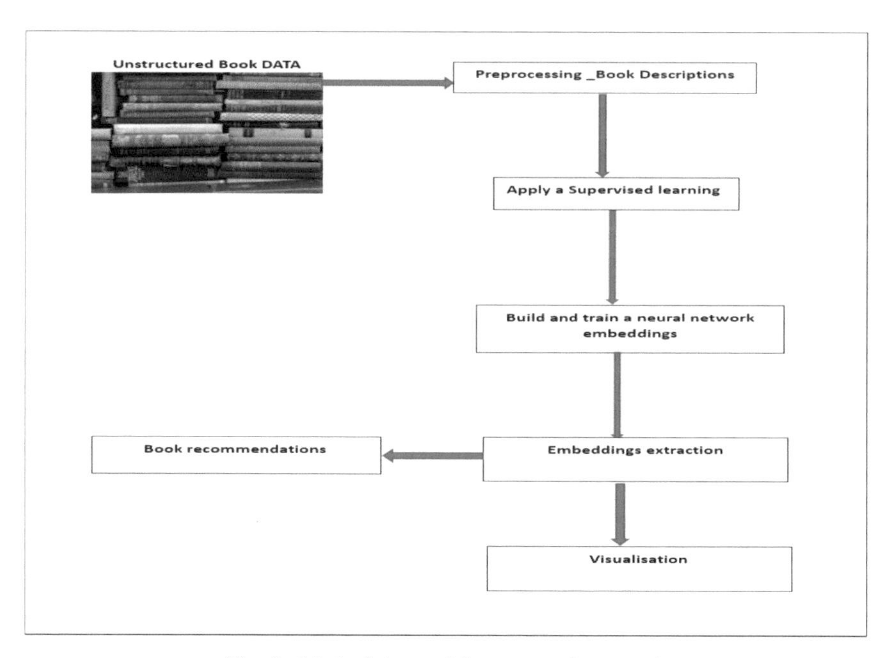

Fig. 1. Methodology of the proposed approach

3.1 Preprocessing

Before beginning the approach procedure, preprocessing is performed, including tokenization, case folding, and cleaning [29]. Tokenization is breaking down each sentence into smaller pieces known as tokens or words [30]. The process of making all of the characters in a sentence text lowercase is known as case folding [30]. Cleaning, on the other hand, employs non-alphabetic letters such as punctuation and digits.

3.2 Supervised Learning

We defined the problem at this point as determining whether or not a specific feature was present in a book article. The training examples will consist of (book, feature) pairs, and it will be up to the network to fine-tune the entity embeddings of the books and features to classify them accurately. Even though we are training for a supervised machine learning task, our ultimate goal is to learn the best entity embeddings and make accurate predictions on new data.

3.3 Embeddings Model

In the case of a word embedding model that generates word representation, we can focus on one hot vector that only contains one in every case [12, 13]. We must build an embedding layer to obtain word representations from large amounts of data [14]. On the other hand, Keras, a Python programming framework, provides a function for constructing an embedding layer with the least effort. The Keras embedding layer is the first layer in a model; positive integers are transformed into dense vectors of fixed size. This phase produces weights to learn on learning because it can be used for information extraction and visualization [15]. The generated embeddings were used to find book entities close in embedding space to one another, potentially allowing us to find the most similar books among tens of thousands of options.

3.4 Visualization

One of the most intriguing aspects of embeddings is that they can be used to visualize concepts like "War and Peace," as well as book titles and biographies. To begin, we must reduce the embeddings from 50 to 3 or 2 dimensions. We used tsne and umap to accomplish this. T-Distributed Stochastic Neighbor Embedding (t-SNE) is a technique for dimensionality reduction that is particularly well suited for visualizing high-dimensional datasets. The technique can be implemented via Barnes-Hut approximations, allowing it to be applied to large real-world datasets.

Uniform Manifold Approximation and Projection (UMAP) is a dimension reduction technique that can be used for visualization similarly to t-SNE, but also for general non-linear dimension reduction. The algorithm is founded on three assumptions about the data:

- The data is uniformly distributed on a Riemannian manifold.
- The Riemannian metric is locally constant.
- The manifold is locally connected.

At this end, we used tsne and umap. TSNE takes much longer and is designed to keep the data's local structure. UMAP is generally faster and designed to achieve a balance of local and global structure in the embedding.

3.5 Book Recommendation

In this case, recommendations are a collection of results from a prediction that includes several books that are similar to one another. It is made by feeding a list of all books into a list of books from a single book, as shown in Fig. 2. To be fitted together, either books or the selected book had to be formatted in Numpy Array. Keras' model was used to fit both Numpy Arrays. function to predict [16]. The results will also be converted to a Numpy Array to make selecting the first entries of the prediction result easier.

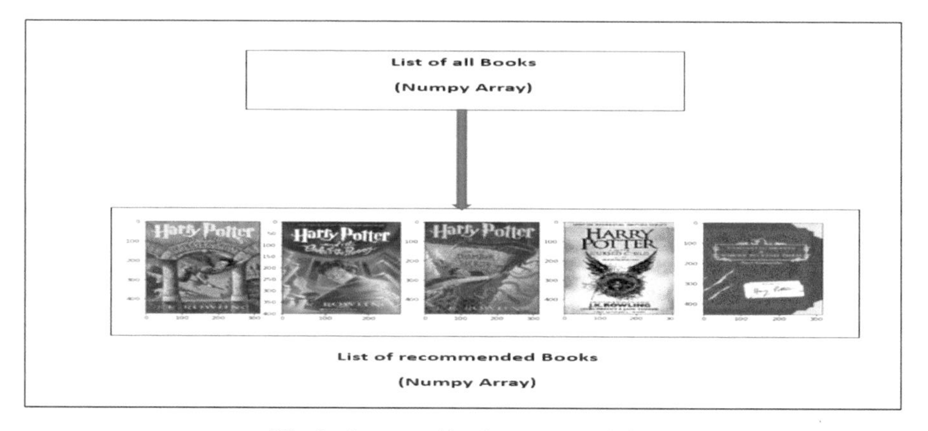

Fig. 2. Process of book recommendation

4 Results and Discussion

This section describes the study's results, and lessons learned and findings. In the training phase of the embedding model, over 37,020 books are used from a Kaggle dataset [17]. It runs for 15 epochs to understand the pattern with a 0.41 loss. The following are the explanations for each phase.

4.1 Preprocessing

The book title and its link are chosen from all columns in the book dataset [17]. To clean the dataset, it is necessary to input rows with null values, as cleaner data yields more accurate results. This phase ends with the dataset being divided into two groups: training and testing data. Training data will be used to teach the model about the data, and testing data will be used to determine the model's success rate when it makes a recommendation. Only mentioning the amount of testing data size is required when using the train test split function from the scikit-learn python library. For testing data, a value of 0.2 is mentioned.

4.2 Embeddings Model

This phase's outcome is the final architecture of an embedding model using Keras. Unlike neural network development, embedding model development necessitates importing functions such as input, embedding, flatten, and dot, as shown in Fig. 3.

Before it flattens to be vectors, books titles and links data are inputted through each input layer and embedded. The model was then compiled using Adam and Mean Squared Error as activation functions, after which the vectors were combined using the dot function. We run the embedding model for 15 epochs. As shown in Fig. 3, the low loss in predicting book recommendations was 0.41.

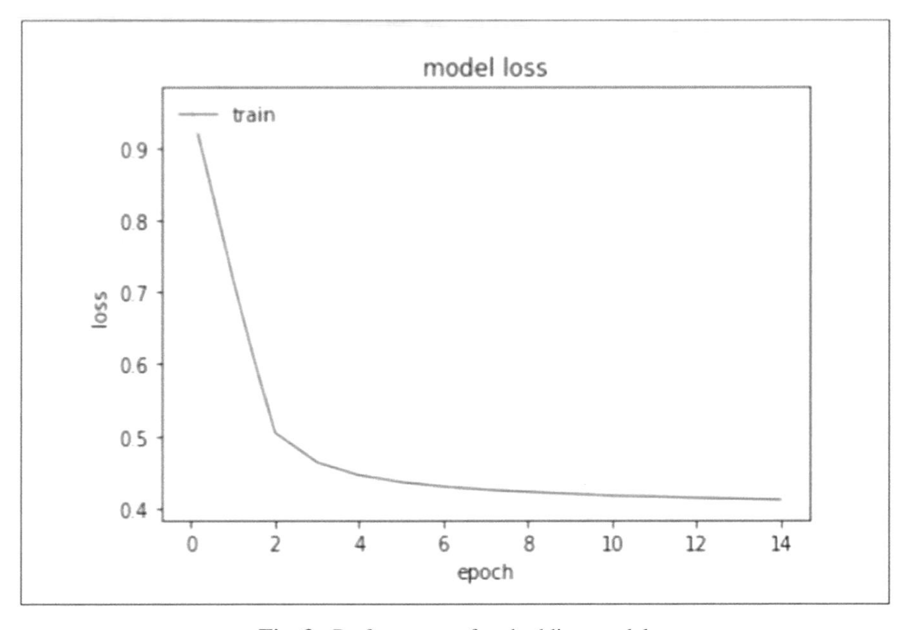

Fig. 3. Performance of embedding model

4.3 Visualization

TSNE (t-Stochastic Distributed Neighbors Embedding) and UMAP (Uniform Manifold Approximation and Projection) are two methods for mapping vectors to a lower-dimensional embedded space that use the concept of a manifold. As a result, in our case of books, we take the 37,000 dimensions, embed them in 50 dimensions with a neural network, and then embed them in 2 dimensions with a manifold. Dimension reduction with a manifold is based on the idea that a lower-dimensional representation of the vectors can still capture the variation between different groups. We want the embeddings to represent similar entities that are close to one another in fewer dimensions so that we can visualize them (Figs. 4, 5 and 6).

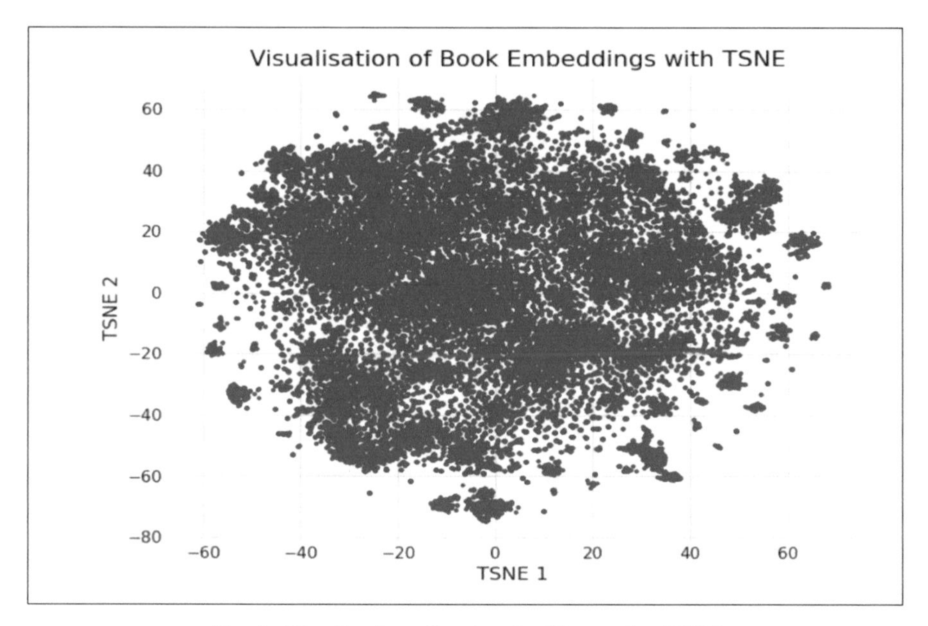

Fig. 4. Visualization of book embeddings using TSNE

The UMAP clustering does not appear to be much separation between the book categories. We decided to stick with TSNE because there are so many parameters to play with in UMAP.

4.4 Book Recommendation

We have trained the model and extracted the embeddings; now that we have them, we can use them to recommend books that are most similar to a given book.

We used a function that takes a book or a link, a set of embeddings, and returns the n books that are the most similar to the query. This is accomplished by computing the dot product of the query and the embeddings. The dot product represents the cosine similarity between two vectors because the embeddings were normalized. In contrast to the Euclidean distance, this is a measure of similarity independent of the vector's magnitude.

We can sort the results to find the closest entities in the embedding space once we have the dot products. Higher numbers in cosine similarity indicate entities closer together, with -1 being the farthest apart and $+1$ being the closest.

If the most similar book is the book itself, our used function works. Because we multiplied the book vector by all of the other embeddings, the item itself with a similarity of 1.0 should be the most similar (Fig. 7).

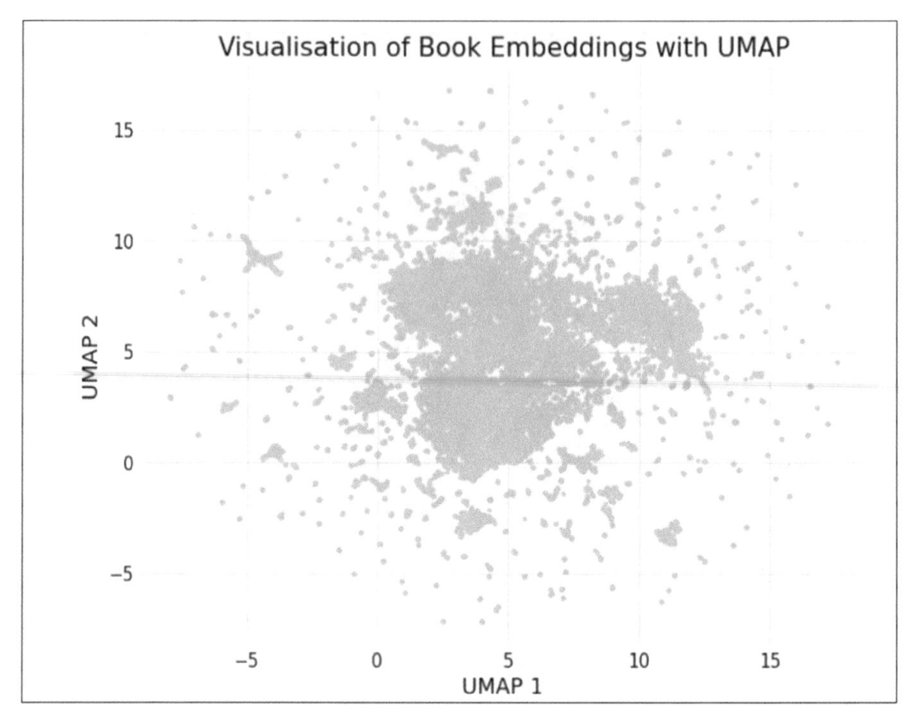

Fig. 5. Visualization of book embeddings using UMAP

5 Discussion

The findings of this study can be used to help with future development. Even though the progress has been gradual, there are still many opportunities for further improvement.

We developed an effective book recommendation system based on the principle that books with similar outgoing links and all of the Wikipedia book articles are similar. We used Keras embeddings to embed both the links and the books. We created a supervised machine learning problem to train the neural network by classifying if a given link was present on a book page. We saw how to thoroughly inspect the embeddings to find the closest books to a given book in the embeddings space, which is more than just training the neural network.

We also learned how to visualize embeddings, which can occasionally reveal interesting clustering. In the process, we reduced the books' original 37,000 dimensions to 50 using a neural network and then to two using a manifold learning method. The neural network embedding helped make recommendations based on nearby entities, whereas the TSNE embedding was primarily valuable for visualization. Another way to improve the model's performance is to create a new model using an Artificial Neural Network (ANN) approach. In the future, the ANN approach could be considered and prioritized in developing book recommendation systems. Reviewing high accuracy models [18–20] reveals that precisely creating recommendations is possible.

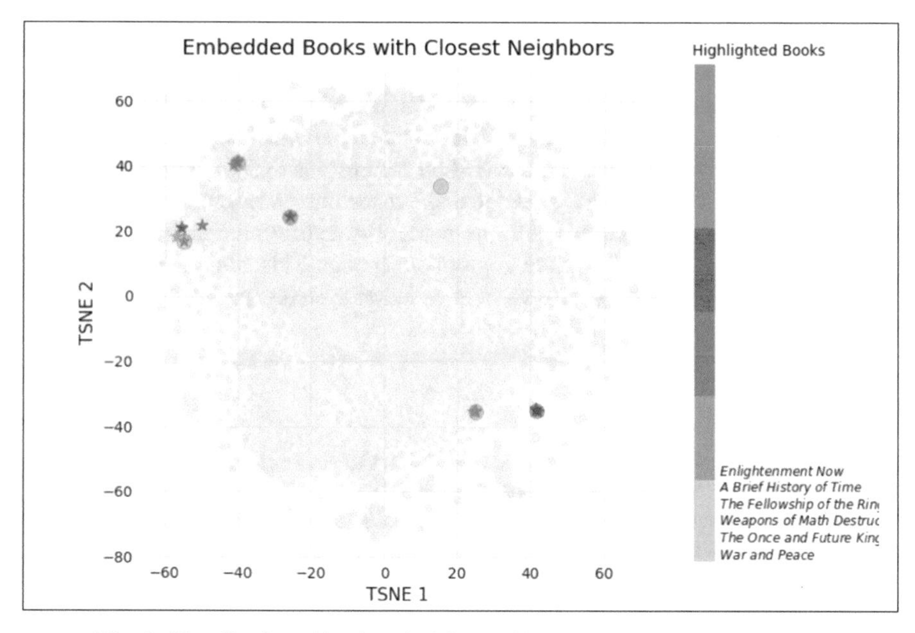

Fig. 6. Visualization of book embeddings with closest Neighbors using TSNE

```
find_similarBooks('War and Peace', book_weights)

Books closest to War and Peace.

Book: War and Peace                  Similarity: 1.0
Book: The Master and Margarita       Similarity: 0.94
Book: Dead Souls                     Similarity: 0.91
Book: Anna Karenina                  Similarity: 0.91
Book: Candide                        Similarity: 0.88
Book: The Idiot                      Similarity: 0.88
Book: Buddenbrooks                   Similarity: 0.88
Book: Demons (Dostoevsky novel)      Similarity: 0.88
Book: The Magic Mountain             Similarity: 0.87
Book: Poor Folk                      Similarity: 0.87
```

Fig. 7. Final result of book recommendation

6 Conclusion

We used textual descriptions of books to create a compelling book recommendation system in this paper, based on the principle that books with similar links are similar. Both the links and the books were embedded using a Keras neural network embeddings. By classifying whether a given link was present on a book page, we created a supervised machine learning problem to train the neural network. More than just training the neural network, we saw how to thoroughly inspect the embeddings to find the closest books to a given book in embedding space. We also learned how to visualize embeddings, revealing interesting clustering on rare occasions. Using a neural network and a manifold learning method, we reduced the books' original 37,000 dimensions to 50 and then to two. The neural network embedding was good for making recommendations based on nearby entities, whereas the TSNE embedding was good for visualization. Finally, this research is expected to continue with a more appropriate model. This study could serve as a model for developing more precise recommendation systems, particularly in systems that archive various innovations in specific fields.

References

1. Seyednezhad, M., Cozart, K., Bowllan, J., Smith, A.: A Review on Recommendation Systems: Context-aware to Social-based, pp. 9–20. IEEE (2018)
2. Portugal, I., Alencar, P., Cowan, D.: The use of machine learning algorithms in recommender systems: a systematic review. Sci. Direct **97**, 205–227 (2018)
3. Alhijawi, B., Kilani, Y.: The recommender system: a survey. Int. J. Adv. Intell. Paradigms **15**(3), 229–251 (2020)
4. Samih, A., Ghadi, A., Fennan, A.: Deep graph embeddings in recommender systems: a survey. J. Theor. Appl. Inform. Technol. **99**(15), 3812–3823 (2022). https://doi.org/10.5281/zenodo.5353504
5. Rahutomo, R., Perbangsa, A.S., Soeparno, H., Pardamean, B.: Embedding model design for producing book recommendation. In: 2019 International Conference on Information Management and Technology (ICIMTech), vol. 1, pp. 537–541. IEEE (2019)
6. Rajpurkar, S., Bhatt, D., Malhotra, P., Rajpurkar, M.S.S., Bhatt, M.D.R.: Book recommendation system. Int. J. Innov. Res. Sci. Technol. **1**(11), 314–316 (2015)
7. Mathew, P., Kuriakose, B., Hegde, V.:. Book Recommendation system through content based and collaborative filtering method. In: 2016 International Conference on Data Mining and Advanced Computing (SAPIENCE), pp. 47–52. IEEE (2016)
8. Das, D., Sahoo, L., Datta, S.: A survey on recommendation system. Int. J. Comput. Appl. **160**(7), 6–10 (2017). https://doi.org/10.5120/ijca2017913081
9. Samih, A., Adadi, A., Berrada, M.:. Towards a knowledge based explainable recommender systems. In: Proceedings of the 4th International Conference on Big Data and Internet of Things, pp. 1–5 (2019)
10. Samih, A., Ghadi, A., Fennan, A.: Translational-randomwalk embeddings-based recommender systems: a pragmatic survey. In: Kacprzyk, J., Balas, V.E., Ezziyyani, M. (eds.) Advanced Intelligent Systems for Sustainable Development (AI2SD'2020): vol. 2, pp. 957–966. Springer International Publishing, Cham (2022). https://doi.org/10.1007/978-3-030-90639-9_77
11. Anwar, K., Siddiqui, J., Saquib Sohail, S.:. Machine learning techniques for book recommendation: an overview. In: Proceedings of International Conference on Sustainable Computing in Science, Technology and Management (SUSCOM). Amity University Rajasthan, Jaipur-India (2019)
12. Gal, Y., Ghahramani, Z.: A theoretically grounded application of dropout in recurrent neural networks. Adv. Neural Inform. Process. Syst. **29** (2016)
13. Géron, A.: Hands-on machine learning with scikit-learn and tensorflow: Concepts. Tools, and Techniques to build intelligent systems. I. Labutov. In: Reembedding words, pp. 489–493, 2013 (2017)

14. Ketkar, N.: Introduction to keras. In: Ketkar, Nikhil (ed.) Deep Learning with Python, pp. 97–111. Apress, Berkeley, CA (2017). https://doi.org/10.1007/978-1-4842-2766-4_7
15. Heidari, E., Balazadeh-Meresht, V., Sharifi-Zarchi, A.: Multivariate Analysis and Visualization using R Package muvis. arXiv preprint arXiv:1810.12184 (2018)
16. Samih, A., Ghadi, A., Fennan, A.: ExMrec2vec: explainable movie recommender system based on Word2vec. Int. J. Adv. Comput. Sci. Appl. **12**(8) (2021). https://doi.org/10.14569/IJACSA.2021.0120876
17. Kaggle datasets: https://www.kaggle.com/datasets
18. Tan, Y.K., Xu, X., Liu, Y.: Improved recurrent neural networks for session-based recommendations. In: Proceedings of the 1st Workshop on Deep Learning for Recommender Systems, pp. 17–22 (2016)
19. Baylari, A., Montazer, G.A.: Design a personalized e-learning system based on item response theory and artificial neural network approach. Expert Syst. Appl. **36**(4), 8013–8021 (2009)
20. Chen, L., Xie, T., Li, J., Zheng, Z.: Graph enhanced neural interaction model for recommendation. Knowl.-Based Syst. **246**, 108616 (2022)

Knowledge Embeddings for Explainable Recommendation

Amina Samih(✉), Abderrahim Ghadi, and Abdelhadi Fennan

Department of Computer Sciences, Data and Intelligent Systems (DIS) Team, FSTT, Abdelmalek Essaadi University, Tetouan, Morocco
amina.samih@etu.uae.ac.ma

Abstract. The explainable recommendation aims to create models that generate high-quality recommendations and intuitive explanations. The explanations may be ad hoc or derived directly from an explainable model (also called interpretable or transparent in some contexts). Explainable recommendation attempts to address the problem of why: by providing explanations to users or system designers, it assists humans in understanding why certain items are recommended by the algorithm, where humans can be either users or system designers. Explainable recommendations help recommendation systems improve transparency, persuasiveness, effectiveness, trustworthiness, and satisfaction. This work aims to use a knowledge embeddings extraction method to improve the explainability of recommender systems.

Keywords: Recommender Systems · Explainable AI · Knowledge Embedding · Machine Learning

1 Introduction

Recommender systems are ubiquitous on the web. Introduced by the digital giants (Google, Amazon, Facebook, Apple, etc.), they are now used by most marketplaces and are intended to increase customer satisfaction while increasing the turnover achieved. Explainability of recommended results is critical in these critical decision-making tasks if users are to understand, appropriately trust, and effectively manage resulted recommendations [1, 2].

Creating explanations for recommendations serves two purposes [3]: (i) It is helpful to explain why a recommendation is made to the user. This helps to persuade users to accept the provided results. As a result, it contributes to increased system trustworthiness and satisfaction, which leads to increased customer loyalty. (ii) From the modelling standpoint, interpretability aids developers in understanding and debugging the system. As a result, its scrutability and efficiency improve.

Explainability, despite its significance, is challenging to achieve. Indeed, with the recent spectacular success of machine learning (ML) and deep neural networks, many recommender systems (RS) rely on these algorithms to improve their performance and prediction at the expense of transparency due to the black-box nature of Machine learning

M. Lazaar et al. (Eds.): BDIoT 2022, LNNS 625, pp. 116–126, 2023.
https://doi.org/10.1007/978-3-031-28387-1_11

algorithms. In the literature, explainability and interpretability are closely related concepts. In general, one approach to achieving explainability is interpretability. Explainable AI (XAI) aims to create models that can explain their (or other models') decisions to system designers or ordinary users [3].

As a result, aware of the importance of addressing explainability issues in recommendation systems and based on Explainable AI, we propose a knowledge Embeddings-based approach for explainable recommendation in this paper. In particular, we extract relationships between entities embedded in the RS to backtrack clear rules that influence the recommendation. Then we match these rules with explanation styles that are human-friendly. The presentation of this paper is structured as follows:

1. We will go over the fundamental concepts.
2. We look at related works in this field.
3. Using an illustrative example, we present our proposed approach and discuss how it aids in resolving explainability issues.
4. We present our findings and future research directions.

2 Background

The main goal of this section is to look into our approach's two main background domains: recommender systems and Explainable AI.

2.1 Recommender Systems

A recommendation system aims to provide users with relevant resources according to their preferences. The latter thus sees his search time reduced and receives suggestions from the system to which he would not have spontaneously paid attention. The rise of the Web and its popularity have notably contributed to establishing such systems in e-commerce. Quote, for example, the popular websites Amazon1 in e-commerce and CiteSeer2, a tool referencing research articles. Recommender systems can initially be seen as a response given to users having difficulty deciding when using a "classic" information retrieval system [1].

The three main recommendation approaches are content-based, collaborative filtering-based, and hybrid recommendations [2].

- Content based recommendation

A system that uses content-based filtering exploits only the representations of documents and the information that can be derived from those documents. Such a type of filtering could, for example, use the similarity of documents in a term-document matrix to determine the relevance of a document. If a user expresses interest in a document, similar documents will also be deemed potentially relevant.

Every system user has a profile that describes their interests. For example, the profile could include a list of themes or preferences the user enjoys or dislikes. When a new item is received, the system compares the item description to the user's profile to predict the item's usefulness for that user [3].

- Collaborative filtering

Collaborative filtering is one of the most popular technologies in recommender systems. It is based on the idea that people seeking information should use what others have already found and evaluated. In daily life, if someone needs information, he tries to generally get information from his friends his colleagues, who will, in turn, recommend articles, films, books, etc. This collaboration between people makes it possible to improve the exchange of knowledge. However, it takes much time since this information resource may not always be available. From here, the idea of collaborative filtering was born, the need to automate and make the exchange of experiences and personal opinions of some people that others can use. Collaborative filtering is the automation of social processes [4].

- Hybrid recommendation

This is another type of recommendation that seeks to overcome the shortcomings of the two preceding approaches. It is a hybrid of two or more distinct recommendation techniques. Hybrid recommender systems combine two or more recommendation strategies in various ways to benefit from their complementary benefits [5]. Content-based and collaborative filtering is the most popular hybrid approach. It uses both the element's content and the ratings of all users [6]. This Recommender employs a wide range of Machine Learning algorithms, including Deep Learning.

Aside from the approaches mentioned above, there are other methods for classifying Recommender Systems, such as Demographic filtering, which uses demographic data such as age, gender, education, address, and so on to build user similarities [7]. Furthermore, Knowledge-based Recommendation is based on existing recommendations in databases or knowledge bases and is thus not dynamically influenced by recent ratings or preferences [8].

The quality evaluation of these valuable tools is one of the most critical issues in the Recommendation field. In general, Recommender Systems are evaluated in various ways, many of which are incomparable. Indeed, Recommendation Systems are one of the most successful applications of Machine Learning techniques [3], and each algorithmic approach has supporters who believe it is superior for some reason. As a result, determining the best algorithm is inherently tricky. Existing work has proposed a wide range of success criteria for a recommendation approach [9], including algorithm accuracy, diversity and novelty, coverage, and serendipity. However, because system-centric criteria are not always correlated with quality, accurate recommendations may not always be the most useful to users.

Thus, evaluation metrics should be more user-centric, considering factors that influence user satisfaction and motivate them to make a recommendation-related decision (see, choose, watch…). In this regard, we believe that a sound Recommender system should be transparent and able to explain (and thus persuade) the active user why items are suggested. Moving beyond accuracy metrics to design and evaluate an explainable recommendation system is a promising but challenging search area. Indeed, because most recommendation approaches are based on Machine Learning algorithms that are "not transparent" by nature, it is difficult to interpret recommendation results.

Fortunately, XAI [10], a new field of study, proposes techniques to make the results of artificial intelligence systems more understandable to humans. We propose then that the results of this field be used, adapted, and projected onto recommendation.

2.2 XAI

Lamy Lamy and colleagues [11] define XAI as the design of intelligent systems capable of explaining their outputs to humans, such as robots, artificial agents, planners, and ML algorithms. Several explanation methods and strategies, particularly for Machine Learning algorithms, have been proposed in the quest to make AI systems explainable in a relatively short period. Explainability methods are classified into three types [12]:

(a) *Interpretability complexity*: it is widely acknowledged that the more complex the model, the more difficult it is to interpret it. There is frequently a trade-off between interpretability and accuracy. As a result, we can divide explainable techniques into two categories based on their complexity:

   i. Algorithms that are inherently and intrinsically interpretable (but less.accurate), such as decision trees.
   ii. Post-hoc explanations consist of building a high complex uninterpretable black-box model with high accuracy and then using a separate set of techniques to perform a reverse engineering process to provide the needed elucidation.

(b) *The scoop of understandability*: Understanding an automated model implies understanding it in two ways, according to the scoop of interpretability: understanding the entire model behaviour or understanding a single prediction. As a result, we distinguish two subclasses: Interpretability on a global and local scale.
(c) *The degree of reliance on the machine learning model used:* Another critical distinction between model interpretability techniques is whether they are model agnostic, which means they can be applied to any artificial intelligence/Machine learning algorithm or model specific, which means they are only applicable to a single type or class of algorithm. The most popular interpretable methods are model agnostic. Indeed, many model-agnostic methods have recently been developed using statistics, machine learning, and data science techniques. These are broadly classified into four technique types: Extraction of knowledge, visualization, influence methods, and example-based explanation.

We are particularly interested in the first type, "Extraction of knowledge" because it fits the context of a recommender system based on data to extract knowledge embeddings, the recommendation in our case.

3 Related Work

In order to identify improvement axes, we propose an analysis of the few existing explainable RS in the literature in this section. The majority of existing explainable recommendations can be divided into four categories: Deep Learning, Topic Modeling, Matrix

Factorization, and Knowledge embeddings-based Models [13]. The following sections examine the most critical proposed works in each class. Seo and al. [14] proposed an interpretable dual attention-based convolutional neural networks model that combines review text and ratings for item rating prediction. Similarly, Damak and al. [15] introduced "SeER," a hybrid deep learning structure that uses ratings and music information to support various music recommendations with short Sequencer segments serving as interpretations.

Tan and colleagues [16] proposed using Rating-Boosted Latent Topics to comprehend users and items. They defined item recommend ability representations and user interest variations in a joint topic space, resulting in better recommendation performance and explainability.

The Explicit Factor Model (EFM) was proposed by Zhang and colleagues [17], a matrix factorization-based model that uses sentence sentiment analysis of user reviews to produce understandable recommendations. They took explicit item attributes and user points of view from feedback. They combined them with user-feature and item-feature relationships and user-item ratings to create a new homogeneous blended matrix factorization template. For both suggested and disrecommended items, the model provides explicit feature-level explanations.

Knowledge bases can help generate rich explanations by storing relevant information about users and items. This is why recent research has focused on developing a knowledge-based, explainable recommendation. Catherine and colleagues [18] propose a method for classifying knowledge graph elements and entities using a personalized PageRank procedure to generate recommendations and explanations. The ripple network was suggested by Wang and colleagues [19] to amplify user preferences on knowledge representation for a recommendation. The framework could also provide explanations based on network hops from the user to the recommended item.

Finally, Ai and colleagues [20] proposed using knowledge graph embeddings to produce explanations, and they built a user-item knowledge graph. Knowledge base embeddings are realized over the graph to get the recommended item. The shortest distance from the user to the recommended item thru the knowledge graph could then be used to provide explanations.

One obvious limitation emerges from examining the approaches mentioned above. Apart from knowledge embeddings-based approaches, existing explainable Recommender systems are primarily based on user feedback, which is not always available. Users prefer to be served with as little interaction with the system as possible. As a result, the lack of user feedback and reviews impacts the quality of existing approaches' explainability. On the other hand, knowledge embeddings-based models can extract a large amount of data with minimal interaction; however, the explanations provided are often complex, unintuitive, and difficult to comprehend by a layperson.

4 Proposed Work

4.1 The Overall Approach

Inspired by existing knowledge-based models and aiming to overcome explainable RS' existing limitations in terms of explanation complexity and the need for user reviews.

We propose a conceptual approach based on extracting knowledge embeddings from various sources of information. It is a model-independent approach that explains the recommendation regardless of the prediction mechanism.

Figure 1 depicts an overview of our modelling approach using a simplified example. The basic idea is to build the user-item knowledge graph to introduce embeddings of both user behaviors and our knowledge of the items. The user-item knowledge graph aims to display various types of user behaviors and item properties over entities in a unified framework.

The goal will then be to reverse engineer the relationship path that connects an item to a user to generate an explanation for the recommendation.

We first build a knowledge-base as a set of triplets N = (ex, ey,r), where ex and ey are head and tail entities, an entity could be a user or an item, and r is the relationship that connects ex and ey. We embed various sources of information in our approach by acquiring a user's actions history, social connections, item tags, and popularity statistics. This allows us to create an extensive list of relationship types based on user-to-user and item-to-item similarities.

The extraction rules that formulate the logic inference sequence from the user to the item in the knowledge graph embeddings are then set up. Finally, we translate each retrieved user-item pair into a natural language sentence because the user should select the item. The proposed model is built around three main extraction rules, which we match with three different explanation styles: social-based, tag-based, and item popularity (Fig. 2).

As a result, our model is primarily composed of two components: (a) knowledge base embeddings and (b) rule extractions.

4.2 Knowledge Embeddings

The more information we have about entities, the rich the explanation will be. Thus, our approach supports the flexibility to incorporate multiple sources of information. We classify acquired data into four classes:

- Users Relationships: it is a social knowledge base that contains users' friendship relationships, demographic information, interests and test data.
- Items Features: contains the items and their associated features such as price, dimension, accessories and tags.
- Trending: contains the most popular Items.
- History: contains items already chosen by users.

We assume that our knowledge data can be represented as a set of triplets N = (eh, et,r), where r is the relation between entities eh and et. Because an entity can be linked to one or more other entities via a single or multiple relations, we propose separating entity and relation modelling for our approach. We project each entity to a low-dimensional latent space, and each relation is treated as a translation function that converts one entity to another.

We model their relationship r as a linear projection from eh to et parameterized by r, namely [1, 21].

$$et = trans(eh, r) = eh + r. \tag{1}$$

To learn the entity embeddings, we can build a movie knowledge graph by connecting entities in the latent space with the translation function. Figure 1 depicts an example of the process of creating such a graph.

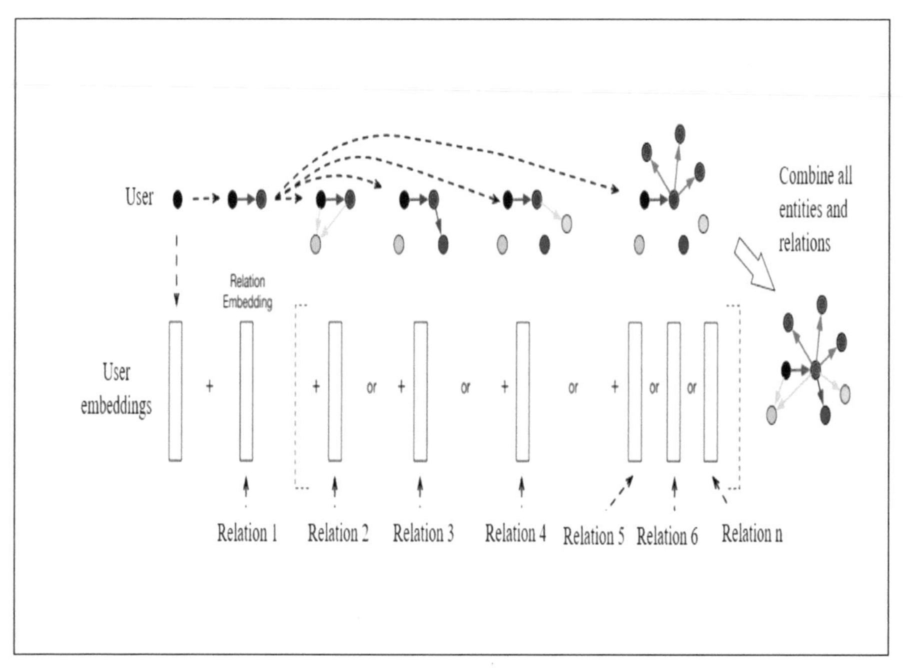

Fig. 1. The process of constructing knowledge embeddings with our proposition. Each entity is represented by a latent vector, and each relation is modeled as a linear translation from one entity to another entity, with the relation embedding parameterized.

We used the IMDB [22] dataset as a data source in our ongoing proof of concept. Our social knowledge base is limited to friendship relationships, and the treated item features are essentially tags.

4.3 Extraction Rules

To model user-item embeddings relationships, we use five extraction rules formulated using the probabilistic soft logic (PSL) framework [23].

The intuition that similar users like similar items are captured by Rule 1. Rule 2 captures the intuition that users will select similar or popular items. Rule 3 uses a transitive friendship rule to calculate user similarity. Rule 4 computes item similarity using a tags-based relationship. Rule 5 is a negative prior, a default fact used in probabilistic logic to adjust the inference.

RULE 1 :(*Connected_User* $(u_n, u_m) \wedge$ *prefer* $(u_n, i_m)) \vee \Rightarrow$ *prefer* (u_m, i_m)

RULE 2 : (*Connected_Item* $(i_m, i_n) \wedge$ *prefer* $(u_n, i_m)) \vee$ *Is-popular* $(i_n) \Rightarrow$ *prefer* (u_n, i_n)

RULE 3 : ((*Friendship* $(u_n, u_s) \wedge$ *Friendship* $(u_s, u_m)) \vee$ *Friendship* $(u_n, u_m) \Rightarrow$ *Connected _User* (u_n, u_m)

RULE 4 :*Tagged* $(i_x, T) \wedge$ *Tagged* $(i_y, T) \Rightarrow$ *Connected_Item* (i_x, i_y)

RULE 5:! *prefer*(u_m, i_n)

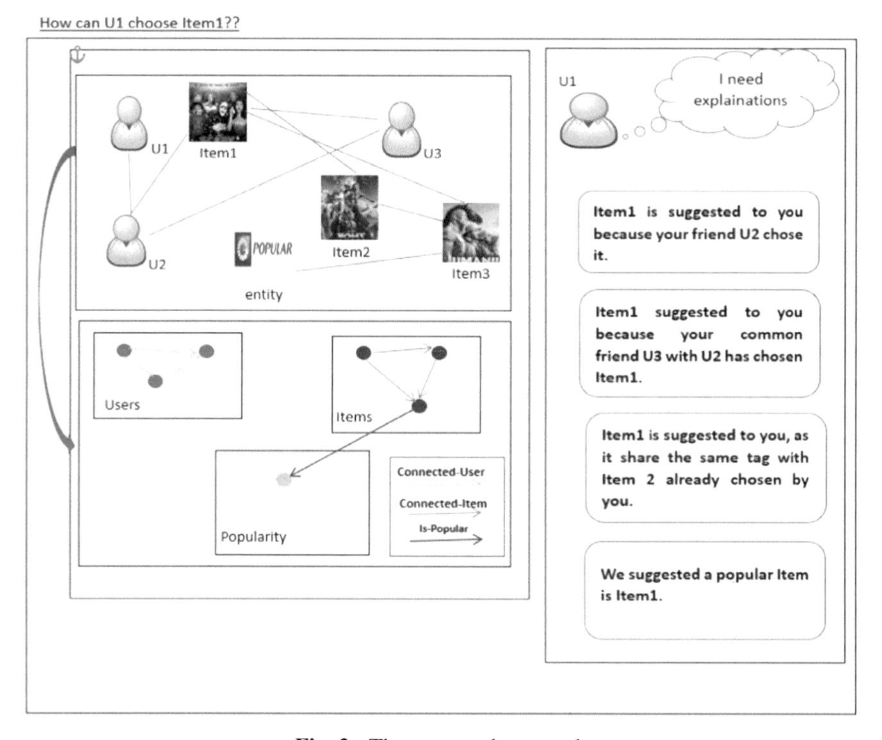

Fig. 2. The proposed approach

Atoms such as Choose (U1, Item1) represent the probability that user U1 will choose the item Item1 and takes values in the interval [0, 1]. To group rules and entities embeddings for classification. The queries are placed in a proof graph by recursively replacing each variable with literals that satisfy the rules. Figure 3 shows an example of a proof graph corresponding to the query "How User U1 can prefer Item1.

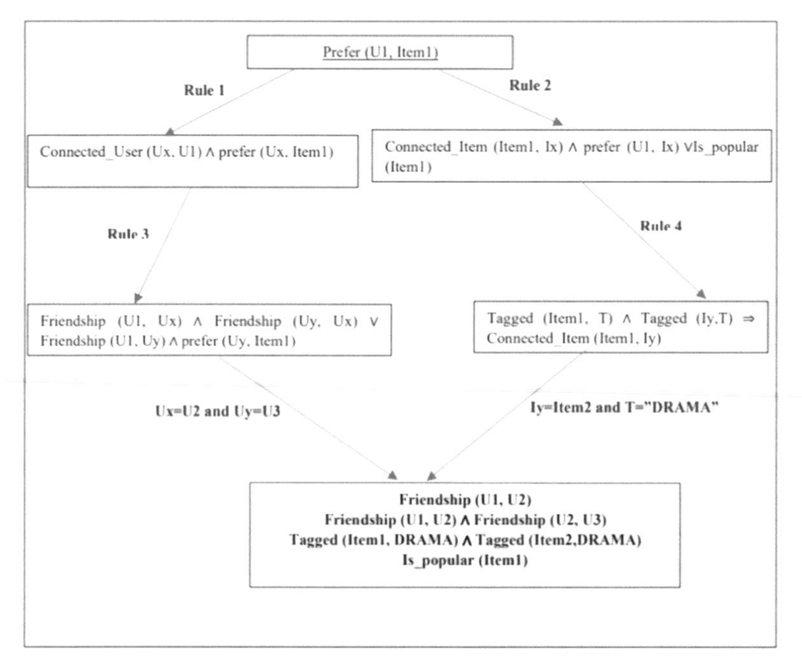

Fig. 3. The proof graph that corresponds to the query "Why will U1 prefer Item 1?"

So, based on the proof graph, we recommend Item1 to U1 because:

- U1's friend, U2, selected Item1.
- Item1 was chosen by U1's common friend U3.
- Item1 is tagged "DRAMA," as is Item2, which has already been chosen by U1.
- Item 3 is well-liked(popular item).

5 Conclusion

In this paper, we propose a method for improving the quality of recommendation systems by including explanations with recommendations. This increases the acceptability of these systems and allows them to be used in a broader range of applications. The proposed method is based on modelling knowledge embeddings for users and items, then extracting explanations using rules.

Aside from completing a proof of concept, we intend to enhance encoded knowledge embeddings in the future by introducing other social relations embeddings based on learned similarity and improving item-item relations using the complementarity concept.

References

1. Samih, A., Ghadi, A., Fennan, A.: Translational-randomwalk embeddings-based recommender systems: a pragmatic survey. In: Kacprzyk, J., Balas, V.E., Ezziyyani, M. (eds)

Advanced Intelligent Systems for Sustainable Development (AI2SD'2020). AI2SD 2020. Advances in Intelligent Systems and Computing, vol. 1418. Springer, Cham (2022). https://doi.org/10.1007/978-3-030-90639-9_77

2. Samih, A., Ghadi, A., Fennan, A.: Deep graph embeddings in recommender systems: a survey. J. Theor. Appl. Inf. Technol. **99**(15) (2022). https://doi.org/10.5281/zenodo.5353504
3. Samih, A., Ghadi, A., Fennan, A.: ExMrec2vec: explainable movie recommender system based on Word2vec. Int. J. Adv. Comp. Sci. Appl. **12**(8) (2021). https://doi.org/10.14569/IJACSA.2021.0120876
4. Katarya,R.: A systematic review of group recommender systems techniques. In: International Conference on Intelligent Sustainable Systems, IEEE, pp. 425 (2017)
5. Seyednezhad,M., Cozart, K., Bowllan, J., Smith, A.: A review on recommendation systems: context-aware to social-based. IEEE, pp. 9–20 (2018)
6. Cano, E., Morisio, M.,: Hybrid recommender systems: a systematic literature review. Intell. Data Anal. **21**(6), 1487–1524 (2017)
7. Portugal, I., Alencar, P., Cowan, D.: The use of machine learning algorithms in recommender systems: a systematic review. Sci. Direct **97**, 205–227 (2018)
8. Prasad , R., Kumari, V.: A categorial review of recommender systems. Int. J. Distrib. Parallel Syst. **3**(5), 73–83 (2012)
9. Alhijawi, B., Kilani, Y.: The recommender system: a survey. Int. J. Adv. Intell. Paradigms **15**(3), 229–251 (2020)
10. Adadi, A., Berrada, M.: Peeking inside the black-box: a survey on explainable artificial intelligence (XAI). IEEE Access **6**, 52138–52160 (2018)
11. Lamy, J.B., Sekar, B., Guezennec, G., Bouaud, J., Séroussi, B.: Explainable artificial intelligence for breast cancer: a visual case-based reasoning approach. Artif. Intell. Med. **94**, 42–53 (2019)
12. Samih, A., Adadi, A., Berrada, M.: Towards a knowledge based explainable recommender systems. In: Proceedings of the 4th International Conference on Big Data and Internet of Things, pp. 1–5 (2019)
13. Zhang, Y., Chen, X.: Explainable recommendation: a survey and new perspectives. Found. Trends Inf. Retr. **14**(1), 1–101 (2020)
14. Seo, S., Huang, J., Yang, H., Liu, Y.: Interpretable convolutional neural networks with dual local and global attention for review rating prediction. In: Proceedings of the eleventh ACM Conference on Recommender Systems, pp. 297–305 (2017)
15. Damak, K., Nasraoui, O.: SeER: An explainable deep learning MIDI-based hybrid song recommender system. arXiv preprint http://arxiv.org/abs/1907.01640 (2019)
16. Tan, Y., Zhang, M., Liu, Y., Ma, S.: Rating-boosted latent topics: understanding users and items with ratings and reviews. In: IJCAI, vol. 16, pp. 2640–2646 (2016)
17. Zhang, Y., Lai, G., Zhang, M., Zhang, Y., Liu, Y., Ma, S.: Explicit factor models for explainable recommendation based on phrase-level sentiment analysis. In: Proceedings of the 37th international ACM SIGIR conference on Research & development in information retrieval, pp. 83–92 (2014)
18. Catherine, R., Mazaitis, K., Eskenazi, M., Cohen, W.: Explainable entity-based recommendations with knowledge graphs. arXiv preprint http://arxiv.org/abs/1707.05254 (2017)
19. Wang, H., et al.: Ripplenet: propagating user preferences on the knowledge graph for recommender systems. In: Proceedings of the 27th ACM international conference on information and knowledge management, pp. 417–426 (2018)
20. Ai, Q., Azizi, V., Chen, X., Zhang, Y.: Learning heterogeneous knowledge base embeddings for explainable recommendation. Algorithms **11**(9), 137 (2018)
21. Zou, W.Y., Socher, R., Cer, D., Manning, C.D.: Bilingual word embeddings for phrase-based machine translation. In: Proceedings of the 2013 conference on empirical methods in natural language processing, pp. 1393–1398 (2013)

22. IMDB dataset: https://www.kaggle.com/c/byuimdbratings. Accessed on 1 Jun 2022
23. Kimmig, A., Bach, S., Broecheler, M., Huang, B., Getoor, L.: A short introduction to probabilistic soft logic. In: Proceedings of the NIPS workshop on probabilistic programming: foundations and applications, pp. 1–4 (2012)

Design of Blended Learning Course Based on SPOC for Primary School Teachers: Case of Soft Skills

Mohammed Chekour[1](✉), Yassine Zaoui Seghroucheni[2], Driss Elomari[3], Nadir El Morabit[4], and El hassan El-hassouny[4]

[1] Ibn Tofail University, Kenitra, Morocco
mohammed.chekour@uit.ac.ma
[2] Faculty of Sciences Mohammed V University in Rabat, Rabat, Morocco
[3] Sidi Mohamed Ben Abdellah University, Fez, Morocco
[4] Abdelmalek Essaadi University, Tetouan, Morocco

Abstract. The contemporary teacher works with a new generation of schoolchildren. They are expected to deal with the unexpected, solve problems, and master educational technologies. Therefore, teachers in the digital age are encouraged to use methods to strengthen soft skills in parallel with hard skills in their students from primary school onwards. Both types of skills are increasingly in demand in the labor market. However, the development of soft skills in pupils is conditioned by the high level of qualification of teachers and adequate further training during their professional careers. The aim of this research is to design a SPOC-based blended learning course for primary school teachers in soft skills. In this training, the pedagogical resources are available online and the practical activities are carried out face-to-face. The pedagogical approaches used in this training are based on problem solving and project work in a collaborative environment.

Keywords: Soft Skills · SPOC · Blended Learning · Primary School Teachers · Collaborative Learning · Continuous Training · Effective Educational Approaches

1 Introduction

In recent decades, educators and educational policymakers have taken cognizance of the importance of education for the development of a knowledge-based economy [1]. Consequently, schools should be committed to providing students with all the skills that enable them to meet the needs of contemporary society and the labor market [2]. Of course, technical, academic, and school skills are essential to build a citizen capable of understanding the world and acting in the future. However, these technical skills are not enough. They must be accompanied by the integration of soft skills in the study courses [3]. Soft skills aim to develop transversal, behavioral, and human competences in students such as communication, creativity, decision-making, problem-solving and teamwork skills [4].

M. Lazaar et al. (Eds.): BDIoT 2022, LNNS 625, pp. 127–137, 2023.
https://doi.org/10.1007/978-3-031-28387-1_12

In Morocco, the Ministry of National Education is opting for a new system of university education. The Bachelor's degree system is being implemented this academic year 2021–2022. This system aims to set up a pathway based on student-centered teaching by reinforcing the "soft skills" of students [5]. However, the acquisition of soft skills at pre-school and primary age will only be beneficial for success in a globalized world [6]. Certainly, the environment in which the modern generation is growing and developing is dynamic [7]. Children are surrounded by an enormous amount of information that comes to them through different channels [8]. Their life skills are essential when the messages and challenges of the social networks and peers are attractive and accessible [9]. Children need to be prepared to deal with life's "pitfalls", to think critically, to control their emotions and relationships with others, to be able to express their ideas and defend their position with arguments [4]. Unfortunately, there is a worldwide tendency for schools to focus mainly on technical and disciplinary knowledge [10]. Therefore, the teaching of soft skills requires high quality for pre- and in-service teachers. Modern education policy in the Kingdom of Morocco is aware of this challenge. Educators stress the importance of high-quality training for graduates in general and future teachers in particular [11].

The aim of this research is to highlight the need for soft skills training of future and in-service teachers. The improvement of the quality of our Moroccan educational system is conditioned by the optimal qualification of teachers, as modern society needs creative and open-minded teachers who are ready to be flexible, react quickly to changing requirements, make creative educational decisions, and conduct research in the field of pedagogical innovation. To achieve this goal, we propose a scenario of a SPOC-Soft skills (Small Private Online Course) for primary school teachers. This SPOC is in a blended learning mode. The choice of blended learning is not arbitrary. It gives a certain degree of flexibility in time and space. Also, the SPOC-Soft skills offers a collaborative environment to help primary school teachers improve their soft skills and to encourage them to produce teaching resources to help their learners develop these skills. This in-service training is based on active pedagogical approaches such as the problem-solving approach and the project-based approach.

The rest of this article is organized as follows: The next section presents the theoretical framework in which this work is embedded. Section 3 is dedicated to the methodology of our research. Section 4 is dedicated to the presentation of the SPOC-Soft skills scenario. The last section concludes the paper.

2 Literature Review

2.1 The Importance of Soft Skills in the Integration of Individuals in the Socio-economic Life

Soft skills help to manage conflict, create ideas, and build inclusive relationships that improve team performance [12, 13]. Yet, most students miss out early opportunities in their careers because they were not exposed to these skills at an early age [14]. Soft skills represent behavioral, transversal, and human skills. They are complementary to the technical skills known as hard skills. These skills are particularly sought after in the job market. In other words, technical skills are no longer enough. The different economic sectors are looking for non-professional qualities such as creativity, open-mindedness, communication skills, complex problem solving, and decision making [15]. The education and training sector is not an exception. These skills have proved themselves to be crucial in the teachers' teaching practices [16]. Indeed, the pedagogical skills of teachers (hard skills) are not enough, they are invited to master a set of personal and social skills allowing the improvement of their professional practices. In addition, teachers must direct and encourage students to develop these skills, taking into consideration the nature of the subject taught and the needs of their learners [17].

2.2 The Impact of In-Service Training on Primary School Teachers' Soft Skills

Globally, the labor market is increasingly looking for graduates with specific soft skills to complement traditional "hard" skills so that they can integrate into a rapidly changing and dynamic world [18]. Therefore, the development of these soft skills implies a change in the educational process as well as in the curricula [19]. In a study conducted by McKinsey & Company [20], which involved more than 8,000 people in eight European countries, employers stated that the lack of soft skills causes serious problems in the economy. Moreover, skill gaps cause the most problems in countries with high unemployment rates. This shows the importance of developing soft skills in teacher training programs to be able to transmit these skills to their learners [21]. Certainly, the modern educational process requires the development of teachers' creativity. Pedagogical dimension is important in our study context. It aims at the development of learners' creative personality from the primary level and the increase of the level of teachers' creative teaching practices [22]. In addition, educational technologies can provide means and tools for teachers to achieve the goals of pedagogical goals.

2.3 Pedagogical Approaches to Teaching Soft Skills

Many educators recommend using an active learning approach to collaboratively teach students basic soft skills that are needed in most careers [13]. The teachers' efforts are expected to be highly rewarding for students. In the collaborative approach, the need for soft skills derives from a real-world problem that results from the application of technical skills in a real-world situation [13]. In addition, collaborative learning helps learners to work in small groups towards a common goal. This method enhances learning because learners can share their knowledge while discussing their opposing viewpoints [23]. Also, it provides exposure to real-life problem-solving techniques to carry out concrete projects. Therefore, we chose to work with the problem-solving and project-based approaches in a collaborative environment.

2.4 The Most Demanded Soft Skills in the Moroccan Job Market

The modern workplace has recently undergone considerable changes. Emerging demands in the workplace have shifted the pendulum toward soft skills [24]. Today, one of the main issues of concern is employment [25]. At this point, the most crucial question to ask in the Moroccan context is: "what soft skills are the most demanded in the labor market? In a recent exploratory research, researchers constructed a taxonomy of soft skills from job advertisements of Moroccan organizations [26]. They found that the most demanded soft skills in the labor market are:

- Creativity
- Leadership
- Decision making
- Curiosity
- Critical thinking
- Problem solving
- Teamwork
- Communication
- Analytical thinking
- Ability to work independently
- Project management
- Organization
- Social skills
- Flexibility
- Skills related to the use of information and communication technologies

3 Methods and Tools

3.1 Choice of Blended Learning

Over the past decade, e-learning has become an important component of education from kindergarten to higher education [27]. However, the absence of human contact and the rate of abandoning the e-learning mode is among the major challenges of virtual classroom-based education [28]. At this point blended learning comes into play. The philosophy of blended learning is quite simple. It combines e-learning and face-to-face learning (F2F). This teaching-learning mode aims at making e-learning more efficient and at increasing the motivation of the learners to reduce the rate of withdrawal in the "e-learning" mode [29]. In addition, it offers the opportunity to personalize learning paths and to intervene in case learners find themselves in cognitive fatigue situations [30]. Several studies have investigated the advantages of blended learning over face-to-face teaching and over 100% online learning (e-learning) [31–34]. Researchers have concluded that blended learning is more effective than pure online learning [34]. The same researchers point out that blended learning has all the capabilities to become the predominant education model in the future. Also, blended learning offers many advantages for learners such as:

- Cost-effectiveness
- The flexibility of learning in terms of time and space
- Easy access to content
- Support from a face-to-face or virtual instructor and
- Contact with peers [35].

In addition, blended learning can be considered an effective approach to distance learning in terms of:

- Learner learning experience,
- Learner-learner interaction
- Learner-instructor interaction [36].

3.2 Choice of SPOCs

MOOCs (Massive Open Online Course) and SPOCs (Small Private Online Course) are two learning modes based on online learning platforms with different characteristics and objectives [37]. The MOOC is suitable for sharing educational resources on a large scale. Also, MOOC is suitable for basic theory education, while SPOC is applicable to professional skill education [38]. In this research, we chose SPOC for the following reasons:

- The educational resources disseminated in SPOC can be in several forms (different versions of the same course) [18]: Texts, Videos, simulations, animations…
- Video lectures offer students great flexibility [39]
- Specific content is made available to a group of learners where specific contents are made available to a group of learners, which offers an individualized follow-up by the tutors [40]
- Learners can check their progress by taking quizzes after the videos [41]
- Learners can view their content at their own pace and review it as many times as they wish before face-to-face lectures [42]
- Teachers can enjoy the benefits of the flipped classroom [41]

3.3 Choice of Microsoft Teams

Microsoft Teams is a proprietary collaborative communication application in SaaS mode. It allows both simplifying teamwork and working remotely [43]. It is an application that is part of the Office 365 suite and allows teams to collaborate in the cloud. It offers several features: messaging, calling, video conferencing and file sharing [44]. In our SPOC, we chose to use the Microsoft Teams platform for the following reasons:

- Microsoft Teams is easy to use [45]
- Microsoft Teams provides a fairly comprehensive learning environment [46]
- The enterprise version is offered by the Moroccan Ministry of Education
- Moroccan teachers and students are trained to work with Microsoft Teams

4 SPOC-Soft Skills Scenario for Primary School Teachers

In the following two sub-sections, we will present the general scenario and the detailed scenario of the SPOC-Soft skills.

4.1 General Scenario SPOC-Soft Skills

The proposed scenario of our SPOC-Soft skills is based on blended learning. In the online part of this training, we will use the Teams platform. In this platform, we will integrate resources and pedagogical activities to help primary school teachers improve their soft skills. The face-to-face sessions (F2F) will be dedicated to the development of workshops and pedagogical resources to develop these skills in students (Fig. 1).

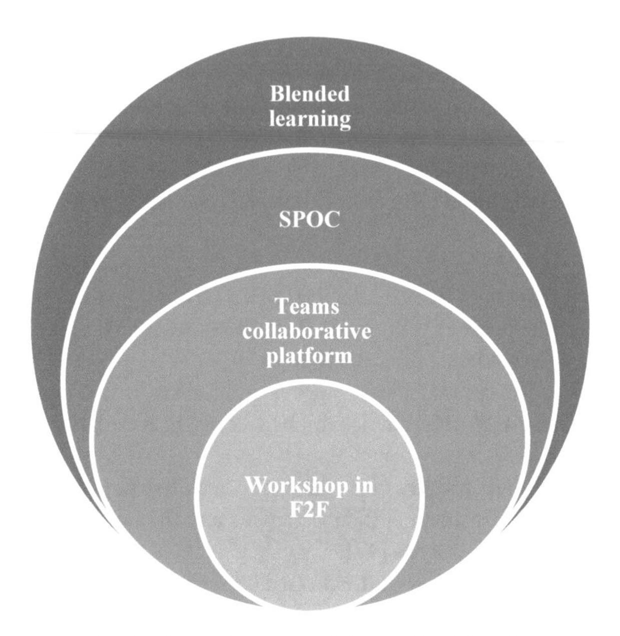

Fig. 1. General scenario of the SPOC-Soft skills

4.2 Detailed Scenario of SPOC-Soft Skills

The SPOC-Soft Skills aims to develop m skills where j represents the skill to be developed and i represents the week number. At the beginning of the training, i and j are initialized to (1) (in other words, we will try to develop SS1 in week 1). Our SPOC-Soft skills is in hybrid mode: the odd weeks will be in face-to-face mode (F2F) and the even weeks will be in online mode (see Fig. 2).

In the first week ($i = 1$) of this training, which takes place in face-to-face mode (F2F), we will build teams according to the level taught. These teams will be disciplinary and categorized to reach the objectives of our training. Then, we will move to video editing software. During the second week ($i = 2$) which takes place in online mode via the Teams platform, the teachers are invited to consult the pedagogical resources to deepen their knowledge at the level of Soft Skills j (with $j = 1$). Moreover, they are invited to take a quiz related to the visualized educational resources. The score in the quiz must be higher than 80% with a coefficient of 25% of the final grade of the teacher in the SPOC-Soft

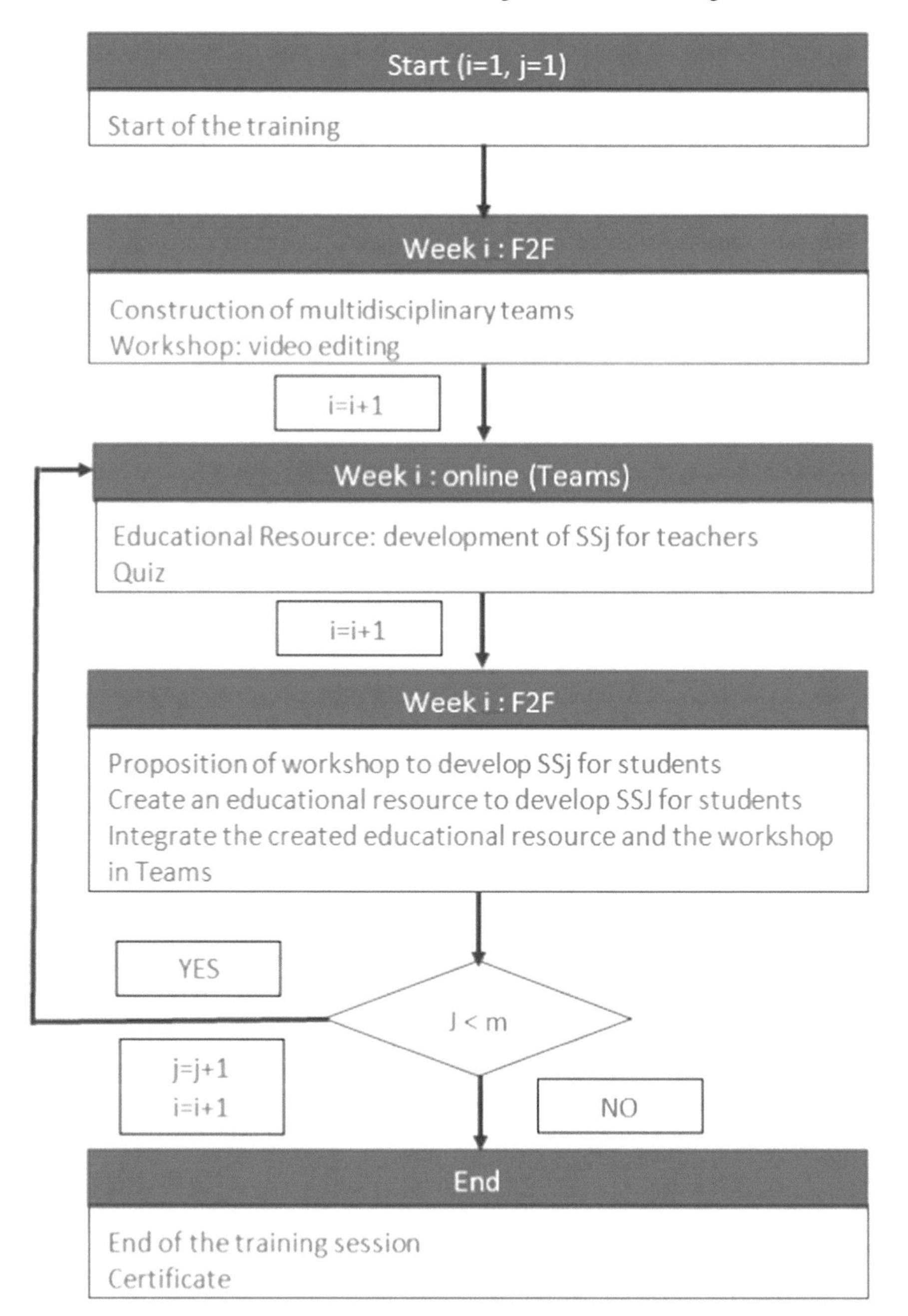

Fig. 2. Detailed scenario of SPOC-Soft skills

skills. The following week ($i = 3$), the teachers will propose, in the framework of the collaborative work, a workshop to develop the SSj (with $j = 1$) for their learners. Then, they start the creation of pedagogical resources to develop the SSj for their learners.

Finally, they will integrate these resources into Teams platform for their classes. The coefficient of the created learning resources is 75% of the teacher's final grade in the SPOC-Soft skills.

The same scenario of week 2 and 3 will be applied for the soft skills J: SSj (with $j = 2$). The training ends when the value of j is higher than m (with m the total number of soft skills to be developed by the learners). A certificate will be awarded at the end of the training.

5 Conclusion

This research aims to identify the weaknesses of soft skills acquisition in the context of the Moroccan educational system. Based on these weaknesses and to contribute to the improvement of the current situation, we propose some solutions. One of these solutions will be in the form of a SPOC to develop teachers' soft skills. In this way, these teachers can improve their soft skills to be able to transmit them to their learners in the best conditions. In addition, the teachers propose pedagogical resources in the form of texts and videos to be integrated into Teams platform for their classes. This research, the first of its kind in Morocco, will provide new data on the place of soft skills in the Moroccan educational system. Its original contribution is to better understand the obstacles that hinder the integration of soft skills in teachers' professional practices. Finally, the results of this research will also open new horizons of reflection for researchers interested in the development of soft skills among teachers and learners.

References

1. Bano, S., Taylor, J.: Universities and the knowledge-based economy: perceptions from a developing country. High. Educ. Res. Dev. **34**(2), 242–255 (2015). https://doi.org/10.1080/07294360.2014.956696
2. Okolie, U.C., Nwosu, H.E., Mlanga, S.: Graduate employability: how the higher education institutions can meet the demand of the labour market. High. Educ. Ski. Work-Based Learn. (2019). https://doi.org/10.1108/HESWBL-09-2018-0089
3. Warrner, J.: Integrating soft skills into an academic curriculum. Am. Assoc. Adult Contin. Educ. (2021). https://files.eric.ed.gov/fulltext/ED611615.pdf
4. Dimitrova, K.: Formation of soft skills in preschool and primary school age - an important factor for success in a globalizing world. Knowl. Int. J. **28**(3), 3 (2018). https://doi.org/10.35120/kij2803909K
5. Tang, K.N.: The importance of soft skills acquisition by teachers in higher education institutions. Kasetsart J. Soc. Sci. **41**(1), 22–27 (2020)
6. Karasheva, Z., Amirova, A., Ageyeva, L., Jazdykbayeva, M., Uaidullakyzy, E.: Preparation of future specialists for the formation of educational communication skills for elementary school children. World J. Educ. Technol. Curr. Issues **13**(3), 467–484 (2021)
7. R. Bruno, L. Oliveira, and P. Ferreira, 'NG2C: Pretenuring N-Generational GC for HotSpot Big Data Applications', *ArXiv Prepr. ArXiv170403764*, 2017. https://doi.org/10.48550/arXiv.1704.03764

8. Groes, S.: Information overload in literature. Textual Pract. **31**(7), 1481–1508 (2017). https://doi.org/10.1080/0950236X.2015.1126630
9. Makhachashvili, R.: Soft skills and ICT tools for final qualification assessment in Universities of Ukraine and India in COVID-19 framework. Psychol. Educ. J. **58**(2), 849–861 (2021). https://doi.org/10.17762/pae.v58i2.1959
10. Sarsekeyeva, Z.Y., Skakova, A. K.: Determinants of formation primary school students' soft skills. ХАБАРШЫСЫ, p. 16 (2020). https://doi.org/10.31489/2020Ped2/16-24
11. Hathazi, A.: Support programs for developing competences of teachers as an essential factor for successful inclusive education. Rev. Educ. E Cult. Contemp. **17**(51), 10–27 (2020). http://periodicos.estacio.br/index.php/reeduc/article/viewArticle/8763
12. Kamin, M.: Soft skills revolution: a guide for connecting with compassion for trainers, teams, and leaders. John Wiley & Sons (2013)
13. England, T.K., Nagel, G.L., Salter, S.P.: Using collaborative learning to develop students' soft skills. J. Educ. Bus. **95**(2), 106–114 (2020). https://doi.org/10.1080/08832323.2019.1599797
14. Solórzano, J., Rojas, Y., Vargas, C., Rueda, O., Palma, H.H.: Soft skills and advantages for learning mathematics at an early age. Indian J. Sci. Technol. **11**(45), 1–7, (2018). https://doi.org/10.17485/ijst/2018/v11i45/137683
15. Y. C. J. Acero, R. B. León, A. M. S. Castrillon, and A. S. Castrillon, 'Soft skills development according to the level of university formation for engineering', *PalArchs J. Archaeol. Egypt-Egyptology*, vol. 19, no. 1, pp. 1776–1791, 2022, from https://archives.palarch.nl/index.php/jae/article/view/10953
16. John, J.: Study on the nature of impact of soft skills training programme on the soft skills development of management students. Pac. Bus. Rev. 19–27 (2009). https://ssrn.com/abstract=1591331
17. Pereira, O.P., Costa, C.A.A.T.: The importance of soft skills in the university academic curriculum: the perceptions of the students in the new society of knowledge (2017). https://doi.org/10.18533/ijbsr.v7i6.1052
18. Ortega-Arranz, A., Bote-Lorenzo, M.L., Asensio-Perez, J.I., Martínez-Monés, A., Gomez-Sanchez, E., Dimitriadis, Y.: To reward and beyond: Analyzing the effect of reward-based strategies in a MOOC. Comput. Educ. **142**, 103639 (2019). https://doi.org/10.1016/j.compedu.2019.103639
19. Stewart, M.: Student perceptions of soft skills as an indicator of workplace success. PhD Thesis, Creighton University (2017). https://www.proquest.com/openview/3953f7a91ce21011c6e1da9e28e64ec2/1?pq-origsite=gscholar&cbl=18750
20. Mourshed, M., Patel, J., Suder, K.: Education to Employment: Getting Europe's Youth into Work. McKinsey Co. (2014). https://www.partners4value.lt/wp-content/uploads/2015/10/Education-to-employment-Getting-Europes-youth-into-work.pdf
21. M. S. Maren, U. K. M. Salleh, and H. Zulnaidi, 'Assessing prospective teachers' soft skills curriculum implementation: Effects on teaching practicum success', *South Afr. J. Educ.*, vol. 41, no. 3, 2021. https://doi.org/10.15700/saje.v41n3a1915
22. Sultanova, L., Hordiienko, V., Romanova, G., Tsytsiura, K.: Development of soft skills of teachers of physics and mathematics. J. Phys: Conf. Ser. **1840**(1), 012038 (2021). https://doi.org/10.1088/1742-6596/1840/1/012038
23. Scager, K., Boonstra, J., Peeters, T., Vulperhorst, J., Wiegant, F.: Collaborative learning in higher education: evoking positive interdependence. CBE—Life Sci. Educ. **15**(4), 69 (2016). https://doi.org/10.1187/cbe.16-07-0219
24. El Messaoudi, M.: Soft skills: connecting classrooms with the workplace–a systematic review. Üniversitepark Bül. Bull. **10**(2), 116–139 (2021) https://link.gale.com/apps/doc/A691005213/AONE?u=anon~7722f1c&sid=googleScholar&xid=43439e5a

25. Kroum, F.Z.: The implementation of soft skills for education and career success: case study of science and technical higher institutions in Morocco. Lingua Lang. Cult. **20**(2), 95–105 (2021)
26. Khaouja, I., Mezzour, G., Carley, K.M., Kassou, I.: Building a soft skill taxonomy from job openings. Soc. Netw. Anal. Min. **9**(1), 1–19 (2019). https://doi.org/10.1007/s13278-019-0583-9
27. Lorenzo, M.F.: Zoom in, zoom out: the impact of the covid-19 pandemic in the classroom. PhD Thesis (2020). https://doi.org/10.3390/su13052531
28. Kumar, P.: Enriching blended learning and teaching through the use of digital marketing and information technology. Int. J. Adv. Res. Educ. Soc. **3**(2), 116–127 (2021) https://myjms.mohe.gov.my/index.php/ijares/article/view/14030
29. Tadlaoui, M.A., Chekour, M.: A blended learning approach for teaching python programming language: towards a post pandemic pedagogy. Int. J. Adv. Comput. Res. **11**(52), 13 (2021). https://doi.org/10.19101/IJACR.2020.1048120
30. Chekour, M., Al Achhab, M., Mohamed, L., El Mohajir, B.: Contribution à l'intégration de l'apprentissage mixte dans le système éducatif marocain. Rev. Int. Technol. En Pédagogie Univ. **11**, 50 (2014). https://doi.org/10.7202/1035613ar
31. Aji, W.K., Ardin, H., Arifin, M.A.: Blended learning during pandemic corona virus: Teachers' and students' perceptions'. IDEAS J. Engl. Lang. Teach. Learn. Linguist. Lit. **8**(2), 632–646 (2020). https://doi.org/10.24256/ideas.v8i2.1696
32. Kaur, M.: Blended learning-its challenges and future. Proc.-Soc. Behav. Sci. **93**, 612–617 (2013). https://doi.org/10.1016/j.sbspro.2013.09.248
33. Ju, S.Y., Mei, S.Y.: Perceptions and practices of blended learning in foreign language teaching at USIM. Eur. J. Soc. Sci. Educ. Res. (2018). https://doi.org/10.33102/mjsl.v8i1.225
34. Tayebinik, M., Puteh, M.: Blended learning or E-learning?' Tayebinik M Puteh M2012 Blended Learn. E-Learn, pp. 103–110 (2013). https://doi.org/10.48550/arXiv.1306.4085
35. Brown, T., et al.: "Learning in and out of lockdown": a comparison of two groups of undergraduate occupational therapy students' engagement in online-only and blended education approaches during the COVID-19 pandemic. Aust. Occup. Ther. J. (2022). https://doi.org/10.1111/1440-1630.12793
36. Abd Gani, N.I., Rathakrishnan, M., Krishnasamy, H.: The effectiveness of Blended Learning on learners' interaction and satisfaction. Opción Rev. Cienc. Humanas Soc. (24), 48–62 (2019). https://dialnet.unirioja.es/servlet/articulo?codigo=8155732
37. Junior, G.L.D.C., Robles, D.C., de la Serna, M.C., Rivas, M.R.: Comparative study SPOC VS. MOOC for socio-technical contents from usability and user satisfaction. Turk. Online J. Distance Educ. **20**(2), 4–20 (2019). https://doi.org/10.17718/tojde.557726
38. Guo, S.: Synchronous versus asynchronous online teaching of physics during the COVID-19 pandemic. Phys. Educ. **55**(6), 065007 (2020). https://doi.org/10.1088/1361-6552/aba1c5
39. Murphy, C.A., Stewart, J.C.: The impact of online or F2F lecture choice on student achievement and engagement in a large lecture-based science course: closing the gap. Online Learn. **19**(3), 91–110 (2015). https://eric.ed.gov/?id=EJ1067503
40. Kaplan, A.M., Haenlein, M.: Higher education and the digital revolution: about MOOCs, SPOCs, social media, and the Cookie Monster. Bus. Horiz. **59**(4), 441–450 (2016). https://doi.org/10.1016/j.bushor.2016.03.008
41. Martínez-Muñoz, G., Pulido, E.: Using a SPOC to flip the classroom. In: 2015 IEEE Global Engineering Education Conference (EDUCON), pp. 431–436. https://doi.org/10.1109/EDUCON.2015.7096007
42. Kunin, M., Julliard, K.N., Rodriguez, T.E.: Comparing face-to-face, synchronous, and asynchronous learning: postgraduate dental resident preferences. J. Dent. Educ. **78**(6), 856–866 (2014). https://doi.org/10.1002/j.0022-0337.2014.78.6.tb05739.x

43. Alabay, S.: Classroom experiences with Microsoft Teams® for foreign language teaching 29. Uluslar. İnsan Ve Sanat Araştırmaları Derg., no. EK 1, pp. 26–29 (2018). https://dergipark.org.tr/en/pub/ijhar/issue/55202/758263
44. Nisrine, S., Abdelwahed, N.: Distance education in the context of the COVID-19 pandemic case of the faculty of sciences Ben M'Sick. Proc. Comput. Sci. **198**, 441–447 (2022). https://doi.org/10.1016/j.procs.2021.12.267
45. Rodriguez-Segura, L., Zamora-Antuñano, M.A., Rodriguez-Resendiz, J., Paredes-García, W.J., Altamirano-Corro, J.A., Cruz-Pérez, M.Á.: Teaching challenges in COVID-19 scenery: teams platform-based student satisfaction approach. Sustainability **12**(18), 7514 (2020). https://doi.org/10.3390/su12187514
46. Pal, D., Vanijja, V.: Perceived usability evaluation of Microsoft Teams as an online learning platform during COVID-19 using system usability scale and technology acceptance model in India. Child. Youth Serv. Rev. **119**, 105535 (2020). https://doi.org/10.1016/j.childyouth.2020.105535

The Impact of the Online Flipped Classroom on the Learning Outcomes and Motivation of Nursing Students in Computer Science

Mouna Hannaoui[1,3,4](✉), Youssra El Janous[1], El Hassan El-Hassouny[1], Jihad El Hachhach[1], Abdelhakim Askam[1], and Issam Habibi[2]

[1] ERIPDS, ENS, Abdelmalek Essaadi University, Tetouan, Morocco
hannaoulmouna@gmail.com
[2] Faculty of Sciene Dhar El Mehraz Chemistry Department – LISAC, Sidi Mohamed Ben Abdellah University, Fes, Morocco
[3] Faculty of Health Sciences, University of Jaén, Andalucía, Spain
[4] High Institute of Nursing Professions and Health Techniques, C.P 93040 Tetouan, Morocco

Abstract. Several research studies dealing with the flipped classroom (FC) model of teaching have pointed out that the decrease in student satisfaction and motivation could severely reduce the effectiveness of this classroom model. Therefore, this study aims to investigate the impact of the online flipped classroom on the learning outcomes and motivation of student nurses in computer science. The study involved 56 students divided into two groups. The first group attended the computer course under traditional conditions, so they were the control group (CG). The second group, the experimental group (EG), undertook the computer course using the online flipped classroom. The research data were obtained through the computer concept test and motivational questionnaire. The results have shown that, with the help of the online flipped classroom, a positive effect on the nursing students' understanding was achieved and a visible increase in the motivation of the experimental group was identified.

Keywords: Computer Science · Flipped Classroom · Motivation · Nurses

1 Introduction

It is clear that nowadays, the computer sciences are becoming an essential element in healthcare where healthcare professionals are responsible for processing a large amount of patient medical data (for example, a record of a patient's symptoms, medical history, previous medical examination reports, test results, doctor's diagnoses and even multiple combinations of these data if necessary) [1, 2]. For this reason, future nurses must master knowledge and computer skills to be able to improve quality, safety and efficiency in health, as well as reduce as much as possible the disparities that may occur [3–5].

In today's learning environments, many teaching strategies have the advantage of integrating technology into them. One of these new active learning strategies is the online flipped classroom [6]. The use of flipped classroom in education has become

M. Lazaar et al. (Eds.): BDIoT 2022, LNNS 625, pp. 138–147, 2023.
https://doi.org/10.1007/978-3-031-28387-1_13

popular in recent years. In a flipped classroom environment, students can access learning content related to the new topics they will learn through media such as lesson videos outside of the classroom where they receive face-to-face instruction. Flipped or "inversed" classroom has been advocated as a way to improve learning outcomes [7–9]. Research findings on the flipped classroom teaching model indicate that it has certain benefits on students' academic performance [7, 8] and engagement [10, 11]. At the same time, it cultivates a positive attitude towards the course in which it is applied [8]. The results also show that this could lead to differentiations in student motivation [6, 12] since; one of the underlying mechanisms that contribute on the increasing performance in flipped classrooms is intrinsic student motivation [13, 14]. According to the theory of cognitive evaluation (a theory of the theory under self-determination), motivation can be reinforced by the awareness of the need for autonomy and competence [15]. With respect to flipped classrooms, autonomy can be supported by the freedom to choose from different study materials when preparing lessons and planning these activities according to students' time and pace [16]. However, to achieve the expected performance of the active learning activities to be carried out in class and to allow students to be motivated in these activities, students must undertake face-to-face lessons with a certain level of prior knowledge and preparation, which is one of the assumptions to ensure the effectiveness of the flipped classroom model [17]. Additionally, students must demonstrate similar motivation and satisfaction in online and in-class course activities to ensure the flipped classroom model is effective. Otherwise, it could reduce the effectiveness of the flipped classroom model. Although research results on the FC teaching model indicate that it has some advantages on learning outcomes [9, 17, 18].

The aim here is to provide online access to learning content and materials and to support students in in-depth and active learning in the classroom. We therefore thought that teaching based on the online flipped classroom could stimulate their motivation and improve the results of nursing students in computer science. These two hypotheses are formulated as follows:

H0: "The use of online flipped classroom has no effect on nursing student outcomes in computer science".
H1: "The use of online flipped classrooms has no effect on the motivation of nursing students in computer science".

2 Methodology

Data for this research were obtained from "Computer Science Concept Tests" and "Computer Science Class Motivation Questionnaires". The different parts forming these tests and questionnaires are presented as follows:

2.1 The Participants

A total of 56 university students participated in this test (28 students in the control group and 28 students in the experimental group). These nursing students were tested in computer science. The Control group consisted of 20 female and 8 male students, while the Experimental group consisted of 17 female and 11 male students.

2.2 Online Learning Platform

In the present study, the Moodle platform was selected as a learning platform at the pre-class and post-class stages. This educational platform is free and it is designed for teachers who want to flip their classes. Moodle allows you to upload pre-existing videos on several sources such as You Tube, Khan Academy, and Personal Computer…, to make them interactive and to share them with students who can benefit from an immediate feedback from the teacher. In order to allow students who have taken the test to familiarize them with Moodle, a short training session has been organized to explain to them how this platform works. The training took place a few days before the start of the experiment.

2.3 Research Design

The research design is given in Table 1. Before the start of the experiment, both the Experimental group and the Control group submitted the same "computing concept test", however, the Experimental group also submitted a test "computing motivation questionnaire", before and after the experiment.

Table 1. Research design.

Groups	Pre-testing	Application	Post-testing
Experimental group	- Computer science test concept - Motivation questionnaire	Flipped classroom	- Computer science test concept - Motivation questionnaire - Semi-structured interview
Control group	- Computer science - Concept test	Traditional teaching	- Computer science - Concept test

In the control group, the lessons were conducted with a traditional approach while in the experimental group; these lessons were conducted with a flipped online classroom approach. At the end of the experimental process, both groups were subjected to the "Computer Science Concept Test" post-test, the experimental group also took the "Computer Science Motivation Questionnaire". We then proceeded to compare the results obtained by the student of the experimental group with those obtained by the control group using recognized statistical methods such as the mean and the student T-test.

2.4 Flipped Classroom Model Inspired by Santikarn and Wichadee

Scripting plays a major role in both traditional and flipped classrooms, so the sequence and progression must be carefully considered. To flip the classroom, this study adopted the model presented in Table 1, which is a model proposed in 2018 by Santikarn and

Wichadee [19].This model was adapted from the one proposed by Estes and colleagues a few years earlier [21] and developed by Hannaoui's team in nursing studies in Morocco for the first time [20]. It consists of three stages, with the first and last stages taking place outside the classroom, while the second stage takes place inside the classroom. This model has been tailored according to the nature of the knowledge to be taught in "computer science" and the didactic tools to be adopted (Table 2).

Table 2. Design of the course scenario based on Santikarn and Wichadee's flipped classroom model and fitted to the approach adopted by Hannaoui's team on nursing studies

Pre-course stage (on the Moodle platform)	Classroom stage (In the classroom)	Post-class stage (on the Moodle platform)
- Training on how to use the Moodle platform. - Exposure of the pedagogical scripting of the course on Moodle platform. - Sending course materials in advance to students (folder of the entire course…).	- Development of the didactic contract. - Explanation of the flipped classroom approach. - Consensus on which course sequences will be addressed in class and which will be addressed in the flipped classroom via E-learning (asynchronous and synchronous). - Learning through problem situations, application exercises, brainstorming discussions, question and answer sessions, etc. (during all face-to-face sessions).	- Consultation of the teaching materials for each teaching sequence. - Productions developed by students on each sequence (formative evaluations). - Audio-visual discussions (between professor and students). - Adjustments of productions (made by students) taking into account the remarks made by the teacher and students. - Re-exhibition of individual works.

2.5 Procedure

This study was carried out in an 8-week period within the academic year 2019–2020. Computer courses were taught to both groups of students.

Prior to the testing process, the experimental group was informed about the online flipped classroom approach. At the beginning of each session, the researchers prepared some activities. And, to ensure that the students followed the content of the online courses correctly, questionnaires were integrated throughout the courses. As the students viewed the course content, the results were automatically marked and sent to the researcher. In addition, participants were provided with answer keys to verify their responses. The instructor did not set aside class time to solve the quiz questions and the concepts on which they were based were discussed in class. During this process, the researcher found that the quizzes further motivated the students to look at the course content, and each quiz counted in the course. This method consisted mainly of two stages: the 'before' class stage and the 'during' class stage.

2.6 Data Collection and Analysis

As mentioned above, a computer science concept test and a motivation questionnaire were used as data collection tools in this work. The analysis of this data was done using arithmetic means, standard deviation tests and the student's T-test.

2.7 Computer Concept Test

- Pre-test: The computer science concept test was used as a pre-test to determine the knowledge levels of both the control and experimental groups. It was also used as a post-test 8 weeks later.
- Post-test: In order to identify the effect of the online flipped classroom approach on students' academic success, a computer science concept post-test was performed by the researchers 8 weeks later for both groups to identify who had better knowledge of computer topics. It consisted of 22 multiple-choice questions and it was applied to 28 students in the experimental group and 28 students in the control group.

2.8 Motivation Questionnaire

In order to identify the effect of the applied learning model on the students' motivations, a motivation questionnaire based on a model developed by Glynn et al. [21] was used. It consists of 22 questions and was used as a pre-test and post-test for the experimental group during the study. The reliability coefficient of the applied motivational questionnaire is equal to 0.0890.

3 Results

In the following section, the results of the data analysis are presented according to the research questions.

3.1 Effect of the Applied Method on Students' Results

The computer science concept test was used as a pre-test applied before the start of the course to identify the knowledge levels of the experimental and control groups. There should be an equal distribution of academic knowledge between the experimental and control groups. As it can be seen in Table 3, the data obtained from the pre-test was applied to the independent T-test; there is no significant difference between the results of the control group and the experimental group ($p > 0.05$). This result indicates that both groups of students have the same level of academic knowledge before starting the experiment.

Table 3. Independent group T-test results regarding pre-test scores of experimental and control groups

R squared		0,005782
Mean of Control group		6,87
Mean of Experimental Group		7,23
Difference between means		0,3571
Error standard mean		0,6373
Confidence interval 95% of the difference	Lower	−0,9206
	Top	1,635
t		0,5604
df		54
P value		0,5775

In order to determine if there are any significant differences between the scores of the experimental and control groups at the end of the study, the "computer science concept test" was used as a post-test, as shown in Table 4. In addition, a T-test was used for the independent groups.

Table 4. Independent groups T-test results for the post-test scores of both the experimental and control groups

R squared		0,2447
Mean of Control group		10,59
Mean of Experimental Group		13,29
Difference between means		−2,696
Error standard mean		0,6447
Confidence interval 95% of the difference	Lower	−3,989
	Top	−1,404
t		4,182
df		54
P value		0,001

The post-test scores of the students in the control group who followed the traditional approach were clearly lower than those in the experimental group who followed the flipped online classroom approach. Although there was an improvement in both groups, the rise in the experimental group was statistically significant with a P-value well below 0.05. From these results, it can be stated that during the teaching of this particular computer course, the online flipped classroom approach shows a more positive effect compared to the traditional approach.

3.2 Effect of the Applied Method on the Students' Motivation

Table 5 contains the results of the questionnaire conducted prior the starting of the course to determine the effect of the online flipped classroom approach on students' motivation, as well as the results of a paired T-test of this questionnaire conducted afterwards. It was observed that there was a statistically significant increase in student motivation ($p < 0.001$). This finding indicates that their motivations increased due to the flipped online classroom approach, which is due to the fact that the computer course has been made more engaging.

Table 5. T-test results of average points in pre-test and post-test motivation questionnaires

R squared		0,7516
Mean of Control group		0,1990
Mean of Experimental Group		0,9336
Difference between means		−1,587
Error standard mean		−2,001
Confidence interval 95% of the difference	Lower	−1,173
	Top	−1,173
t		7,971
P value		<0,0001

For the motivation questionnaire, the following five elements showed the greatest increases. For the element "I like to learn about computer topics", the mean pre-test score was $X = 0.62$, while the mean post-test score was $X = 4.60$. For the element "The computer science I learned was relevant to my personal goals", the mean pre-test score was $X = 0.83$, while the mean post-test score increased to $X = 4.57$. Also, for the item "I like to do better than other students on computer tests" the mean pre-test score was $X = 0.47$, while the mean post-test score was determined as $X = 4.82$. Similarly for the element "I am not nervous about how I will do the computer science tests", the mean pre-test score was $X = 0.87$, while the mean post-test score increased to $X = 4.64$., for the item "It's my fault that I don't understand computers" the mean pre-test score was $X = 0.97$, finally, the mean post-test score was determined as $X = 3.92$.

4 Discussion

This study has been conducted in order to assess the effects of the online flipped classroom on the learning outcomes and motivation of computer science nursing students. Prior to the implementation, the results of the pre-test computer concept tests indicated that the knowledge levels of both groups were basically equal. However, the post-test results, in contrast to the pre-test results, indicated a significant improvement in the learning of the students in the experimental group that exceeded that of the students in the control group.

The reasons behind this improvement in the experimental group's results can be interpreted by the fact that the transmission of knowledge acquired through the course content was done outside the classroom, so these students were able to assimilate the information at their own learning pace and could also retake the course as many times as they wanted. In addition, the transfer of data was carried out through the course content and the class time was divided between discussions and problem-solving activities. The use of exercises also helped to increase the success.

The flipped classroom approach encourages students to acquire cognitive skills such as the acquisition of knowledge prior to lessons, while focusing on higher level cognitive skills such as application, analysis, synthesis and evaluation with the support of their friends and teachers, which is in line with the Bloom's taxonomy.

In the literature, as observed in this study, the flipped classroom approach has been found to significantly increase students' academic ability [7–9]. However, this is not always the case, as the results of a study conducted by Frydenberg [23] are found to differ from the results of this study. Where they found that teaching Excel using a flipped classroom approach, showed no increases in the students' understanding compared to the traditional approach. In another study by Winter [24] involving the flipped classroom approach in a physics course, once again, no increase in student achievement was observed.

The other important result of this study was the significant increase in motivation that appears in the questionnaire given to the experimental group. It is believed that the online flipped classroom approach builds active learners by giving them better chances to participate in classroom discussions. The result of this study therefore supports the idea that the students' motivation levels increase under active learning conditions [25–27]. The results of this study are consistent with other studies regarding the flipped classroom approach [8–14, 20–27].

Finally, the present study will be contributing to the literature as there are limited studies available regarding the online flipped classroom approach for computer science courses in Universities.

5 Conclusion

The results of this study indicated that the flipped classroom had a positive effect on the grades and motivation of nursing students with respect to computer science.

Thus, this approach could be an alternative to the traditional course in nursing and more precisely in scientific subjects. Also, it's a solution to the insufficient time volume of certain courses such as informatics, since the content of the courses would be taken at home and the class would be intended for application exercises, problem situations, etc…

Therefore, other studies should be adapted to apply this approach to other subjects and to other cohorts.

References

1. Demiris, G., Zierler, B.: Integrating problem-based learning in a nursing informatics curriculum. Nurse Educ. Today **30**(2), 175–179 (2010)
2. Kim, H.N.: A conceptual framework for interdisciplinary education in engineering and nursing health informatics. Nurse Educ. Today **74**, 91–93 (2019)
3. Buerck, J., Feig, D.: Knowledge discovery and dissemination: a curriculum model for informatics. In: Working group Reports on ITiCSE on Innovation and technology in Computer Science Education, pp. 48–51 (2006)
4. Dixon, B.E., Newlon, C.M.: How do future nursing educators perceive informatics? Advancing the nursing informatics agenda through dialogue. J. Prof. Nurs. **26**(2), 82–89 (2010)
5. De Gagne, J.C., Bisanar, W.A., Makowski, J.T., Neumann, J.L.: Integrating informatics into the BSN curriculum: a review of the literature. Nurse Educ. Today **32**(6), 675–682 (2012)
6. Yilmaz, R.: Exploring the role of e-learning readiness on student satisfaction and motivation in flipped classroom. Comput. Hum Behav. **70**, 251–260 (2017)
7. Freeman, S.: Active learning increases student performance in science, engineering, and mathematics. Proc. Natl. Acad. Sci. **111**(23), 8410–8415 (2014)
8. Yilmaz, R.: Knowledge sharing behaviors in e-learning community: exploring the role of academic self-efficacy and sense of community. Comput. Hum. Behav. **63**, 373–382 (2016)
9. Wang, F.H.: An exploration of online behaviour engagement and achievement in flipped classroom supported by learning management system. Comput. Educ. **114**, 79–91 (2017)
10. Jayawardena, P.R., Van Kraayenoord, C.E., Carroll, A.: Factors that influence senior secondary school students' science learning. Int. J. Educ. Res. **100**, 101523 (2020)
11. Hu, R., Gao, H., Ye, Y., Ni, Z., Jiang, N., Jiang, X.: Effectiveness of flipped classrooms in Chinese baccalaureate nursing education: a meta-analysis of randomized controlled trials. Int J Nurs Stud **79**, 94–103 (2018)
12. Davies, R.S., Dean, D.L., Ball, N.: Flipping the classroom and instructional technology integration in a college-level information systems spreadsheet course. Educ. Technol. Res. Dev. **61**(4), 563–580 (2013)
13. Presti, C.R.: The flipped learning approach in nursing education: a literature review. J. Nurs. Educ. **55**(5), 252–257 (2016)
14. Sergis, S., Sampson, D.G., Pelliccione, L.: Investigating the impact of flipped classroom on students' learning experiences: a self-determination theory approach. Comput. Hum. Behav. **78**, 368–378 (2018)
15. Tian, Z.J., Jin, R.H., Liu, C.F.: Application of flipped instruction in the course of Geriatric Nursing. Chin. J. Nurs. Educ. **12**(5), 333–335 (2015)
16. Mikkelsen, T.R.: Nursing students' experiences, perceptions and behavior in a flipped-classroom anatomy and physiology course. J. Nurs. Educ. Pract. **5**(10), 28–35 (2015)
17. Huang, Y.-N., Hong, Z.-R.: The effects of a flipped English classroom intervention on students' information and communication technology and English reading comprehension. Educ. Tech. Res. Dev. **64**(2), 175–193 (2015). https://doi.org/10.1007/s11423-015-9412-7
18. Deslauriers, L., Schelew, E., Wieman, C.: Improved learning in a large-enrollment physics class. Science **332**(6031), 862 (2011)
19. Santikarn, B., Wichadee, S.: Flipping the classroom for english language learners: a study of learning performance and perceptions. Int. J. Emerg. Technol. Learn. **13**(9), 123 (2018)
20. Hannaoui, M.: The effect of the flipped classroom on nursing students' learning and motivation outcomes in health education. Turk. J. Comput. Math. Educ. (TURCOMAT) **12**(6), 1213–1223 (2021)

21. Estes, M.D., Ingram, R., Liu, J.C.: A review of flipped classroom research, practice, and technologies. Int. HETL Rev. **4**(7), 1–8 (2014)
22. Glynn, S.M., Taasoobshirazi, G., Brickman, P.: Science motivation questionnaire: Construct validation with nonscience majors. J. Res. Sci. Teach. **46**(2), 127–146 (2009)
23. Frydenberg, M.: Flipping excel. In: Proceedings of the Information Systems Educators Conference ISSN, vol. 2167, p. 1435 (2012)
24. Winter, J.B.: The effect of the flipped classroom model on achievement in an introductory college physics course (2013)
25. Hannaoui, M., Madrane, M., El Hassouny, E.H., Janati-Idrissi, R., Zerhane, R., Grande-Gascón, M.L.: Towards exploring midwifery students' misconceptions of reproductive health education: Case of Sexually transmitted infections. Elementary Educ. Online **19**(4), 898 (2021)
26. Estes, M.D., Rich, I., Juhong, C.L.: Un examen de la recherche, de la pratique et des technologies en classe inversée. Revue Internationale HETL **4**(7), 1–8 (2014)
27. Evans, L., et al.: Flipping the classroom in health care higher education: a systematic review. Nurse Educator **44**(2), 74–78 (2019)

Digitizing Writing: An Empirical Exploratory Study Towards Students' Writing Skill Fluency Improvement in 'Padletosphere'

Nadir El Morabit[1](✉) and Najemeddin Soughati[2]

[1] Abdelmalek Essaadi University, Tetouan, Morocco
ndrmorabet@gmail.com
[2] Ibn Tofail University, Kenitra, Morocco

Abstract. Boosting students' writing skill is thorny and of utmost challenge to EFL teachers. Paper and pencil method has proved itself myopic in the era of widespread technology where the new generation of learners are characterized by digital affinity to learning rather than a mere conventional method. This article aims at gauging the impact of the smart application Padlet as an integral part of a digitized syllabus intervention on the learners' writing skill rather than a mere conventional paper and pencil common method. The research design for this paper is mixed method. The qualitative part is embodied in the in-class discussions about the app and the open questions conducted via an electronic questionnaire where structured open-ended questions represent most items learners should respond to. On the other hand, the quantitative part of the questionnaire aims at measuring the learners' degree of motivation, method of work, degree of commitment to working on the app, helping features and affordances, etc. The results obtained from the collected data are appealing and encouraging to us as practitioners to restructure our teaching methods towards updating and digitizing them to cater for students' learning needs and language skills delivery in the classroom.

Keywords: Padlet · Writing Skill · Instructional Technologies · New Affordances · Mobile Mediated Learning (MML)

1 Introduction

Technology today has prevailed all aspects of life. These digital literacies are widespread nowadays in many countries' educational systems forming an integral part of the curricula. The Moroccan educational system is still low-tech. Therefore, students' expectations are torn between the reality as being digital natives and the incompatible classroom environment with that reality. Moroccan EFL teachers concede that the students' writing fluency is characterized by low-proficiency level let alone accuracy. One of the interesting questions raised in this context is whether instructional digital literacies may be conducive to motivating and stimulating students with poor performance in the writing skill. Accordingly, we argue in this paper that implementing Padlet as a smart application is helpful and appealing to both practitioners and students as well. To achieve this

M. Lazaar et al. (Eds.): BDIoT 2022, LNNS 625, pp. 148–159, 2023.
https://doi.org/10.1007/978-3-031-28387-1_14

goal, a mixed-method study is used to measure our students' experiences (i.e., primary data) with Padlet, degree of motivation, the changes affecting their writing, the app's affordances, etc. An electronic questionnaire is used as a post-tool to measure these variables.

The primary goal of this mixed-method study is to demonstrate the potential linkage between the smart app Padlet as an independent variable and writing fluency as a dependent variable. This study aims to fill a gap in the few studies that have been conducted in the field by reinforcing research regarding the little inquiry that has been directed towards this topic and its importance on the Moroccan EFL high school students, who are characterized by low performance in writing due to poor teaching methods and the conventional practices with younger generations known as digital natives.

The study's significance aims at adding to the body of knowledge about the effects of technology on low-proficiency students' motivation and writing ability. The study is conducted to see how learners of today view the usage of technology whether in the classroom or at home as a follow-up. These findings may inform a wide range of people, particularly instructors to revisit their teaching practices and approaching the writing skill differently to cope with the new demands of the twenty first century quality education.

Two main research questions will be addressed to examine the impact of digital mobile technologies (Padlet in this case) on the writing performance (i.e., fluency) of low-ability students:

- How frequently do students use Padlet?
- Does Padlet implementation have positive impact on their writing skill?
- How can Padlet enhance the writing skill for low-ability students?

2 Literature Review

Since the stepwise emergence of the term educational technology in the 1940s as investigated by the leading figure in the history of educational technology Paul Saettler (as cited in [1]), the term has been redefined several times to establish a grounded 'theory' of the 'concept' by the AECT (Association for Educational Communications and Technology) since the 1963 definition [2]. Many critics had a say in the concept and each writer has been trying to approach the concept differently. Yeaman considers this as a "propaganda" with "naïve optimism" [1]. Establishing the theoretical background is of utmost importance to the concept yet spending decades trying to do so is a matter of unproductive controversy. The world is moving fast towards digitizing the educational system for a new generation which was born in the era surrounded by technology available in their pockets, wrists, ears, clothes, on desks and walls, in streets, classrooms, means of transportation, kitchen, etc. however, the criticism is still partially oriented towards focusing on the theoretical foundations of the concept [3, 4], thorny underestimation of instructional technology [5]; motivation, the importance and the frequency use of technology [6–8], and the basics of e-learning [9, 10].

Bax (2003) came with the idea of normalizing the instructional modalities, however the new concept, regardless of its importance to make a remarkable shift from the concept-focus to concept-implementation in the classroom, remains limited. The unintelligibility lies in the classical shift to either the theoretical side where the critics are still striving to establish the conceptual framework of the term, or the current trend of the recent descriptive papers which focus on the pseudo practicality of the educational technological modalities per se. In other words, many papers have been published prescribing apps to be implemented in the classroom. These apps are unfortunately prescribed artificially without any measurement of their impact on students' language skills as in Smith [11] where he listed several free and payable apps recommended for teachers.

Blake [12] reviewed and listed some new technologies considering the affordances they offer to improve the four skills following the task-based language teaching (TBLT) and Dewey's ideas on experiential learning. Despite Blake's intention to highlight the merits of the affordances offered by the new technologies of Web 2.0, his study did not focus on measuring the impact of the listed apps on students' productivity and language fluency especially in writing. The works cited above focus more on the question of what apps to assign for each skill rather than how each app may improve L2 given the techniques proposed, such as creating digital stories, engage in fanfiction, translate, or subtitle movies or series, etc.

In [13], Wasniewski and Boechler published a paper stressing the importance of the role of "the potential impact [of] assistive technology" (p. 2491). The results were very encouraging in language development, learning strategies (self-reported), learning autonomy, lexical knowledge, and "strategic support for differentiated instruction" (p. 2495). Wasniewski and Boechler's findings were harmonious with the proposed definition by Oxford (1990): "learning strategies are specific actions taken by the learner to make learning easier, faster, more enjoyable, more self-directed, more effective, and more transferable to new situations" [14]. Hence, Wasniewski and Boechler with their concise study have made a remarkable shift towards a strategic implementation of instructional technology.

In [15], Al Sharqi and Abbasi underscored the importance of the writing skill. Their study emphasized the role of technology (social media as an example) to sustain writing even if it was conducive to "a new shorthand English dialect called Text-speak" (p.1). Text-speak, which is also known as Digi-talk and Tech-speak by technology savvy users, should not be underestimated nor considered it to be "infiltrating students' assignments [and] blurring the distinction between formal and informal writing" [15]. However, the writers justified their conviction by a study conducted by Merritt in 2013 which accentuated the lack of any correlation between Text-speak and poor grammar. Moreover, Merritt pointed out that "these children were able to develop arguments, write thesis statements, and structure their thought appropriately, which may serve as a reassurance to parents who fear that Text speak may be degrading their children's grasp of Standard English" [15]. Al Sharqi & Abbasi reached the conviction that "technology is here to stay and Text-speak is the main mode of communication used by the net-generation [and] accepting Text-speak as a skillful language seems to be the only appropriate choice" [15].

Alsulami conducted quantitative research to measure the impact of digital technologies on learning English as a foreign language (EFL) for freshmen, sophomores, juniors, and senior students. According to the writer, the results were highly positive and encouraging for "students to be more proficient in the English language" [16]. Alsulami aimed at measuring three main areas: frequency use, type of technologies, and students' motivation and impression. As far as the writing skill is concerned, she found that "(70%) of students agreed that word processing software (i.e., Google Docs and Microsoft Word) can be very useful in developing their writing skill" (p. 12), and "(88.9%) used [technology] for chatting on social networking websites as a way to improve their writing skills." (p. 10). Although Alsulami's study is important at measuring the impact of digital technologies on the students' language skills, the lack of measuring the real impact with pre-established criteria to assess the students' performance was critical. On the other hand, the study did not introduce any developed strategies by the students (self-reported & self-directed).

Like Alsulami, Nomass [17], Bloch [18] emphasized the role of technology with the myriad of tools provided by the development of Web 2.0 in improving the writing skill. Although he did not establish a concrete demarcation between his positive stand based on the different digital modalities in favor of teachers and learners, Bloch's approach is still important in the sense that he focuses more on the social aspect of writing rather than only on the bunch of modalities with a plethora of affordances per se. He maintained that:

> One of the areas where convergence of writing theory and technology was most evident in the development of what were called social-cognitive theories of writing. The original process approaches were criticized for ignoring social aspects of the writing process. The development of ever-expanding networks, from the classroom to the world, would facilitate this social process so that writing could be shared within the classroom, outside the classroom, and eventually anywhere in the world. The later development of technologies such as wikis or file-sharing programs such as Google Docs, would further facilitate social interactions [18].

Block, as many other writers, is not only concerned with the plethora of modalities and its unquestionable appealing affordances, but also concerned with how CALL (Computer-Assisted Language Learning) and how teachers respond to the pedagogical goals in the sense that "teaching with technology requires a great amount of careful planning and evaluation" [19]. Bloch published a paper highlighting the importance of blogging and how Abdullah, a 1.5 generation Somali student, developed his academic writing via blogosphere. Bloch's attention was not only to explore how blogging can be used in the context of L2 composition classroom, but also how students (Abdullah as an example) took advantage of blogging "as a tool in the acquisition of more academic forms of" [19].

Nomass argued that the challenge resulted from the writing skill in terms of "generating ideas, organization, and perfect use of grammar and vocabulary" [17] can adeptly be dealt by the aid of word processing programs. Nomass argued that "there [was] a weak tendency for encouraging students to use technology in the classroom lessons" (p. 115), she found that (96%) of the students questioned "believe that using computer will help them to develop their writing skill. Despite Nomass's strive to accentuate the role of technology in the learning process, her attempt was myopic in the sense that writing does not only entail some electronically mediated communication softwares or apps, but "writing potentially entails both production (writing itself) and decoding (reading) [...] [where] we have few studies" [20].

Another attempt towards drawing the attention to digital literacies was made by Li and coauthors [21]. The authors counted several affordances of the listed web-based softwares and apps to help teachers and students develop the writing skills effectively since new information and communication technologies are significantly affecting writing processes and practices. In this context, Li and coauthors argue that "all teachers, schools, and colleges are challenged to respond to the changing nature of writing" [21]. The authors listed some studies where the implementation of Automated Writing Evaluation (AWE) and web-based softwares were conducive to successful empirical results. Li and coauthors unlike many above studies, tried to shift from the traditional mainstream of instructional technology to focusing on measurable benefits of these tools on the writing skill. In this sense, they argued that:

"technology-supported peer feedback is characterized by motivating interaction, flexible discourse patterns and language usages, effective shifts in teachers' and students' roles in the process of providing peer feedback [and] improvement in English writing in terms of content, organization, structure, vocabulary and spelling, and genre awareness" [21].

As far as Li et al. and other studies mentioned above concerned, we argue in this study that the proliferation of digital tools require further examination of the real impact on students' writing fluency far from the conventional approach to the concept of educational technology in terms of the theoretical framework, frequency of use, motivation, students' preferences, tools' affordances, strategy development, etc. The investigation should target the real impact of these technologies on students' language skills (writing in our case) within the framework of a digitized course design with clear established criteria and measurable variables in comparison to the traditional pencil and paper approach with pre- and post-tools of measurement.

2.1 Defining Padlet: An Outlook on Affordances

Padlet is a simple online collaboration platform that allows students to publish text, images, documents, links, videos, and voice recordings. It was founded by Nitesh Goel (CEO) and Pranav Piyush in 2012. Nowadays, Padlet is credited a remarkable fame as:

"a tool that teachers can exploit with any level of learner. As learners add their own posts, what skill they develop depends on what task given is to them" [22].

What characterizes Padlet is the multifaceted affordances appealing to students to partake in the activities guided by the teacher. For instance, this software provides the educator's dashboard with many features, such as evaluating students' outcomes numerically or via stars as preferred by many educators, filtrating profanity, controlling the deadline of each posted task via approval option, backchannel for discussion board, or comments' spaces where peer feedback takes place. Synchronous and asynchronous interaction are also possible to prompt more ideas among learners. Moreover, Padlet helps teachers to monitor the learners' outcomes via providing corrective feedback and trace back the learners' progress either by teachers or learners' themselves by looking back at the prior Padlets.

2.2 The Role of the Teacher

In Padlet, the role of the teacher is multi-faceted and depends on the nature of the task, yet the instructor plays the role of an assessor, facilitator, and prompter. Tasks assigned to students by the teacher are either related to the post-stage where one hour session is challenging if not impossible to finish a particular lesson, let alone those pertaining to the writing skill, or as part of the production stage of other skills as well. To help students improve their writing fluency, production-stage for the reading skill takes the form of writing (i.e., summary, critique, analysis of the author's stand, students' impressions, etc.). This process is part of the digitized course design of the English syllabus [23] to be conducive to boost the Moroccan high school EFL students' writing fluency. Furthermore, students are encouraged to use Padlet and make it their digital platform where they are supposed to post their writing tasks whether they are related to the reading (production stage) or the writing skill. As students seem skeptic about the process, we were much obliged to motivate students and target their psychological readiness adeptly as Nycz and Cohen point out: "to understand what knowledge is, we need to understand psychological aspects of the individual" [10]. The first task the teacher asked students to introduce themselves to their friends by providing as much information about them as possible. Our intention was to prepare students psychologically from the easiest task to the prospective challenging ones. Once the process was done, students reacted positively to the first experience using Padlet. Hence, the teacher took advantage of the students' approbation to work on this digital tool to disseminate the goal of using the app in boosting their writing skill which was immediately approved by them enthusiastically. It should be noted that the most asynchronous processes of the interactions taken place in Padlet are teacher-student, student-teacher, and student-student. This process was also helpful in the sense that two hours are saved every two or three weeks. These hours were dedicated by the teacher to share some feedback about the students' progress from each Padlet (task) to another. Finally, students' outcomes are filtrated using online plagiarism checkers to make sure that students stick to their learning agency. Finally, the best outcomes are read and commented constructively in class before being copied collectively. This process generated a fierce competition among learners, where each student tries to make a remarkable progress to be the quintessential follow suit model.

2.3 The Role of the Learner

The student's role in Padletosphere (i.e., the interactive atmosphere in Padlet) is empowered by the shift from being passive receiver to active contributor. Students should stick to the deadline assigned to each task as a code of conduct agreed on between the instructor and the learners. The mode of work varies from individual to pair-work, and small groups. Students must communicate their knowledge authentically via the use of all the available digital tools (softwares, web-based sites, smart apps, etc.) from their own choice to achieve optimal learning experiences. What is more, students are strongly advised to avoid any act of word-to-word plagiarism without any reference to their authors. Furthermore, the student's role does not only lie in posting their tasks per se, but also interacting with each other for the sake of peer learning and prompting as many ideas as possible.

3 Methodology

The purpose of this research was to examine the positive effects of the smart app Padlet on EFL students' writing skills. To achieve this goal, a mixed-method study is used to measure the variable. To implement the process, a quantitative mixed-method study is used to measure the impact of the digital smart tool Padlet in Al Imam El Ghazali High School in Tetouan, Morocco. The process started one year before using graded readers and Flipgrid as an integral part of the digitized syllabus process. Therefore, this research section is structured into three parts. The first section was designed for data collection, the second section discusses the targeted population, and the third section discusses the instruments or measures.

3.1 Data Collection

To measure the effectiveness of Padlet on students' writing skill, an electronic questionnaire is designed to collect primary data. It is structured into four parts. The first part measures the demographic data, the second part collects data about Padlet use experience, the third part gauges the students' motivation, and the last part aims to collect data about the app's affordances. The three last sections contain close, semi-close, and open questions. The closed items are primarily meant to collect numerical data and the open ones were intended to collect qualitative data.

3.2 Participants

To collect primary data, the electronic questionnaire was administered to 104 students studying et al. Imam El Ghazali High School in Tetouan, Morocco. The respondents' demographic data are documented in the following Table 1:

Table 1. Participants' demographic information.

Age	Number	%	Level	%
15–17	74	71.1	Common Core	39.4
18–23	30	28.9	Second Bac	60.6

3.3 Instruments/Measures

Different types of questions were designed to measure the variable (i.e., impact of Padlet on students' writing skill). The first part consists of demographic items (gender, age, and level), the second part consists of four closed questions targeting the participants' experience using Padlet. Participants are also encouraged to justify their answers. The third part included two questions, a closed and open item to justify the degree of motivation by using linear scale type (0 to 5) where zero represents unmotivated and 5 highly motivated. Eventually, the fourth section of this research is made up of four multiple-choices questions and eight open questions to elicit the participants' impressions about Padlet's features, affordances, and recommendations.

4 Results

This part aims at presenting the main findings after analyzing the data. Having introduced the demographic items above, the findings concerning this section are overlooked due to the focus of the study. The main findings of part two are included in the following Table 2:

Table 2. Participants' experience using Padlet.

Questions	%		Type
	Yes	No	
Did you like Padlet ?	92.3	07.7	closed
Were you committed to using it?	70.2	28.8	closed
Do your regret not using it?	71.9	21.9	closed
Did you face any problems using the app?	73.1	26.9	closed
Mean	**76.8**	**21.3**	***

Most students who responded with NO oversaw the justification of their choices. Aggregately, students' responses may be structured into technical problems where students, whose smart phones are old fashioned, found it difficult to download and install the app. Moreover, those who succeeded claimed that the app failed in the notifications feature. Others admitted neglecting the app due to lack of time, oblivion, and the internet access.

The third part of the questionnaire targeted the degree of motivation using Padlet. The following figure describes all the students' responses as generated automatically by the web-based software (Fig. 1).

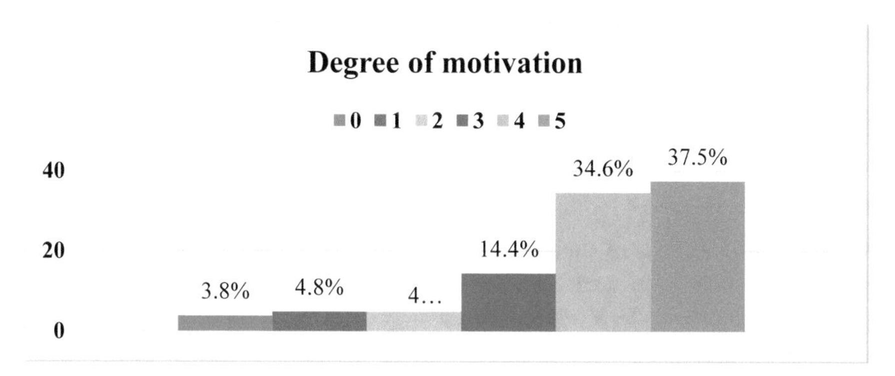

Fig. 1. Degree of motivation

Having used the linear scale to measure the degree of motivation, the results were very encouraging, 72.1% of students chose (4 to 5 as highly motivated). Students whose degree of motivation is either low or null referred to the above-mentioned reasons in the second section.

The third section targeting the affordances of Padlet, and students' points of view was rich in recommendations (see recommendations section). Hence, the following table summarizes this part (Table 3):

Table 3. Padlet's affordances and recommendations.

Questions		%		Type
		Yes	No	
How do you find using educational technology in teaching?		93.3 (good)	6.7 (bad)	closed
Do you support using technology in teaching?		95.2	4.8	closed
Which mode of work do you prefer?	**Individual** 44.2%	**Pair work** 29.8%		**Group work** 26%
Has Padlet helped you improve you writing skill?		84.6	15.4	closed
Mean excluding item 3		**91**	**8.9**	***

Since the central open question of this section was the one related to whether Padlet has helped the participants improve their writing skill. The following reactions are the students' impressions structured to participants *for* and *against*. Anonymity is maintained as agreed on with the participants. It should be noted that the participants answers are translated from Arabic to English without any further additions (Table 4).

5 Discussion

This section examines the primary data collected holistically except for some points in need of an in-depth analysis. It has become clearer from the data collected that the higher mean number (percentage) in all the sections that Moroccan students, most importantly the younger generations known as digital natives, are eager for creative and innovative learning practices implementing the new digital technologies.

5.1 Padlet Experience

The higher mean of 76% of students favoring Padlet as a tool improving their writing skills denotes the students' tendency towards the digital method of work at the expense of the common traditional one prevailing the Moroccan public schools. What consolidates this conviction is that 71.9% of students regretted having missed the opportunity to participate in the app. These students, in open discussions, admitted that their classmates who were committed to posting their tasks have made remarkable progress in their writing whereas they did not the way they wanted. Furthermore, many participants in open questions concede that the traditional method undergone by teachers in delivering writing lessons demotivates them.

Table 4. Padlet's role in the writing skill improvement.

Participants for	Participants against
- "it makes the work easier and takes less time to do than without technology." - "it feels like it pushed me towards practicing my writing more." -"using it pushed me to look for information from other resources like translation softwares." - "I couldn't write well, but using Padlet made me write on paper, cast a look on my classmates outcomes, then I re-write my paragraph as well which helped me so much to improve." - "because of it I started to wake up early, look up words in the dictionary and write." - "I learned so much due to the type of tasks." - "I learned a lot and most importantly from the interaction with the teacher in the comments." - "I learned great deal of vocabulary." - "we've time to search, translate, and much freedom at home."	- "the problem of basics" - "I feel obliged to translate all the time." - "time restrictions." - "we can do the same on our copybooks." - "the topics posted on Padlet are very basic."

5.2 Motivation

86.5% is the mean number depicting the aggregate percentage of students who feel motivated to participate in Padlet. Again, this percentage denotes the shift from the traditional method in delivering writing lessons to the newer method students admitted having never worked with. What is more, this higher figure questions the traditional methods implemented in the classroom delivering writing lesson on the one hand, and other skills on the other for this generation where technology prevails their surrounding starting from their smart phones in their own pockets. We also deduce from the participants' reactions the inclination towards improving their writing skills if the tools meet their expectations, namely using instructional technologies, especially smart apps.

5.3 Padlet's Affordances

There are many features which make Padlet preferable to students in contrast to the pencil and paper method. In-class method requires students to finish their writing tasks within the allocated time. Students find it challenging to brainstorm their prior knowledge adeptly. Moreover, they are mostly not allowed to use their mobile phones due to the lack of trust between teachers and their students in fear of ending up using social media rather than staying on task. On the contrary, using Padlet motivates students to work creatively and differently. Students may resort to translation softwares, prompt each other, learn from one another (peer learning), create thematic discussion boards themselves, newer mode of work (e.g., GPS), reinforce their answers with clickable links, images, and videos. Moreover, students can comment on each other constructively and gain higher self-esteem which is very recommended in the classroom. In addition to that, students are granted more time and freedom to come up with quality work which targets higher levels of thinking instead of the lower ones where students are asked to analyze (e.g., writers' points of view), synthesize (i.e., integrating resources and ideas, timeline, etc.), and evaluate (e.g., the education system).

6 Conclusion

In summary, the findings of this study show that Padlet, as a smart app implemented to improve the writing skill of the low-ability students, proves itself an effective and promising tool. The focus was to measure the students' degree of motivation, frequency use, aspects of change, and the helping affordances conducive to writing improvement in relation to the digitized English syllabus where Padlet is forming an integral part. Accordingly, the study reveals that students prefer using technology to enhance their writing rather than pencil and paper-based method. As far as students concerned, Padlet can enhance authentic learning experiences, exploited for communication facility, peer learning engagement, communicate their knowledge differently and adeptly, compare their *Padlets* (assignments), check their progress, and receive their peer and teacher's feedback. Furthermore, it is an app where the sense of authority shifts from the teacher to the student as the teacher plays the role of facilitator, prompter, and assessor. Most importantly, it is an app which stimulates the use of other softwares (vocabulary builder apps and translation softwares) as revealed by students' impressions and recommendations.

References

1. Januszewski, A.: Educational technology: The development of a concept. Libraries Unlimited (2001). https://books.google.com/books/about/Educational_Technology.html?id=mlZsIIoOaSYC
2. El Morabit, N.: Educational technology: from a historical perspective to an empirical exploration of moroccan learners' EFL speaking fluency. Global J. Hum.-Soc. Sc. **21**, 39–48 (2021). https://doi.org/10.34257/GJHSSGVOL21IS11PG39
3. Thorne, S.L., Smith, B.: Second language development theories and technology-mediated language learning. CALICO J. **28**, 268–277 (2011). http://www.jstor.org/stable/calicojournal.28.2.268
4. Levy, M.: Computer-Assisted Language Learning: Context and Conceptualization. Oxford University Press (1997). https://books.google.co.ma/books?id=RRGgrjteVjUC&lpg=PR8&ots=8s7dI9oFw5&lr&pg=PR8#v=onepage&q&f=false
5. Clark, R.E.: Reconsidering research on learning from media. Rev. Educ. Res. **53**, 445–459 (1983). https://doi.org/10.3102/00346543053004445
6. Bolkan, J.: Report: Schools not meeting students' technology needs. Campus Technol. (2012). https://thejournal.com/articles/2012/09/13/report-schools-not-meeting-students-technology-needs.aspx
7. Sailer, M., Murböck, J., Fischer, F.: Digital learning in schools: what does it take beyond digital technology? Teach. Teach. Educ. **103**, 103346 (2021). https://doi.org/10.1016/j.tate.2021.103346

8. Francis, J.: The effects of technology on student motivation and engagement in classroom-based learning (2017). https://dune.une.edu/theses/121
9. Jamiai, A.: Measuring master students' online learning perceptions and satisfaction during Covid-19 crisis in Morocco. Int. J. Lang. Lit. Stud. **3**, 1–11 (2021). https://doi.org/10.36892/ijlls.v3i1.488
10. Nycz, M., Cohen, E.: The basics for understanding e-learning. Princ. Eff. Online Teach. 1–17 (2007). https://www.academia.edu/17666101/Principles_of_Effective_Online_Teaching
11. Smith, B.: Technology in language learning: An overview (2015). https://doi.org/10.4324/9781315673721
12. Blake, R.: Technology and the four skills. Lang. Learn. Technol. **20**, 129–142 (2016). http://llt.msu.edu/issues/june2016/blake.pdf
13. Wasniewski, E., Boechler, P.: The Impact of Technology on Language Strategy Development. In: Society for Information Technology & Teacher Education International Conference, pp. 2491–2496. Association for the Advancement of Computing in Education (AACE) (2013). https://www.learntechlib.org/p/48477
14. Rebecca, O.: Language learning strategies: What every teacher should know. N.Y. (1990). https://www.pdfdrive.com/language-learning-strategies-what-every-teacher-should-know-e161545126.html
15. Al-Sharqi, L., Abbasi, I.S.: The Influence of Technology on English Language and Literature. Engl. Lang. Teach. **13**, 1–7 (2020). https://doi.org/10.5539/elt.v13n7p1
16. Alsulami, S.: The effects of technology on learning English as a foreign language among female EFL students at Effatt College: an exploratory study. Stud. Lit. Lang. **12**, 1–16 (2016). https://doi.org/10.3968/7926
17. Nomass, B.B.: The impact of using technology in teaching English as a second language. Engl. Lang. Lit. Stud. **3**, 111 (2013). https://doi.org/10.5539/ells.v3n1p111
18. Bloch, J.: Technology for teaching english as a second language (ESL) writing. In: The TESOL Encyclopedia of English Language Teaching, pp. 1–8 (2018). https://doi.org/10.1002/9781118784235.eelt0440
19. Bloch, J.: Abdullah's blogging: a generation 15 student enters the blogosphere. Lang. Learn. Technol. **11**, 128–141 (2007). http://llt.msu.edu/vol11num2/bloch/
20. Baron, N.S.: Do mobile technologies reshape speaking, writing, or reading? Mob. Media Commun. **1**, 134–140 (2013). https://doi.org/10.1177/2050157912459739
21. Li, Z., Dursun, A., Hegelheimer, Volker: Technology and L2 writing. In: Chapelle, C.A., Sauro, S. (eds.) The Handbook of Technology and Second Language Teaching and Learning, pp. 77–92. Wiley (2017). https://doi.org/10.1002/9781118914069.ch6
22. The Digital Teacher | Cambridge English https://thedigitalteacher.com/. Accessed 29 May 2022
23. El Morabit, N.: Graded readers: an empirical study measuring the impact on low-proficiency EFL students' writing fluency in Morocco. Int. J. Linguist. Lit. Transl. **4**, 237–244 (2021). https://doi.org/10.32996/ijllt.2021.4.6.28

AraBERT with GANs for High Performance Fine-Grained Dialect Classification

Ibtissam Touahri(✉)

Department of Computer Science, Superior School of Technology, University Moulay Ismail, Meknes, Morocco
Ibtissamtouahri555@gmail.com

Abstract. Arabic nations share the same mother tongue, Arabic. However the used vernacular language is different, that, in turn, may vary from one region to another. In this paper, we aim to identify various dialects by performing text classification. We distinguish between Moroccan, Algerian, Tunisian, Egyptian, and Lebanese Arabic dialects denoted respectively MDA, ADA, TDA, EDA and LDA. Aside from explaining the collecting process of system resources, we identify linguistic specificities that characterize each dialect. We build models from the preprocessed text using a combination of the pretrained model, AraBERT; and Generative Adversarial Networks (GANS). The work establishes the foundation of a dialect identification system by gathering freely available corpora and reaching the state-of-the-art.

Keywords: Dialect Identification · Arabic Language · AraBERT · GANs

1 Introduction

Arabic is a Semitic language that has three linguistic forms: Classical Arabic, the language of the Holy Qur'an; Modern Standard Arabic (MSA), the official language in Arabic countries and Dialectal Arabic (DA) that is the vernacular language enhanced by loanwords. It is characterized by the absence of written standards and does not adhere to grammatical rules or strict word order [1]. Furthermore, it disregards the dual number and feminine plural. The main barriers to written dialect are linguistic shortcuts and short vowels collapse. Many factors influence dialects, including location, lifestyle, education, gender; personal and professional conditions. There are several varieties of MSA that reflect dialectal languages including Egyptian, Gulf, Levantine, Iraqi, and Maghrebi [2].

Many influences, including Islamic conquest, colonialism, and immigration, have resulted in substantial linguistic similarities amongst Maghrebi languages. Furthermore, there is a geographic adjacency factor, particularly between spoken local dialects in border areas (such as Moroccan East and Algerian West, Algerian East and Tunisian North). Thus, Moroccan and Algerian dialects may have a significant degree of similarity, but there is a large linguistic dialectal separation between Moroccan and Tunisian dialects.

There are 92 million Egyptians who speak Egyptian Arabic as their native language. Egypt is the Arab world's most populated country, hence its Arabic is the most frequently

M. Lazaar et al. (Eds.): BDIoT 2022, LNNS 625, pp. 160–170, 2023.
https://doi.org/10.1007/978-3-031-28387-1_15

spoken Arabic dialect. It has several characteristics in common with MSA and has been impacted by a variety of other languages, including Coptic, Turkish, French, and, English.

Lebanon's rich past has had a significant influence on its language. Lebanon has been a part of several empires, including the Armenian and Arabic Empires, which made people speak a variety of languages. Besides Arabic, English, Armenian, French, and Kurdish are among the spoken languages in Lebanon.

Dialect identification, which is required for the development of additional language-dependent systems that handle each language individually, continues to be a difficulty. When we try to separate dialectal words from MSA terms, that may have the same graphical shape and different pronunciation, the work gets more difficult; which is also the case when separating each dialect from the other.

The proliferation of material on the web in dialectal written form is widespread. As a result, dialects suffer analytical issues linked to natural language processing (NLP) activities since NLP tools are dedicated to MSA and rely heavily on its specificities [3]. As a result, unless in common language contexts, they may not deal with DA. Distinguishing between MSA and dialects, also known as language identification, is critical for later usage of specialized and sophisticated dialectal tools and systems such as morphological analyzers. Dialect identification, in turn, is also interesting since it may be used as a preprocessing step for the tools that treat each dialect separately.

Dialect identification is difficult due to the resemblance between dialects [4]. Furthermore, code swapping complicates matters. Code-switching is common on the Internet since internet users move between several languages and dialects by utilizing online Arabic, namely Arabizi, Franco-Arabe, and Arabish, or by switching between various dialects which causes their utterances to be composed of multiple languages, making language identification difficult.

The knowledge-based technique computes the total importance of dialectal phrases in the studied utterances using a lexicon of dialectal words [5]. Likewise, the machine learning approach, which produces a model from labeled vectors that represent training data, uses the resulting model to classify unlabeled chunks of text; supervised learning may also be based on pretrained models [6]. Finally, the deep learning technique, a form of artificial intelligence that produces functionality automatically, is applied in this field [7]. Classification can be binary, which defines the corresponding dialect from labeled dialectal phrases, tripolar, which examines the separation of three dialects, or fine-grained, which refines the tripolar classification by adding more dialects [8].

This paper collects a freely available corpus, whose expressions are labeled using five tags MDA, ADA, TDA, EDA and LDA, the collected corpus is used for language identification. Afterward, we build a dialect identification system that separates the Moroccan dialect from the Algerian, the model gathers their specificities in a unique model in order to test the separability of adjacent dialects. The model is enriched by other dialects in an upward direction of dissimilarity till reaching the five-class model built on the herein studied dialects.

The rest of this paper is structured as follows. In the next part, we outline relevant studies, then provide the collected dataset along with its statistics. Section four describes our system's methodology. Following that, we present the results of the system evaluation on the dialects under consideration. Then, we end by providing a recall of our system findings. Furthermore, we provide some perspectives.

2 Related Works

The task of automatically recognizing the variety of a given text or audio fragment is known as dialect identification. Dialect recognition is a prerequisite for gaining lexical and morphological information of a language variety that might be useful for NLP and possible AI downstream tasks. In this context, [2] described the development of a new Arabic resource with dialect annotations based on crowdsourcing. [9] proposed an open dataset collection containing social data material in several Arabic dialects. The corpora were collected from Twitter and comprise ~50 K tweets in Morocco, Algeria, Tunisia, Egypt and Lebanon national dialects. These data were also labeled for numerous applications, including sentiment analysis, topic detection, and dialect recognition. This dataset was used to build several models. [10] offered a technique for automatically identifying fine-grained dialects in Arabic text. In addition to Modern Standard Arabic, the method differentiates the dialects of several Arab towns. [2] utilized the data to train and assess artificial classifiers for dialect identification. They found that classifiers trained on dialectal data outperform baselines trained on MSA-only data, obtaining near human classification accuracy. [11] conducted empirical comparative research to test the effect of preprocessing such as stemming, lemmatization, and part of speech tagging on dialect identification. They have used many classifiers to perform the identification. They have also used a variety of oversampling strategies to solve the issue of unbalanced data. [12] performed a sentence-level Arabic dialect identification using the Multidialectal Parallel Arabic Corpus (MPCA). [13] have based their study on the AOC dataset, a large-scale dataset of Arabic dialects with hand labeling for four variants of the language. The dataset was used as a benchmark for classical and deep learning classification focusing on dialect identification. [8] provided a detailed explanation of the technique they used to perform the EACL WANLP-2021 Shared Task 1. The proposed models are built on pretrained Transformer-based networks, AraBERT and AraELECTRA. [14] gave a comprehensive analysis of dialect similarity and confusability; and contributed fascinating insights. [15] utilized the supplied dataset by the MADAR Shared Task on Arabic Fine-Grained Dialect Identification, subtask 1. Their model comprises three classifiers and then ensembles the classifier predictions to predict the outcome. [16] discussed their contribution to NADI 2021 Shared Task for Arabic Dialect Identification. They have used five classical machine learning classifiers for dialect identification. [17] reported work on a sentence-level Saudi Dialect recognition problem in which they have trained and evaluated three classifiers using datasets gathered from Twitter. They have trained models on three classical machine learning classifiers. [18] presented the findings of the NADI 2021 shared task. They reported that determining the linguistic variety of short texts based on limited geographical regions of origin is doable, albeit difficult. [4] investigated MADAR and NADI, two well-established datasets containing Arabic dialects from various cities and regions. They reported that, depending on their geographical proximity, cities in different nations may have more dialectal similarities than cities within the same country. [19] have described their findings in the MADAR Shared Task on Arabic Fine-Grained

Dialect Identification. They made a city-level dataset as well as a new country-level dataset publicly available. [20] demonstrated that using noisy web crawled data rather than structured data is more practical for such non-standardized language. Furthermore, their results have shown that a very small web crawled dataset yields results that are comparable to those produced with bigger datasets. They reported that their finetuned TunBERT model gave pertinent results. [21] reported that good dialect identification necessitates accounting for dialect-specific lexical, morphological, and phonological occurrences.

3 Linguistic Resources

We have collected a multidialectal dataset made publically available by [9]. The information was acquired by scraping tweets at random from active users in a preset range of Arab nations, including Algeria, Egypt, Lebanon, Tunisia, and Morocco. There were no restrictions on the date of the tweets or their precise location in the country. The online navigation was automated using the Selenium Python package, and the tweets were scraped using BeautifulSoup. Table 1 shows the five dialect samples extracted from the studied dataset.

The collected dataset contains 49,306 tweets in total. Table 2 shows the distribution of tweets per nation.

We have visited site[1] that gives social media reports in Arabic countries. The five dialects we are interested in are labeled by their respective names Egypt, Lebanon, Morocco, Algeria and Tunisia. Hence, we found out that the distribution of the collected tweets from Table 2 has nothing to do with the statistics depicted in Fig. 1; since Egypt is characterized by the most active users; However, it has a minimal number of the collected tweets, whereas, for Lebanon which Twitter users are not too active as those of Egypt, the collected data related to it is two times more than those related to Egypt comments.

4 System Approach

We have explored the aforedescribed dataset to perform dialect classification based on the combination of the state-of-the-art model AraBERT [22] for classification tasks and Generative Adversarial Networks (GANs) [23]. AraBERT is a pretrained model which ultimate vocabulary size is 64k tokens. Our data are represented by a vector of IDs that define the overlap between the studied comments and AraBERT vocabulary terms. On the other hand, GANs, enable the learning of deep representations from the training material. They do this by generating backpropagation signals through a competitive process involving a pair of networks. The generator in GAN architecture generates fake data and forwards them to the discriminator that will distinguish whether the given comment can be tagged with a specific dialect or it is fake.

We build our system on two experiments to investigate the best model for dialect identification and then we explore the effect of dialect adjacency as well as the data balance on the classification.

[1] https://arabsocialmediareport.com/twitter/.

Table 1. Dialect samples

Dialects	Samples
Algerian	لدكتورة نجوى قامة في القصف الصاروخي لك ان ترى منشوراتها وتدخلاتها /Ldktwrp njwY qAmp fy AlqSf AlSArwxy lk An trY mn$wrAthA wtdxlAthA/ Dr. Najwa legend in the missile bombing, you can see her publications and interventions
Algerian	والفو بالتزوير والتزييف /wAlfw bAltzwyr wAltzyyyf/ They are accustomed to forgery and counterfeiting
Morocco	🌹🌹🌹🌹 صباح الخير و التيسير... يوم جديد /SbAH Alxyr w Altysyr … ywm jdyd/ Good morning and facilitation... New day 🌹🌹🌹🌹
Morocco	.. نشوفو هادشي فين غايخرجنا /.. n$wfw hAd$y fyn gAyxrjnA/
	Let's see, where this will get us ..
Egypt	ربنا يحفظهم من كل سوء يارب /rbnA yHfZhm mn kl sw' yArb/ May God protect them from all evil, Lord
Egypt	😂😂😂ايه ده انت بتتعامل مع نفس البنك /Ayh dh Ant bttEAml mE nfs Albnk 😂😂😂/ What are you dealing with the same bank 😂😂😂
Lebanon	💔💔 مبروك /Mbrwk 💔💔/ Congratulations 💔💔
Lebanon	□♀□ الله اعلم.. بلكي لانها عفوية /Allh AElm .. blky lAnhA Efwyp □♀□/ God knows..perhaps because it is spontaneous □♀□
Tunisian	قراءة كتاب /qrA'p ktAb/ reading a book
Tunisian	مغرفة /mgrfp/ spoon

Table 2. Distribution of tweets per country

Arabic dialect	MDA	ADA	TDA	EDA	LDA
Number of Tweets	9,965	13,393	8,044	7,519	14,482

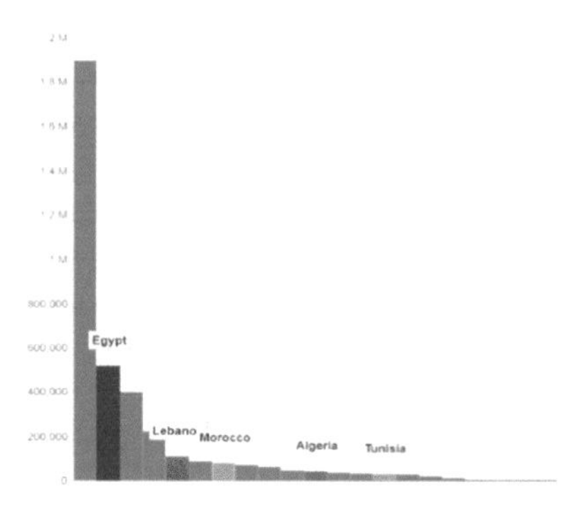

Fig. 1. Active users per country (See footnote 1)

We preprocessed the data by deleting all non-Arabic letters and then transferred learning from the domains covered by the pre-trained models to our system. In the first experiment, we perform two setups. In the first setup, we finetune AraBERT on the training data based on the configuration in Table 3. In each test, we use 80% of the data for training and the remainder for testing. We use seed 42 for reproducibility. We set a batch size of 64 and a maximum length of 128 so that our data would not be truncated. The hyperparameters proposed by AraBERT are kept. Adam, who has a learning rate of 2e−5, is utilized as an optimizer. Based on the macro averaged F1-score, we saved the best model among those created across ten epochs.

Table 3. System parameters

Parameter	Value
Batch size	64
Maximum length	128
learning rate	2e−5
optimizer	Adam

In the second setup, we enhance AraBERT with GANs (Fig. 2). GANs help to generate noisy vectors that have high similarity to real data used for training. Afterward, a discriminator, fed with the generated fake data and the vectors corresponding to real data generated by AraBERT, is used to distinguish between real and fake data, namely the noisy ones generated by the generator. Then, after calculating the loss, the discriminator sends its feedback backward to improve the classification. By annotation, if a comment is repeated, it will be tagged with the same dialect tag which means that the characteristics vectors generated by AraBERT, that identify the corresponding Ids of comments terms within the 64k terms lexicon, are different between dialects. However, if only common terms between dialects within a specific comment are present in the lexicon, then our system will decide which dialect to be affected based on the highest probability given to each dialect among five. The separation of real and fake data that are characterized by their high similarity helps the model to differentiate between the studied 5 dialects, in our case, by building a fine and similarity-wise separation approach. In other words, the distinction between vectors with close features is very helpful to distinguish between dialects that share the same characteristics or graphical terms.

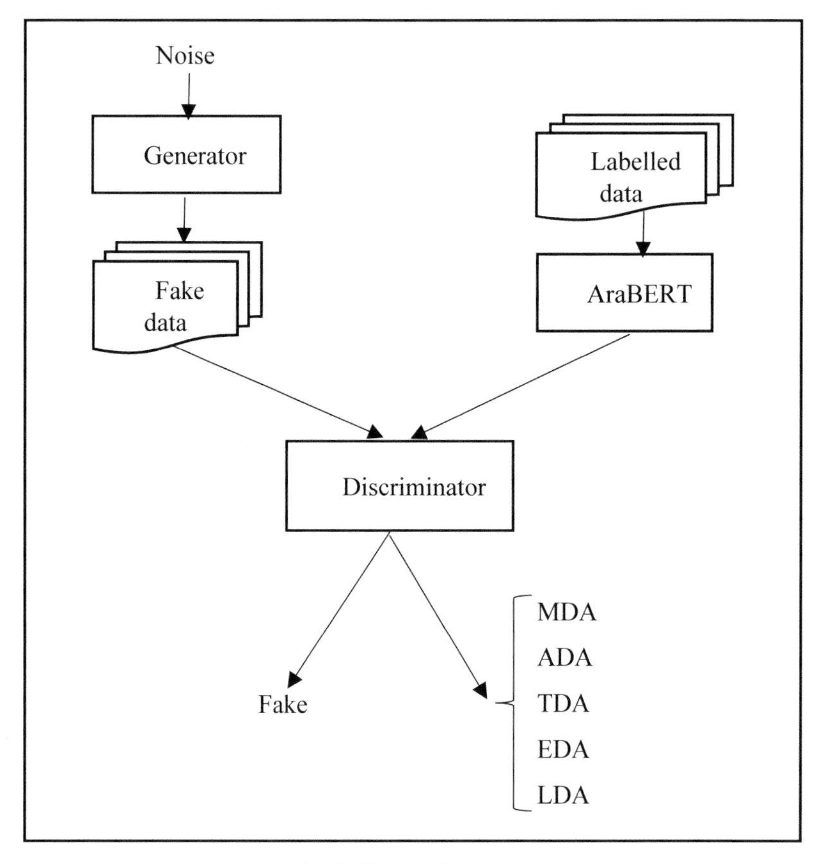

Fig. 2. System description

The second experiment used the model AraBERT + GANs that proved its efficiency in the first experiment. In this experiment, we study the separability of dialects that are adjacent at a certain rate (Moroccan, Algerian). Then we add other dialects respectively in downward adjacency order (Tunisian, Egyptian, Lebanese). Since the aforetests are based on unbalanced data, we perform an undersampling to the dialects subsets to match the one of the minimal size, namely the Egyptian. Hence our balanced dataset consists of 37,595 dialectal expressions where each dialect is represented by a size of 7,519.

Our system is evaluated using accuracy, precision, recall and F-measure metrics. The metrics are defined by:

$$\text{Accuracy} = \frac{\text{Number of correctly classified tweets}}{\text{Number of tweets in the test corpus}}$$

$$\text{Precision } C_i = \frac{\text{Number of correctly classified tweets}}{\text{Number of tweets classified as } C_i}$$

$$\text{Recall } C_i = \frac{\text{Number of correctly classified tweets}}{\text{Number of tweets of the class } C_i}$$

$$\text{F} - \text{measure } C_i = \frac{2 \times \text{Precision} \times \text{Recall}}{\text{precision} + \text{Recall}}$$

5 System Evaluation and Analysis

In this section, we follow the instructions described in our system approach to perform the first experiment based on the two setups to build the AraBERT and AraBERT + GANs models. We present in Table 4 the obtained results using accuracy, precision, recall, and F-measure metrics.

Table 4. Statistics of the first experiment results

Model	Metrics	MDA	ADA	TDA	EDA	LDA
AraBERT	Accuracy	80				
	Precision	0.82	0.86	0.88	0.70	0.71
	Recall	0.86	0.81	0.78	0.77	0.71
	F-measure	0.84	0.83	0.82	0.74	0.71
AraBERT + GANs	Accuracy	86				
	Precision	0.83	0.87	0.76	0.87	0.90
	Recall	0.82	0.86	0.76	0.90	0.90
	F-measure	0.82	0.87	0.76	0.89	0.90

In the first experimental results given in Table 4, the Moroccan, Algerian, and Tunisian dialects identification have reached comparable F-measure. However, the F-measure obtained for both Egyptian and Lebanese dialects classification is degraded, which may be explained by the small size of the Egyptian dataset in comparison to others, and hence, small amount of data to train on. Moreover, for the Lebanese dialect classification, even though the data to train on is of large amount, the test set contains many data to test on which causes the overall F-measure degradation. The second experiment based on AraBERT + GANs has turned the scales where the model major focus was on the last two dialects that suffer from degraded results. The model has improved their F-measures respectively which is also the case for the Algerian dialect. Even though, the identification of the Moroccan and Tunisian dialects have more or less preserved their rank and degraded respectively, the precision and recall for each dialect identification have become very close. Moreover, the overall accuracy and average F-measure have been improved in comparison to the previous model which proves the efficiency of GANs in improving classification.

We perform the second experiment based on dialects combination and data balance. The results are given in Table 5 using only accuracy and Average F-measure metrics in order to lighten the paper.

Table 5. Statistics of the second experiment results

Model	MDA	ADA	TDA	EDA	LDA	Accuracy	F-measure
AraBERT + GANs	√	√				93.4	0.93
	√	√	√			87.2	0.87
	√	√	√	√		85.8	0.86
	√	√	√	√	√	86	0.85
	√	√	√	√	√	86.4	0.86

The results of Table 5 prove that the Moroccan and Algerian dialects are easily separable by our system even though they may be adjacent. The results of the second, third and fourth tests are comparable which enhances the strength of our proposed system that deals with fine-grained classification. However, the second experiment shows that adding Tunisian dialect degraded the results which may be explained by the low representation of TDA in comparison to MDA and ADA. The comparison between balanced and imbalanced datasets (the fourth and fifth tests respectively), in turn, gives comparable results which proved that our system can deal with balanced and imbalanced data. When analyzing our system errors, the main sources were related to the presence of MSA expressions that are common between the five dialects (Table 1).

In Table 6, we compare our system to the corpus owner [9]. The comparison is unfair because their method produced only baseline findings, which were surpassed by our AraBERT-based approach. However, since the dataset was recently released and AraBERT was state-of-the-art in many papers and was utilized in many shared tasks, we may use it as a baseline for our comparison.

Hence, from the obtained results, we believe that our system that enhances AraBERT with GANs gives better results much more than the ones obtained using AraBERT only which proves the efficiency of our system.

Table 6. System Comparison

System	Dataset	Features	Accuracy	Average F-measure
[9]	[9]	TF-IDF vectorization - BOW + classical classifiers	75	0.75
Our system		AraBERT	80	0.79
		AraBERT + GANs	86	0.85

6 Conclusion and Further Work

The herein described system helped to reach pertinent results measured by accuracy and average F-measure. The system is built on a freely available corpus based on which many models have been created. The models are mainly based on the pretrained AraBERT and a combination of it with GANs. We have performed several experiments based on various setups to deal with dialect alteration as well as data size and balance. The model enhanced with GANs gave better results than the one based on AraBERT only which proved the efficiency of using GANs that help to improve the system performance by the integration of the generator and discriminator fed with real and fake data which makes it aware when dealing with various dialect even adjacent ones. Our system reaches an accuracy of 93.4 for the Moroccan, and Algerian dialects identification and 86.4 for fine-grained dialect classification based on a balanced dataset. As further work, we intend to apply our system to other freely available corpora as well as enhance it with linguistic features that characterize each dialect.

References

1. Habash, N., Diab, M., Rambow, O.: Conventional Orthography for Dialectal Arabic. no January 2012 (2015)
2. Zaidan, O.F., Callison-Burch, C.: Arabic dialect identification. Comput. Linguist. **40**(1), 171–202 (2014)
3. Guellil, I., Saâdane, H., Azouaou, F., Gueni, B., Nouvel, D.: Arabic natural language processing: an overview. J. King Saud Univ. – Comput. Inform. Sci. **33**(5), 497–507 (2021). https://doi.org/10.1016/j.jksuci.2019.02.006
4. Abdul-Mageed, M., Zhang, C., Elmadany, A., Bouamor, H., Habash, N.: Nadi 2021: The second nuanced arabic dialect identification shared task. arXiv preprint arXiv:2103.08466 (2021)
5. Guellil, I., Azouaou, F.: Arabic dialect identification with an unsupervised learning (based on a lexicon). application case: Algerian dialect. In: 2016 IEEE Intl Conference on Computational Science and Engineering (CSE), pp. 724–731. IEEE (2016)
6. Qwaider, C., Chatzikyriakidis, S., Dobnik, S.: Pre-trained models or feature engineering: the case of dialectal arabic. In: Proceedings of the OSACT 2022 Workshop @LREC2022, pp. 41–50. Marseille, 20 June 2022
7. Issa, E., AlShakhori, M., Al-Bahrani, R., Hahn-Powell, G.: Country-level Arabic dialect identification using RNNs with and without linguistic features. In: Proceedings of the Sixth Arabic Natural Language Processing Workshop, pp. 276–281 (2021)

8. Wadhawan, A.: Dialect identification in nuanced arabic tweets using farasa segmentation and AraBERT. arXiv preprint arXiv:2102.09749 (2021)
9. Boujou, E., Chataoui, H., Mekki, A.E., Benjelloun, S., Chairi, I., Berrada, I.: An open access NLP dataset for Arabic dialects: Data collection, labeling, and model construction. arXiv preprint arXiv:2102.11000 (2021)
10. Obeid, O., Salameh, M., Bouamor, H., Habash, N.: ADIDA: Automatic dialect identification for Arabic. In: Proceedings of the 2019 Conference of the North American Chapter of the Association for Computational Linguistics (Demonstrations), pp. 6–11 (2019)
11. Lichouri, M., Abbas, M., Lounnas, K., Benaziz, B., Zitouni, A.: Arabic dialect identification based on a weighted concatenation of tf-idf features. In: Proceedings of the Sixth Arabic Natural Language Processing Workshop, pp. 282–286 (2021)
12. Malmasi, S., Refaee, E., Dras, M.: Arabic dialect identification using a parallel multidialectal corpus. In: Hasida, K., Purwarianti, A. (eds.) Computational Linguistics. CCIS, vol. 593, pp. 35–53. Springer, Singapore (2016). https://doi.org/10.1007/978-981-10-0515-2_3
13. Elaraby, M., Abdul-Mageed, M.: Deep models for arabic dialect identification on benchmarked data. In: Proceedings of the Fifth Workshop on NLP for Similar Languages, Varieties and Dialects (VarDial 2018), pp. 263–274 (2018)
14. Salameh, M., Bouamor, H., Habash, N.: Fine-grained arabic dialect identification. In: Proceedings of the 27th International Conference on Computational Linguistics, pp. 1332–1344 (2018)
15. Younes, M.B., Younes, M.B., Al-khdour, N.: Identify Arabic dialect using ensemble model (2022)
16. Nayel, H., Hassan, A., Sobhi, M., El-Sawy, A.: Machine learning-based approach for arabic dialect identification. In: Proceedings of the Sixth Arabic Natural Language Processing Workshop, pp. 287–290 (2021)
17. Aseri, Y., Alreemy, K., Alelyani, S., Mohanna, M.: Meeting challenges of modern standard arabic and Saudi dialect identification (2022)
18. Alsudais, A., Alotaibi, W., Alomary, F.: Similarities between Arabic dialects: investigating geographical proximity. Inf. Process. Manage. **59**(1), 102770 (2022)
19. Bouamor, H., Hassan, S. and Habash, N.: The MADAR shared task on Arabic fine-grained dialect identification. In: Proceedings of the Fourth Arabic Natural Language Processing Workshop, pp. 199–207 (2019)
20. Messaoudi, A., et al.: TunBERT: pretrained contextualized text representation for Tunisian dialect. In: Bennour, A., Ensari, T., Kessentini, Y., Eom, S. (eds.) Intelligent Systems and Pattern Recognition: Second International Conference, ISPR 2022, Hammamet, Tunisia, March 24–26, 2022, Revised Selected Papers, pp. 278–290. Springer International Publishing, Cham (2022). https://doi.org/10.1007/978-3-031-08277-1_23
21. Darwish, K., Sajjad, H., Mubarak, H.: Verifiably effective arabic dialect identification. In: Proceedings of the 2014 Conference on Empirical Methods in Natural Language Processing (EMNLP), pp. 1465–1468 (2014)
22. Antoun, W., Baly, F., Hajj, H.: Arabert: Transformer-based model for arabic language understanding. arXiv preprint arXiv:2003.00104 (2020)
23. Creswell, A., White, T., Dumoulin, V., Arulkumaran, K., Sengupta, B., Bharath, A.A.: Generative adversarial networks: an overview. IEEE Signal Process. Mag. **35**(1), 53–65 (2018)

Genetic-Novelty Oversampling Technique for Imbalanced Data

Hajar Ait Addi(✉), Redouane Ezzahir, and Nouhaila Boukhlik

MAISI/ENSA, University of Ibn Zohr, Agadir, Morocco
hajaraitaddi@gmail.com, r.ezzahir@uiz.ac.ma

Abstract. Imbalance data is in important topic vexed researchers in practice of classification problems. A data is imbalanced if the distributions of categories are not approximately equally represented. The class with small samples is called minority class, while the other classes form the majority class. Standard learning classifiers tend to misclassify the minority samples; they assume that the distribution of data is relatively balanced. However in real world application, the corrected prediction of minority samples is more valuable than correctly classify samples belonging to the majority class. In this paper, we propose GNOT a novel oversampling strategy that combines algorithm genetic concept and novelty detection technique to generate consistent with the original distribution of the minority class while avoiding outliers. We tested GNOT on seven real-world imbalanced datasets. Our experimental analysis shows that GNOT can effectively improve the performance of classifiers in terms of G-mean and F1-measure.

1 Introduction

In real world applications like sentiment analysis [20], medical diagnostic [22], detection of oil spills [17], and fraud detection [26], the classification datasets suffer from the class imbalance problem. A dataset is imbalanced if the distributions of categories are not approximately equally represented [29]. The class with small samples is called minority class, while other classes assemble the majority class. Traditional classification algorithms cannot deal with imbalance data; they assume that the data is well represented. Thus, traditional classifier ignores the difference between classes on the training task that leads to increase the misclassification rate for the minority class [28]. For some applications, misclassifying the minority class may be more expensive and serious. For instance, in fraud detection, if the non-fraud transaction is regarded as majority class and fraud transaction as minority class, then missing fraud transaction is much more serious than missing a non-fraud transaction.

To tackle imbalanced dataset problem, many approaches have been proposed at both data and algorithmic levels [10,16,25]. Data level includes sampling techniques that resize the original dataset either by over-sampling the minority class or by under-sampling the majority class, until the classes are balanced. Algorithmic level methods directly modify existing learning algorithms to relieve the bias towards majority class and adapt them to imbalanced distribution. In this work, we focused on sampling techniques at data level. Some of the powerful sampling technique is Synthetic Minority

M. Lazaar et al. (Eds.): BDIoT 2022, LNNS 625, pp. 171–185, 2023.
https://doi.org/10.1007/978-3-031-28387-1_16

Over-sampling Technique (SMOTE) [7]. This method generates synthetic samples of the minority class to generalise the decision region of the minority class. It creates synthetic minority samples by interpolating between two randomly selected neighbouring samples.

The drawback of SMOTE-based algorithms is that the accuracy decreases for dataset with high imbalance ratio [6], because many new samples may be irrelevant to the classifier (see Sect. 2 for more details). To generate samples that incorporate feasible information, we propose a novel pre-processing strategy (GNOT) that combines algorithm genetic concept and novelty detection technique. As presented in Sect. 3, we first use a novelty technique to model the distribution of the minority class. Then, we apply crossover and mutation operators on the minority class to create new samples. Next, we select the most relevant samples using the generated model. Experimental results that we reported in Sect. 5, shows that GNOT can effectively improve the accuracy of classifiers for datasets with high imbalance ratio.

2 Related Work

Different approaches for sampling techniques have been proposed, which are mainly divided into undersampling and oversampling. Undersampling methods remove some samples from the majority class. The most renowned technique is Random Undersampling (RU) that randomly eliminates samples of the majority class until the distribution becomes balanced [18]. This technique may lead to loss of information. To tackle this drawback, several overlapping simple filter techniques have been introduced. Edited Nearest Neighbours (ENN) removes any sample from the majority class whose is different from at least two of its three nearest [31]. In [11], Repeated Edited Nearest Neighbor applies ENN repeatedly until ENN can't remove any further instance. Condensed nearest neighbor rule (CNNr) has been proposed to remove redundant samples of majority class for a training set [31]. Tomek Links (TL) [2], a modification of CNNr, is proposed to eliminate not only redundant samples, but also the noisy and borderline samples of majority class. Undersampling could result in loss of important information [33], which leads to not consider it as standalone technique.

For oversampling technique, the first popular method was Random Oversampling that replicates the minority class samples. Despite its simplicity and safety, this technique yields to increase the likelihood of occurring over-fitting. To avoid the problem of over-fitting, Synthetic Minority Oversampling Technique (SMOTE) has been proposed [7]. SMOTE generates synthetic samples using a linear process among the samples that are near in term of distance. In [12], SMOTE-Borderline, an extend version of SMOTE, oversamples the borderline samples. Although the best performance of SMOTE based approach, the accuracy of classification decreases for high imbalance ratio.

To overcome this bottleneck, some researchers have attempted to improve the quality of oversampling using metaheuristic algorithms. In [14], GASMOTE algorithm has been proposed. GASMOTE uses genetic algorithm (GA) to discover the optimized sampling rates of the minority class. After, it performs SMOTE technique to oversample the dataset using these samples. In [27], GA-based oversampling engine has been introduced. This algorithm also applies GA to generate a set of samples that increase the accuracy of learning algorithm. Another approach that adapts the same idea is GenSample algorithm [15]. The performance of both techniques GA and GenSample depends on the training set, the performance metric, and the classifier. For these reasons, we propose a preprocessing method that attempt to generate samples before classification stage and incorporates the genetic approach.

3 Proposed Solution Approach: GNOT

In this section, we propose a Genetic-Novelty oversampling technique (GNOT) to deal with imbalanced data. The idea behind our approach is to generate new samples, using genetic operators (GO), consistent with the original distribution of the minority class while avoiding outliers. According to [13], an outlier is an observation that deviates so much from other observations as to arouse suspicion that it was generated by a different mechanism. Hence, the proposed GNOT algorithm uses a novelty detection technique to identify samples. Novelty detection attempts to identify outliers that are generated by GO and do not match the expected behavior of the original (or normal) data [23].

Most studies on novelty detection were statistical approaches [5]. Statistical approach models data based on its statistical proprieties and uses the trained model to estimate whether a simple belongs to the same distribution or not. These studies can be mainly classified into two categories: parametric approach and no-parametric approach [8]. Parametric approach assumes that data follow a standard distribution (e.g. Normal, Poisson, Exponential, etc.), and can be modelled based calculating certain parameters, such as mean and covariance for Gaussian distribution. No-parametric approach makes no assumption on the statistical form of data distribution [21]. Thus, the density function can be estimated using either nearest neighbour based or clustering based. For these approaches, a threshold is fixed to estimate whether a new sample comes from the same distribution or not. For nearest neighbour based technique, a similarity measure (distance or density) is required between two samples that belong to the original data. An outlier is detected when the similarity score to its neighbours surpasses the predefined threshold. Clustering based assumes that original data forms large and dense clusters, while outlier do not belong to any cluster. For each new data, a membership degree to each cluster is assigned. If the degree of membership exceeds the predefined thresholds associated to each cluster, then the new sample is an outlier.

For our approach, we used Local Outlier Factor (LOF) as it's a powerful non parametric technique. In [5], the authors assign to each sample a degree of being a novel, this score is called LOF. Given a sample, the LOF score is equal to ratio of average local density of the k nearest neighbours of the sample and the local density of the sample itself. To compute the local density for a sample, the authors search the radius of the smallest hyper-sphere that centers the sample and contains its k nearest neighbours. The local density is then calculated by dividing k by the volume of this hyper-sphere. The local density of a normal sample is similar to that of its neighbours, while for a novel sample, its local density is lower than that of its nearest neighbours.

The flowchart 1 illustrates the steps of the algorithm. First, GNOT models the data distribution using a novelty technique and drive an estimator f (Fig. 1.(a)). To generate samples, GNOT starts by representing samples of minority class as chromosomes

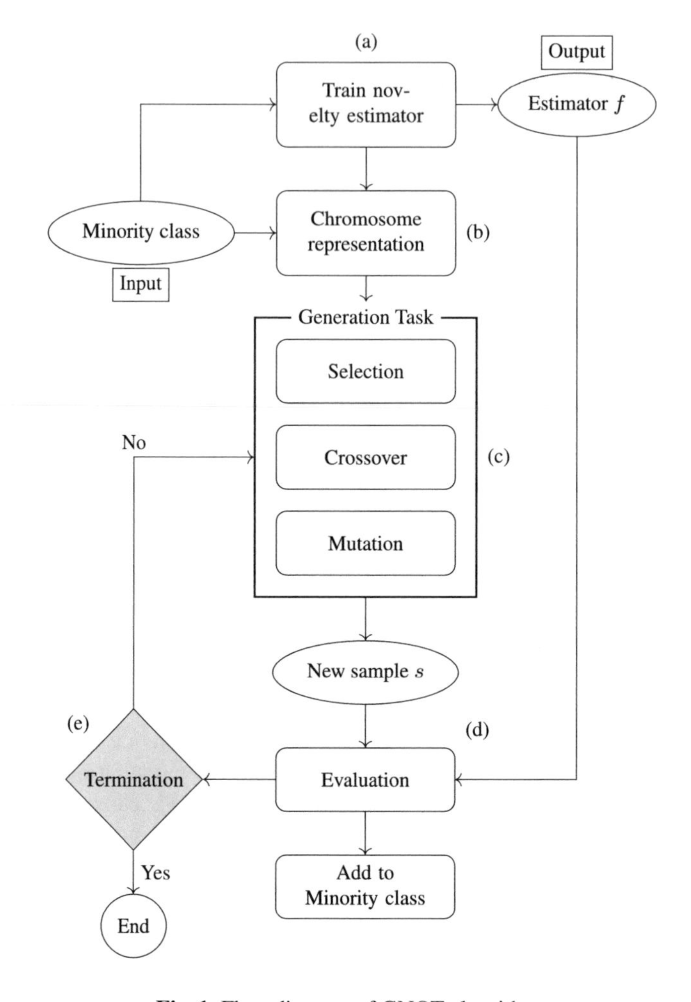

Fig. 1. Flow diagram of GNOT algorithm

(Fig. 1.b). Next, GNOT selects two chromosomes and apply crossover and mutation operator to generate new samples (Fig. 1.c). If f identifies the generated sample as normal (not outlier) (Fig. 1.d), GNOT adds this sample to the minority class. GNOT repeats these steps (generation and evaluation) until the class distributions are balanced. Selection, crossover and mutation operators are elaborated below.

Individual Representation. We represent each minority class as a chromosome, where features play gene role. Let x_i, x_j two chromosomes in a problem of d features:

$$x_i = (x_{i1}, \ldots, x_{id})$$
$$x_j = (x_{j1}, \ldots, x_{jd})$$

where x_{ki} is the value of feature k in the sample i of a minority class.

Selection. First we create a set *pop* of size N, in which we store original samples of the minority class. Next, we select randomly two parents for crossover (x_i, x_j), $i,j \in \{1,..., N\}$). To avoid choosing a redundant pair of parents, we use a memory set to save the indices of the selected pair. A pair of parents is redundant if it has already been picked in a previous iteration.

Crossover. The aim of crossover operator is to produce new chromosomes based on two parent chromosomes. Many strategies are available for both discrete and continuous spaces. As we currently only handle datasets with continuous features, we suggest to apply linear crossover [32] and barycentric crossover [4] in GNOT. These methods are more adopted to continuous data. Given two parents $parent_1$ and $parent_2$, linear crossover generates three children using the following formulas:

$$Child_1 = 0.5parent_1 + 0.5parent_2$$

$$Child_2 = 1.5parent_1 - 0.5parent_2$$

$$Child_3 = 0.5parent_1 - 1.5parent_2$$

Example 1. *Let assume that in selection step, GNOT picks up two samples:* $(9, 3, 6, 12, 0)$ *and* $(-3, 0, 12, 9, 0)$*. Then, the outputs of linear crossover be as follow:*

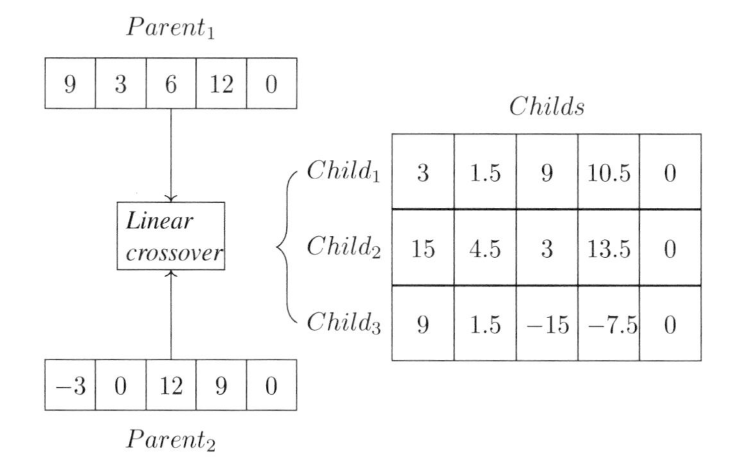

Barycentric crossover generates two children using a random parameter α in $[0, 1]$:

$$Child_1 = \alpha parent_1 + (1 - \alpha)parent_2$$

$$Child_2 = (1 - \alpha)parent_1 + \alpha parent_2$$

The parameter α decides how much information about the distribution should be propagated from the parents to the new samples. This parameter can be fixed at the beginning.

Example 2. *Let take the same samples of Example 1, and* $\alpha = \frac{2}{3}$. *The outputs of barycentric crossover are:*

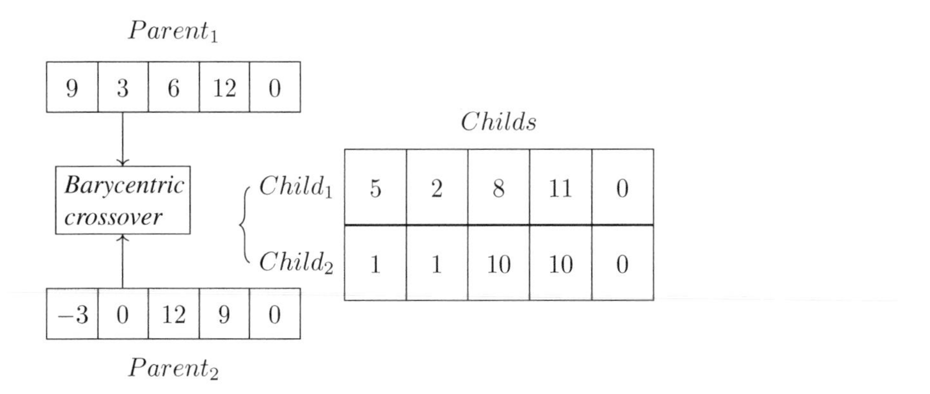

Mutation. In GNOT, the mutation mechanism is designed to create more larger decision regions. There several mutation techniques in the literature. In GNOT, we apply DE/rand/1 strategy [3] that generates a new child using three individuals (chromosomes). This mutation requires a fix parameter F in $[0, 2]$. Given Three individuals x_1, x_2, x_3, the new child x_{child} is calculated as follow:

$$x_{child} = x_1 + F(x_2 - x_3)$$

Example 3. *Let take the generated samples of linear crossover in Example 1. For* F, *we choose the same value mentioned in the original paper:* 0.8. *So, the output of mutation is:*

$Child_1$	3	1.5	9	10.5	0
$Child_2$	15	4.5	3	13.5	0
$Child_3$	9	1.5	−15	−7.5	0

Mutation

$Child_4$	7.8	3.9	23.4	27.3	0

For barycentric crossover of Example 2, the new childs are:

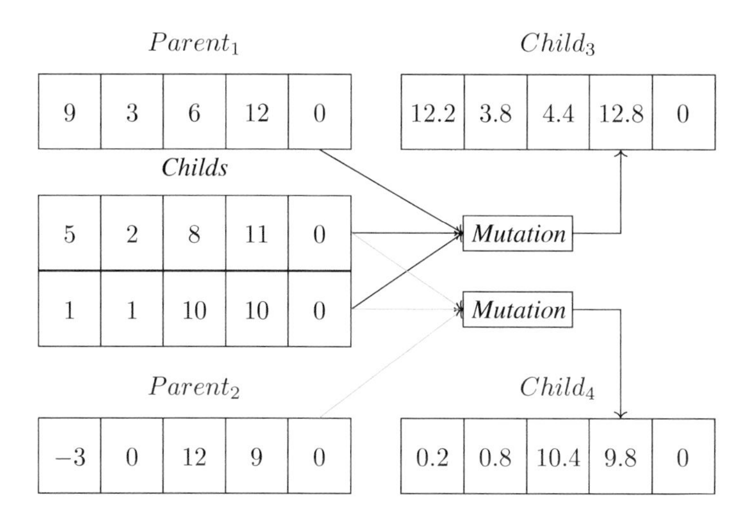

GNOT uses $Child_1$, $Child_2$, and $Child_3$ ($parent_1$, $Child_2$, and $Child_3$) for the mutation when a linear crossover (barycentric crossover) operator is applied. We denote the version of GNOT that applies linear crossover by $GNOT_L$, and the version that applies barycentric crossover by $GNOT_B$.

4 Performance Metrics

Generally, studies on imbalanced dataset classification mainly focus on binary-class problem as the multi-class problem can be transformed into binary-class problems. By convention, the minority class and the majority class are labelled as positive and negative, respectively.

To evaluate the performance of classifier, we use confusion matrix (see Table 1). It is a two dimensional matrix, where the first column represents the actual class label and the first row represent the predicted class label. In the confusion matrix, True Negative (**TN**) and True Positive (**TP**) are the number of negative and positive samples, respectively, that are correctly predicted. False Negative (**FN**) and False Positive (**FP**) are the number of positive and negative samples, respectively, that are misclassified as negative. **FP** (False Positive) is the number of negative samples that are misclassified as negative. Overall accuracy is defined as:

Table 1. Confusion matrix

Actual label	Predicted label	
	TN	FP
	FN	TP

$$Accuracy = \frac{TP + TN}{TP + TN + FP + FN}$$

Accuracy is not suitable when the data is imbalanced [30] Therefore, there are other metrics to evaluate the performance of classifiers, which we present in Table 2.

Table 2. Evaluation metrics

Metric	Formula
$Recall$	$\frac{TP}{TP+FN}$
$Precision$	$\frac{TP}{TP+FP}$
$F1$	$F1 = 2 \times \frac{Precision \times Recall}{Precision + Recall}$
Geometric-mean (GM)	$\sqrt{TP_{rate} \times TN_{rate}}$

Where

$$TP_{rate} = \frac{TP}{TP + FP} \quad and \quad TN_{rate} = \frac{TN}{TP + FP}$$

Each metric can be briefly explained as follow:

- $Recall$: is defined as the number of samples correctly predicted as positive among TP ones. Recall is very important when the cost of false negative is high. To illustrate, if the presence of a nuclear missile corresponds to the minority class, we would not want risk people's life by not detecting a nuclear missile.
- $Precision$: is defined as the positive samples that are correctly predicted out of the ones predicted as positive. Precision is very important when the cost of false positive is high. To illustrate, if a problem involves the detection of tumours, a low precision leads to told many patients that they have a tumours. This will generate extra tests and stress.
- $F1$: is defined as the harmonic mean between recall and precision. F1 score is a good discriminator metric to create a balanced model.
- GM: is defined as the geometric mean between true positive rate and true negative rate. This metric maximizes the TP rate and TN rate, with keeping both rates approximately balanced.

$F1$ and GM are reported as an excellent metrics and performed better than the total accuracy [30]. To evaluate the performance of a classifier when the datasets are imbalanced, we should use $F1$ to measure the performance of classification model on the minority class and GM to measure the overall accuracy.

5 Experiment and Analysis

5.1 Datasets

We experimented on seven different datasets. These datasets are taken from KEEL [1] dataset repository and UCI Machine Learning repository [9]. They are elaborated below:

1. **The Pima Indian Diabetes**: this data is used to predict whether or not a patient from near Phonex, Arizona has diabetes. There are 8 features, 2 classes and 768 samples. The class distribution is 500 negative samples and 268 positive samples.
2. **The Vehicle Silhouettes**: the aim of the dataset is to classify whether or not a given silhouette is from Saab company. It has 18 features, 846 samples with 217 samples belong to minority class.
3. **The Thyroid Disease**: the dataset has 5 features, two classes and 215 samples. The data is used to identify positive thyroid cases. The number of positive class is only 35 samples.
4. **Image Segmentation**: The images were hand segmented to create a classification for every pixel. There are 10 features and 2, 324 samples. The number of minority samples is 324.
5. **The Satimage dataset**: the database consists of the multi-spectral values of pixels in 3×3 neighbourhoods in a satellite image, and the classification associated with the central pixel in each neighbourhood. The aim is to predict this classification, given the multi-spectral values. The dataset has 36 features, 6 classes and 6, 431 samples. We chose the smallest class as positive samples, and collapsed the other five into negative samples. Hence, we got 625 minority samples and 5, 806 majority samples.
6. **The Dermatology dataset**: uses 34 features to differentiate patients with erythemato-squamous disease from those without. The dataset is skewed with 20 minority and 338 majority samples.
7. **The Oil dataset**: the objective of this data is to detect oils spills from radar images of the sea surface. There are 49 features, 41 of oil slick (minority class) and 896 of non-oil slick (majority class).

The Table 3 summarizes these datasets. For each dataset, we present: the dataset name (Datasets), the number of minority samples $\#Min$, the number of majority samples $\#Maj$, and the imbalanced ratio IR.

5.2 Experimental Set-up

We tested our proposed sampling techniques using standard classifiers; Naive Bayes (NB), Support Vector Machine (SVM), and Decision Tree (DT). To compare the performance of our approach with SMOTE technique, we combine each classifier with $GNOT_B$, $GNOT_L$, and SMOTE, respectively. We took an average of 200 runs for each experiment. All tests have been realized using Python libraries: Scikit-learn [24] and imbalanced-learn [19], in a personal laptop of 8 GB in memory and i7 core processor with a speed up to 3.5 GHz.

Table 3. Experimental datasets, ordered in increasing level of IR.

Datasets	$\#Min$	$\#Maj$	IR
Pima	268	500	1.86
Vehicle	217	629	2.9
Thyroid	35	180	5.14
Segmentation	324	2,000	6.02
Satimage	625	5,806	9.28
Dermatology	20	338	16.9
Oil	41	896	21.85

5.3 Results and Analysis

Tables 4 and 5 respectively present the results of the experiments on datasets with a low imbalance ratio (IR<6), and those with high imbalance ratio (IR>6). For each classifier, the best performance metric is bolted. The purpose of our first analysis is to compare the performance of $GNOT_B$ to that of $GNOT_L$.

Based on data in both Tables, the classification performance using $GNOT_B$ is obviously superior to $GNOT_L$ in all datasets except Vehicle dataset. Thus, the first conclusion we can draw is that barycentric crossover is better than linear. The reason is that barycentric method leads to generate data inside the original region of minority class, while linear creates new samples a little bit far from this region. This has the effect that barycentric technique forms a dense region that is easy to discriminate. Thus, We limit our syudy to $GNOT_B$.

Let focus now on the comparison between the performance of $GNOT_B$ with SMOTE to determine under which setting resorting to $GNOT_B$ approach pays off. We observe from the results in Tables 4 and 5 that $GNOT_B$ shows good performance for almost datasets, especially in those with high imbalance ratio. Overall accuracy ($Accu$) is better with $GNOT_B$ for all datasets in Table 5. Besides, the combination $GNOT_B$ and NB is the best technique for all metrics on Oil dataset.

Let us now investigate the performance $GNOT_B$ in terms of F1 and GM. As mentioned before, F1 is index to evaluate the performance of minority class and GM is a measure of overall accuracy. We report the improvements of $GNOT_B$ in terms of both GM and $F1$ in Table 6. These improvements are calculated as follow:

$$Value_{improvement} = \frac{Value_{GNOT_B} - Value_{SMOTE}}{Value_{SMOTE}}$$

The performance of oversampling techniques in terms of F1 depends on learning algorithm. $GNOT_B$ is the best technique over all datasets, except Vehicule dataset, when using NB. For SVM and DT, SMOTE is better in three datasets out of 7. Even the F1 score is not the highest for most datasets, is always slightly less than the best performer. This behaviour is not ensured by SMOTE.

Table 4. Experimental results for Datasets with $IR < 6$

Dataset	**Classifier**		$Accur$	$Prec$	Rec	$F1$	GM
Pima	NB	$GNOT_B$	**78**	**76**	**81**	**78**	**78**
		$GNOT_L$	77	75	79	77	75
		SMOTE	74	75	80	72	76
	SVM	$GNOT_B$	**79**	**77**	**82**	**79**	**79**
		$GNOT_L$	78	77	80	78	78
		SMOTE	74	75	71	73	74
	DT	$GNOT_B$	**75**	**75**	**75**	**75**	**75**
		$GNOT_L$	73	73	73	73	73
		SMOTE	74	73	**75**	74	74
Vehicule	NB	$GNOT_B$	70	77	58	66	69
		$GNOT_L$	**73**	**77**	**65**	**70**	**72**
		SMOTE	68	67	72	69	68
	SVM	$GNOT_B$	**82**	87	74	80	81
		$GNOT_L$	**82**	**88**	75	81	82
		SMOTE	81	77	**88**	**82**	**83**
	DT	$GNOT_B$	78	78	78	78	78
		$GNOT_L$	81	81	80	81	81
		SMOTE	**82**	**81**	**84**	**83**	**83**
Tyroid	NB	$GNOT_B$	**97**	94	**99**	**97**	**97**
		$GNOT_L$	96	**98**	93	96	96
		SMOTE	**97**	94	**99**	**97**	**97**
	SVM	$GNOT_B$	96	95	97	96	**96**
		$GNOT_L$	96	95	97	96	**96**
		SMOTE	**99**	**98**	**99**	**99**	94
	DT	$GNOT_B$	97	97	97	97	**97**
		$GNOT_L$	95	94	96	95	95
		SMOTE	**98**	**98**	**98**	**98**	**97**

The Table $GNOT_B$ gives the most accurate results for all datasets except Vehicle dataset. The best improvement is 15.67%, and it's enjoyed by SVM in Oil dataset, where the imbalance ratio is very high.

Table 5. Exprimental results for Datasets with $IR > 6$

Dataset	**Classifier**		$Accu$	$Prec$	Rec	$F1$	GM
Segment	NB	$GNOT_B$	**94**	**97**	91	**94**	**94**
		$GNOT_L$	91	97	83	90	90
		SMOTE	89	83	**98**	89	**94**
	SVM	$GNOT_B$	99	99	99	99	99
		$GNOT_L$	99	99	99	99	99
		SMOTE	99	99	99	99	99
	DT	$GNOT_B$	99	99	99	99	99
		$GNOT_L$	99	99	99	99	99
		SMOTE	99	99	99	99	99
Satimage	NB	$GNOT_B$	**89**	**91**	87	**89**	**89**
		$GNOT_L$	87	89	85	87	87
		SMOTE	85	82	**88**	85	**89**
	SVM	$GNOT_B$	**78**	**89**	65	75	**77**
		$GNOT_L$	75	73	74	75	74
		SMOTE	74	66	**95**	**78**	75
	DT	$GNOT_B$	93	**93**	92	93	**93**
		$GNOT_L$	92	92	92	92	92
		SMOTE	**94**	92	**95**	**94**	91
Dormo	NB	$GNOT_B$	**97**	**99**	94	**97**	**96**
		$GNOT_L$	**96**	99	92	96	**96**
		SMOTE	96	92	**99**	96	94
	SVM	$GNOT_B$	99	99	99	99	99
		$GNOT_L$	99	99	99	99	99
		SMOTE	99	99	99	99	99
	DT	$GNOT_B$	**99**	**99**	**99**	**99**	**99**
		$GNOT_L$	98	98	98	98	98
		SMOTE	**99**	**99**	**99**	**99**	98
Oil	NB	$GNOT_B$	**82**	**91**	**70**	**80**	**81**
		$GNOT_L$	57	80	19	30	42
		SMOTE	58	54	98	70	78
	SVM	$GNOT_B$	**97**	**97**	97	**97**	**97**
		$GNOT_L$	84	81	88	84	83
		SMOTE	95	92	**98**	95	84
	DT	$GNOT_B$	**96**	**97**	**97**	**96**	**97**
		$GNOT_L$	**96**	**96**	96	**96**	96
		SMOTE	95	94	**97**	95	**97**

Table 6. Improvements (%) in F1 and GM of $GNOT_B$ over SMOTE

Datasets	NB		SVM		DT	
	$F1$	GM	$F1$	GM	$F1$	GM
Pima	8.30	2.63	8.22	6.76	1.35	1.35
Vehicule	−4.35	1.47	−2.44	−2.41	−6.02	−6.02
Tyroid	0	0	−3.30	2.13	−1.02	0
Segment	5.62	0	0	0	0	0
Satimage	4.70	0	−4	2.67	−1.06	2.19
Dormo	1.04	2.13	1.04	2.13	0	1.02
Oil	14.28	3.85	2.10	15.67	1.05	0

6 Conclusion

In this paper, we proposed a genetic-novelty oversampling technique (GNOT) to handle class imbalance problem. GNOT employed genetic operators (crossover and mutation) to generate new samples, and then applied a novelty detection technique to select the relevant samples that are consistent with the original distribution of the minority class. We oversampled seven real-world imbalanced datasets using GNOT algorithm. Three standard classifiers (Naive Bayes, Support Vector Machine, and Decision Tree) trained the resulted datasets. Then, we compared the results with SMOTE oversampling approach. Based on the results of our experiments, we observed that GNOT outperformed SMOTE in terms of G-means for 6 out of 7 datasets, and F1 value was at par and sometimes higher with GNOT than its value with SMOTE, for 13 out of 21 times.

References

1. Alcalá-Fdez, J., et al.: KEEL: a software tool to assess evolutionary algorithms for data mining problems. Soft. Comput. **13**, 307–318 (2008). https://doi.org/10.1007/s00500-008-0323-y
2. At, E., Aljourf, M., Al-Mohanna, F., Shoukri, M.R.: Classification of imbalance data using Tomek link(T-Link) combined with random under-sampling (RUS) as a data reduction method (2016)
3. Baatar, N., Zhang, D., Koh, C.: An improved differential evolution algorithm adopting λ -best mutation strategy for global optimization of electromagnetic devices. IEEE Trans. Magn. **49**(5), 2097–2100 (2013)
4. Bernard, T., Nakib, A.: Adaptive ECG signal filtering using Bayesian based evolutionary algorithm. In: Metaheuristics for Medicine and Biology, pp. 187–211 (2017). https://doi.org/10.1007/978-3-662-54428-0_11
5. Breunig, M.M., Kriegel, H., Ng, R.T., Sander, J.: LOF: identifying density-based local outliers. In: Proceedings of the 2000 ACM SIGMOD International Conference on Management of Data, 16–18 May 2000, Dallas, Texas, USA, pp. 93–104 (2000)
6. Cervantes, J., Li, X., Yu, W.: Using genetic algorithm to improve classification accuracy on imbalanced data. In: 2013 IEEE International Conference on Systems, Man, and Cybernetics, pp. 2659–2664, October 2013

7. Chawla, N.V., Bowyer, K.W., Hall, L.O., Kegelmeyer, W.P.: SMOTE: synthetic minority over-sampling technique. J. Artif. Intell. Res. **16**, 321–357 (2002)
8. Desforges, M.J., Jacob, P.J., Ball, A.D.: Fault detection in rotating machinery using kernel-based probability density estimation. Int. J. Syst. Sci. **31**(11), 1411–1426 (2000)
9. Dua, D., Graff, C.: UCI machine learning repository (2017). http://archive.ics.uci.edu/ml
10. Estabrooks, A., Jo, T., Japkowicz, N.: A multiple resampling method for learning from imbalanced data sets. Comput. Intell. **20**, 18–36 (2004)
11. Guan, D., Yuan, W., Lee, Y., Lee, S.: Nearest neighbor editing aided by unlabeled data. Inf. Sci. **179**(13), 2273–2282 (2009)
12. Han, H., Wang, W.-Y., Mao, B.-H.: Borderline-SMOTE: a new over-sampling method in imbalanced data sets learning. In: Huang, D.-S., Zhang, X.-P., Huang, G.-B. (eds.) ICIC 2005. LNCS, vol. 3644, pp. 878–887. Springer, Heidelberg (2005). https://doi.org/10.1007/11538059_91
13. Hawkins, D.M.: Identification of Outliers. Monographs on Applied Probability and Statistics, Springer, Cham (1980). https://doi.org/10.1007/978-94-015-3994-4
14. Jiang, K., Lu, J., Xia, K.: A novel algorithm for imbalance data classification based on genetic algorithm improved smote. Arab. J. Sci. Eng. **41**, 3255–3266 (2016)
15. Karia, V., Zhang, W., Naeim, A., Ramezani, R.: Gensample: a genetic algorithm for over-sampling in imbalanced datasets. CoRR abs/1910.10806 (2019)
16. Kotsiantis, S.B., Kanellopoulos, D., Pintelas, P.E.: Handling imbalanced datasets: a review (2006)
17. Kubat, M., Holte, R.C., Matwin, S.: Machine learning for the detection of oil spills in satellite radar images. Mach. Learn. **30**, 195–215 (1998)
18. Laza, R., Pavón, R., Reboiro-Jato, M., Fdez-Riverola, F.: Evaluating the effect of unbalanced data in biomedical document classification. J. Integr. Bioinform. **8**(3), 105–117 (2011)
19. Lemaître, G., Nogueira, F., Aridas, C.K.: Imbalanced-learn: a python toolbox to tackle the curse of imbalanced datasets in machine learning. J. Mach. Learn. Res. **18**(17), 1–5 (2017). http://jmlr.org/papers/v18/16-365.html
20. Li, Y., Guo, H., Zhang, Q., Mingyun, G., Yang, J.: Imbalanced text sentiment classification using universal and domain-specific knowledge. Knowl.-Based Syst. **160**, 1–15 (2018)
21. Markou, M., Singh, S.: Novelty detection: a review - part 1: statistical approaches. Sig. Process. **83**(12), 2481–2497 (2003)
22. Mena, L.J., Gonzalez, J.A.: Machine learning for imbalanced datasets: application in medical diagnostic. In: FLAIRS Conference (2006)
23. Miljkovic, D.: Review of novelty detection methods. In: The 33rd International Convention MIPRO, pp. 593–598, May 2010
24. Pedregosa, F., Varoquaux, G., Gramfort, A., Michel, V., Thirion, B., Grisel, O., Blondel, M., Prettenhofer, P., Weiss, R., Dubourg, V., Vanderplas, J., Passos, A., Cournapeau, D., Brucher, M., Perrot, M., Duchesnay, E.: Scikit-learn: machine learning in python. J. Mach. Learn. Res. **12**, 2825–2830 (2011)
25. Rout, N., Mishra, D., Mallick, M.K.: Handling imbalanced data: a survey. In: Reddy, M.S., Viswanath, K., K.M., S.P. (eds.) International Proceedings on Advances in Soft Computing, Intelligent Systems and Applications. AISC, vol. 628, pp. 431–443. Springer, Singapore (2018). https://doi.org/10.1007/978-981-10-5272-9_39
26. Phua, C., Alahakoon, D., Lee, V.C.S.: Minority report in fraud detection: classification of skewed data. SIGKDD Explor. **6**, 50–59 (2004)
27. Saladi, P.S.M., Dash, T.: Genetic algorithm-based oversampling technique to learn from imbalanced data. In: Bansal, J.C., Das, K.N., Nagar, A., Deep, K., Ojha, A.K. (eds.) Soft Computing for Problem Solving, pp. 387–397. Springer Singapore, Singapore (2019). https://doi.org/10.1007/978-981-13-1592-3_30

28. Tomasev, N., Mladenic, D.: Class imbalance and the curse of minority hubs. Knowl.-Based Syst. **53**, 157–172 (2013)
29. V., C.N.: Data mining for imbalanced datasets: an overview. In: Maimon, O., Rokach, L. (eds.) Data Mining and Knowledge Discovery Handbook. Springer, Boston, pp. 853–867. Springer, Boston (2005). https://doi.org/10.1007/978-0-387-09823-4_45
30. VALUATIONS, E.: A review on evaluation metrics for data classification evaluations (2015)
31. Wilson, D.L.: Asymptotic properties of nearest neighbor rules using edited data. IEEE Trans. Syst. Man Cybern. **2**(3), 408–421 (1972)
32. Wright, A.H.: Genetic algorithms for real parameter optimization. In: Proceedings of the First Workshop on Foundations of Genetic Algorithms. Bloomington Campus, Indiana, USA, 15–18 July 1990, pp. 205–218 (1990)
33. Zewdu, T., HiLCoE, T.B.: Prediction of HIV status in Addis Ababa using data mining technology (2015)

Neural Network for Arabic Text Diacritization on a New Dataset

Zubeiri Iman[1,2](✉), Souri Adnan[1,2], and El Mohajir Badr Eddine[1,2]

[1] New Trend Technology Team, Abdelmalek Essaadi University, Tetouan, Morocco
imane.zubeiri@gmail.com, B.elmohajir@ieee.ma

[2] National School of Applied Sciences, Abdelmalek Essaadi University, Tetouan, Morocco

Abstract. Arabic language is one of the most spoken languages in the world, it's the official language of many countries and the fourth most used language on the internet. Arabic texts are often written without diacritic marks. However, those marks are important to clarify the sense and meaning of words. Automatic diacritization is the process of assigning diacritics to letters, and it's an important field in Arabic Natural Language Processing (ANLP). In this work, we try to find the effect of increasing the training dataset on the diacritization error rate (DER) by building a new dataset and concatenating it with the Tashkeela dataset. We trained a deep learning model based on bidirectional long short-term memory BLSTM that transcribes undiacritized sequences of Arabic letters and produces an output sequence of the same length fully diacritized. Our model shows significant results on the new dataset in terms of DER and validation loss.

Keywords: Automatic diacritization · Sequence transcription · BLSTM

1 Introduction

Arabic language is one of the most spoken languages in the world and a native language of 27 countries. Arabic is ranked the fourth most used language on the internet and the fastest growing in the last few years [1]. Arabic language has many special features; for example, Arabic is written from right to left; there is no capitalization in Arabic, and the letters could change their shape according to their position in the word. Moreover, The Arabic writing system is composed of a letter and a diacritic mark, above or below the letters, representing the phonetic information. The Arabic alphabet consists of 28 letters (see Fig. 1). Arabic texts are splitted into Classical Arabic (CA), like the Holy Quran, old books, and old poetry. And Modern Standard Arabic (MSA), used for writing news, letters, and formal speeches [2].

Arabic texts can be partially or fully diacritized, the major function of diacritics is to determine and clarify the meaning of words and sentences and improve reading comprehension. However, MSA texts are often written without those marks, so it becomes difficult to determine the correct diacritic for each letter, this could lead to ambiguity even for natives. The biggest challenge is that the same word in the undiacritized form may have many interpretations depending on the diacritization of each letter (see Table 1) [3].

M. Lazaar et al. (Eds.): BDIoT 2022, LNNS 625, pp. 186–199, 2023.
https://doi.org/10.1007/978-3-031-28387-1_17

The diacritization process consists of adding the correct diacritics to the corresponding letters. A robust automatic diacritization tool should have overcome some challenges such as:

Data Sparsity challenge: High-quality diacritized datasets tend to be quite small and there is a lack of linguistic resources available for research. These factors result in diacritized Arabic being very data sparse.

Ambiguity challenge: Could be summarized in two main types:

- Ambiguity within the Part-of-speech (POS) tagging: words with the same spelling and POS tag but a different lexical sense, and homographs where words have the same spelling but different POS tags and lexical senses.
- Ambiguity at grammatical level: when sentences and phrases could be interpreted in more than one way [4].

Automatic diacritization task has been discussed in several works in various NLP areas. The existing state-of-the art approaches that address the automatic discretization are splitted into two main categories, rules-based approaches and machine learning approaches.

Rule-based methods: Rely on well-formed rules and the knowledge of Arabic morphology and syntax to give the correct diacritization of a given sentence. Those methods consistently require a language expert as a reference in writing the rules [5].

Machin learning methods: learn from diacritized texts to predict the diacritization of the output text. They have been useful in different domains including Arabic NLP, they offer many advantages:

- ML algorithms can automatically focus on common cases.
- ML algorithms can produce models for unfamiliar data.
- ML can be accurate by merely increasing the input data [6].

However, those models need a large corpus of fully diacritized texts and they may suffer from data sparsity.

In this work, we focus on finding the effect of increasing the training dataset on the diacritization error rate by building new datasets. We proposed also a deep learning model that will be trained firstly on the Tashkeela subset presented in [3] and secondly on the concatenation of the Tashkeela subset and our dataset.

أ	ب	ت	ث	ج	ح	خ	د	ذ	ر
ز	س	ش	ص	ض	ط	ظ	ع	غ	ف
ق	ك	ل	م	ن	ه	و	ي		

Fig. 1. Arabic letters

Table 1. One Arabic word has multiple meanings depending on diacritics

A written word without diacritics signs	Manning	Written word with diacritics signs
سلم	Peace	سِلْمٌ
سلم	Submit	سَلَّمَ
سلم	Stairs	سُلَّمٌ
سلم	Survive	سَلِمَ
سلم	He say's hi	سَلَّمَ
سلم	without Dispute	سَلَمٌ

The rest of this paper is organized as follows; in Sect. 2, we review the existing diacritization tools and their categories and we focus on ML techniques. In Sect. 3 we address Arabic diacritization as a sequence transcription problem. Section 4 illustrates the experimental setup and presents the obtained results. We conclude in Sect. 5.

2 Literature Review

We start this section by reviewing important related work in Arabic text diacritization. The earliest approaches were rule-based approaches [7, 8], which relied on morphological analyzers, dictionaries, and grammar modules. The major drawbacks of rule-based diacritization are the high development cost and the reliance on parsed corpora that are difficult to create. Additionally, the rules must be continuously maintained as new words and terms are generated in living languages. More recently, there have been several statistical, machine-learning approaches. Authors in [9] used hidden Markov models (HMMs) to capture the contextual correlation between words. His approach restores only short vowels. In [10] they used statistical n-gram language modeling of a large corpus. They used the Tashkila diacritized Arabic text corpus. The possible diacritized word sequences of an undiacritized input sentence are assigned probability scores using the n-gram models. Then using a dynamic programming algorithm, the most likely sequence is found. Smoothing techniques were used to handle unseen n-grams in the training data. The accuracy of n-gram models depends on the order n. larger order gives higher accuracy as it incorporates longer linguistic dependencies. However, larger order results in larger models and requires larger training data. In [11] authors used n-gram models of order three (trigram), and thus, do not exploit long-range context dependencies. However, this system requires the availability of speech input as it combines acoustic information from the speech input to complement the text-based conditional random fields model.

Most current work relies on hybrid approaches that combine rule-based and statistical modules. In [12] researchers investigated the effect of combining several knowledge sources (acoustic, morphological, and contextual) to automatically diacritize Arabic text. Authors in [7] treated diacritization as an unsupervised tagging problem where each word is tagged as one of the many possible forms provided by the Buckwalter Arabic morphological analyzer (BAMA). They also tested the use of Arabic dialectal speech they used two different corpora: the Foreign Broadcast Information Service (FBIS) corpus of MSA speech and the LDC CallHome Egyptian Colloquial Arabic (ECA) corpus. However, they did not model the Shadda diacritic. In [13] authors proposed a weighted finite state machine algorithm to restore the missing diacritics. Their basic module consists of a standard trigram language model that chooses the most probable diacritized word sequence that could have generated the undiacritized text. They also used several other transducers.

The most important work is [14], the open-source project Shakkala for Arabic text diacritization; it used B-LSTM networks in addition to character embedding. The authors trained the model on Tashkeela Corpus several times while removing the data with a negative influence on the training process. The system provides the diacritization service through an interactive web interface, without providing an API.1. The website allows users to diacritize text containing up to 490 symbols. More recently, Deep Neural Networks (DNN) are used in Arabic text diacritization and show significant improvements over the previous techniques. Many deep learning models are simple sequential models consisting of Recurrent Neural Networks (RNNs) and fully connected (FC) layers. In [15] authors designed a system based on DNN and Confused Sub-Classes Resolution (CSR). In [16] they proposed a deep learning model based on long short-term memory (LSTM) layers. The system significantly improved the diacritization over the previous works. In [17] they examine several deep learning models with different architectures and different numbers of layers. In authors [18] used two deep learning approaches: Feed-Forward Neural Network (FFNN) and RNN, with several techniques such as 100-hot encoding, embedding, CRF, and Block-Normalized Gradient (BNG).

In [19] they presented a character-level sequence-to-sequence deep learning model. The model used a Neural Machine Translation (NMT) setup on overlapping windows of words. In [20] authors proposed a novel architecture for labeling character sequences. Their model contains a Two-Level (Hierarchical) recurrence that operates on the word and character levels separately enabling faster training and inference than comparable traditional models. In addition, a cross-level attention module further connects the two and opens the door for network interpretability. Researchers in [21] proposed three deep learning models to recover Arabic text diacritics. First a baseline model: consists of an embedding layer (512 dimensions), followed by three bidirectional 256 cells LSTM layers, then a fully connected layer. Second an encoder/decoder model: uses location-sensitive attention and the output of the decoder is a sequence of diacritics instead of a sequence of frames. Third The CBHG model has 14 Million trainable parameters, much more than the encoder-decoder model. The proposed scheme in [22] was the LSTM model; the authors trained the model on the LDC ATB3 dataset as an example of MSA and a clean subset of the Tashkeela dataset as an example of CA. They tuned the model by going for a deeper network and applying dropout. Their best results were: 2.46% and 1.97% for ATB3 and Tashkeela, respectively. Authors in [4] proposed a novel multitask learning model for diacritization, which trains a model to both diacritize and translate. They demonstrate that training a diacritics model to both diacritize and translate substantially outperforms a model trained on the diacritization task alone. They manually analyse the errors and obtained 4.79% on DER as the best rate.

In [23], the authors used the ML approach to classify and diacritize Arabic poetry. The proposed model is a deep and narrow recurrent neural network that has an embedding layer at the input, four hidden bidirectional LSTM layers, and a softmax output layer. The suggested model tuned and achieved an average accuracy much higher than previous work (97.27%) The model achieves higher accuracy on diacritized verses compared with the accuracy on undiacritized verses. Authors in [24] implemented a model for diacritics restoration based on the BERT. They annotated all reported missing predictions in Czech and found out that more than one correct variant is sometimes possible. They elaborated on these phenomena using morphological annotations and utilized them to further analyze real confirmed errors of the systems. Researchers in [25] presented results from the investigation of the effect of diacritic quality of training corpus on the performance of some simple and commonly used machine learning algorithms for diacritic restoration tasks. The results show that the completeness and correctness of diacritics have a significant effect on the performance of the algorithms, but it is most important for the decision tree algorithm. The aim of this paper was to identify a simple algorithm that achieves state-of-the-art results on a relatively small training corpus since existing research has used a larger corpus. Authors present in [26] a diacritic restoration by a model that considers the output distributions for different related tasks to improve the performance of diacritic restoration. The results show significant improvements across all evaluation metrics. This shows the importance of considering additional linguistic information at morphological and/or syntactic levels. Including semantic information through pre-trained words, embedding within the diacritic restoration model also helped boost the diacritic restoration performance.

3 Sequence Transcription and Neural Networks

3.1 Sequence Transcription

Sequence Transcription is the process of transforming an input sequence into the corresponding output sequence. Arabic text diacritization is considered as a sequence transcription problem the input sequence X consists of T characters $x_1, x_2, x_3, \ldots, x_t$ that represent the undiacritized sequence. The corresponding output sequence is $y_1, y_2, y_3, \ldots,$, where y_i is the diacritic mark of x_i. There are four categories of Sequence transcription solved via Recurrent Neural Networks (RNN) [22]:

- One-to-one networks take an input sequence and produce an output sequence of the same length.
- Sequence-to-vector networks transcribe input sequences into one final output by ignoring all previous outputs.
- Vector-to-sequence networks take one input vector and produce an output sequence.
- The general sequence-to-sequence network has an output sequence that is not of the same length as the input sequence.

3.2 Neural Networks

Neural networks obtained excellent results in many complex tasks such as neural machine translation, text-to-speech synthesis, sentiment analysis, image captioning, and speech recognition. Multi-stage or cascaded architectures are models that can be used for accurate prediction and classification. In these models, the results of the early stages are fed to the next stages as part of the input vector. The results of such stages can be either extracted features or classification results [23].

Recurrent Neural Networks (RNNs):
RNNs are powerful extensions of feed-forward layers that can be used with sequence transcription (see Fig. 2). This is because the cell's hidden states are functions of all previous states with respect to time.

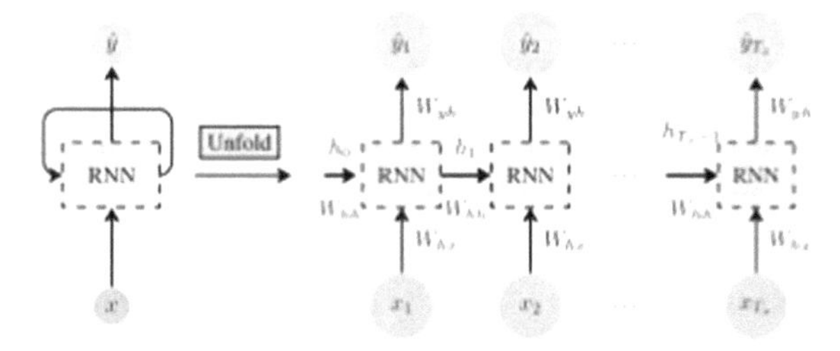

Fig. 2. RNN layer's structure: This figure shows one-unit RNN, and its unfold version, the input state is represented by x, the output is y, and the hidden state is represented by h, while w is the weight of the network.

Fully-Connected Layer (FC):
FC is a type of feed-forward layer in which each neuron is connected to all neurons in the next layer (see Fig. 3). These are powerful layers that are used in almost all deep learning. Models. In many cases, a fully-connected layer is required as a projection layer to change the output dimension.

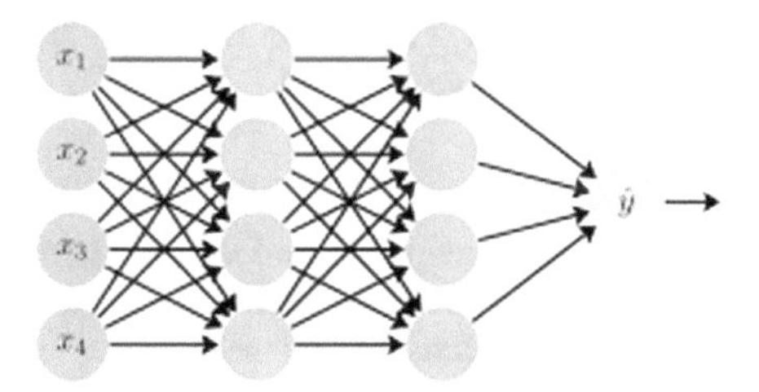

Fig. 3. FC layer's structure: In this type of network the inputs (represented by x) are connected to every activation unit of the next layer to produce an output y.

Long Short-Term Memory (LSTM):
In the LSTM RNN, each memory cell has two states, the short-term state (also used as the cell output) h(t) and a long-term state c(t). These states are updated using an input gate, a forget gate, an output gate, and a cell activation unit. This collective operation enables

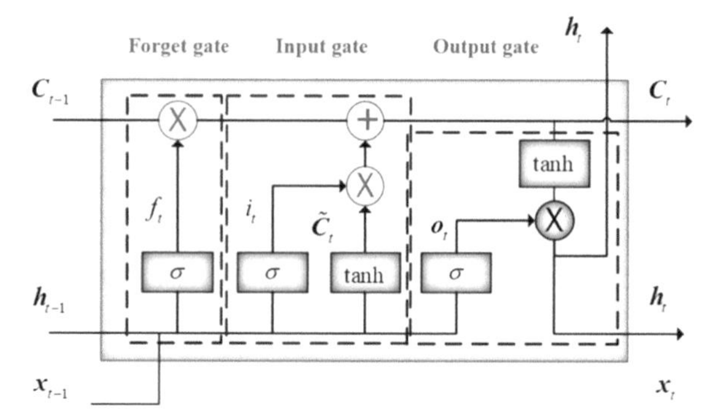

Fig. 4. The structure of LSTM Cell. As we can see in this figure the memory cell can be driven by three control gates; the input gate decides if the content of the cell should be modified; the forget gate decides whether to reset the content of the cell to 0; the output gate decides if the content of the cell should influence the output of the neuron.

the LSTM cell to capture long-term patterns by recognizing important inputs, preserving them if they are needed, and extracting them whenever they are needed (Fig. 4).

Bidirectional Recurrent Neural Networks (BRNN):
BRNN use information from both past and future inputs. BRNN layers achieve this by comprising two unidirectional layers that process the sequence in both time directions producing two hidden vectors (Fig. 5).

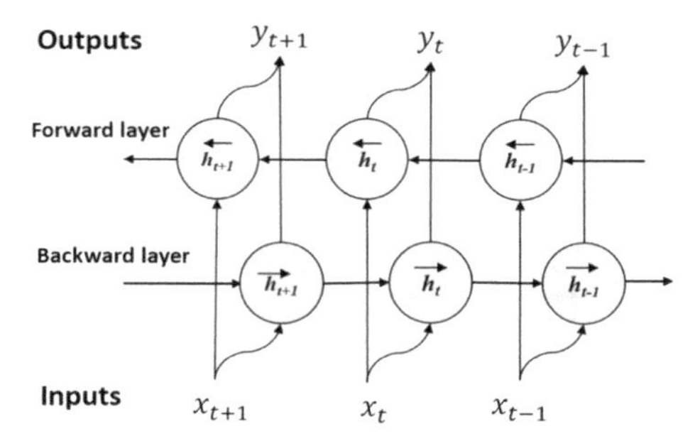

Fig. 5. The general structure of the bidirectional neural network: in the BRNN architecture the same output is both connected to two hidden layers in the opposite direction so the output layer can get the information from the past and future simultaneously.

In our work, we are going to deal with automatic Arabic diacritization as a one-to-one sequence transcription problem since for each input sequence of characters; the output sequence of diacritics is of the same length. First, we are going to train our model on an existing dataset, and second we are going to extend this dataset by adding more unseen data to the model in order to evaluate the impact on the validation loss and the diacritization error rate.

4 Experiments and Discussion

In this section, we provide details of our experiments. We illustrate the methodology used, how datasets were extracted, and the structure of our model.

4.1 Methodology

Our experiments went through two steps:

- The first one is training the model.
- The second one is testing its diacritization Errors rate and the validation loss.

In the training step, diacritics are removed from the training sequence X to generate undiacritized sequence Y. Sequence Y represents the model input whereas X sequences are the model target sequences. Both sequences X and Y are fed to the model after being encoded.

In the testing step, diacritics are removed from the testing sequences. The trained model takes the obtained undiacritized sequences as input to predict their diacritics.

Finally, we compare the model diacritized sequences with the target sequences in measures of DER rate (Figs. 6 and 7).

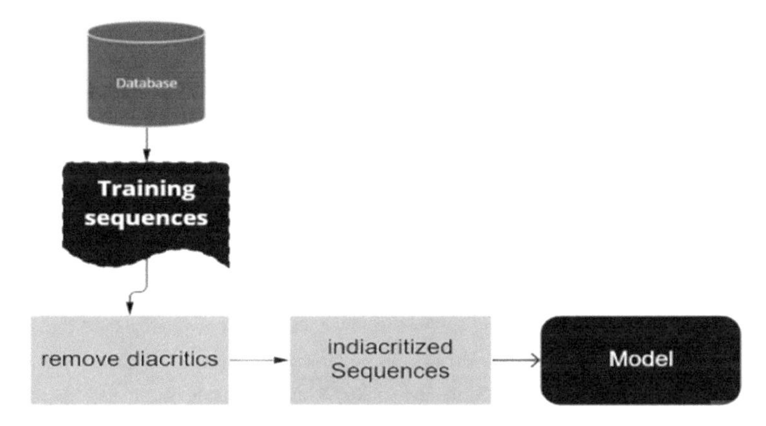

Fig. 6. This figure shows the steps performed in the training phase of the performed experiments.

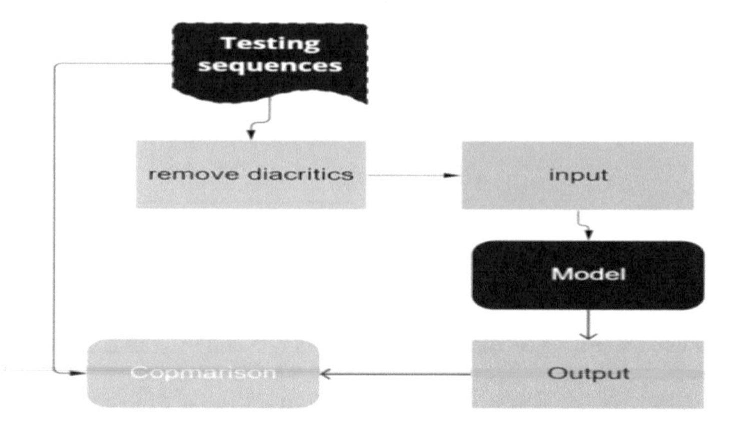

Fig. 7. This figure shows the steps performed in the testing phase of the performed experiments.

4.2 Datasets

We first tested our model on the subset of 55,000 sequences extracted from the Tashkeela dataset by [3]. The dataset was cleaned by removing English letters and extra whitespaces and separating numbers from words. One of the main goals of this work is to study the effect of incrementing the training data size on the diacritization accuracy by adding more unseen sequences to the model. In order to reach this goal, we extract and clean new datasets from several books in the scientific and medical fields from Shamela library, which is an electronic library. The new dataset contains 27497 sequences (Fig. 8).

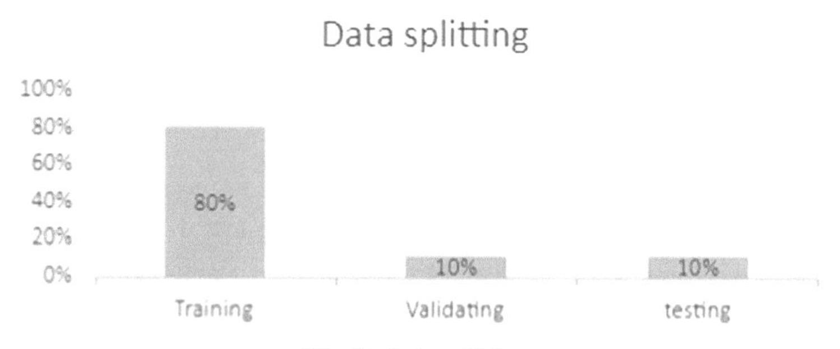

Fig. 8. Data splitting.

4.3 Model

Our model is a BiLSTM that consists of an embedding layer of 25 dimensions, followed by two bidirectional 256 cells LSTM layers, then two fully connected layers. Lastly, a softmax layer is used to output a probability distribution over the output diacritics (see Fig. 9).

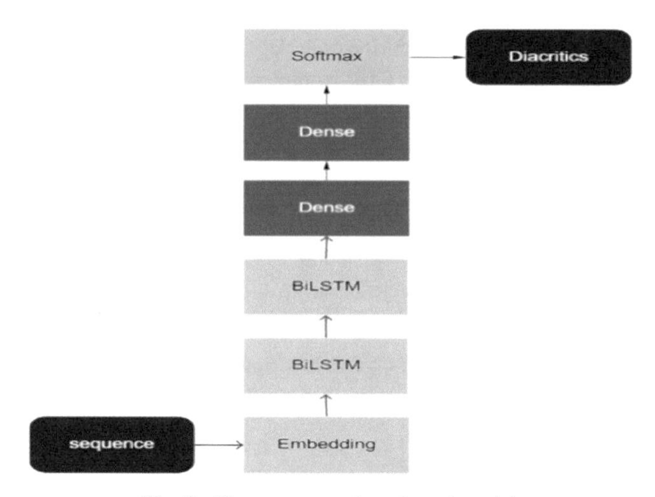

Fig. 9. The structure of our based model.

4.4 System Evaluation

To evaluate the diacritization system we use one of the popular metrics the Diacritic Error Rate (DER), this metric calculates the percentage of characters that were not correctly diacritized. We calculate this metric by extracting diacritics from both the original and the predicted files.

$$DER = \frac{\mathrm{Dw}}{\mathrm{Dw} + \mathrm{Dc}} \times 100 \tag{1}$$

This metric can be calculated either with Case-Ending or without Case-Ending. The calculation without CE excludes each word's last character from error calculation since they mostly depend on grammatical rules.

4.5 Experiment and Results

In the first experiment, we trained our model on the Tashkeela subset presented in [24]. The subset contains 50,000 sequences. To reduce the training time and memory usage, we wrap sequences as such to have maximum lengths of 400 characters. We obtained 300,138 sequences. We report the performance of the model during the training in terms of validation loss and DER. In the second experiment, we increased the number of the sequence to study the effect of incrementing the training data size on the diacritization error rate and validation loss by concatenating the subset presented in [24] with our dataset to obtain a total number of 328,136 sequences that we call the incremented dataset.

We can clearly see that the validation loss decreased in the function of the size of the data set. Figure 11 shows that the DER improves as the number of epochs increases. We can also see that the DER decreased in the function of the size of the data set the best DER reported is 0.78% without case ending and 1.02% with the case ending for the incremented dataset (Fig. 10).

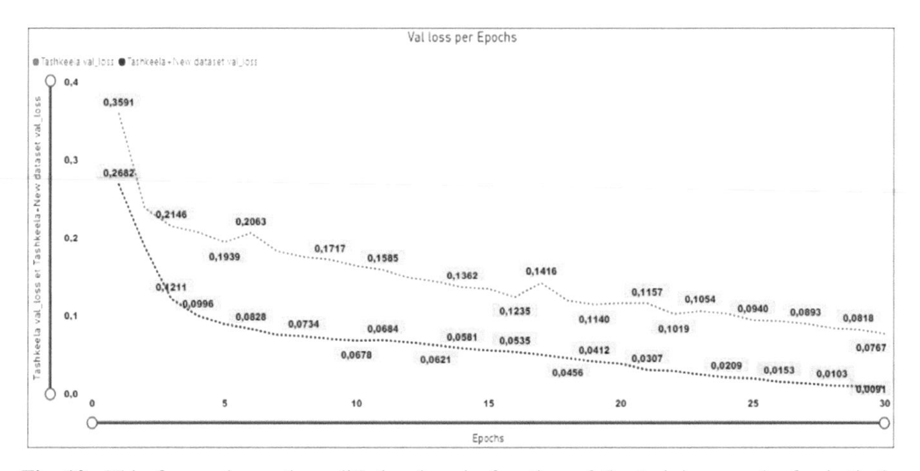

Fig. 10. This figure shows the validation loss in function of the training epochs for both the Tashkeela subset and the incremented dataset, the best validation loss achieved for the first data set is 0,0767 and for the incremented dataset is 0,0091.

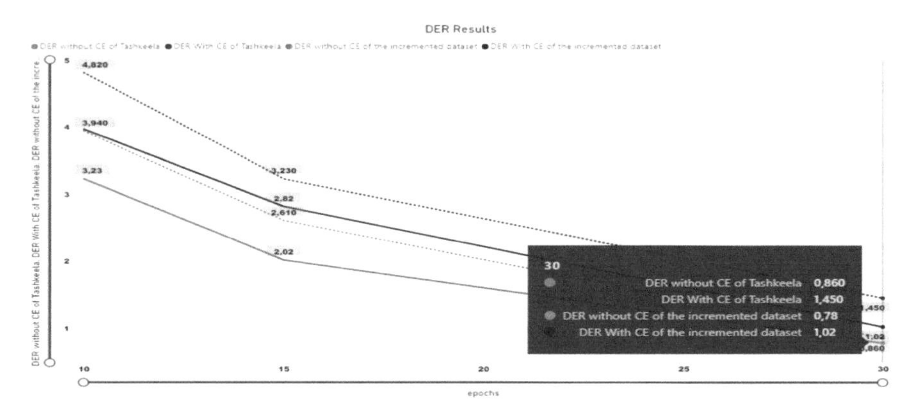

Fig. 11. In this figure we illustrate the obtained DER with and without case ending on the tashkeela data set and the extended dataset.

The best DER reported after 30 epochs was on one hand, 0.860 without case ending and 1.450 with case ending on the Taskeela dataset, on the other hand, 0.78 without case ending 1.02 on the extended data set (Fig. 12).

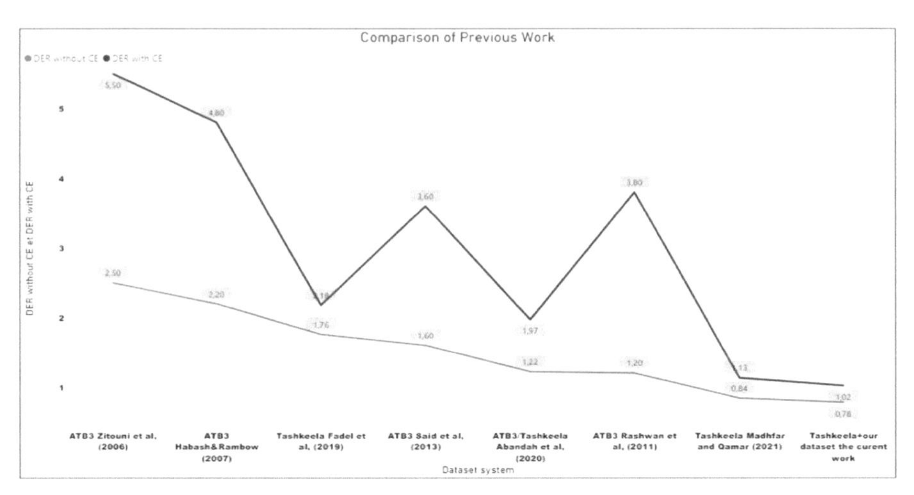

Fig. 12. This figure summarizes the comparison of our work results and the best results of the published system. For each system, the figure shows the dataset used and the publication year.

5 Conclusion

Automatic diacritization is one of the most important fields in Arabic NLP. In this paper, we reviewed some previous works using ML techniques to deal with the automatic diacritization, we proposed a deep learning model based on BLSTM to automatically recover diacritics. In order to find the effect of increasing the training dataset on the diacritization error rate we extend the training dataset by adding more unseen data to our model. We conducted two experiments, the first one on the Tashkeela dataset of 300,138 ssequences and the second one on a concatenated dataset of 328,136. Our model achieves a DER of 0.78% without case ending and 1.02 with case ending. As expected, the model showed significant results on the largest dataset in terms of DER and validation loss.

References

1. Boudad, N., Faizi, R., Thami, R.O.H., Chiheb, R.: Sentiment analysis in Arabic: a review of the literature. Ain Shams Eng. J. **9**(4), 2479–2490 (2018)
2. Farghaly, A., Shaalan, K.: Arabic natural language processing: challenges and solutions. ACM Trans. Asian Language Inform. Process. **8**(4), 1–22 (2009)
3. Fadel, A., Tuffaha, I., Al-Jawarneh, B., Al-Ayyoub, M.: Arabic text diacritization using deep neural networks. In: 2019 2nd International Conference on Computer Applications & Information Security (ICCAIS), Riyadh, Saudi Arabia, pp. 1–7 (2019). https://doi.org/10.1109/CAIS.2019.8769512
4. Thompson, B., Alshehri, A.: Improving Arabic Diacritization by Learning to Diacritize and Translate. https://arxiv.org/ftp/arxiv/papers/2109/2109.14150.pdf
5. Almanea, M.M.: Automatic methods and neural networks in Arabic texts Diacritization: a comprehensive survey. IEEE Access **9**, 145012–145032 (2021). https://doi.org/10.1109/ACCESS.2021.3122977

6. Larabi, S., Marie-Sainte, S., Alalyani, N., Alotaibi, S., Ghouzali, S., Abunadi, I.: Arabic natural language processing and machine learning-based systems. IEEE Access **7**, 7011–7020 (2019). https://doi.org/10.1109/ACCESS.2018.2890076
7. El-Sadany T., Hashish M.: Semi-automatic vowelization of Arabic verbs. In: 10th National Computer Conference, pp. 725–732 (1988)
8. Al-Sughaiyer, I.A., Al-Kharashi, I.A.: Arabic morphological analysis techniques: a comprehensive survey. J. Am. Soc. Inf. Sci. Technol, **55**(3), 189–213 (2004)
9. Gal, Y.: An HMM approach to vowel restoration in Arabic and Hebrew. In: ACL-02 Workshop on Computational Approaches to Semitic Languages, pp. 1–7 (2002)
10. Hifny, Y.: Smoothing techniques for Arabic diacritics restoration. In: 12th Conf. on Language Engineering, pp. 6–12 (2012)
11. Dahl, G., Yu, D., Deng, L., Acero, A.: Context-dependent pre-trained deep neural networks for large-vocabulary speech recognition. IEEE Trans. Audio Speech and Language Process. **20**(1), 30–42 (2012)
12. Vergyri, D., Kirchhoff, K.: Automatic diacritization of Arabic for acoustic modeling in speech recognition. In: Workshop on Computational Approaches to Arabic Scriptbased Languages, pp. 66–73 (2004)
13. Nelken, R., Shieber, S.M.: Arabic diacritization using weighted _nite-state transducers. In: ACL Workshop on Computational Approaches to Semitic Languages, pp. 79–86 (2005)
14. Barqawi, A., Zerrouki, T.: Shakkala, arabic text vocalization. https://github.com/Barqawiz/Shakkala (2017)
15. Al Sallab, M., Rashwan, H., Raafat, M., Rafea, A.,: Automatic Arabic diacritics restoration based on deep nets. In: Proceedings of the MNLP Workshop Arabic Natural Lang. Process. (ANLP). Association Computational Linguistics, Doha, Qatar, pp. 65–72. https://www.aclweb.org/anthology/W14-3608 (2014)
16. Abandah, G.A., Graves, A., Al-Shagoor, B., Arabiyat, A., Jamour, F., Al-Taee, M.: Automatic diacritization of Arabic text using recurrent neural networks. Int. J. Document Anal. Recogn. (IJDAR) **18**(2), 183–197 (2015). https://doi.org/10.1007/s10032-015-0242-2
17. Belinkov, Y., Glass, J., : Arabic diacritization with recurrent neural networks. In: Proceedings of the Conference Empirical Methods Natural Language Processing Lisbon, pp. 2281–2285. Association Computational Linguistics, Portugal. https://www.aclweb.org/anthology/D15-1274 (2015)
18. Fadel, A., Tuffaha, I., Al-Jawarneh, B., Al-Ayyoub, M.: Neural arabic text diacritization: state of the art results and a novel approach for machine translation. In: Proceedings 6thWorkshop Asian Translation, pp. 215–225. Association Computational Linguistics, Hong Kong (2019)
19. Mubarak, H., Abdelali, A., Sajjad, H., Samih, Y., Darwish, K.: Highly effective Arabic diacritization using sequence-to-sequence modeling. In: Proceedings of the Conference North American Chapter Association Computational Linguistics, Human Language Technologies, vol. 1, pp. 2390–2395. Association Computational Linguistics, Minneapolis, MN, USA (2019)
20. AlKhamissi, B., ElNokrashy, N., Gabr, M.: Deep Diacritization: Efficient Hierarchical Recurrence for Improved Arabic Diacritization. arXiv:2011.00538v1 (2020)
21. Madhfar, M.A.H., Qamar, A.M.: Effective deep learning models for automatic Diacritization of Arabic Text. IEEE Access **9**, 273–288 (2021). https://doi.org/10.1109/ACCESS.2020.3041676
22. Abandah, G., Abdel-Karim, A.: Accurate and fast recurrent neural network solution for the automatic Diacritization of Arabic text. Jordanian J. Comput. Inform. Technol. **06**, (02) 1 (2020)
23. Abandah, G.A., Khedher, M.Z., Abdel-Majeed, M.R., Mansour, H.M., Hulliel, S.F., Bisharat, L.M.: Classifying and diacritizing Arabic poems using deep recurrent neural networks. J. King Saud Univ. – Comput. Inform. Sci. **34**, 3775-3788 (2022)

24. Náplava, J., Straka, M., Straková, J.: Diacritics restoration using BERT with analysis on Czech language. Prague Bull. Math. Linguist. **116**(1), 27–42 (2021)
25. Ayogu, I.I., Abu, O.: Automatic diacritic recovery with focus on the quality of the training corpus for resource-scarce languages. In: 2020 IEEE 2nd International Conference on Cyberspac (CYBER NIGERIA), pp. 98–103. Abuja, Nigeria (2021)
26. Alqahtani, S., Mishra, M., Diab M.: A Multitask Learning Approach for Diacritic Restoration arXiv:2006.04016v1 (2020)
27. Abdel Karim, A., Abandah, G.: On the training of deep neural networks for automatic Arabic-text Diacritization. Int. J. Adv. Comput. Sci. Appl. **12**(8), 276–286 (2021)
28. Abandah, G.A., Suyyagh, A.E., Abdel-Majeed, M.R.: Transfer learning and multi-phase training for accurate diacritization of Arabic poetry. J. King Saud Univ. – Comput. Inf. Sci. **34**(6), 3744–3757 (2022). https://doi.org/10.1016/j.jksuci.2022.04.005

Processing Time Performance Analysis of Scheduling Algorithms for Virtual Machines Placement in Cloud Computing Environment

Hind Mikram[1], Said El Kafhali[1](✉), and Youssef Saadi[2]

[1] Faculty of Sciences and Techniques, Computer, Networks, Modeling, and Mobility Laboratory (IR2M), Hassan First University of Settat, B.P. 539, 26000 Settat, Morocco
{h.mikram,said.elkafhali}@uhp.ac.ma

[2] Data Science For Sustainable Earth Laboratory, Faculty of Sciences and Techniques, Sultan Moulay Slimane University, Beni Mellal, Morocco
y.saadi@usms.ma

Abstract. In recent years, a large range of applications have been migrated from traditional computing environments to cloud systems. On the other hand, organizations with existing infrastructure investments leverage the server consolidation in the cloud to serve the upcoming load from various clients around the world. Server consolidation techniques in cloud computing are applied at infrastructure levels and spilled into three main steps; load detection, selection and placement of migrated virtual machines. The placement process can be divided to initial placement and task scheduling while allowing fair load and better resource utilization. The challenge is executing several tasks with the available shared computing resources. Further, the cloud provider should apply a good task strategy depending on the customer's need. To maintain the best performance of the cloud system, the optimal resource utilization and the shortest completion time for task execution are the essential keys. The heuristics and metaheuristics methods are applied to achieve this goal. In this work, some of the proposed scheduling algorithms will be compared through CloudSim as a simulation tool and through processing time as a performance metric. More specifically, the comparison is made by proposing various simulation scenarios, using the processing time to classify algorithms according to time-shared and space-shared. The evaluation results show that SJF (Short Job First) and FCFS (First Come First Serve) outperforms other baseline algorithms according to different simulations set up.

Keywords: Server Consolidation · Cloud Computing · Resource Efficiency · Virtual Machines Placement · Energy Consumption · scheduling algorithms · CloudSim

M. Lazaar et al. (Eds.): BDIoT 2022, LNNS 625, pp. 200–211, 2023.
https://doi.org/10.1007/978-3-031-28387-1_18

1 Introduction

Data centers and cloud computing are rapidly evolving, opening up new software and hardware resources as services through the Internet [1]. The energy consumed by data centers results in the production of carbon dioxide (CO2) emissions which is an issue within the field of Green Computing [2,3]. The infrastructure level uses virtualization to address this issue, allowing a single physical machine (PM) to group several virtual machines (VMs) [4]. Hence, grouping several instances risk degrading the service quality (QoS), which leads to applying the consolidation algorithms [5].

The server consolidation algorithm allows finding a solution for the energy consumption taking into account the use of the computer resources to avoid any performance degradation of services [6]. It is divided into three main steps; each step has its own policy. The first policy detects the load of the PM, the second one selects the VMs that should migrate to another PM, and the last policy searches for a new PM for the migrated VMs. Within the consolidation process, scheduling is a fundamental step for VMs and task planning. Especially with the growth of user demands that increased the problem complexity.

Scheduling may be characterized as a collection of policies that control how processing tasks is organized in cloud computing systems. In addition, the assignment capacity of the VM determines the task's execution time. Due to the enormous number of users, cloud computing tasks may be of massive size. Without breaching the priority limitations, the scheduling algorithm distributes the tasks among the appropriate and readily accessible resources in the cloud environment. In other words, task scheduling is one of the key elements of a cloud environment since it relates user tasks to the proper resource consumption [7,8]. If task scheduling techniques are effective, then a positive impact on the performance of the entire cloud computing system is reached [9,10].

Several metaheuristic and heuristic methods for task scheduling and load balancing in cloud resource management have been suggested in the literature to find the optimal solution [11]. Heuristic algorithms look to approximate the solution while speeding up to achieve it. However, they do not ensure the discovery of the best solution. As a result, they are considered more general approximate solution rather than precise ones. On the other side, meta-heuristic algorithms seek to effectively investigate the search space for solutions that are close to optimum. But these algorithms have significant temporal complexity compared to heuristic algorithms because the solution iterates until the stop criterion or the maximum number of iterations is achieved [12].

In this paper, different metaheuristic and heuristic methods are compared to find the appropriate algorithm for the scheduling of VMs as well as the tasks on suitable PM. Particle Swarm Optimization (PSO) and Ant Colony Optimization (ACO) are the most popular and effective metaheuristics [13]. First Come First Serve (FCFS), Shortest Job First (SJF), MinMin, and MaxMinx are the heuristic methods that will be compared in this paper across processing time using the CloudSim toolkit. In addition, several comparative studies treat allocating and scheduling tasks in the cloud to save energy. However, there is a need to consider the processing time as a metric of comparison.

The following list summarizes our significant contributions:

- We will introduce some heuristic and metaheuristic methods for scheduling problems and define scheduling types to solve load balancing and reallocation of resources.
- Based on scheduling policy types, we will compare SJF, FCFS, MinMin, MaxMin, PSO, and ACO using the CloudSim toolkit.
- In order to determine the optimal strategy for processing time, we will discuss the results by varying simulation parameters such as the number of data centers, PMs, VMs and Tasks.
- We will compare the simulation performance of the different algorithms and identify what conditions perform better.

The remaining sections are organized as follows. Section 2 reviews the literature. In Sect. 3, we formally present the scheduling policies. The simulation of compared algorithms is described in Sect. 4. We will present the conclusion of our comparison in Sect. 5.

2 Related Work

The workload distributes among the VMs via scheduling algorithms, which assign requests to the relevant VMs based on the resources needed. These algorithms apply to maintain the effectiveness and stability of the energy in cloud computing [14]. Research on load balancing and task scheduling has used several methods like heuristic and metaheuristic. This work aims to compare some of these algorithms. In [15], the authors compared several heuristic methods such as FCFS, SJF, RoundRobin, and others, where they used average turnaround time and average waiting time as performance parameters. Moreover, another comparison is introduced in [16], where it compared five methods using DISSECT-CF as a simulator, simulation duration, total power consumption, and the number of migrations as performance metrics. The results show that population-based metaheuristics do not provide a significant improvement in solution quality to compensate for the longer simulation duration. However, numerous researchers have suggested population-based metaheuristic algorithms. They are frequently used to solve optimization issues and achieve results close to optimum in a reasonable time.

ACO and PSO are popular population-based metaheuristics [16]. PSO method resolve scheduling and routing issues. The dimensions of this issue, the inertia weight, the number of particles, the range of iterations, the random values of the social and cognitive components, and the acceleration coefficients are the fundamental parameters for PSO [13]. Furthermore, the result in [13] shows that the choice of random initial parameters is insufficient for scheduling the performance of the evolutionary process on a large scale. The suggested technique in [17] achieves load balancing in virtual machines by maintaining high availability and minimizing downtime problems when a data center experiences significant traffic. Its goal is to improve the speed of data transfer and remote server functioning performances. The related scheme has been evaluated with the other algorithms against three server broker policies.

To determine data center processing time (DCPT) and total response time (TRT) in a cloud system, the authors in [18] analyze the performance of four load balancing algorithms that are nature inspired. Whereas the assignment of VMs to incoming tasks differs significantly depending on the nature of the service's attributes. In addition, the basis for calculating processing time is interval data, which offers a finite range of equally possible processing time values for each activity [19]. Each task has two characteristics in the scheduling problem, the size and the processing time. Due to the machines' limited and varied capabilities, different assignments of tasks have a major impact on processing times, which results in solutions of variable quality [20].

Maintaining the QoS at an acceptable level for cloud clients requires an accurate and well suitable performance analysis model. However, Hanini *et al.* [21] developed a mathematical model to control the incoming request's arrival to enhance the QoS in the cloud system. The proposed model is cross-validated using MATLAB software implementation. The obtained analysis results showed the positive impact of the system under control in comparison with a system with no control on the QoS in terms of the loss probability, the mean number of requests in the system, and the mean requests delay while varying the incoming request arrival rate. For load balancing on the virtual machines and scheduling tasks with a low computational cost, the algorithm in [22] is based on a hybrid algorithm that combines Max-Min and genetic algorithms. In the experimental simulation, the proposed algorithm outperformed Max-Min and other compared algorithms in terms of makespan, transmission time, resource usage, average waiting time, and degree of imbalance. Ouammou *et al.* [23] proposed an analytical technique to improve the QoS of the cloud system, maximize the utilization of cloud resources and reduce the overall energy consumption in the cloud data center. The goal is to manage resource utilization by exploiting the virtual machine migration concept in the cloud data center. Two techniques are presented and studied, namely, Combinations of Migrations (CM) and High Priority (HP). The obtained numerical results show the effectiveness of the proposed techniques in terms of makespan and energy efficiency while ensuring the QoS.

3 Types of Scheduling Policy

This section introduces four policy types: VM space shared and task space shared, VM time-shared and task space shared, VM space shared and task time-shared, and VM time-shared and task time-shared.

In the Space-Shared scheduling strategy, each VM schedule one task at a particular time until finished. After that, another one will map to the same VM. Under the Time-Shared scheduling strategy, on the virtual machine simultaneously schedules and splits the time between all the tasks.

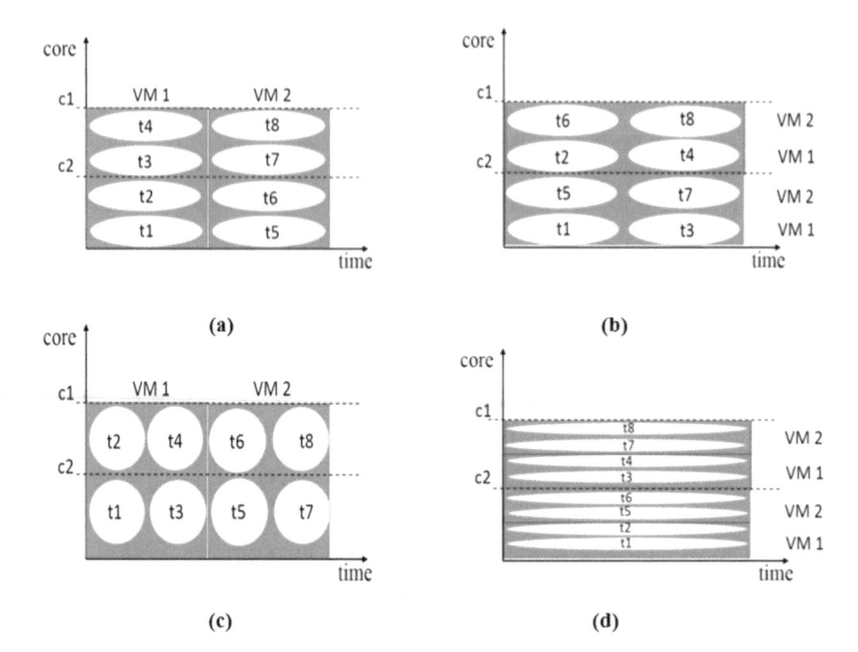

Fig. 1. Types of scheduling policy; VM space-shared and task space shared (a), VM space-shared and task time-shared (b), VM time-shared and task space-shared (c), and VM time-shared and task time-shared (d)

The scheduling types are shown in Fig. 1. In which a PM with two CPU cores receive a request to host two VMs, each of which needs two cores to perform four task units: t1, t2, t3, and t4 in VM1, and t5, t6, t7, and t8 in VM2 [24].

4 Simulations Results and Analysis

4.1 Simulation Environment

Scheduling algorithms are used to run many tasks on VM efficiently. This paper will compare some existing algorithms, using processing time as a metric. We implemented the compared techniques in the CloudSim toolkit, using the configuration shown in Table 1.

Table 2 presents four scenarios, analyzing processing times for heuristic and metaheuristic approaches. We vary the number of tasks, the number of VMs, the number of PMs, and the number of data centers.

For each scheduling type, we test the different scenarios as shown in Table 2 by using the processing time as a performance metric. The results for transmitting the tasks, processing them, then obtaining the answers from the compared algorithms are shown in milliseconds in the in the graphs below (Figs. 2, 3, 4, and 5) based on scheduling policies.

Table 1. Configuration of the simulation environment

Server	System architecture	X86
	Operating System	Linux
	VM Monitor (VMM)	Xen
	Time Zone	10
	Second Cost	3
	Memory Cost	0.05
	Bandwidth Cost	0.0
	Storage Cost	0.001
	Broker	2
Host	Mips	500000
	Processing Elements	5
	Memory (MB)	500000
	Bandwidth (MB/S)	100000
	Storage	1000000
VM	Mips	2000
	Processing Elements	5
	Memory (MB)	512
	Bandwidth (MB/S)	1024
	Storage	10000
Task	Length	10000
	Processing Elements	1
	Filesize	1
	Output size	1

Table 2. Scenarios for heuristic and metaheuristic methods

Scenarios	Data center	Physical machines	Virtual machines	Tasks
Scenario 1	1	3	10	{100, 500, 900}
Scenario 2	1	3	{15, 30, 60}	2000
Scenario 3	1	{3, 5, 10}	30	2000
Scenario 4	{1, 2, 3}	10	30	2000

4.2 Results and Discussion

Figure 2 shows the simulation of Scenario 1, where a significant difference between the heuristic and metaheuristic methods, especially FCFS and SJF, in all the scheduling types.

For example Fig. 2 (a), in the case of VM space shared and task space shared, the small values of processing time in SJF and FCFS were 6 and 8 ms, respectively. On the other hand, the PSO and ACO values in 100 tasks were 70 and

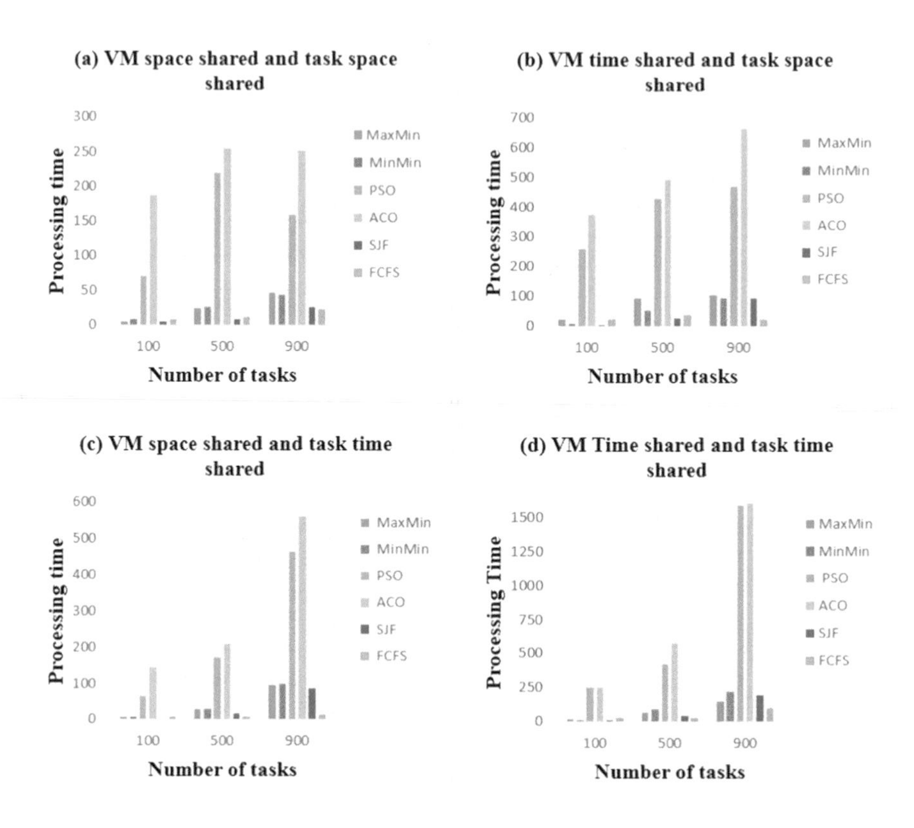

Fig. 2. Processing time performance for each scheduling policy in scenario 1

188 ms, respectively. Furthermore, overall the processing time increases when we augment the number of tasks. We observe in Fig. 2 (b) (The VM time-shared and task space shared) that SJF outperforms other algorithms in 100 and 500 tasks where the processing time were 2 and 26, respectively. FCFS has a smaller value in 900 tasks which is 24 ms. The small values of PSO and ACO in 100 tasks were 260 and 376, respectively. In the VM space shared and Task time-shared in Fig. 2 (c), we observe approximately a zero value in 100 tasks for SJF while the processing time values for FCFS were 8 and FCFS was 8 and 12 ms in 500 and 900 tasks, for PSO and ACO in 100 tasks, they were 64 and 144 ms, respectively. In Fig. 2 (d), the VM time-shared and task time-shared showed increased processing time when the number of the tasks increased too. MinMin and MaxMin demonstrate the smaller values in 100 tasks. We observed respectivley 8 ms and 16 ms for processing time.

Figure 3 presents different scheduling policies as presented in Scenario 2.

In Fig. 3 (a), where VM space is shared and task space is shared, we observe that FCFS outperforms other algorithms when the number of VMs is 15 and 60, and SJF outperforms when this number is 30 VMs. The smaller processing time

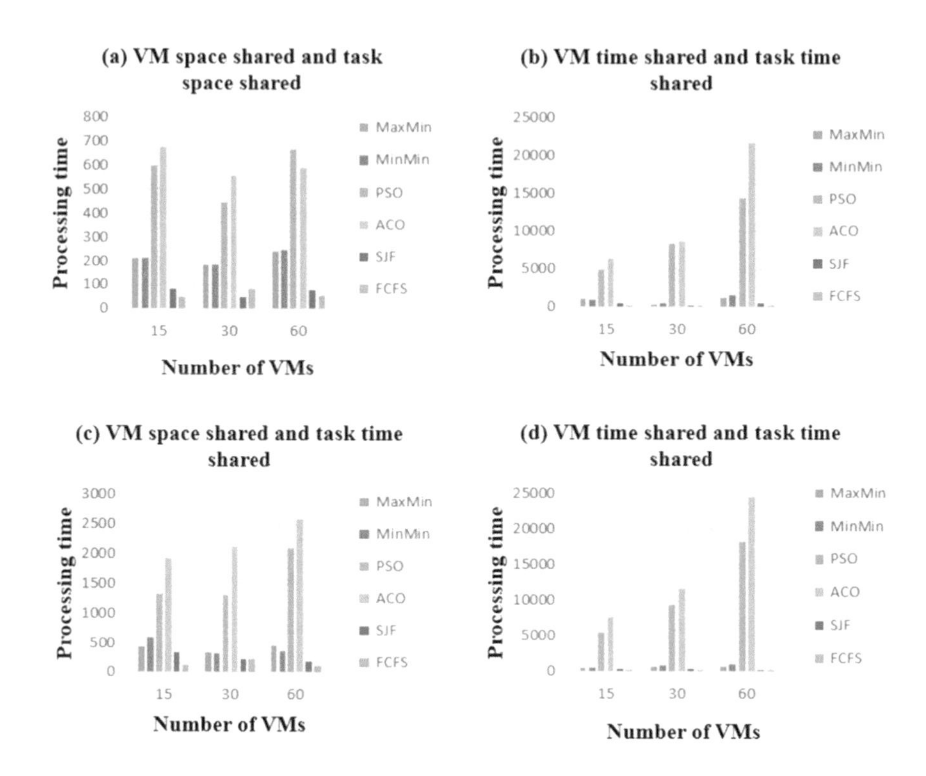

Fig. 3. Processing time performance for each scheduling policy in scenario 2

value was 48 ms for both FCFS (in the case of 15 VMs) and for SJF (in the case of 30 VMs). The lowest processing time values observed for data centers of size of 30 VMS were 446, 556, 184, and 182 ms, for PSO, ACO, MinMin, and MaxMin, respectively. The second type is when VMs shared the time and tasks shared the space in Fig. 3 (b), FCFS outperforms other algorithms in case of 15, 30, and 60 VMs. The small values for heuristic methods namely, FCFS, SJF, MaxMin, and MinMin in the case of 30 VMs were 94, 114, 344, and 430 ms, respectively. However, the processing time for PSO and ACO were 4854 and 6402 ms, respectively. In VMs space shared and tasks time shared in Fig. 3 (c), FCFS outperforms other algorithms in the cases of 15, 30, and 60 VMs. The small values for FCFS and SJF according to 60 VMs were 102 and 188 ms, whereas MinMin, MaxMin, and PSO showed the lowest processing time values in the case of 30 VMs which were 322, 332, and 1308 ms, respectively. In the last case as illustrated in Fig. 3 (d), the FCFS values outperform other algorithms in the cases 15, 30, and 60 VMs.

Figure 4 presents Scenario 3, where the number of PMS is varied.

The first policy is the VM space-shared and task space-shared presented in Fig. 4 (a), the processing time increases when the number of PMs increases,

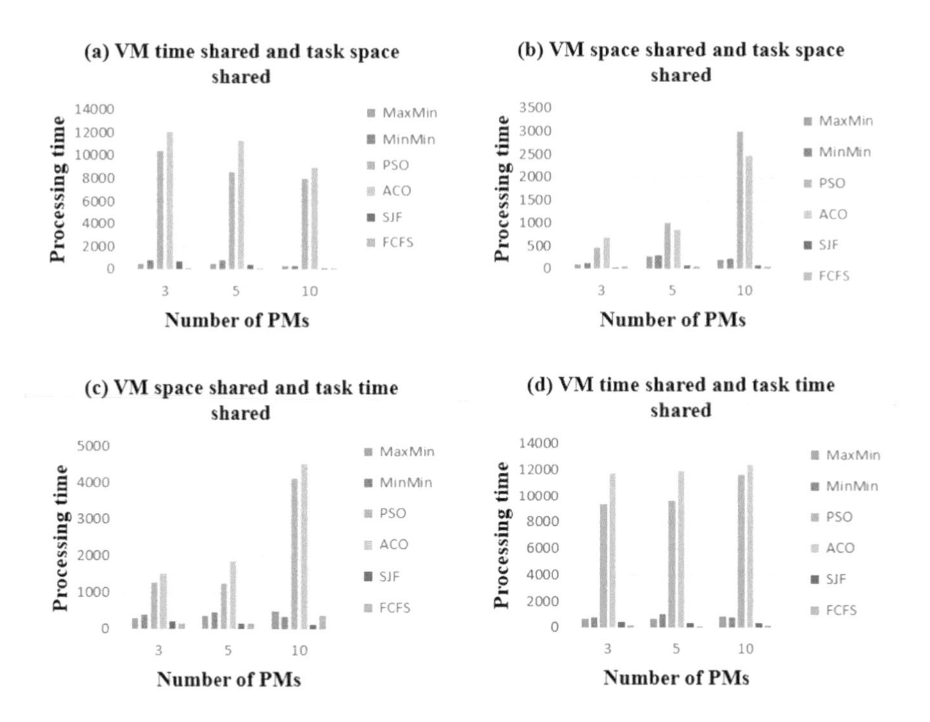

Fig. 4. Processing time performance for each scheduling policy in scenario 3

except for MinMin and MaxMin, which are higher in case of 5 PMs. SJF outperforms other algorithms in case of 3 PMs, and FCFS is better in cases of 5 and 10 PMs. In Fig. 4 (b), the processing time decrease when the PMs' number increases in PSO, ACO, and SJF. FCFS outperforms others in cases of 3 and 5 PMs, but SJF outperforms other algorithms in case of 10 PMs. Figure 4 (c) shows the case of VM space shared and tasks time shared where the highest value for the most algorithms was achieved in the case of 10 PMs, except for SJF and MinMIn where the highest values were obtained in the cases of 3 and 5 PMs, respectively. FCFS outperforms other algorithms in 5 and 3 PMs, and SJF showed better values in the cases of 5 and 10 PMs. The VM time shared and task time shared are presented in Fig. 4 (d), where FCFS outperforms other algorithms (cases of 3, 5 and 10 PMs). When the number of PMs increases, the processing time increase for PSO, ACO, and MaxMin in contrast to SJF, where the value of processing time decreases. The lower value was 100 ms in 5 PMs' scenario for FCFS, and the highest processing time was 12488 ms for the ACO algorithm in 10 PMs scenario.

Figure 5 shows the evaluation of the processing time, that occurred by several methods under various scheduling algorithms for scenario 4.

For space shared of VMs and time shared of tasks presented in Fig. 5(a), the lower value was 40 in 2 data centers for SJF and the highest value was 3272

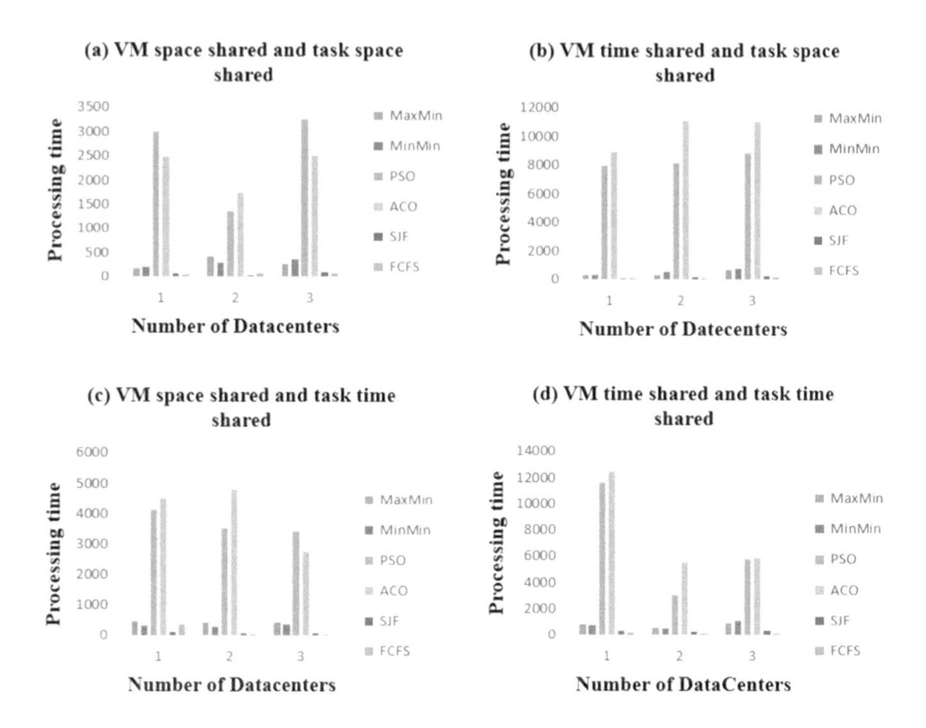

Fig. 5. Processing time performance for each scheduling policy in scenario 4

in the case of 3 datacenters for the ACO algorithm. We observed that FCFS outperforms other algorithms in 1 and 3 datacenters and SJF in 2 datacenters. Figure 5 (a) presents the case of VMs time shared and tasks space shared. In SJF, MinMin, and MaxMin, the processing time increases when the number of data centers augment. The lowest processing time was 30 ms in SJF, and the highest value was 11116 ms in 2 datacenters configuration. FCFS outperforms other algorithms in the cases 2 and 3 data centers and SJF in the case of 1 datacenter. Figure 5 (c) presents the case of VM space shared and tasks time shared, where the smaller value was 36 ms in the case of 2 datacenters for FCFS, and the highest value was 4804 ms in the case of 2 datacenters for ACO. FCFS outperforms other algorithms in the cases of 2 and 3 datacenters and SJF in 1 datacenter. The last case is VM time shared and task time shared in Figure 5 (d), the highest value was 12448 ms in the case of datacenter for ACO, and the lowest processing time was 78 ms in the case of 2 datacenters for FCFS. FCFS outperforms other algorithms in the cases of 1 and 3 datacenters and SJF in the case of 2 datacenters.

According to the Figures above, FCFS shows better performance when the size of datacenters increases, unlike JSF which is better in small and medium sizes. Also, VM space shared shows better results compared with time shared, especially for metaheuristic algorithms. Going through each scenario, we observe that the rate of processing time increases. The more we share the tasks with other components the more the processing time increases.

5 Conclusion and Future Work

In this paper, we compared traditional heuristic and some metaheuristic methods for scheduling problems, using processing time as a performance metric in cloud computing environment. We defined four scenarios and applied various scheduling types for VMs and tasks as constraints for these algorithms. In each scenario, we varied the number of components in the infrastructure, such as data centers, PMs, VMs, and tasks, in order to seek for the better algorithm according to processing time. Furthermore, we classified the algorithms through processing time using various types of scheduling policies. Thus, we determined more precisely that the space-shared scheduling approach performs better than time-shared scheduling policy. Moreover, we observed a significant difference between the heuristic and metaheuristic methods in processing time, especially FCFS and SJF, against the deployed scheduling policies in this study. In future work, we will address the scheduling issue with a variety of objectives and a combination of different algorithms. Moreover, we will take into consideration other performance parameters as well as the problem of energy consumption in the cloud data center.

References

1. El Kafhali, S., El Mir, I., Hanini, M.: Security threats, defense mechanisms, challenges, and future directions in cloud computing. Arch. Comput. Methods Eng. **29**(1), 223–246 (2022). https://doi.org/10.1007/s11831-021-09573-y
2. Mandal, R., Banerjee, S., Islam, M.B., Chatterjee, P., Biswas, U.: QoS and energy efficiency using green cloud computing. In: Ghosh, U., Chakraborty, C., Garg, L., Srivastava, G. (eds.) Intelligent Internet of Things for Healthcare and Industry. Internet of Things, pp. 287-305. Springer, Cham (2022). https://doi.org/10.1007/978-3-030-81473-1_14
3. El Kafhali, S., Salah, K.: Modeling and analysis of performance and energy consumption in cloud data centers. Arab. J. Sci. Eng. **43**(12), 7789–7802 (2018)
4. Tissir, N., El Kafhali, S., Aboutabit, N.: Cybersecurity management in cloud computing: semantic literature review and conceptual framework proposal. J. Reliable Intell. Environ. **7**(2), 69–84 (2021)
5. Chaurasia, N., Kumar, M., Chaudhry, R., Verma, O.P.: Comprehensive survey on energy-aware server consolidation techniques in cloud computing. J. Supercomput. **77**(10), 11682–11737 (2021). https://doi.org/10.1007/s11227-021-03760-1
6. Saadi, Y., El Kafhali, S.: Energy-efficient strategy for virtual machine consolidation in cloud environment. Soft. Comput. **24**(19), 14845–14859 (2020). https://doi.org/10.1007/s00500-020-04839-2
7. El Kafhali, S., El Mir, I., Salah, K., Hanini, M.: Dynamic scalability model for containerized cloud services. Arab. J. Sci. Eng. **45**(12), 10693–10708 (2020)

8. Hanini, M., Kafhali, S.E., Salah, K.: Dynamic VM allocation and traffic control to manage QoS and energy consumption in cloud computing environment. Int. J. Comput. Appl. Technol. **60**(4), 307–316 (2019)
9. Alworafi, M.A., Dhari, A., Al-Hashmi, A.A., Darem, A.B.: An improved SJF scheduling algorithm in cloud computing environment. In: 2016 International Conference on Electrical, Electronics, Communication, Computer and Optimization Techniques (ICEECCOT), pp. 208-212. IEEE (2016)
10. Salah, K., El Kafhali, S.: Performance modeling and analysis of hypoexponential network servers. Telecommun. Syst. **65**(4), 717–728 (2017). https://doi.org/10.1007/s11235-016-0262-3
11. Ziyath, S., Senthilkumar, S.: MHO: meta heuristic optimization applied task scheduling with load balancing technique for cloud infrastructure services. J. Ambient. Intell. Humaniz. Comput. **12**(6), 6629–6638 (2021)
12. Mapetu, J.P.B., Chen, Z., Kong, L.: Low-time complexity and low-cost binary particle swarm optimization algorithm for task scheduling and load balancing in cloud computing. Appl. Intell. **49**(9), 3308–3330 (2019). https://doi.org/10.1007/s10489-019-01448-x
13. Kumar, N., Mishra, A.: Comparative study of different heuristics algorithms in solving classical job shop scheduling problem. Mater. Today Proc. **22**, 1796–1802 (2020)
14. Liu, Z., Qiu, X., Zhang, N.: ACPEC: a resource management scheme based on ant colony algorithm for power edge computing. Secur. Commun. Netw. **2021**, 1–9 (2021)
15. Pirani, M., Ranpariya, D., Vaishnav, M.: A comparative review of CPU scheduling algorithms. Int. J. Sci. Res. Eng. Trends **7**(4), 2446–2452 (2021)
16. Ponto, R., Kecskeméti, G., Mann, Z.Á.: Comparison of workload consolidation algorithms for cloud data centers. Concurr. Comput. Pract. Exp. **33**(9), e6138 (2021)
17. Banerjee, C., Roy, A., Roy, A., Saha, A., De, A.K.: A time efficient threshold based ant colony system for cloud load balancing. In: Mandal, J.K., Mukhopadhyay, S., Dutta, P., Dasgupta, K. (eds.) CICBA 2018. CCIS, vol. 1030, pp. 206–219. Springer, Singapore (2019). https://doi.org/10.1007/978-981-13-8578-0_16
18. Arulkumar, V., Bhalaji, N.: Performance analysis of nature inspired load balancing algorithm in cloud environment. J. Ambient. Intell. Humaniz. Comput. **12**(3), 3735–3742 (2021)
19. Silva, M., Poss, M., Maculan, N.: Solution algorithms for minimizing the total tardiness with budgeted processing time uncertainty. Eur. J. Oper. Res. **283**(1), 70–82 (2020)
20. Jia, Z., Yan, J., Leung, J.Y., Li, K., Chen, H.: Ant colony optimization algorithm for scheduling jobs with fuzzy processing time on parallel batch machines with different capacities. Appl. Soft Comput. **75**, 548–561 (2019)
21. Hanini, M., El Kafhali, S.: Cloud computing performance evaluation under dynamic resource utilization and traffic control. In Proceedings of the 2nd International Conference on Big Data, Cloud and Applications, pp. 1-6. ACM (2017)
22. Kodli, S., Terdal, S.: Hybrid max-min genetic algorithm for load balancing and task scheduling in cloud environment. Int J Intell Eng Syst. **14**(1), 63–71 (2021)
23. Ouammou, A., Tahar, A.B., Hanini, M., El Kafhali, S.: Modeling and analysis of quality of service and energy consumption in cloud environment. Int. J. Comput. Inf. Syst. Ind. Manag. Appl. **10**, 098–106 (2018)
24. Sidhu, H.S.: Comparative analysis of scheduling algorithms of Cloudsim in cloud computing. Int. J. Comput. Appl. **97**(16), 8887 (2014)

Business - IT Alignment in Cloud Environment Comparative Study

Manal Bouacha and Hanaa Sbai(✉)

Laboratory of Mathematics, Computer Science and Applications (LMCSA), University Hassan II of Casablanca, Mohammedia, Morocco
sbai.hanae87@gmail.com

Abstract. In recent years, much research has focused on Business Process reuse to reduce the cost and time of process development. However, in a competitive market where business services are provided in a cloud environment, there is a need to develop non-installable applications or so-called "multi-tenant" applications. BPM practitioners and researchers are involved in improving the BPaaS (Business Process as Service) model, which today defines business process management in the context of the cloud. However, the problem of aligning IT operations in the cloud, bridging the gap between business requirements and technical cloud solutions, is appearing as an emerging issue in digital transformation. This is because cloud solutions are described technically, making it difficult for business stakeholders, who typically understand business language, to properly evaluate the most appropriate cloud solution. Identifying the most appropriate cloud solutions in this context, therefore, requires specifying the requirements and capabilities of service, both at the business and IT levels. In this paper, we present a comparative study to evaluate approaches that explore the problem of IT-Business alignment in the cloud to analyze how the alignment has been modeled and then extract the limitations and weaknesses of each approach, to propose an approach that will be able to address most of these critical points.

Keywords: Business - IT alignment · Business Process (BP) · Business Process Management (BPM) · Business Process as a Service (BPaaS) · Cloud computing · process mining

1 Introduction

Nowadays, with the lack of control and flexibility as well as the automation of processes in data-centric information systems, companies are increasingly motivated and driven to adopt a new generation of process-aware information systems, known as PAIS (Process-Aware Information System) where the main unit of IS is the business process model.

In parallel, the cloud computing paradigm has emerged and evolved from its initial computing-centric service models (IaaS, PaaS, SaaS) [18, 19] to more fine-grained, multi-dimensional service models [20] that encompass, among other things, the storage and network dimension by becoming an important driver of business models and innovation.

M. Lazaar et al. (Eds.): BDIoT 2022, LNNS 625, pp. 212–224, 2023.
https://doi.org/10.1007/978-3-031-28387-1_19

The Cloud computing adopters were originally IT-oriented companies, while SaaS allows non-technology companies to benefit from the cloud as well. However, SaaS offerings are generally inflexible, and their providers expect them to be one-size-fits-all solutions for their customers, often ignoring the fact that they must integrate with users' existing business processes. To overcome this inflexibility, business processes are now seen as an appropriate instrument to manage the digitization of companies.

Outsourcing all or part of business processes is a common approach to reduce costs and increase flexibility [2]. New and innovative cloud service models are needed that support non-IT customers with a business-oriented view [21] and thus support existing processes. In addition, the proposed solutions must be able to grow (and shrink) with the size of the business.

The basis for determining IT-Business alignment in the cloud environment is the idea of Business Process as Service (BPaaS), which today describes business process management in the context of the cloud, where conceptual models and semantics are applied to align business processes with workflows deployed in the multi-cloud.

BPaaS is a business process-based concept. A business process consists of a set of activities that serve the specific objectives of an organization [22]. While activities perform operations that are intended to produce the results desired by an organization. Therefore, the idea of BPaaS is not to focus on another combination of cloud services, but to support the design of a domain-specific service and its alignment with IT services [2].

However, the problem of IT-Business alignment in the cloud and in particular the alignment of the BPaaS layer with cloud services (IaaS, PaaS, or SaaS) is addressed in a single modeling/architecture which is the CloudSocket approach. This approach, although has enabled the alignment of IT-Business with Cloud services, the methods used in this approach are still informal; automation is absent, business process modeling is still from scratch, and service selection is done manually by a broker… And synchronization between services and the BPaaS layer is managed manually, which is an error-prone task.

In this context, our study consists in analyzing the different points of view on the existing approach and comparing its limitations and strengths to propose a more complete approach and then respond to the observed missing needs.

The recap of this paper is organized as follows. In Sect. 2, we summarize some basic concepts on BPM, Process Mining, BPaaS, and Business-IT alignment in the cloud. In Sect. 3, we present an overview of related work; we mainly discuss the proposed approaches to Business-IT alignment in a traditional non-cloud approach and then in a cloud environment (CloudSocket architecture). Section 4 introduces the different criteria used in our comparative study and presents and discusses our results. Then, in Sect. 5, we present our proposed framework to address the observed limitations, and finally, Sect. 6 concludes the paper with guidelines for future research.

2 Background and Foundation

In this section, we present the three basic concepts related to our work: BPM, Business Process as a Service, IT-Business Alignment.

2.1 Business Process Management

Classical Approach. A business process is a sequence of activities designed to achieve specific organizational objectives by aligning them with all actions performed within the enterprise and ensuring that resources are optimally used, regardless of who performs them.

Business process modeling consists of structuring and representing the activities of an organization using standard modeling languages - such as BPMN (Business process model and notation) - and based on the designer's opinion and experience to meet the customer's needs to the highest level.

Process Mining Based Approach. Process mining is a technical suite of methods that link IS domains to business process management (BPM) to support business process analysis based on automatic process discovery from event logs. The objective of process mining is to transform event data into information and actions.

Process mining is considered the second derivation of data mining that focuses on log-based business processes. This concept aims to analyze business processes through the data that runs through them by analyzing business process data, independently of each software.

The three basic types of process mining are explained in terms of input and output: (a) discovery, (b) compliance checking, and (c) improvement. Process mining techniques are capable of extracting knowledge from the event logs commonly available in today's information systems [7].

2.2 Business Process as a Service (BPaaS)

Business Process as a Service (BPaaS) introduces a cloud-based service model for BP. This is a new type of cloud service that delivers configurable and executable business processes to customers over the Internet [6]. It sits on top of the other three fundamental cloud services: SaaS, PaaS, and IaaS. Typically, BPaaS is defined as workflows in which Workflows are seen as orchestrations of IT services in the cloud [3].

The BPaaS layer comes to define business process management (BPM) in a cloud environment. Thus, as a cloud-oriented approach, the size of the IT infrastructure can grow and shrink as needed, and the potential third-party services used to implement and automate a business process are not necessarily statically linked [4].

2.3 Business- IT Alignment in Cloud Environment

Business –IT Alims to use IT in the most efficient way possible to achieve business objectives and improve financial performance or market competitiveness [5].

In addition, IT-Business alignment enables domain-driven controls to be transformed into workflow-driven technology by leveraging automation [4].

Business-IT alignment is a concept that already existed before cloud computing and several research works have implemented solutions and frameworks that support IT-Business alignment in non-cloud contexts [8, 15–17], it includes the mapping of BPs to workflows. Workflows orchestrate the services that need to be selected and allocated. In other words, IT business alignment involves mapping IT layer services to business layer processes to maintain consistency between the two and to facilitate change management.

In a cloud environment, BPaaS implementation requires flexibility for many steps in business-IT alignment [4].

IT-Business alignment to transform a domain-specific BP into a cloud-compatible BP is applied in five levels of transformation: Level I (business process), Level II (technical workflows), Level III (executable workflows), Level IV (deployed workflows), Level V (Instance workflows) [2, 4].

3 Related Works

In a classical framework and a traditional non-cloud approach [8, 15–17], research around PAIS has highlighted the need to align business processes and the application layer, to manage the traceability of business requirements expressed at the business layer and their realization as a service in the application layer.

Thus, alignment enables the implementation of business processes as a service while developing collaborative solutions, i.e., solutions that involve both the business and IT layers.

This is the work that was deconstructed and implemented in this thesis [8, 15–17], where the authors were able to establish an IT-Business alignment that aligns the BPMN models with the application layer (web services), by establishing mapping rules that map the different elements of the BPMN model as the source modeling language and the elements of SOAML as the target modeling language.

In a cloud environment, all current research that addresses the alignment of the BPaaS layer with the lower layers (IaaS, PaaS, and SaaS) is based on the CloudSocket architecture [2–4, 9]. As mentioned in [2], CloudSocket considers BP management as a holistic approach by considering design, assignment, execution, operation, and evaluation as a continuous management lifecycle. The implementation of the CloudSocket approach in this work [2] also includes vertical integration using business-IT alignment and introduces deployment aspects (e.g., country-specific restrictions, business rules, strategies, preferred payment model) into the business process and IT alignment.

Even if the approach is the same (CloudSocket), the angles of treatment differ from one to another given the richness of this architecture whether it is at the level of modeling and implementation languages or functional aspects and business requirements. Each study found focuses on one of the environments defined by the CloudSocket architecture: I - Design environment [3, 10, 11], I - Allocation environment [12, 13], III - Execution environment [12, 13], IV - Evaluation environment [14], V - MarketPlace.

Existing works, while addressing the problem of IT-Business alignment in the context of the Cloud, but their limitations are also clear and require additional work to evolve and improve the alignment approach, including automating the approach and proposed services using the event log-based discovery process, and then making the approach formal by applying the MDA (Model Driven Architecture).

4 Comparative Study

In this section, we will first define the different comparison criteria on which we have based our study, and then we will group all the comparisons made according to these criteria in a comparative table and discuss it.

4.1 Comparison Criteria

As mentioned earlier, all existing studies that address the issue of alignment between IT and business process are based on the CloudSocket architecture as a mapping of BPMS (Business Process Management System: Design, Modeling, Execution, Monitoring, and Optimization) to BPaaS environments, while projecting the different predefined BPaaS levels (standard levels) onto the different environments of the CloudSocket architecture.

BPaaS Environment. This criterion presents the BPaaS environment of the Cloud-Socket architecture that the work studied addresses by mapping BPaaS levels to the BPaaS environments. The BPaaS environments as defined above are I - Design environment, II - Allocation environment, III - Execution environment, IV - Evaluation environment, V - MarketPlace.

Alignment Approach. It allows us to classify the approach used in the study into two categories: Formal Approach or Informal Approach. An informal alignment approach is defined as a method that is not applied in any formal alignment framework or approach. This allows us to distinguish the methods and to think about implementing the proposed solutions in a formal framework if this is not the case; optimize and innovate the existing ones.

Type of Mining Process. It specifies the type of process mining domain (discovery, conformance, or enhancement) used in the work studied. This allows us to evaluate the use of process mining types and their utilities in the study.

Type of Mapping. This criterion is considered the most important because it defines the mapping approaches used to achieve alignment between models of different BPaaS levels as well as between the BPaaS layer and lower layers (cloud services). The mapping can be horizontal (between models of the same level) or vertical (from one level to another in a top-down or bottom-up manner).

Modeling Language. It determines the modeling language(s) used in the work. In the design environment, the modeling languages used are BPMN and DMN (BPMN + Ontology) and their extensions. Otherwise, for the other environments, the modeling language most used in the CloudSocket approach is the CAMEL language and its extensions.

Tools. Indicates the tool used to implement the proposed solution. This could allow us to propose more innovative tools in our prototype.

4.2 Comparison Results

The following table contains the results of the comparative study carried out concerning the various articles found and studied on our problem.

Certainly, this research subject is quite new and the works that deal with it can be counted on our fingers, but this does not prevent us from saying that the study carried out was very beneficial and enriching in the way that we were able to extract several elements in common between the approaches and thus make a global criticism of the existing and find the limits on which we can work to improve them (Table 1).

BPaaS Environment. Given the richness and complexity of the alignment process used in each environment, most works dealing with the CloudSocket approach focus on one or two levels to deepen the research on it and present more complete alignment realizations.

However, some articles deal with the whole CloudSocket architecture, without detailing the technical aspect of each environment, to present a global explanation of the CloudSocket approach and introduce its different stages as well as the proposed modeling languages.

Type of Mining Process. The presence of process mining methods is almost absent, especially in the first 3 environments. In the evaluation environment, we can read in [9] that the use of the "Conformance" part of process mining is a perspective on which studies are still in progress. Otherwise, the work [14] mainly uses the Evaluation Ontology of which process mining is an essential component, especially for the discovery and evaluation part, as well as the use of logs and the process mining algorithms.

Type of Mapping. The type of mapping differs from one environment to another, but if we compare work that focuses on the same levels, we can find commonalities between several approaches, including their alignment approaches and the mapping methods used.

Modeling Language. The modeling languages used in the implementation of the alignments designed by the CloudSocket approach are numerous and there this criterion allows to distinguish well the studied works. The most used languages, which can be found in all the approaches read, are BPMN, CAMEL, and their extensions.

Table 1. Comparative Study Results.

	Article	BPaaS Environment	Alignment approach		Type of mining process			Type of mapping	Modeling languages	Tools
			Formel	Informel	Discovery	Conofrmance	Enhancement			
CloudSocket Approach	[2]	I II III IV	Semantic Lifting Ontology	Yes	No	No	No	Smart Business-IT alignment (Ontology) DMN to CAMEL Mapping	DMN CAMEL SRL OWL-Q	DevOps tools Eclipse EMF Eclipse XText Eclipse RAP OCL CAMEL editors ADOxx REST API Semantic KB
	[3]	I	FODA approach (SD)	Yes	No	No	No	Horizontal BPMN and DMN Vertical BPMN cloud-specific Enrichment Vertical Alignment with Semantic Lifting	FDMM BPMN BPD CD CM DDM WE DMN DRD STF STM	ADOxx INTERREFS OMS ArchiMEO REST services
	[4]	I II III IV	No	Yes	No	No	No	–	BPMN DMN SML CloudML CAMEL CAMP TOSCA	Adonis ADOxx Bonita BPM Camunda modeler DevOps Tools Juju COTs Brooklyn Scalr Cloudiator InfluxDB

(continued)

Table 1. *(continued)*

Article	BPaaS Environment	Alignment approach		Type of mining process			Type of mapping	Modeling languages	Tools
		Formel	Informel	Discovery	Conofrmance	Enhancement			
[9]	I II III IV	Semantic Lifting FODA approach	Yes	No	Yes	No	Smart Business-IT alignment. DMN to CAMEL Mapping	UML BPMN RDF DMN OWL-Q CAMEL	REST Services ADOScript Javascript MFB Web-GUI
[11]	I	Semantic Lifting	Yes	No	No	No	Horizontal BPMN and DMN & KPI Weaving Vertical BPMN cloud-specific Enrichment Vertical Alignment with Semantic Lifting BPaaS Alignment Model as a Packaging Role	BPMN RDF OWL-Q ArchiMEO -Ontology ArchiMate UML DMN FDMM BPD, CM, DDM, WE, DRD, CEM, INDM, SDM, STM, BPAM SPARQL SPIN FEEL XML XSLT	ADOxx OMiLAB TOGAF SEC framework CCM framework INTERREF

(continued)

Table 1. *(continued)*

	Article	BPaaS Environment	Alignment approach		Type of mining process			Type of mapping	Modeling languages	Tools
			Formel	Informel	Discovery	Conofrmance	Enhancement			
	[12]	II III	FORTH Approach	Yes	No	No	No	DMN to CAMEL Mapping	BPMN CAMEL OWL-Q WSMO USDL TOSCA CloudMLSoaML WSBPEL SPARQL SLA languages	DevOps tools COAPS API ADOxx PaaS-API REST-ful API PaaSHopper Cloudiator Apache Brooklyn OpenStack Heat Scalr Stratos Brooklyn Cloudify CLOUDQUAL SMI index
	[13]	II III	FORTH's framework UULM's framework	Yes	No	No	No	DMN to CAMEL Mapping CAMEL to PaaS/SaaS CAMEL extension	CAMEL PaaS/SaaS CAMEL extension Q-SLA OWL-Q OWL-S DMN	Tomcat DevOps tools Java REST API REST SDS service UI component ADOxx Cloudiator VM PaaS Unified -Library (PUL) COAPS API Heroku OpenShift v2 CloudFoundry v2

(continued)

Table 1. *(continued)*

	Article	BPaaS Environment	Alignment approach		Type of mining process			Type of mapping	Modeling languages	Tools
			Formel	Informel	Discovery	Conofrmance	Enhancement			
	[14]	IV	No	Yes	Yes	Yes	No	SQL queries, OLAP and event-based metric formula calculation to WSML rules and SPARQL queries OWL-Q to SPARQL Transformation	BPMN TOSCA CAMEL OWL-Q Score-ML WS-QoSPolicy UML Q-SLA SPARQL PMDM OLAP SQL queries	HIGO framwork REST APIs TSDBs iBOM platform FAPE CDOSim SelCSP QuARAM ProM framework

5 Proposed Framework

The thorough reading of existing works as well as the comparative study carried out allowed us to understand the limitations, we need to work on to create a more innovative and improved IT-Business alignment solution in the Cloud that meets all the needs raised by our study.

In this sense, the framework we propose is as follows (Fig. 1):

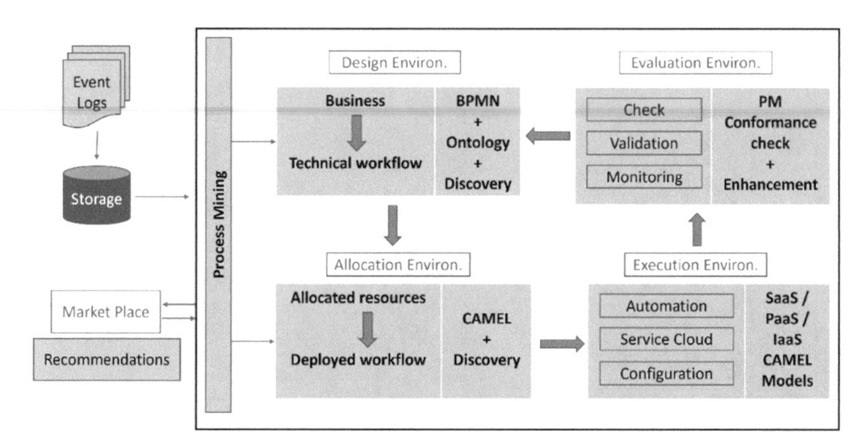

Fig. 1. Proposed framework for business-IT alignment in a cloud environment.

The framework we propose is based on using the discovery part of process mining, especially the use of event logs and service automation.

We propose to keep the same CloudSocket architecture with its different BPaaS environments but apply MDA to make the alignment approach more formal and well-designed.

6 Conclusion

The problem of alignment between IT and business has become the main focus of many current research studies. The modeling of business processes in the cloud computing environment and the application of the BPaaS layer make establishing the alignment between the different layers of cloud computing as well as its services with the business layer a primary need and a hot topic.

In this paper, we examine different approaches to address the IT-business alignment problem, especially in the context of cloud computing. We found that most existing work is based on the CloudSocket architecture which lacks automation, selection, and automatic synchronization of services with the BPaaS layer.

We are currently working on validating the proposed framework, and from a perspective, we intend to improve the alignment approach proposed by the CloudSocket architecture and make it more formal using MDA.

References

1. Woitsch, R.: Bpaas modelling: business and IT-cloud alignment based on adoxx. Modellierung 2016 Workshopband, In: Informatics (LNI), Gesellschaft für Informatik, 141, Bonn (2016)
2. Griesinger, F., et al.: BPaaS in multi-cloud environments - the cloudsocket approach. In: European Space Projects: Developments, Implementations, and Impacts in a Changing World, pp. 50–74, EPS Porto (2017)
3. Woitsch, R., Hinkelmann, K., Juan Ferrer, A.M., Yuste, J.I.: Business process as a service (BPaaS): the smart BPaaS design environment. Paper Presented at the Advanced Information Systems Engineering Workshops: CAiSE 2016, Ljubljana, Slovenia (2016)
4. Domaschka, J., Griesinger, F., Seybold, D., Wesner, S.: A cloud-driven view on business process as a service. In: Proceedings of the 7th International Conference on Cloud Computing and Services Science, pp. 739–746. CLOSER (2017)
5. Woitsch, R., Karagiannis, D., Plexousakis, D., Hinkelmann, K.: Business and IT Alignment: The IT-Socket. In: e & i Elektrotechnik und Informationstechnik (2009)
6. Chiranjeevi, M., Jaya Krishna, K.: An efficient BPaaS modeling approach for three steps configuration and transactional behavior verification process. Pramana Res. J. **9**(4), 868–875 (2019)
7. van der Aalst, W., et al.: Process mining manifesto. In: Daniel, F., Barkaoui, K., Dustdar, S. (eds.) BPM 2011. LNBIP, vol. 99, pp. 169–194. Springer, Heidelberg (2012). https://doi.org/10.1007/978-3-642-28108-2_19
8. Sbai, H.: PAIS (process aware information system) orienté services: modélisation et evolution processus configurables (2015)
9. CloudSocket Project, D4.5 Final CloudSocket Architecture, https://www.cloudsocket.eu/deliverables. Accessed 30 Sept 2016
10. CloudSocket Project, D3.2 Modelling Prototypes for BPaaS. https://www.cloudsocket.eu/deliverables. Accessed 31 Aug 2016
11. CloudSocket Project, D3.1 Modelling Framework for BPaaS. https://www.cloudsocket.eu/deliverables. Accessed 31 Aug 2016
12. CloudSocket Project, D3.3 BPaaS Allocation and Execution Environment Blueprints. https://www.cloudsocket.eu/deliverables. Accessed 31 Aug 2016
13. CloudSocket Project, D3.4 BPaaS Allocation and Execution Environment Prototypes. https://www.cloudsocket.eu/deliverables. Accessed 31 Dec 2016
14. CloudSocket Project, D3.5 BPaaS Monitoring and Evaluation Blueprints. https://www.cloudsocket.eu/deliverables, accessed 31 Dec 2016
15. Sbai, H., Fredj, M., Chakir, B.: Generating services supporting variability from configurable process model. J. Teor. Appl. Inf. Technol. **72**(2), 111–124 (2015)
16. Sbai, H., Fredj, M., Kjiri, L.: A pattern-based methodology for evolution management in business process reuse. IJCSI Int. J. Comput. Sci. Issues **11**(1), 211–220 (2014)
17. Faquih, L., Sbai, H., Fredj, M.: Semantic variability modeling in business processes: a comparative study. In: 9th International Conference for Internet Technology and Secured Transactions (ICITST-2014), pp. 131–136, 8–10 December 2014, London, UK (2014)
18. Mell, P., Grance, T.: Draft NIST Working Definition of Cloud Computing (2009)
19. Foster, I., Zhao, Y., Raicu, I., Lu, S.: Cloud computing and grid computing 360-degree compared. In Grid Computing Environments Workshop (GCE 2008), Austin, Texas, USA, pp. 1–10 (2008)
20. Kachele, S., Spann, C., Hauck, F. J., and Domaschka, J.: Beyond iaas and paas: An extended cloud taxonomy for computation, storage and networking. In UCC, 2013 IEEE/ACM 6th International Conference (2013)

21. Smith, F., Missikoff, M., Proietti, M.: Ontology-based querying of composite services. In: National Research Council, IASI "Antonio Ruberti", Viale Manzoni 30, 00185 Rome, Italy (2012)
22. Leymann, F., Roller, D.: Production Workflow: Concepts and Techniques. Prentice Hall, Upper Saddle River (2000)
23. Woitsch, R., Utz, W.: Business process as a service (BPaaS). In: 14th Conference on e-Business, e-Services and e-Society (I3E), Delft, Netherlands, pp. 435–440 (2015)
24. Taher, Y., Haque, R., Heuvel, W., Finance, B.: BPaaS - a customizable BPaaS on the cloud. In: Proceedings of the 3rd International Conference on Cloud Computing and Services Science (CLOSER-2013), pp. 290–296 (2013)

Reduce Cheating in e-Exams Using Machine Learning: State of the Art

Ilhame Khabbachi(✉), Abdelhamid Zouhair, Aziz Mahboub, and Nihad Elghouch

DSAI2S Research Team, Faculty of Sciences and Technologies, Abdelmalek Essaâdi University (UAE), Tetouan, Morocco
ilhame.khabbachi@gmail.com

Abstract. Cheating on online exams becomes a black spot in distance learning environments. On the one hand, it threatens the credibility of these exams by violating the principle of equality and success on merit. On the other hand, it also has negative repercussions on the reputation of the institutions. Without a doubt, in the Covid-19 health crisis and following the recommendations of the World Health Organization to respect social distancing, the majority of establishments have adopted the distance learning system, including online exams. However, the difficulty of monitoring learner activity in remote settings characterizes this type of assessment by inequity. In practice, each establishment has relied on a monitoring solution adapted according to certain criteria in order to guarantee a fair passage of the exams and to control them well. AI-assisted proctoring tools add a layer of protection to online exams. In this article we will discuss and compare the different uses of Artificial Intelligence tools to reduce cheating in online exams, based on the use of Machine Learning techniques.

Keywords: Machine Learning · supervised learning · unsupervised learning · reinforcement learning · e-exam · e-learning · cheating · Covid19

1 Introduction

No one can deny the major importance of remote (online) learning in teaching due to the development of information and communication technologies and the Internet, this technique has long been used by universities in parallel with the birth of the e-learning concept. Indeed, following the COVID 19 health crisis, several universities around the world have resorted to adopting distance education in an emergency as a substitute for face-to-face mode, which requires ensuring the integrity of exams in online environments. Guaranteeing fair and secure passage of exams is becoming a major challenge for many educational institutions.

In this regard, various remote exam solutions are now available and can solve this problem. These methods must be tested and evaluated according to their cost, their security, their resistance to fraud, the stress induced for the students and the conditions of implementation. [1].

M. Lazaar et al. (Eds.): BDIoT 2022, LNNS 625, pp. 225–238, 2023.
https://doi.org/10.1007/978-3-031-28387-1_20

Many of these institutions outsource the exam aspect of their course to online proctoring service providers who can hold the exams close to where the student lives. [1].

Hylton, Levy and Dringus (2016) state that « deception and dishonesty in online exams are believed to be related to their unsupervised nature where users appear to have the opportunity to collaborate or use unauthorized resources during these evaluations» the main objective of their study was to investigate the deterrent effect of webcam surveillance on misconduct during online exams [2].

Today the evolution of technological advances in the field of artificial intelligence, particularly in Machine Learning and Deep Learning, allows a renewal of these traditional monitoring methods and poses certain scientific solutions to obtain developed and more effective results in order to reduce the risk of cheating in remote exams and ensure an impartial and protected environment. And this is how this article was designed to study and compare the different solutions proposed in the context of monitoring online exams against monitoring tools assisted by Artificial Intelligence.

The rest of this paper is organized as follows: we discuss the different approaches of machine learning indicating the three general types of ML (Supervised learning, Unsupervised learning and Reinforcement learning), then, in Sect. 3 we present a paragraph that discusses distance exams and the extent of its dependence in distance learning especially in the Covid19 period, as well as the cheating phenomenon in this type of exams, In Sect. 4 we discuss in detail the phenomenon of cheating, then in the next section we talk about the studies that have focused on the fight against cheating in distance exams, in Sect. 6 we propose a comparative study between the algorithms and techniques used for the detection of cheating in distance examinations and finally in Sect. 7 we present the conclusion and our future work.

2 Machine Learning

Nowadays automatic learning or Apprentissage Automatique in French has become an essential thing. According to its inventor ARTHUR Samuel (1954), it is defined as follows: «Machine Learning is the science of giving a machine the ability to learn, without programming it explicitly».

Its objective is to define methods and algorithms making it possible to extract relevant information from data or to learn a behavior from the observation of a phenomenon. It allows computers to use previously collected data to predict behaviors, outcomes and future developments/trends. It is considered a field of study of artificial intelligence (AI) where predictions are made from learning techniques. It is characterized by a set of rules used to solve data processing and analysis problems, mathematical calculation or automated deduction. In several areas of interest, these techniques can make apps and devices smarter. For example, when we shop online, machine learning can recommend other products that may be of interest to us based on our purchase history. Or when a credit card is used to make a transaction, machine learning compares the outstanding transaction to a database of transactions and helps the bank detect fraud. As well as for pattern recognition, anomaly detection, facial detection and many other applications [3].

At a general level, learning algorithms can be classified into large families among which the main ones are: supervised learning, unsupervised learning, reinforcement learning, as shown in Fig. 1. In the following, we briefly define each type of learning.

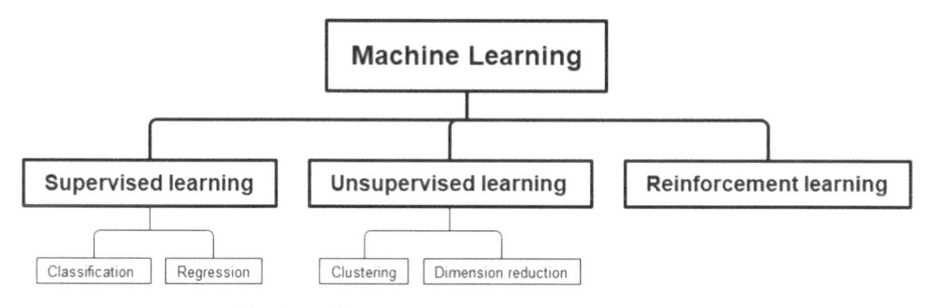

Fig. 1. Different types of machine learning [43].

- Supervised learning: supervised learning is generally the task of machine learning to learn a function that maps inputs to outputs based on samples of input-output pairs [43], it is the system that provides both the input data and the expected output data, these data are labeled for classification by a person expert in his field in order to establish a learning base for further processing of data [4]. In supervised learning, the machine already knows the answers it is expected to give. The most common supervised tasks are "classification" which separates the data and "regression" which matches the data [43].

The Fig. 2 below in the form of Mind Mapping made in EdrawMind that summarize the different approaches to Supervised learning including the algorithms adapted for each type of problem, the areas of their uses, and the types of data.

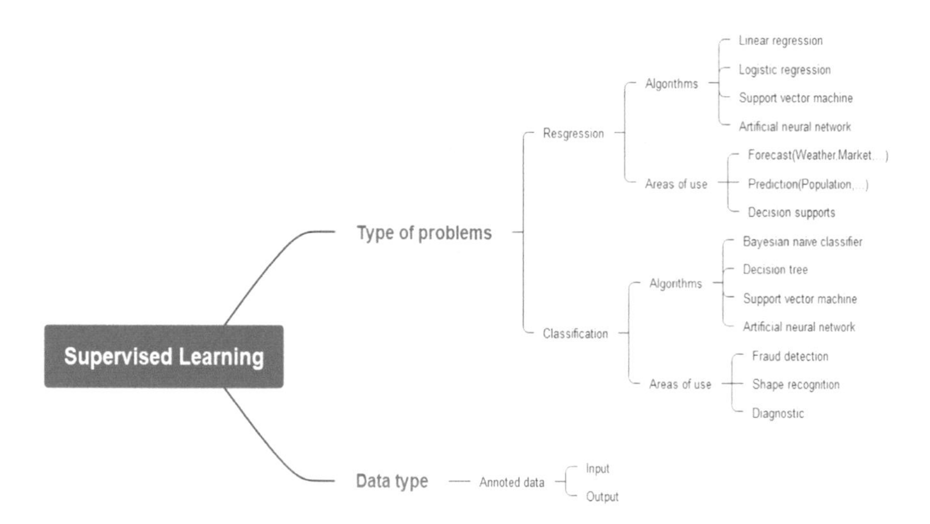

Fig. 2. Different approaches to Supervised learning [44].

- Unsupervised learning: unsupervised learning analyzes unlabeled data sets without human intervention, where only data is available and where the model will highlight a more or less hidden structure present in this data [5]. The most common unsupervised learning tasks are clustering, density estimation, feature learning, dimensionality reduction, association rule finding, anomaly detection, etc. [43].

The Fig. 3 below in the form of Mind Mapping made in EdrawMind that summarize the different approaches to Unsupervised learning including the algorithms adapted for each type of problem, the areas of their uses, and the types of data.

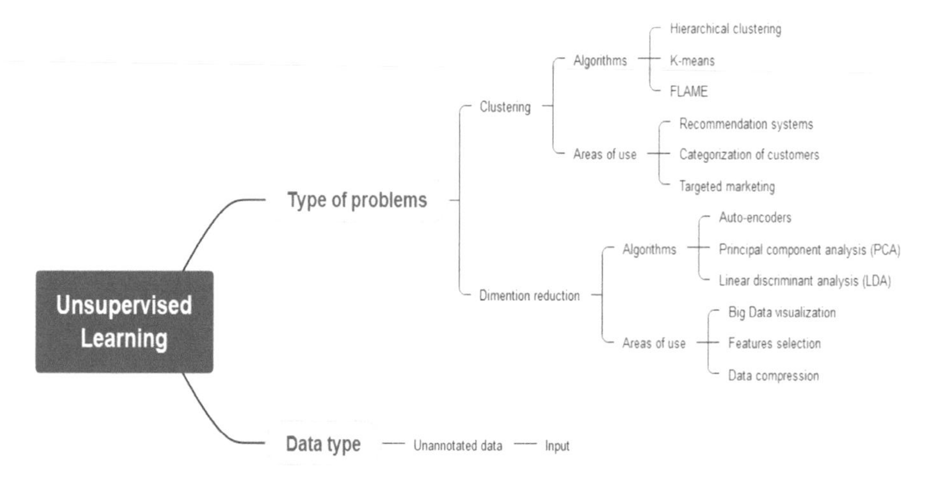

Fig. 3. Different approaches to Unsupervised learning [44].

- Reinforcement learning: specifies a set of methods that allow agents to learn to choose actions to take autonomously. Immersed in a given environment, the agent learns by receiving rewards or punishments based on their actions. Using his experience, the agent tries to find the optimal decision-making strategy that allows him to maximize the accumulated reward over time [6]. It is a powerful tool for training AI models that can help increase automation or optimize the operational efficiency of sophisticated systems such as robotics, autonomous driving tasks, manufacturing, and supply chain logistics, but it is not best used to solve basic or simple problems [43].

The Fig. 4 below in the form of Mind Mapping made in EdrawMind that summarize the different approaches to Supervised learning including the algorithms adapted for each type of problem, the areas of their uses, and the types of data.

.

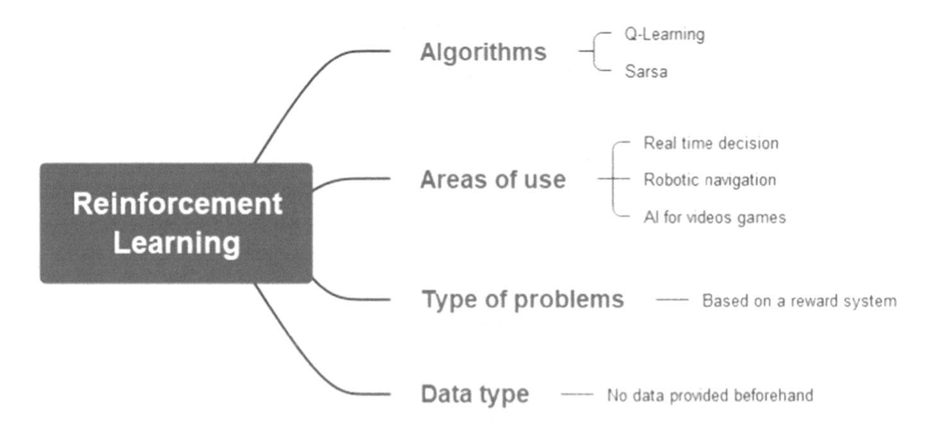

Fig. 4. Different approaches to Reinforcement learning [44].

It should be noted that there is a wide variety of Machine Learning algorithms. Some are more commonly used than others, depending on the type of problem.

3 Online Exams

Since the end of the 20th century the new e-learning concept has started to acquire a certain popularity in academic circles and among training professionals [12]. This period saw the birth of an entire industry that specializes in the creation of technological platforms for managing educational content, called Learning Management System (LMS) such as Moodle and Skaia. And since every aspect of education requires some kind of assessment; the assessment of distance learning on online platforms (such as Lms) can enable individual and collective activities, however, these activities are not supervised. For examinations, conditions and identity verification must be as rigorous as if the test were carried out at the establishment [13]. And at the disposal of all kinds of technological tools, there are many methods of cheating at online spaces, It could have been collaborative fraud via discussion groups with classmates, on platforms like Discord or WhatsApp. Others have outright paid more advanced candidates to pass the exams for them. Or, more simply, took their exams online with a second computer with internet access [11].

Online exams have begun to become the preferred method of assessment in online and even traditional learning environments. When used appropriately in an e-learning program, they offer various benefits to the learning process and to the learner [21]. It should be noted that the COVID-19 pandemic has also led to major changes in educational assessment worldwide, requiring that all assessments be conducted by online methods [22].

Online exams are an integral part of the e-Learning solution for fair and equitable evaluation of student performance. Designing and executing online exams is the most difficult aspect of e-Learning. Indeed, online exams are often administered on online learning platforms, where students and teachers do not physically exist in the same place [21]. This creates integrity and security loopholes, the thing that will increase the possibility of cheating and violating the principle of equality between learners during these online exams.

In this respect, many researchers have been interested in this subject and have proposed different solutions, including the use of software capable of proving identity through facial recognition, regular recording of students to analyze suspicious movements, and even allows for the exploration of the room where students take the exam [23]. While other institutions have made e-proctoring tools available to their educational staff to support these online exams. E-proctoring tools facilitate remote monitoring of the process using telematics resources [24][22].

Several attributes are associated when conducting an online exam. For example, an important attribute is candidate verification/authentication during online exams [21].

These attributes can be classified based on the various studies that target online reviews into four important classes (categories): 1) Verification and abnormal candidate behavior are very important features of online testing. There are two types of verification [25], static and continuous. In static verification, candidates are verified only once at the beginning of the online exam. In continuous verification, candidate authentication/verification is performed continuously after a certain period of time during the online exam. Thus, preventing cheating by detecting abnormal behavior is important to ensure the fairness of online exams. 2) The security of online examinations is an important feature that ensures unauthorized access to the various components of the system (for example, user management, question banks, etc.). 3) Automatic generation of question banks and evaluation of candidates' answers are very important in online examinations. 4) Usability which this present in simplicity and user-friendliness are also important characteristics of online examination systems [21].

3.1 Online Exams During the Covid19 Period

Following the emergence of coronavirus 2019 (COVID-19), educational institutions worldwide have been forced to adopt the distance learning mode. In addition, electronic distance exams (E-exams) have been considered as a mode of assessment [26]. In this context, several means have been adopted to ensure the continuity of studies in the face of this crisis. In the epidemic situation caused by COVID-19, the adoption or implementation of a distance learning and resource sharing platform including online exams becomes increasingly urgent and important in the case of a global emergency [27]. Like traditional classroom-based exams, the e-exam provides access to multiple test formats. They can be classified according to the purpose of the exam and the evaluation criteria, such as the "Closed-book" or the knowledge control exam, where the student has to answer a series of QSM or open-ended questions within a time limit without any help [28]. The evaluation during which the student has an aide-memoire document prepared by him/her or provided on the day of the exam (also called "cheat sheet"). The objective of the cheat sheet is to check the mastery of knowledge and the ability to apply it, rather than its acquisition [28]. And there is the "open book" exam, which mobilizes critical thinking and creativity. The student can use any support he or she wishes to answer the questions [28]. The principle of e-exams is not so new in practice, and many institutions around the world are already using them. E-exams are part of the tools that some universities are making available as part of their digital transition [28]. But in the circumstances of the Corona pandemic and to limit the spread of the virus, the World Health Organization issued recommendations of social distancing and the obligation to wear the mask and a continuous sterilization, as well as the imposition of quarantine in all countries attacked by the virus. As a result, students and teachers at all levels have been forced to adapt quickly to online learning and testing [29]. For its part, the DGESIP (Directorate-General for Higher Education and Professional Integration) of the Ministry of Higher Education, Research and Innovation offers advice on organizing distance learning exams [30]. Different learning management platforms (Moodle, Canvas, Blackboard, etc.) offer evaluation tools such as quizzes that can take the form of multiple choice questionnaires, single choice questionnaires (between true and false, for example), or open-ended questions whose answers are limited to a few words [31]. On the other hand, these tools do not allow for the assurance of the learner's identity or for monitoring against cheating attempts. Only strategies such as random posting of questions, the use of question banks, limiting the time of taking the quiz and the number of attempts per quiz can limit the risks of cheating [31]. Tracking distance learners is not easy. Monitoring systems exist, but the costs associated with their use are high compared to the expected benefits, such as ProctorExam, Managexam and TestWe, which can use several forms of remote monitoring: human remote monitoring, called "synchronous monitoring", and algorithmic remote monitoring, which relies on images of the candidate to detect anomalies, called "asynchronous monitoring" [30].

4 Cheating

With the rapid growth of technology and the availability of several electronic tools in the markets, educational institutions increasingly suffer from cheating. Research by L. McCabe et al. shows that cheating is accepted and that some forms of cheating have increased over the past 30 years [7]. Statistics on cheating on exams (often including plagiarism in assignments) vary between 30% and 80% [8].

Cheating methods differ, Fontaine.S identifies three categories of methods most used to cheat, the first consists of exchanging information that should not be shared such as the disclosure of exam questions on social networks. The second category includes the use of prohibited material during exams, for example, the use of notes written on the forearms or information encoded in smart watches. And finally the third category of methods applies after the exam, for example, when a student claims that the teacher made mistakes in his correction when he himself changed his answers when he returned from the exam in class [9]. A survey conducted by Michaut.C reveals that in a sample of 1909 high school students reveal that nearly 20% of them have already used a mobile phone to cheat during their schooling [10].

These cheating issues come back to nature that students have to fight to get the best grades possible especially when it comes to certification exams on the one hand. On the other hand, when the learner is under family pressure and there is no room for error, cheating may seem the only way to meet expectations. We can also mention that a student can start cheating just to join a group of peers who appreciate this practice, contrary to the rules of the school system [11].

Cheating in an online environment is more convenient than traditional offline testing. More recently, survey-based research has increasingly reported that academic dishonesty - whether perceived or self-reported behavior - is more common in online environments [32]. Therefore, detection and prevention of online cheating is crucial for online assessment. Therefore, this issue is one of the biggest challenges facing e-learning platforms [33].

In fact, in addition to traditional cheating methods that can also be used to cheat on online exams, there are various techniques and tools that can be used to more easily cheat on online exams [33], For example, using remote desktop and split screen, searching for solutions on the internet, using social networks, etc. A study by Chirumamilla, Sindre, and Nguyen-Duc concluded that students and teachers perceived cheating to be easier with electronic tests, especially with the use of a personal device, especially with the "bring your own device" system [34].

Choosing the right methodology to use to prevent cheating in online exams is a very important thing that requires a lot of time and effort, we find the following protocols: (i) a single set of exam times for all students; (ii) the computer available for the online exam will be available for a short period of time (assume 15 min on the window); (iii) a random sequence of questions; (iv) a question will be presented in a timely manner; (v) a drawing or template for online review for a limited time; (vi) single access to the exam; (vii) require students to use the Blackboard Answer Lock Browser (ALB) only when the exam is complete (this limits exit/return, copy/paste, use of electronic device computers, etc.), (viii) change one-third of the multiple choice questions each time they take an exam [35].

5 Cheat Detection Systems in Online Exams

Several researches have been raised regarding cheating in online exams and with the emergence of the Covid-19 epidemic, this problem has become the focus of attention of many researchers Hylton, Levy and Dringus suggested adopting webcam surveillance to detect potential students who cheat and then reduce these behaviors in order to tackle online cheating, the main objective of this study was to investigate the deterrent effect of webcam surveillance on misconduct in online exams [2].

However Korman focuses on investigating a computerized methodology for detecting cheating on exams based on behavioral measures and machine learning. The detection potential of this approach is mainly represented by the signals leakage theories, whose topics can be covered using pattern recognition and anomaly detection theory, through a biometric behavior-mental approach [14].

While Alotaibi emphasizes the paramount importance of candidate authentication for online exams to ensure fairness during remote exam taking with a proposed method to ensure that unauthorized individuals are not allowed to take a test [15].

Another monitoring tool that checks the browsing activity of students during the online exam is developed by Kasliwal to detect cheating in real time with internet usage enabled [16].

In the same context Khlifi has proposed a new, slightly advanced method to secure electronic evaluations, the program collects information about students and their behavior in classroom activities and then allows it to be used to monitor unethical behavior during electronic assessment, the proposed method randomly and periodically generates a number of questions throughout the assessment from existing knowledge that students must answer to ensure their continued authentication [17].

In the same direction, another method based on CNN technologies was provided by Nitesh Kumar et al. in order to counter cheating attempts on electronic exams, this technology allows to identify each unknown situation during online tests [18].

Advancements in Machine Learning techniques are expanding time-consuming tasks that previously required significant effort from educators. Renzella et al. in this regard were developing a system that integrates an online oral assessment tool, Real Talk, with Deep Speaker, which is a speaker recognition and verification system that verifies with great precision whether two audio samples resemble the same person speech [19].

While Haytom has come up with a solution that tries to address the need for remote exam monitoring by combining solutions based on current technologies and techniques with different biometric modalities and implementing an automated monitoring system to monitor and verify learner identity during online exams with two modalities: dynamic typing and facial recognition to verify learner identity during remote exams [20].

6 Comparative Study of Machine Learning Algorithms and Techniques Applied for Detecting Cheating in Remote Examination

In examinations, monitoring methods to detect and reduce the possibility of cheating are important to ensure academic equity. A variety of methods have been proposed to ensure effective and comfortable tracking of online examinations [21].

Some researchers are interested in machine learning-based methods for verifying learners, improving remote testing, and detecting abnormal behavior. Thus, many studies propose different techniques/algorithms for facial recognition, head pose detection and behavior detection [21].

The work of Hadian and Bandung will propose a method for continuous user verification based on facial verifications by implementing an incremental training process using images captured from m-learning online course sessions as the training dataset to increase robustness against pose and lighting variations. The algorithm is trained each time a user completes their course session [36]. Senbo hu builds a system that uses a webcam to monitor images of the candidates' heads, and then inputs the head posture and mouth state information into a rule-based reasoning system. This system can detect the respondent's behavior and determine whether the behavior is abnormal [37]. Another system created by Garg, Verma and Patidar has the general objective of detecting/tagging/recognizing the face, as it can monitor the student's behavior and look for any malpractice during the online assessment using the webcam [38].

One method of cheating in online environments reported in the literature is CAMEO (Copying Answers using Multiple Existences Online) [39], where a user creates several accounts, one of which is the main account through which he/she will eventually obtain the certificate. The other ghost accounts are used to obtain the answers to the different evaluation quizzes and thus feed the original account [42]. In this study, the researchers will build a random forest classifier that detects submissions that have been cheated using the CAMEO method [39]. Other researchers have focused their work on biometrics such as Haytom who has integrated a solution based on an innovative multimodal biometric technique with two modalities: keystroke dynamics and facial recognition in order to verify the identity of learners during a remote exam [20].

The Table 1 below represents the comparison between some studies that are interested in the detection of cheating in online exams according to the use of Machine Learning techniques and algorithms.

Table 1. Comparison between different cheat detection systems in online exams based on Machine Learning.

N°	Online Cheating Detection Systems	BT issue/ Improvement	Techniques	Algorithms	Exam type
1	Continuous user verification	Improve the robustness of pose and lighting variations by performing an incremental training process using the training dataset obtained from m-learning online course sessions [36]		Convolutional neural networks CNN	
2	Abnormal behavior detection	Obtain information about the candidate's head posture and oral status via webcam and discriminate abnormal behavior of examiners during the online examination [37]	Facial recognition Detection of head placement	Convolutional neural networks CNN	

(*continued*)

Table 1. (*continued*)

N°	Online Cheating Detection Systems	BT issue/ Improvement	Techniques	Algorithms	Exam type
3	Abnormal behavior detection	Detect and identify students' faces using the Haar Cascade classifier and deep learning and will apply some constraints to stop these practices (for example, multiple face detection) [38]	Facial recognition	Convolutional neural networks CNN	Multiple choice
4	Improve the quality of teaching	Propose an evaluation method that takes place throughout the curriculum process introducing online examinations [40]	Generation and evaluation of question banks	K-means clustering	True/false Single choice Multiple choice
5	Identification of cheaters	Identify and classify submissions that have been cheated using the cameo method [39]		Random forest classification model	Multiple choice
6	Biometric system	Improve learner identity verification and detect unusual events without violating learner privacy [20]	Behavior detection Facial recognition		

(*continued*)

Table 1. *(continued)*

N°	Online Cheating Detection Systems	BT issue/ Improvement	Techniques	Algorithms	Exam type
7	Improve the supervision of remote examinations	To provide assistance and support in order to secure large-scale examinations [41]	Security	Generative adversials networks GANs Recurrent neurals networks RNN Convolutional neural networks CNN	

It is important to mention that appropriate information regarding the technique, algorithm, or type of examination proposed is not available in some studies. While, other authors did not provide any substantial information about the underlying techniques and algorithms used for the development of the system. Therefore, these studies are not included in Table 1 above.

7 Conclusion and Perspectives

Academic misconduct in online exams remains a major challenge for institutions. For any institute, maintaining academic integrity is of paramount importance in order to build and ensure the credibility of the tests. A kind of fairness in these types of examinations can only be achieved with a guarantee of strict and complete supervision. In this article we have discussed the different solutions proposed to counter deception and dishonesty online. Solutions based on machine learning techniques remain more effective than any other solution.

Most studies have relied on CNN algorithms as a solution to detect abnormal learner behavior, using several techniques such as authentication, facial recognition, head pose detection etc. Biometric systems also pose important solutions in reducing cheating in distance learning exams.

In our future work we plant to develop a new generic and efficient approach based on machine learning for the detection of cheating during online exams. The approach should be independent of the exam subject and e-learning domain.

References

1. Lafleur, F., Samson, G.: Online Training and Learning. Press of the University of Quebec, Quebec (2019)

2. Hylton, K., Levy, Y., Dringus, L.P.: Utilizing webcam-based proctoring to deter misconduct in online exams. Comput. Educ. **92**, 53–63 (2016)
3. Elhadji Ille Gado, N.: Random methods for learning data in high dimension: application to shared learning. In: Hal Open Science, University of Technology of Troyes, France (2017)
4. LEMAGIT. https://www.lemagit.fr/definition/Apprentissage-supervise. Accessed 03 June 2022
5. Duda, R., Hart, P., Stork, D.: Pattern Classification. 2nd edn. (2000)
6. DataScientest. https://datascientest.com/reinforcement-learning. Accessed 01 June 2022
7. McCabe, D.L., Kleb Trevino, L., Butterfield, D.: Cheating in academic institutions: a decade of research. Ethics Behav. **11**, 219–232 (2010)
8. Williams, M.W.M., Williams, M.N.: Academic dishonesty, self-control, and general criminality: a prospective and retrospective study of academic dishonesty in a New Zealand university. Ethics Behav. **22**, 89–112 (2012)
9. Fontaine, S.: Exam cheating: a research snapshot. Educ. Occup. 139–141 (2020)
10. Michaut, C.: The new tools of school cheating in high school. Hal open-archives (2013)
11. GoStudent. https://insights.gostudent.org/fr/consequences-tricherie-examens. Accessed 16 June 2020
12. Riyami, B.: Analysis of the effects of ICT on higher education in Morocco in a context of training in collaboration with a French university (2019)
13. Beust, P., Duchatelle, I., Cauchard, V.: Exams taken at the student's home. Hal Open Science (2019)
14. Matus, K.: Behavioral detection of cheating during the online exam (2010)
15. Alotaibi, S.: Using biometrics authentication via fingerprint recognition in e-Exams in e-Learning environment. In: The 4th Saudi International Conference. The University of Manchester (2010)
16. Kasliwal, G.: Cheating detection in online examinations. San José State University (2015)
17. Khlifi, Y.: An advanced authentication scheme for E-evaluation using students behaviors over E-learning platform. In: International Journal of Emerging Technologies in Learning (iJET), pp. 90–111 (2020)
18. Sharma, N.K., Gautam, D.K., Rathore, S., Khan, M.R.: CNN implementation for detect cheating in online exams during COVID-19 pandemic: a CVRU perspective. Elsevier Direct Science (2021)
19. Renzella, J., Cain, A., Schneider, J.-G.: Verifying student identity in oral assessments with deep speaker. Comput. Educ. Artif. Intell. **3**, 100044 (2022)
20. Haytom, M.A.: Behavioral analysis remote exam monitoring (2021)
21. Muzaffar, A.W., Tahir, M., Anwar, M.W., Chaudry, Q., Rasheed Mir, S., Rasheed, Y.: systematic review of online testing solutions in e-learning: techniques, tools and global adoption. IEEE Access **9**, 32689–32712 (2021)
22. Balderas, B., Caballero-Hernández, J.A.: Analysis of learning records to detect student cheating on online exams: case study during COVID-19 pandemic. In: 8th International Conference on Technology Ecosystems to Enhance Multiculturalism, pp.752–757. ACM Digital Library (2021)
23. EducationalInnovation. https://innovacioneducativa.wordpress.com/2020/04/15/adaptacion-de-la-evaluacion-presencial-a-evaluacion-online/. Accessed 29 Aug 2022
24. González-González, C., Infante-Moro, A., Infante-Moro, J.: Implementing electronic monitoring in online education: a study of motivational factors. MDPI (12) (2020)
25. Moukhliss, G., Belhadaoui, H., Filali Hilali, R.: A new model for automatic and continuous monitoring of online reviews. IEEE (2019)
26. Elsalem, L.: Stress and behavioral changes with remote electronic exams during the Covid-19 pandemic: a cross-sectional study of undergraduate medical students. Ann. Med. Surg. **60**, 271–279 (2020)

27. Yao, S., Li, D., Yohannes, A., Song, H.: Exploring a networked distance learning and resource sharing system for higher education in a COVID-19 epidemic, pp. 807–813 (2021)
28. Sillard, B.: Digital assessment methods of learning. In: Taking Ownership of and Learning About Digital Technology, pp. 62–72. Annales des Mines, Mars (2022)
29. Chand, A.: Remote Learning and Online Teaching in Fiji During COVID-19: The Challenges and Opportunities. Elsevier, Amsterdam (2020)
30. Campusmatin Homepage. https://www.campusmatin.com/numerique/pedagogie/pratiques/covid-19-comment-organiser-les-examens-a-distance.html. Accessed 04 July 2022
31. Catia, P., Gérin-Lajoie, S., Hebert, M.H..: Getting closer to remote assessment: ten ways to respond. Int. J. Educ. Res. Train. 201–206 (2020)
32. Dendir, S., Maxwell, R.S.: Cheating in online courses: evidence from online proctoring. Comput. Hum. Behav. Rep. **2**, 100033 (2020)
33. Noorbehbahani, F., Mohammadi, Z., Aminazadeh, A.: A systematic review of research on cheating on online exams from 2010 to 2021. Educ. Inf. Technol. **27**, 8413–8460 (2022)
34. Chirumamilla, A., Sindre, G., Nguyen-Duc, A.: Cheating in e-exams and paper exams: the perceptions of engineering students and teachers in Norway. Assess. Eval. High. Educ. **7**(45), 940–957 (2020)
35. Sharma, N.K., Gautam, D.K., Rathore, S., Khan, M.R.: CNN implementation for detect cheating in online-exams during COVID19 pandemic: a CVRU perspective. Mater. Today Proc. (2021)
36. Asep, H.S.G., Bandung, Y.: A design of continuous user verification for online exam proctoring on M-learning. In: 2019 International Conference on Electrical Engineering and Informatics (ICEEI). IEEE, Bandung (2020)
37. Hu, S., Jia, X., Fu, Y.: Research on abnormal behavior detection of online examination based on image information. In: 2018 10th International Conference on Intelligent Human-Machine Systems and Cybernetics .IEEE, Hangzhou (2018)
38. Garg, K., Verma, K., Patidar, K., Tejra, N., Patidar, K.: Convolutional neural network based virtual exam controller. In: 2020 4th International Conference on Intelligent Computing and Control Systems (ICICCS). IEEE, Madurai (2020)
39. Ruiperez-Valiente, J.A., Munoz-Merino, P.J., Alexandron, G., Pritchard, D.E.: Using machine learning to detect 'multiple-account' cheating and analyze the influence of student and problem features. IEEE Trans. Learn. Technol. **1**(12), 112–122 (2017)
40. Chen, Q.: An application of online exam in discrete mathematics course. In: TURC 2018: ACM Turing Celebration Conference, pp. 91–95. ACM Digital Library, Chine (2018)
41. Pierre, S.: Aide intelligente pour la surveillance d'examens. Corpus UL, pp. 88–96 (2020)
42. The student. https://www.letudiant.fr/educpros/actualite/sur-edx-mit-et-harvard-font-la-chasse-aux-fraudeurs.html. Accessed 30 Aug 2022
43. Sarker, I.H.: Machine learning: algorithms, real-world applications and research directions. SN Computer Science **2**, 160 (2021)
44. Projeduc. https://projeduc.github.io/intro_apprentissage_automatique/introduction.html#i-3-types-des-algorithmes-dapprentissage. Accessed 31 Aug 2022

Comparative Study of Bayesian Optimization Process for the Best Machine Learning Hyperparameters

Fatima Fatih[1(✉)], Zakariae En-Naimani[2], and Khalid Haddouch[1,2]

[1] Laboratory LISA, ENSA University of Sidi Mohamed Ben Abdellah, Fez, Morocco
fatih.fatima@gmail.com

[2] Laboratory SSDIA, ENSET University of Hassan II Casablanca, Mohammedia, Morocco

Abstract. Bayesian optimization is important algorithm that uses two essential components, namely the surrogate model and the acquisition function. They are used to approximate the unknown objective function. This optimization is used as a hyperparameter tuning technique for the four machine learning algorithms to increase their performance. In this work, we applied Bayesian optimization to choose the best hyperparameters for a set of ML algorithms namely RF, SVM, KNN and LR. For this, we used a heart disease dataset. In this context, we obtained the best hyperparameters with accuracy for each machine learning algorithm optimized by BO-GP and BO-TPE. The results demanstrate the highest accuracy in BO-GP and BO-TPE are respectively LR is 89.01% and SVM is 89.01%. Then, the setting of hyperparameters allows to find the best hyperparameters that is improved accuracy for each algorithm.

Keywords: Bayesian optimization · Machine learning · Gaussian process · Tree structured parzen estimator · Hyperparameter optimization

1 Introduction

Hyperparameter tuning [8] is used to test the combination of hyperparameters that are randomly chosen to improve machine learning problems. It is difficult to manually choose the best values of hyperparameters because the choice of hyperparameters affects the performance of the model. So, the performance of the learning model depends on better choice of hyperparameters. In this regard, there existe an important technique for tuning hyperparameters are random search, grid search, particle swarm optimization, genetic algorithm and Bayesian optimization [8,9,16,17]. Bayesian optimization is one of the good technique for tuning hyperparameters in automatic learning models.

Bayesian optimization [4,11,13] is a method used to solve objective functions that are costly to evaluate and also to find the global maximum of this function.

M. Lazaar et al. (Eds.): BDIoT 2022, LNNS 625, pp. 239–249, 2023.
https://doi.org/10.1007/978-3-031-28387-1_21

It is based on the Gaussian process, Random forest and Tree structure parzen estimator (TPE) which constitutes two density functions (good density and bad density) and the observations are divided in to together [1]. In TPE the ratio must be maximized for minimize the expected improvement of the acquisition function which allows to find the new configuration of the hyperparameters [1]. All three surrogate models are used to approximate the objective function. Haw-ever, the most of the time, the Gaussian process is the most used in Bayesian optimization.

In order to explain this, we use the heart disease dataset. This dataset is on of the serious diseases that threaten human life. ML machine learning algorithms playing a key role in predicting heart disease based on different symptoms such as age, gender etc. The main objective is to detect the patient in its early stages where it can be treated and save lives from death in order to reduce the morality rates by heart disease. In this work, we applied Bayesian optimization to find the best hyperparameters that improved the accuracy for each algorithm namely RF, SVM, KNN and LR as the performance of the learning models depends on better hyperparameters. So, our result shows that the highest accuracy in BO-GP and BO-TPE are of LR and SVM respectively.

This paper is structured as follows. We present the related Bayesian optimization work in Sect. 2. Section 3 illustrates two components of Bayesian optimization. Then the steps of Bayesian optimization to find the global maximum of an objective function which are costly to evaluate are presented in Sect. 4. The optimization process which consists of three surrogate functions is explained in Sect. 5. In Sect. 6, we present experimental results. And finally, we end with a conclusion.

2 Related Work

Bayesian optimization (BO) [4,11,13,17] is an efficient method that consists of two essential components namely the surrogate models and the acquisition function to determine the next hyperparameters configurations that allows to find an approximation of a costly objective function to be evaluated. The surrogate models are: Tree structure parzen estimator (TPE) [1], random forest [15,17], and Gaussian process [4,7,8,10,13]. The acquisition functions are the expected improvement (EI) [2], the probability of improvement [11] and the upper confidence bound (UCB) [11]. The most used acquisition function in Bayesian optimization is the expected improvement [11]. But, according to [2] shows that there is a better acquisition function than EI is the acquisition function E^3I which balances exploitation and exploration in BO.

The concept of Bayesian optimization was introduced in [11] with two experiments. The first experiment is determining the global maximum of the objective function f(x, y). The second experiment is compared between Bayesian optimization and random search in the SVM machine learning algorithm, which shows that there is no difference in the performance of these two methods. There are more works as [6,8,9,16,17] show that Bayesian optimization is one of the most effective hyperparameter optimization techniques for tuning hyerparameters in machine learning models.

In this article [9] presents a comparative study between three HPO methods: grid search, random search, Bayesian optimization. This comparison is used to find the best method that can be used to obtain the highest accuracy in a short time simulation. The results of [9] show that the method of Bayesian optimization is more efficient than the other methods.

This work [6] presents a comparative analysis of various hyperparameter tuning techniques, namely Grid Search, Random Search, Bayesian Optimization, Particle Swarm Optimization (PSO), and Genetic Algorithm (GA). They are used to optimize the accuracy of six machine learning algorithms, namely, Logistic Regression (LR), Ridge Classifier (RC), Support Vector Machine Classifier (SVC), Decision Tree (DT), Random Forest (RF), and Naive Bayes (NB) classifiers. These algorithms are used to solve the tree sentiment classification problem. The results of [6] shows that the performance for each machine learning algorithm before and after setting the hyperparameters shows that the highest accuracy was given by SVC before and after setting the hyperparameters with the highest scores obtained when using Bayesian optimization.

We have seen in [17] a comparative study between eight different hyperparameter optimization methods that are implemented on three machine learning models (KNN, RF, SVM) of classification and regression. First, it compares accuracy and computation time (CT) for classification problems that evaluated on the MNIST dataset. Secondly, it is compared MSE and computation time for the regression problem which is evaluated on the Bosten-houssing dataset. In this paper it is shown that using the default hyperparameter settings does not give the best model performance. So, it is important to use HPO methods to determine the best hyperparameters.

3 The Components of Bayesian Optimization

Bayesian optimization uses the following two important components:

3.1 Surrogate Functions

The surrogate model [11] is a probability model that gives a representation of the objective function that is expensive to evaluate. We will see in Sect. 5 three Surrogate models namely: gaussian process (GP), random forest (RF), tree structure parzen estimator (TPE). However, in the most of the time, GP [3] is good tool used in Bayesian optimization. The main idea of these surrogate models are used to approximate the unknown objective function and to search the global optimization of this function.

3.2 Acquisition Functions

The acquisition function is an essential technique in Bayesian optimization. Mathematically, the point that maximizes the acquisition function is used to propose the next sampling point for the next iteration. The most commonly used acquisition functions in Bayesian optimization are:

1. **Probability of Improvement (PI)**
   We can define the improvement $I(x)$ as follows:

$$I(x) = \max((f(x) - f(x^*), 0) = \begin{cases} f(x) - f(x^*) & \text{if} \quad f(x) > f(x^*) \\ 0 & \text{if} \quad f(x) < f(x^*) \end{cases}$$

   The probability of improvement is defined as follows

$$\begin{aligned} PI(x) &= \mathbb{P}[I(x) > 0] \\ &= \mathbb{P}[f(x) > f(x^*)] \\ &= \Phi(\frac{\mu(x) - f(x^*)}{\sigma(x)}) \end{aligned}$$

   where
   - The mean μ and the variance σ,
   - Φ is the cumulative distribution functions (CDF):

$$\Phi(z) = \int_{-\infty}^{z} \varphi(z)dz$$

   with $\varphi(z) = \frac{1}{\sqrt{2\pi}} exp^{\left(\frac{-z^2}{2}\right)}$ is the probability density function (PDF) of the normal distribution $\mathcal{N}(0,1)$.
2. **Expected Improvement (EI)**
   The expected improvement is defined as follows:

$$EI(x) = \begin{cases} (\mu(x) - f(x^*))\Phi(\frac{\mu(x)-f(x^*)}{\sigma(x)}) + \sigma(x)\varphi(\frac{\mu(x)-f(x^*)}{\sigma(x)}) & \text{if} \quad \sigma(x) > 0, \\ 0 & \text{if} \quad \sigma(x) = 0 \end{cases}$$

   where Φ and φ are the cumulative distribution functions (CDF) and probability density function (PDF).
3. **The Upper Confidence Bound (UCB)**
   The Upper Confidence Bound is defined as the sequence [11]:

$$UCB(x) = \mu(x) + \beta\sigma(x)$$

   where $\beta > 0$ is a user-selected parameter that is used to balance exploration and exploitation [11].

4 Bayesian Optimization Steps

To find the new hyperparameters to approximate an unknown objective function f, we have the following Bayesian optimization steps [17]:

1. Build a surrogate model of the objective function and we used almost all the time the Gaussian process to approximate the true objective function.
2. Find the optimal values of hyperparameters on the substitution model. In this step we used the acquisition function to choose the next hyperparameters. The hyperparameters that maximizes the acquisition function is an hyperparameter chosen to use as the first sample in the graph of the substitution function.
3. We compute the true objective function of this new hyperparameter that we obtained in step 2 and obtain a score.
4. Update the substitution probability model with the new results. In this step we compute the substitution function to determine the mean and the variance of this hyperparameter, then we define the value of μ in new iteration.
5. Repeat steps 2 through 4 until the maximum iteration pattern is reached.

Finally we find an approximation of the real objective function that allows us to find the global maximum from the previously evaluated samples.

5 Optimization Process

There are three following substitution models in Bayesian optimization:

5.1 Bayesian Optimization - Gaussian Process (BO-GP)

The Gaussian process [11,17] is a surrogate model most commonly used in Bayesian optimization to approximate the objective function $f : X \longrightarrow \mathbb{R}$ with X is a finite set of N points, and the values of the objective function $f = [f(x_1), \ldots, f(x_n)]$ [11] are distributed according to a multivariate Gaussian distribution. Thus the Gaussian process is given by [9,12,14]:

$$f \sim \mathcal{GP}(\mu(x), K(x, x^{'}))$$

where μ is a mean vector and K is a covariance matrix. Predictions following a normal distribution [17]:

$$P(y|x, D) = \mathcal{N}(y|\tilde{\mu}, \tilde{\sigma}^2)$$

where D is the configuration space of the hyperparameters, and $y = f(x)$ is the result of the evaluation of each hyperparameter value X [17]. We assume $\mu(x) = 0$ so the new means and variances are [14]:

$$\tilde{\mu} = K(x)^T K^{-1} y,$$
$$\tilde{\sigma}^2 = K(x, x) - K(x)^T K^{-1} K(x).$$

These new means and variance will be used in the acquisition function to find the next evaluation point of the true objective function f.

5.2 Sequential Model-Based Algorithm Configuration (SMAC)

Bayesian optimization using RF as a surrogate model. It's also called sequential model based algorithm configuration (SMAC) [17]. Assuming that there is a Gaussian model $\mathcal{N}(y|\tilde{\mu}, \tilde{\sigma^2})$, which $\tilde{\mu}$ and $\tilde{\sigma^2}$ are the mean and variance of the regression function $r(x)$, respectively [5,15,17]:

$$\tilde{\mu} = \frac{1}{|B|} \sum_{r \in B} r(x)$$

$$\tilde{\sigma^2} = \frac{1}{|B|-1} \sum_{r \in B} (r(x) - \tilde{\mu})^2$$

where B is a set of regression trees in the forest.

5.3 Tree Structure Parzen Estimator (TPE)

The tree structured parzen estimator (TPE) is another common surrogate model for Bayesian optimization [17]. It creates a model applying the Bayes rule to calculate $p(y|x)$ following:

$$P(y|x) = \frac{P(x|y)P(y)}{P(x)}$$

But this method takes a different approach, since Bayesian optimization try to determine $p(y|x)$ [1], Tree structure parzen estimator models $p(x|y)$ and $p(y)$, i.e. (TPE) does not directly model $p(y|x)$ but rather $p(x|y)$ et $p(y)$, and the likelihood probability is defined as follows [1]:

$$P(x|y) = \begin{cases} l(x) & \text{if } y < y^* \\ g(x) & \text{if } y > y^* \end{cases}$$

where $l(x)$ is the probability density function formed using the observed variables x such that the objective function value is less than the threshold y^*. Then $l(x)$ models the density of the best observations, and $g(x)$ is the density function using the remaining observations such that the objective function value is greater than the threshold y^*. Then $g(x)$ models the density of bad observations [1]. TPE uses the following expected improvement [1]:

$$\begin{aligned} EI_{y^*}(x) &= \frac{\gamma y^* l(x) - l(x) \int_{-\infty}^{y^*} P(y)\, dy}{\gamma l(x) - (1-\gamma) g(x)} \\ &\propto (\gamma + \frac{g(x)}{l(x)}(1-\gamma))^{-1} \end{aligned}$$

The expected improvement is proportional to the ratio $\frac{l(x)}{g(x)}$ the Tree structure parzen estimator works by drawing x values from $l(x)$ based only on x values that give scores below the threshold, and not $g(x)$ to increase the EI, then to maximize the expected improvement, we must maximize this ratio [1].

6 Experimental and Results

The heart disease dataset, we downloaded from github, contain 76 attributes but all published experiments refer to the use of a subset of 14 of them. So the 14 features are detailed as follows:

1. Age: Age of the patient,
2. Sex: Sex of the patient,
3. exang: exercise induced angina (1 = yes, 0 = no),
4. ca: number of major vessels (0–3),
5. cp: Chest Pain type (1: typical angina, 2: atypical angina, 3: non-anginal pain, 4: asymptomatic),
6. trtbps: resting blood pressure (in mm Hg),
7. chol: cholestoral in mg/dl fetched via BMI sensor,
8. fbs: (fasting blood sugar >120 mg/dl) (1 = true; 0 = false),
9. rest-ecg: resting electrocardiographic results (0: normal, 1: having ST-T wave abnormality (T wave inversions and/or ST elevation or depression of >0.05 mV), 2: showing probable or definite left ventricular hypertrophy by Estes' criteria),
10. thalach : maximum heart rate achieved,
11. target : (0 = less chance of heart attack 1 = more chance of heart attack contenu).

Based on this dataset, we categorized patients as 1 indicating the presence of heart disease and 0 indicating the absence of heart disease. Then we used 70% training data and 30% used data to test the obtained model.

6.1 Results

A comparative analysis between BO-GP and BO-TPE are used to determine process gives the highest accuracy for different ML algorithms such as RF, KNN, SVM and LR applying to heart disease prediction (Table 1 and Figs. 1, 2, 3, 4 and 5):

Table 1. The accuracy of ML algorithms without using BO.

ML	Accuracy	Time (s)
RF	0.8461	0.2447
KNN	0.6593	0.0210
SVM	0.5714	0.0230
LR	0.8791	0.0601

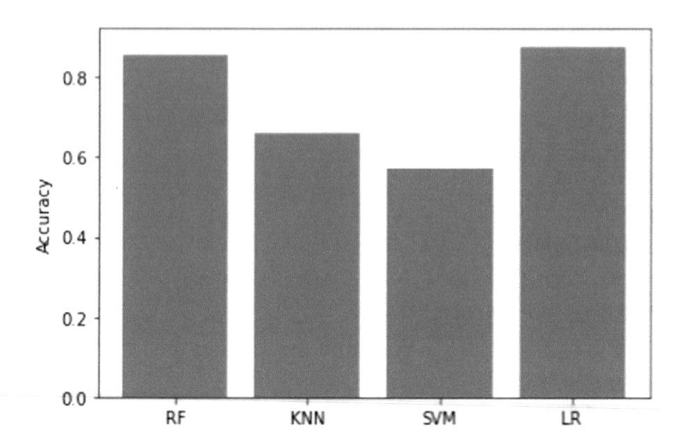

Fig. 1. Accuracy.

Table 2. Evaluation of BO-GP performance.

ML	Best hyperpameters	Precision	Recall	F1_score	accuracy	Time (s)
RF	criterion='entropy', max_depth=61, max_features='sqrt', min_samples-leaf=11, min_samples-split=14 n_estimators=205	0.8	0.9756	0.8791	0.8791	80.36
KNN	n-neighbors=17	0.8	0.9756	0.8791	0.8791	15.34
SVM	C=13.273723936150628 kernel='linear'	0.8	0.9756	0.8791	0.8791	169.41
LR	C=3.131333053846714 penalty=l2, solver= 'liblinear'	0.8163	0.9756	0.8888	0.8901	25.87

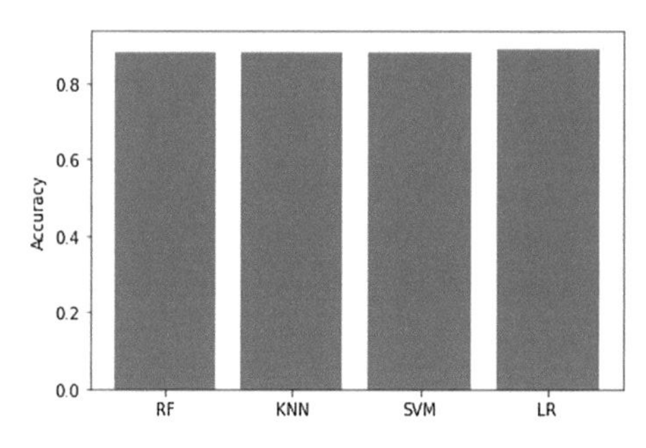

Fig. 2. Accuracy

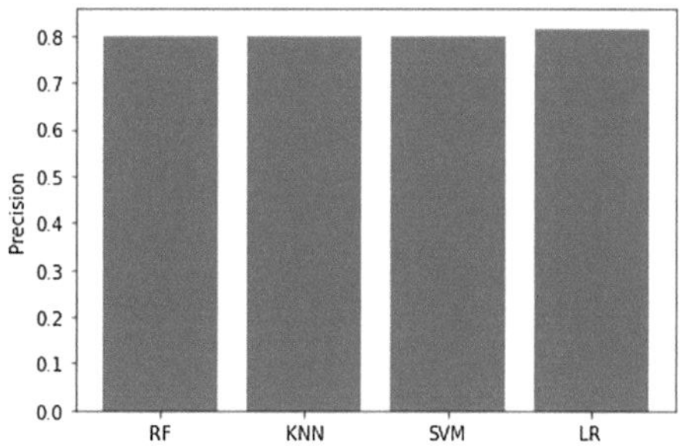

Fig. 3. Precision

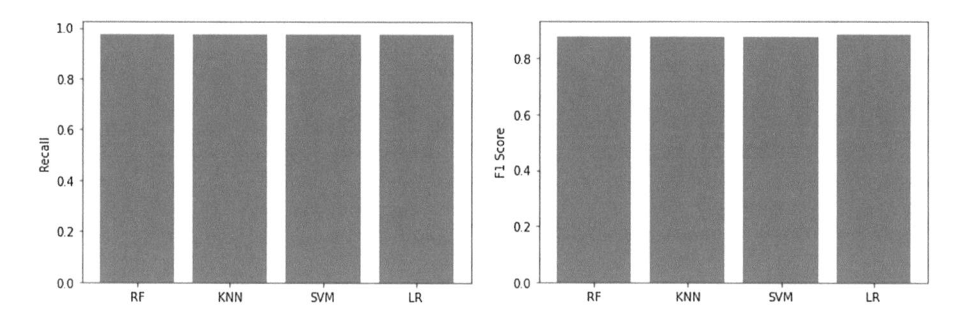

Fig. 4. Recall

Fig. 5. F1-score

Table 3. Evaluation of BO-TPE performance.

ML	Best hyperpameters	Precision	Recall	F1_score	accuracy	Time (s)
RF	criterion='gini,', max_depth=80.0, max_features='auto', min_samples-leaf=0.1232, min_samples-split=0.2559 n_estimators=210	0.7692	0.9756	0.8602	0.85714	53.283
KNN	n_neighbors=24.0	0.6078	0.7560	0.6739	0.6703	0.5642
SVM	C= 6.609156 kernel='linear'	0.8297	0.9512	0.8863	0.8901	42.917
LR	C =1.1082953781569425 penalty=l1, solver= 'liblinear'	0.8	0.9756	0.8791	0.8791	0.5868

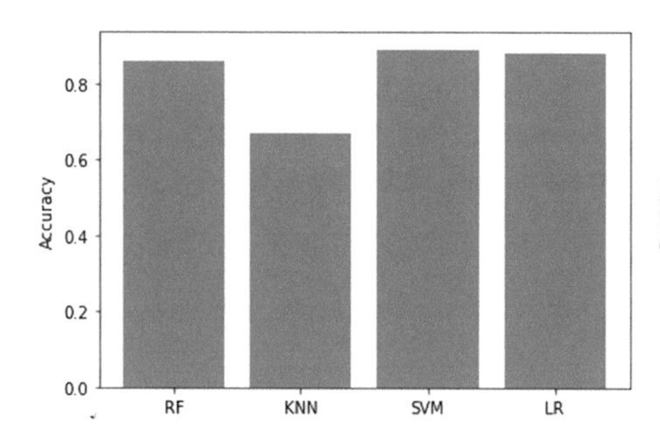

Fig. 6. Accuracy

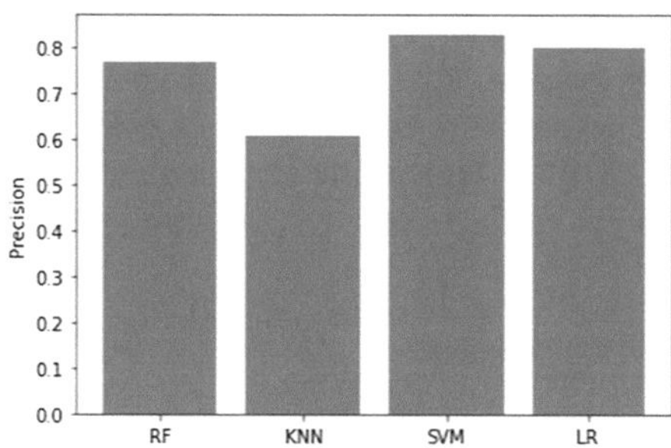

Fig. 7. Precision

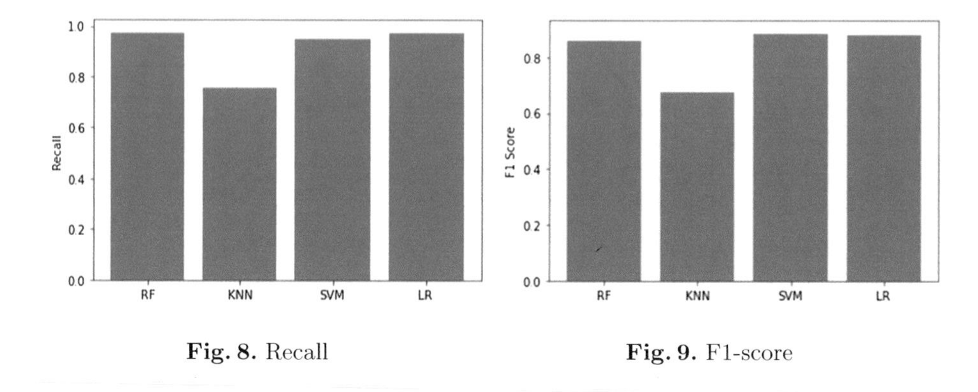

Fig. 8. Recall

Fig. 9. F1-score

6.2 Discussion of Results and Comparisons

Automatic learning algorithms are used to measure the performance of RF, SVM, KNN and LR. According to Table 2 the highest accuracy is of LR with a score of 87.91%, while the lowest accuracy we obtained for SVM is 57.14% (Figs. 6, 7, 8 and 9).

BO-GP shows that the highest accuracy for LR of 89,012% compared to other learning algorithms. The BO-TPE results show that the highest accuracy is from SVM with a score of 89.01%, and the lowest accuracy is from KNN which gives a score of 67.03%. Then, the accuracy of LR in BO-GP and the accuracy of SVM in BO-TPE are larger than the accuracy of Table 3. From these results we deduce that the hyperparameters setting allows to find the best parameters that helped to improve the accuracy for each learning model.

7 Conclusion

Bayesian optimization is an efficient hyperparameter tuning technique to improve machine learning problems. Experience shows that BO-GP and proposed BO-TPE can find the best hyperparameters with accuracy for machine learning models namely RF, KNN, SVM and LR.

The results we obtained from the highest accuracy in BO-GP and BO-TPE are respectively LR is 89.01% and SVM is 89.01%. Then, hyperparameter tuning allows to find the best hyperparameters that improve the accuracy for each algorithm. Our study shows that the right choice of hyperparameters depends on the performance of machine learning models.

References

1. Bergstra, J., Bardenet, R., Bengio, Y., Kégl, B.: Algorithms for hyperparameter optimization. Adv. Neural Inf. Process. Syst. **24** (2011)

2. Berk, J., Nguyen, V., Gupta, S., Rana, S., Venkatesh, S.: Exploration enhanced expected improvement for Bayesian optimization. In: Berlingerio, M., Bonchi, F., Gartner, T., Hurley, N., Ifrim, G. (eds.) Machine Learning and Knowledge Discovery in Databases. ECML PKDD 2018. LNCS, vol. 11052, pp. 621-637. Springer, Cham (2019). https://doi.org/10.1007/978-3-030-10928-8_37
3. Bodin, E., Kaiser, M., Kazlauskaite, I., Dai, Z., Campbell, N., Ek, C.H.: Modulating surrogates for Bayesian optimization. In: International Conference on Machine Learning, pp. 970–979. PMLR (2020)
4. Brochu, E., Cora, V.M., De Freitas, N.: A tutorial on bayesian optimization of expensive cost functions, with application to active user modeling and hierarchical reinforcement learning (2010). arXiv preprint arXiv :1012.2599
5. Dewancker, I., McCourt, M., Clark, S.: Bayesian optimization for machine learning : a practical guidebook (2016). arXiv preprint arXiv :1612.04858
6. Elgeldawi, E., Sayed, A., Galal, A.R., Zaki, A.M.: Hyperparameter tuning for machine learning algorithms used for Arabic sentiment analysis. Informatics **8**, 79. Multidisciplinary Digital Publishing Institute (2021)
7. Hoffman, M., Brochu, E., De Freitas, N., et al.: Portfolio allocation for Bayesian optimization. In: UAI, pp. 327–336. Citeseer (2011)
8. Joy, T.T., Rana, S., Gupta, S., Venkatesh, S.: Hyperparameter tuning for big data using Bayesian optimisation. In: 2016 23rd International Conference on Pattern Recognition (ICPR), pp. 2574–2579. IEEE (2016)
9. Kim, H.-C., Kang, M.-J.: Comparison of hyper-parameter optimization methods for deep neural networks. J. IKEEE **24**(4), 969–974 (2020)
10. Li, D., Kanoulas, E.: Bayesian optimization for optimizing retrieval systems. In: Proceedings of the Eleventh ACM International Conference on Web Search and Data Mining, pp. 360–368 (2018)
11. Matosevic, A.: On Bayesian optimization and its application to hyperparameter tuning (2018)
12. Nguyen, V., Gupta, S., Rana, S., Li, C., Venkatesh, S.: Regret for expected improvement over the best-observed value and stopping condition. In: Asian Conference on Machine Learning, pp. 279–294. PMLR (2017)
13. Nomura, M., Abe, K.: A simple heuristic for Bayesian optimization with a low budget (2019). arXiv preprint arXiv :1911.07790
14. Rasmussen, C.E., Nickisch, H.: Gaussian processes for machine learning (GPML) toolbox. J. Mach. Learn. Res. **11**, 3011–3015 (2010)
15. van Hoof, J., Vanschoren, J.: Hyperboost: hyperparameter optimization by gradient boosting surrogate models (2021). arXiv preprint arXiv :2101.02289
16. Wu, J., Toscano-Palmerin, S., Frazier, P.I., Wilson, A.G.: Practical multifidelity Bayesian optimization for hyperparameter tuning. In: Uncertainty in Artificial Intelligence, pp. 788–798. PMLR (2020)
17. Yang, L., Shami, A.: On hyperparameter optimization of machine learning algorithms: theory and practice. Neurocomputing **415**, 295–316 (2020)

The Effect of Normalization and Batch Normalization Layers in CNNs Models: Application to Plant Disease Classifications

Saloua Lagnaoui[1](✉), Zakariae En-Naimani[2], and Khalid Haddouch[1]

[1] Laboratory LISA, ENSA University of Sidi Mohamed Ben Abdellah, Fez, Morocco
saloua.lagnaoul@usmba.ac.ma

[2] Laboratory SSDIA, ENSET University of Hassan II Casablanca, Mohammedia, Morocco

Abstract. In fact, although deep learning has been very successful in many areas such as computer vision, natural language processing and information retrieval. Among the most important deep learning models, we mention convolutional neural networks (CNNs). This type of network is the most effective for performing image classification in real life domains such as medicine, industry, education, agriculture, etc. Research on the acceleration and the regularization of convolutional neural networks is limited, i.e., research on the effect of normalization layers and batch normalization layers on the model. In this work, we discuss the sensitivity of a CNN model to the normalization layer and batch normalization layer on the one hand. And on the other hand, we present how to accelerate and regularize the CNNs models. After that, we apply these CNNs models in a very interesting field which is agriculture, particularly plant disease classifications. And at the end of this work, we synthesize the results and discuss them.

Keywords: Deep Learning · Convolutional Neural Networks · Regularization · Normalization · Plant Disease Classification

1 Introduction

Since 2000, deep learning has emerged as a branch of artificial intelligence that deals with different types of artificial neural networks [1]. After 10 years, this field has experienced a great revolution with the appearance of a type of deep neural network with at least two hidden layers [2]. This revolution is related to the increasing power of computers, which makes it possible to create and train neural networks with dozens of hidden layers. Deep learning has been invested in different fields such as computer vision, natural language processing, information retrieval, and known a big success in all fields.

The convolutional neural network (CNN) is a type of direct multilayer artificial neural network. This specific model of neural network appeared in 1990 by

M. Lazaar et al. (Eds.): BDIoT 2022, LNNS 625, pp. 250–262, 2023.
https://doi.org/10.1007/978-3-031-28387-1_22

Yann Lecun [3]. CNNs are the best known deep neural networks, thanks to their computational power and performance, which constitute a revolutionary evolution, especially in fields related to pattern recognition, image processing and speech recognition [4]. This type is called convolutional thanks to the importance of the convolution layer, it is always at least their first layer, which is based on the mathematical principle of convolution and allows the detection and recognition of features regardless of their position in the image. More precisely, the most beneficial aspect of CNNs is the reduction of the number of parameters compared to other types of artificial neural networks. Another important aspect of CNNs is to obtain abstract features when the input propagates to deeper layers. A CNN is essentially composed of four layers: convolutional layers, non-linear layers, pooling layers and fully connected layers (see Fig. 1). There are several architectures of CNNs, the best known are: LeNet, AlexNet, VGGNet, GoogLeNet, ResNet, MobileNet, and EfficientNet.

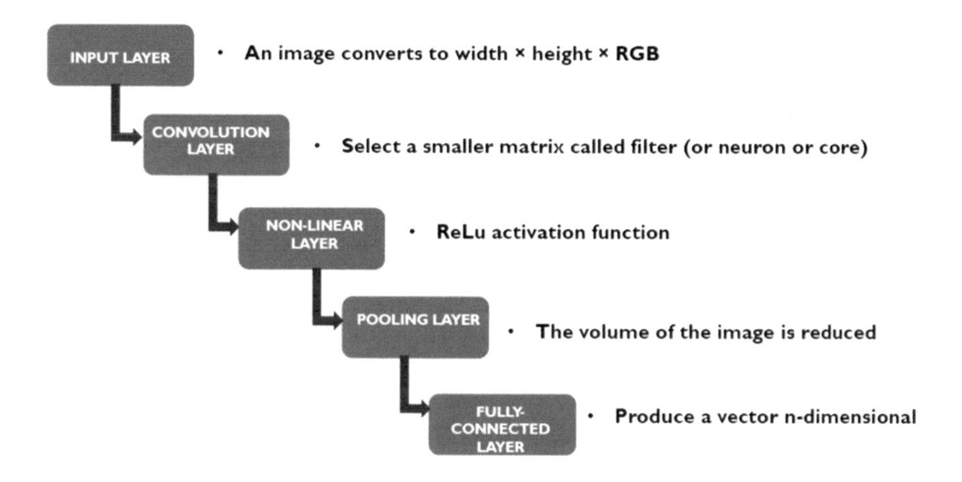

Fig. 1. The Structure of Convolutional Neural Networks.

In recent years, plant production has been greatly reduced due to various diseases, which seriously endanger food safety [5]. Therefore, there is an urgent need for the accurate detection of plant diseases. Traditional classification methods, such as naked-eye observation and laboratory tests, have many limitations, such as a long time for disease detection. Nowadays, deep learning methods, especially those based on convolutional neural networks (CNN), have been successful in most of the applications in plant disease classification. They have solved or partially solved the problems of traditional classification methods [6].

In this paper, we will study convolutional neural networks, and we are interested to add two layers Normalization layer and Batch Normalization layers in the CNNs models, and we present the effect of this addition on one hand. In another hand, do we define how the Batch Normalization layer accelerates and regularizes the CNNs models?

In the first section, we present a preliminary on the field of Deep Learning, the CNNs in many features, and an idea on the field of application to the plant disease classification, in section two, we define the convolutional neural networks with different layers and we define the Forward and Backpropagation, in section tree, we presented the normalization layer, in section four, we presented the batch normalization layer. For finished this work, we study the application of CNNs to the domain of agriculture. We see the experiments on three datasets (PlantVillage, Plant Pathology 2020 and Plant disease recognition) that we define in section five and we also see the different results and discuss several modifications at the levels of a CNN model that we create in section six and seven. And finally, in section eight we have a conclusion and a vision for research in this context in the future.

2 Convolutional Neural Networks

2.1 Different Layers of CNNs

In this section, we will see the different layers of the convolutional neural networks [7,8] and [9].

1. Convolutional Layer
   This layer is used to cut the image into sub-regions and then analyzed by a convolution kernel. The analysis of the features of the image by the convolution kernel is a filtering operation with an association of weights to each pixel. The application of the filter to the image is called convolution. After this operation, a feature map is obtained, it is an abstract representation of the image. Its values depend on the parameters of the applied convolution kernel and the pixel values of the input image. Its purpose in general is to identify the presence of a set of features in the images received at the network input. The characteristics are extracted using the following formula:

$$Z_k^i = \sigma\left(\sum_{l=1}^{N_{i-1}} K_{l,k}^i * X_l^{i-1}\right) \tag{1}$$

   where $*$ denotes the convolution product, σ is the activation function and K is the trained kernel.
2. Non-Linearity Layer
   The correction or activation layer is the application of a non-linear function to the feature maps output from the convolution layer. Making the data non-linear, facilitates the extraction of complex features. This layer is applied to saturate the output or limit the generated output. The most used nonlinear functions are Sigmoid, Tanh, ReLu, Softmax, PReLU, LRelu, GELU, ELU, ... And for CNN the activation function ReLu is used for the hidden layers and Softmax for the last layer.

3. Pooling Layer
   Pooling consists in reducing the size of the images. It reduces the number of hyperparameters, and the computations in the network and decreases the risk of over learning. The pooling layer makes the network less sensitive to the position of the features. The most common pooling operations are Max-pooling, Average pooling, ...
4. Fully-Connected Layer
   This layer is used to make the classification of the image from the extracted features, and it is Fully-Connected. For access to all input information. Each neuron assigns to the image a probability value of belonging to class i among the C possible classes (Fig. 2).

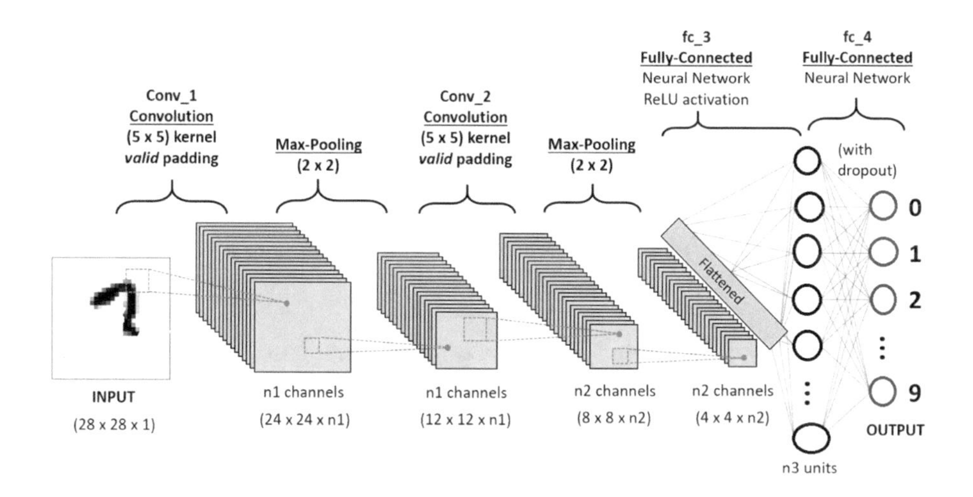

Fig. 2. The example of CNN with 2 hidden layers.

2.2 Forward Propagation and Back-Propagation

The direct propagation in CNN is done layer by layer. For the input layer we receive the image as input X we associate each input by a weight W and each neuron by a bias b then we go to the second layer which is the convolution layer at the level of this layer we will apply the convolution product between the convolution filter and the scanned part of the image by the effect of filter sliding by a stride that we choose after that we will apply an activation function the most part is the function Relu that we note σ then we have an output.

$$Z = \sigma\left(\sum_i X^i W^i + b^i\right) \tag{2}$$

where σ activation function, X Input of CNN model, W weight, b bias and Z output of CNN model.

And we now move to the layer Pooling on which we will reduce the size of the image of several manners often Max-Pooling then we have the Fully- Connected layer to have the probability of each image belonging to the different classes. And finally, we have the output layer on which we have the final results Y [10].

Back-propagation is the essence of neural network training. It is the method of adjusting the weights of a neural network according to the error rate obtained in the previous epoch. Proper adjustment of the weights allows you to reduce the error rates and make the model reliable by increasing its generalization. A neural network is an abbreviated form of back-propagation of errors. It is a standard method of training artificial neural networks. This method calculates the gradient of a loss function concerning all the weights of the network. Back-propagation algorithm: allows the neural network to quickly and efficiently compute the gradients that will then allow adjusting the parameters of these neural networks to better explain the data [11].

The Idea of back-propagation in CNNs is as the different types of artificial neural networks of the direct type are to adjust the different parameters of the network (weights, bias) where different layers of CNN (Convolution, Pooling, Fully-Connected) whose goal is to present the input data well where the loss function is minimal [12].

3 Normalization Layer

Normalization is a technique typically used to prepare data before model training. The main purpose of normalization is to provide a uniform scale for the numerical values. If the Dataset contains numerical data that vary over a wide range, the training process will be biased, resulting in a poor model. The normalization method ensures that there is no loss of information and that the range of values is not affected. Normalization is performed by the formula below, subtracting the mean and dividing by the standard deviation.

$$\hat{X} = \frac{X - mean(X)}{std(X)} \tag{3}$$

where $\hat{X}$ Normalized output, X input data, $mean(x)$ mean of data and std(X) standard deviation.

In spite of, the normalization of the input data, the value of activation's of some neurons in the hidden layers may start to vary over a wide range during the training process. This means that the input of neurons in the next hidden layer will also vary over a large scale, resulting in instability [13].

Normalization Layer why it is important?
Training state-of-the-art deep neural networks are computationally expensive. One approach to reducing training time is to normalize the activities of the neurons. For this, the normalization layer is very important for convolutional neural networks which are deep neural networks due to their computational complexity, learning, and computation time. So, CNN with a normalization layer

is different from CNN without a normalization layer, for the first one, we have two advantages: we have homogeneity of the data and a low error, these two things influence the learning of the net-works, so we have a CNN plus two benefits, and for the second one, we have an error rate that is very high during the learning process [14].

This normalization layer is placed after the input layer of a convolutional neural network to normalize the inputs and then we have the convolution layer to start the training of a CNN. Although the CNN gives good results in the field of image processing either with the normalization layer or without, it is better to use a CNN with a normalization layer to adjust the error during the training. Hence the normalization layer is an added value for convolutional neural networks.

4 Batch Normalization Layer

Batch Normalization layer (BN) has achieved great empirical success for deep network training, but today there is still no general consensus on why BN is useful to the training process of CNNs, and the effectiveness of BN is still not well understood by researchers, and its mathematical explanation has been researched by researchers. Training convolutional neural networks are complicated by the fact that the distribution of inputs at each layer changes during training based on the parameters of previous layers. This slows down training by requiring lower learning rates and careful initialization of parameters and makes it notoriously difficult to train models with saturating Non-Linearity. We call this phenomenon an internal covariate shift (ICS). And we address the problem through the Bach normalization layers [15].

The Batch normalization layer[1] allows for full normalization of the model architecture and that is performed for each mini-batch of training (see Fig. 3). The Batch normalization layer allows us to use much higher learning rates and be less conservative with initialization. It also acts as a regularized, eliminating the need for Dropout regularization in some cases. Batch normalization also achieves the same accuracy with 14 times fewer learning steps. And gains a more efficient model. And also enables faster CNNs [16].

The general principle of a BN layer is applied for neural networks where the learning is done by mini-batches. We divide the data into batches of a certain size and then run them through the network. BN is applied to the activation of the neurons for all samples in the mini-batch so that the mean of the output is close to 0 and the standard deviation is close to 1.

The advantages of BN used in CNNs [17]:

- BN mitigates to some extent the problem of gradient disappearance and explosion.
- Accelerates the convergence of the network formation.

[1] Usually inserted after Fully-Connected or Convolutional layers, and before Non-Linearity.

- Allows for deeper neural network training and prevents over-fitting.
- Also allows the network to use a higher learning rate and reduces the need for parameter tuning.

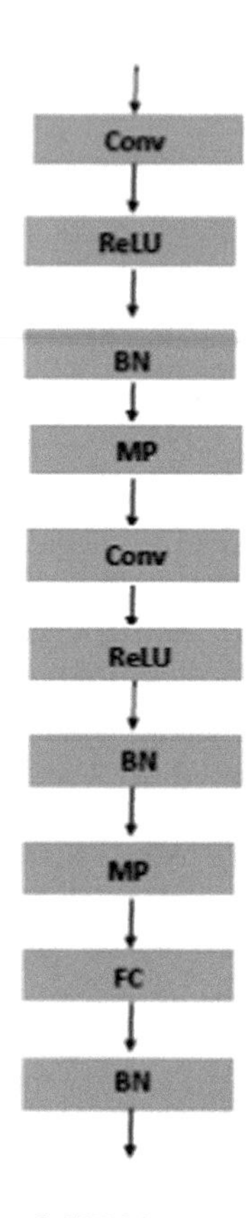

Fig. 3. The architecture of CNN has two hidden layers with BN.

5 Experimental Setup

In this section, we present an overview of the data sets, the models used for the experiments and the procedure used for this experiment.

5.1 Datasets

The different datasets used for this experiment are color image datasets used in many experiments concerning the agriculture domain and they are well structured and prepared.

1. Dataset PlantVillage [18].
2. Dataset Plant-Pathodology-2020 [19]
3. Dataset plant-disease-recognition [20] (Table 1 and Fig. 4)

Table 1. the Characteristics of each Datasets.

Name	Number of Images	Classes	Task	training set	test set
PlantVillage	54305	38	Image Classification	43 444	10 861
Plant-Pathodology-2020	3624	4	Image Classification	1456	365
plant-disease-recognition	1532	3	Image Classification	1322	150

(a) (b) (c)

Fig. 4. Some pictures of the Datasets PlantVillage (a), plant-pathology-2020 (b), and plant disease recognition (c).

5.2 Models

we will define a CNN model, neither complex nor simple, containing four hidden layers. We will try to add a normalization layer and batch normalization layer. The objective is to:

1. Detect the effects of these different modifications on the architecture of the CNNs.
2. To determine how I will choose the best model.
3. To understand the importance of the normalization layer and the batch normalization layer.

The Networks Structures used of Plant-Pathodology-2020 and plant-disease-recognition-dataset are:

1. Model: is the CNNs models with four hidden layers.
2. Model+ Normalization layer
3. Model + Batch Normalization layer
4. **Normalization Model:** is the CNNs models with Normalization layer and Batch Normalization layers (Table 2).

Table 2. Convolutional neural network (CNN) model description

Layer	Description
input	32 × 32 color
2 × convolution	64 filters
maxpool	output size: 16 × 16
2× convolution	128 filters
maxpool	output size: 8 × 8
2× convolution	256 filters
maxpool	output size: 4 × 4
2× convolution	512 filters
maxpool	output size: 2 × 2
reshape	flatten
2× linear	512 hidden units
softmax	c-way softmax

To have the effect of the normalization layer and the Batch normalization layer we will use a large Plantvillage dataset and deeper CNN models VGG16 and VGG19 to test also **Normalization Model**.

The Hyperparameters Initializations of Model: epochs= 50, image_size= $(150, 150)$, batch_size= 100, learning_rate= 10^{-2}, Droptout with $p = 0.3$, kernels_size= $(3, 3)$, pool_size= $(2, 2)$.

6 Experimental Results

In this section, we will view the results associated with each CNNs models and each Dataset (PlantVillage, Plant Pathodology 2020, and Plant disease recognition).

Configuration of the Computer: The hardware configuration used in our implementation is:

- Lenovo i7 CPU 3.00 GHZ laptop
- RAM size 8 GO
- Windows 10, 64 bit operating system (Tables 3, 4, 5 and Figs. 5, 6 and 7)

Table 3. The different results of Datasets Plant-Pathodology-2020 for many modifications.

Datasets	Method	Accuracy	Compilation time (GPU)
Plant-Pathodology-2020	Model	71% (mean)	1 min 21 s
	Model+ Normalization layer	74% (mean)	2 min 3 s
	Model + Batch Normalization layer	80% (mean)	1 min
	Normalization Model	80% (mean)	2 min 9 s

Table 4. The different results of Datasets plant-disease-recognition-dataset for many modifications.

Datasets	Method	Accuracy	Compilation time (GPU)
plant-disease-recognition	Model	86% (mean)	2 h 29 min
	Model+ Normalization layer	60% (mean)	3 h 26 min
	Model + Batch Normalization layer	88% (mean)	1 h
	Normalization Model	88% (mean)	1 h 42 min

Table 5. The different results of Datasets PlantVillage.

Datasets	Method	Accuracy	Loss	Compilation time (GPU)
PlantVillage	VGG16	10% (mean)	3, 3	2 h 24 min
	VGG19	10% (mean)	3, 3	2h 24 min
	Normalization Model	75% (mean)	0, 33	1 h 30 min

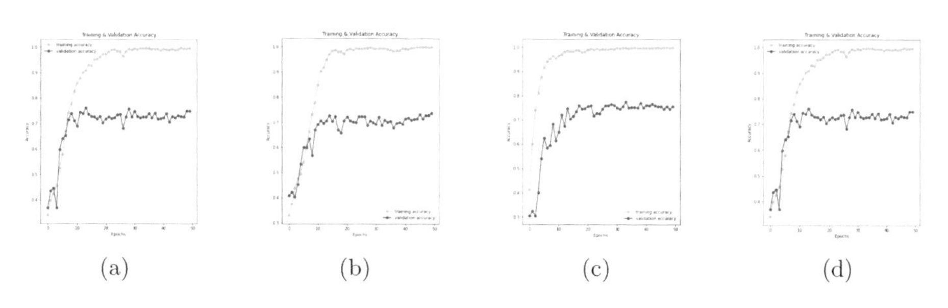

(a) (b) (c) (d)

Fig. 5. The results of Datasets Plant-Pathodology-2020 for different models: Model (a), Model+ Normalization layer (b), Model + Batch Normalization layer (c), **Normalization Model** (d).

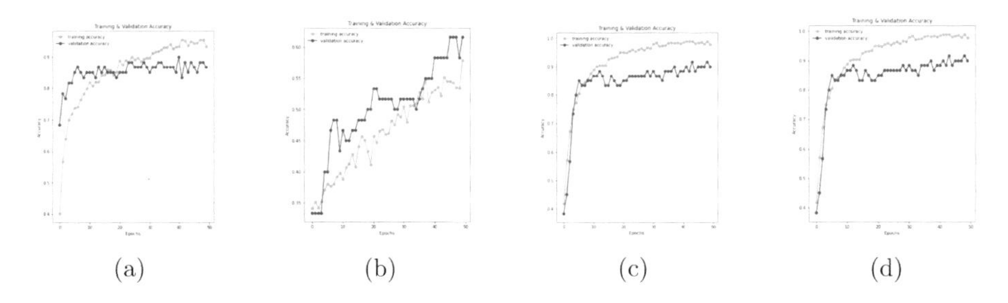

(a) (b) (c) (d)

Fig. 6. The results of Datasets plant-disease-recognition-dataset for different models: Model (a), Model+ Normalization layer (b), Model + Batch Normalization layer (c), **Normalization Model** (d).

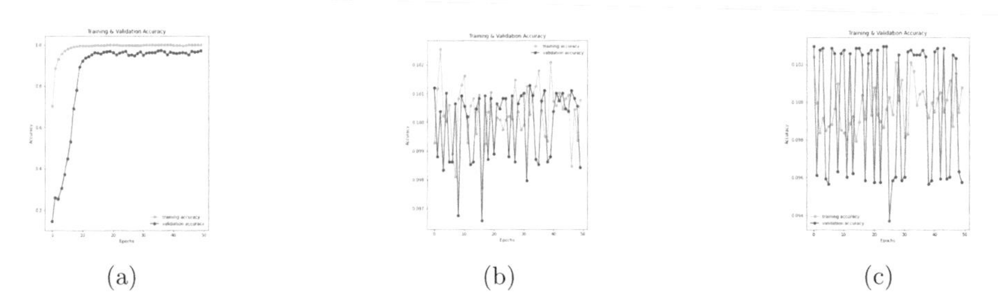

(a) (b) (c)

Fig. 7. The results of Datasets PlantVillage: **Normalization Model** (a), VGG16 (b) and VGG19 (c).

7 Discussion

The result of the plant-pathology-2020 and plant-disease-recognition datasets is beneficial for two things: the first is the necessity of normalization and batch normalization layers. For the normalization layer in general, we do not have the best result, but the purpose of this addition is to reduce the error and homogeneity of the data. And for the batch normalization layers are the acceleration of CNN and the regulator managed to get the best result in a short time of the model without BN [17]. The second is the effect of the batch normalization layer in model CNNs is accelerated and regularized and this is well justified by the results we found. So the Batch normalization has resulted in two big problems concerning the CNNs that is the problem of regularization and the problem of acceleration. Hence the use of a BN layer implies that it is not really used as a regularization method and also BN can be considered the most efficient acceleration method for the CNNs [21]the effect of the BN layer is well on convolutional neural networks but the justification of this effect is between two things, the reduction of the ICS that we have shown in section five. And we also have the justification to make the optimization smooth [22]. Based on this, we will create the **Normalization Model** that we will apply to PlantVillage datasets and compare it with the architecture of VGG16, and VGG19 CNNs. From the results, we find that the best model created is more efficient than the VGG16, and VGG19 architectures. These results assure that we need the normalization layer, and batch normalization layer.

In the sense of comparison, there are many articles in this field that deal with the problem of normalization in convolutional neural networks, we can compare this work with the following works [16,21] and [22] these three articles confirm that the effect of the addition of the Batch Normalization layer in the architecture of CNNs is acceleration and regularization and this is very clear from the results obtained in these articles or in this work But the problem is what is the justification of these effects of BN? The answer to this question is different in these three articles one is justifying these effects by making the optimization smooth [22] and the other one says that the justification is the reduction of ICS [16], in this term the goal of my work in the future is to get the right answer to this question with justification based on mathematical relations and real results.

8 Conclusion

In spite of all the advantages and revolutions that have known the convolutional neural networks in different fields. There are disadvantages in the research for CNNs, among these disadvantages we have Batch Normalization, the effect of adding a layer of Batch Normalization on a CNNs and accelerate and regularize the model. The question is: what is the justification of this acceleration, is it to reduce the ICS or to make the optimization smooth?

From this study we can conclude that using a CNNs with layers of Normalization, Batch Normalization gives better results than a usual CNNs architectures. And finally, the questions about Batch Normlization remain open for future research.

References

1. LeCun, Y., Bengio, Y., Hinton, G.: Deep learning. Nature **521**(7553), 436–444 (2015)
2. Shrestha, A., Mahmood, A.: Review of deep learning algorithms and architectures. IEEE Access **7**, 53040–53065 (2019)
3. LeCun, Y., Denker, J.S., Henderson, D., Howard, R.E., Hubbard, W., Jackel, L.D.: Handwritten digit recognition with a back-propagation algorithm. In: Proceedings of NIPS (1990)
4. Gu, J., et al.: Recent advances in convolutional neural networks. Pattern Recogn. **77**, 354–377 (2018)
5. Lu, J., Tan, L., Jiang, H.: Review on convolutional neural network (CNN) applied to plant leaf disease classification. Agriculture **11**(8), 707 (2021)
6. Arunnehru, J., Vidhyasagar, B.S., Anwar Basha, H.: Plant leaf diseases recognition using convolutional neural network and transfer learning. In: Bindhu, V., Chen, J., Tavares, J.M.R.S. (eds.) International Conference on Communication, Computing and Electronics Systems. LNEE, vol. 637, pp. 221–229. Springer, Singapore (2020). https://doi.org/10.1007/978-981-15-2612-1_21

7. Wu, J.: Introduction to convolutional neural networks. Natl. Key Lab Novel Softw. Technol. Nanjing Univ. **5**(23), 495 (2017)
8. Albawi, S., Mohammed, T.A., Al-Zawi, S.: Understanding of a convolutional neural network. In: 2017 International Conference on Engineering and Technology (ICET), pp. 1–6. IEEE, Turkey (2017). https://doi.org/10.1109/ICEngTechnol.2017.8308186
9. Aloysius, N., Geetha, M.: A review on deep convolutional neural networks. In: 2017 International Conference on Communication and Signal Processing (ICCSP), pp. 0588–0592. IEEE, India (2017). https://doi.org/10.1109/ICCSP.2017.8286426
10. Glorot, X., Bengio, Y.: Understanding the difficulty of training deep feedforward neural networks. In: Proceedings of the Thirteenth International Conference on Artificial Intelligence and Statistics, pp. 249–256. JMLR Workshop and Conference Proceedings (2010)
11. Zhang, Z.: Derivation of backpropagation in convolutional neural network (CNN). University of Tennessee, Knoxville (2016)
12. LeCun, Y., Bottou, L., Orr, G.B., Müller, K.-R.: Efficient backprop. In: Orr, G.B., Müller, K.-R. (eds.) Neural Networks: Tricks of the Trade. LNCS, vol. 1524, pp. 9–50. Springer, Heidelberg (1998). https://doi.org/10.1007/3-540-49430-8_2
13. Shao, J., Hu, K., Wang, C., Xue, X., Raj, B.: Is normalization indispensable for training deep neural network? Adv. Neural. Inf. Process. Syst. **33**, 13434–13444 (2020)
14. Ba, J.L., Kiros, J.R., Hinton, G.E.: Layer normalization. arXiv preprint arXiv:1607.06450 (2016)
15. Bjorck, N., Gomes, C.P., Selman, B., Weinberger, K.Q.: Understanding batch normalization. Adv. Neural Inf. Process. Syst. **31** (2018)
16. Ioffe, S., Szegedy, C.: Batch normalization: accelerating deep network training by reducing internal covariate shift. In: International Conference on Machine Learning, pp. 448–456. PMLR (2015)
17. Zhang, H., Feng, L., Zhang, X., Yang, Y., Li, J. : Necessary conditions for convergence of CNNs and initialization of convolution kernels. Digit. Signal Process. 103397 (2022)
18. PlantVillage Dataset - Kaggle. https://www.kaggle.com/datasets/emmarex/plantdisease. Accessed 30 Oct 2018
19. Plant Pathology 2020 - FGVC7 - Kaggle. https://www.kaggle.com/c/plant-pathology-2020-fgvc7. Accessed 2020
20. Plant disease recognition dataset - Kaggle. https://www.kaggle.com/datasets/rashikrahmanpritom/plant-disease-recognition-dataset. Accessed 2021
21. Schilling, F.: The effect of batch normalization on deep convolutional neural networks (2016)
22. Santurkar, S., Tsipras, D., Ilyas, A., Madry, A.: How does batch normalization help optimization?. Adv. Neural Inf. Process. Syst. **31** (2018)

The Role of Ground Truth Annotation in Semantic Image Segmentation Performance for Autonomous Driving

Ihssane Bouasria(✉), Walid Jebrane, and Nabil El Akchioui

Research and Development Laboratory in Engineering Sciences (LRDSI), Abdelmalek Essaâdi University, Al Hoceima, Morocco

ihssane.bouasria@gmail.com, walid.jebrane@etu.uae.ac.ma

Abstract. The key to reaching full autonomous vehicles lies in the degree of robustness and generalization of the visual perception system for any prompt scenario during driving or under any weather condition. Regardless of all the advent in semantic segmentation and the numerous datasets existing, they still fail to perform well when the input images are unclear (fog, rain, snow, or nighttime), or small-scale. In this study, we will highlight the impact of choosing the adequate coarse ground truth annotation on achieving the best outcome possible out of the preparation and training of the dataset to increase the accuracy of the perception system in driverless cars. Our experiment has shown that indeed the accuracy values of semantic image segmentation using coarse ground truth annotation outperform the fine ground truth for the principal classes (car, road) which implies faster datasets preparation and tuning before pragmatic applications.

Keywords: Semantic Segmentation · Autonomous Vehicles · Computer Vision · Perception · Deep learning · Accuracy · Object Detection

1 Introduction

The application of Deep learning in combination with computer vision for tasks, such as scene understanding, and recognition is a key choice for accomplishing its utmost performance. However, these systems rely on large amounts of labeled training data, expensive data collection, and labeling to generate fine ground truth (GT) annotations that are utilized later for the training and testing of deep learning networks [1, 2]. Semantic segmentation is a revolutionary idea to maintain satisfactory performance for scene understanding. Generally, this technique assigns each pixel of an image to a corresponding class label from a predefined set of categories [3]. in spite of the fact that this problem has been deemed to be a pixel-level classification within the pixels of an image, it is known to be a much more sophisticated approach, contrary to the typical classification which aims to predict the label of the entire image.

Although autonomous driving vision-based applications strictly require both effectiveness and efficiency it is possible to investigate the ability of some functionalities e.g.,

M. Lazaar et al. (Eds.): BDIoT 2022, LNNS 625, pp. 263–278, 2023.
https://doi.org/10.1007/978-3-031-28387-1_23

braking and anti-collision systems, to rely only on the coarse ground truth (GT) annotations without diminishing the accuracy. Accordingly, we reduce the labor-intensive task of high-quality datasets preparation by annotating selective main classes, for instance, road, car, pedestrian, and traffic signs, etc. instead of going through the time-consuming process of annotating the whole and each existing object class to run the image semantic segmentation tasks. By labeling each pixel with the class of the item or region it surrounds, semantic picture segmentation creates a segmentation mask. This task has drawn a lot of attention from professionals across a variety of disciplines, including autonomous driving, robot navigation, scene analysis, and medical imaging, due to its broad applicability. Deep learning has established itself as the standard method for semantic image segmentation as a result of its enormous success. Deep CNNs are typically trained for image analysis tasks including classification, object identification, and semantic picture segmentation using vast amounts of labeled training data. This is particularly true for semantic picture segmentation, where each training image's pixel must be labeled or annotated in order to infer the labels of each test image's pixel. It is difficult to get enough densely annotated photos, especially in fields like material science, engineering, and medicine where it takes a lot of effort and significant user knowledge to annotate images. Therefore, creating high-performance deep segmentation networks that can learn from sparsely annotated training data is very advantageous. The assumption that coarse object shapes learning could to some extent address the problem of generalization for segmentation to novel categories, has been evaluated as a key solution, especially in instance segmentation using Shape Mask [4], this type of annotation can then be adapted to common object identification and localize them faster in a crowded urban environment [5–8].

Contribution. We will tackle the issue of how beneficial is the use of coarse ground truth annotation for semantic image segmentation, with emphasis on accuracy and performance. The current trend of numerous applications of datasets in the AVs is focusing mainly on applying the fine ground truth annotations, the issue with this type of annotation is the long hours of labeling to provide a single fine annotated image, thus multiplies the difficulty of preparing and training the model for a problem within a highly changeable and dynamic environment such as for the self-driving cars scenarios. The suggestions about avoiding annotating each objects classes used for semantic segmentation tasks have been speculated on many occasions as a solution to reduce the labeling and data preparation time, with the assumption that coarse object shapes are more favorable for identification and localization on the road, therefore speeding up the preparation of the datasets for model training. Our contribution will investigate this proposal's reliability by comparing the results of fine ground truth annotation and coarse annotation trained on the PSPNet network with regard to the accuracy of semantic image segmentation, based on the mean intersection over union (mIoU) metric of various object classes. This study will demonstrate the margin between the performance quality of semantic image segmentation in an urban area crowded with traffic using high-quality fine GT annotation and inferior images with coarse annotation.

The rest of the paper is organized as follows: we will highlight a few relevant works and we present an exhaustive list of the state-of-the-art datasets and their characteristics in Sect. 2. The approaches, datasets, and metrics used in the evaluation in this work are elaborated in Sect. 3. In Sect. 4, we present our analysis results with discussion and interpretation, and conclusions are drawn in Sect. 5.

2 Background and Related Work

Machine learning applications, particularly in the deep learning domain, are rapidly diversifying and expanding to be effective, deep learning methods must employ a large amount of data for training. To date, a deficiency of information has made preparing profound neural systems (DNNs) challenging. Further, extra data is needed to validate a trained DNN to confirm that its predictions are trustworthy [23]. Since many real-world situations do not have ideal data configurations, numerous strategies have been developed to train models effectively despite a scarcity of data. Understanding that semantic segmentation is not an independent field but rather a logical step in the path from coarse to fine inference is crucial to understanding how semantic segmentation is approached by contemporary deep learning architectures. The starting point could be classified, which entails creating a prediction for the entire input, such as identifying the object in an image or even providing a ranked list if there are several of them. The next step towards fine-grained inference is localization or detection, which provides not only the classes but also extra details about their spatial location, such as centroids or bounding boxes. Given that, it follows naturally that semantic segmentation is the next stage in achieving fine-grained inference; its objective is to make dense predictions inferring labels for every pixel, labeling each pixel with the class of the surrounding item or region.

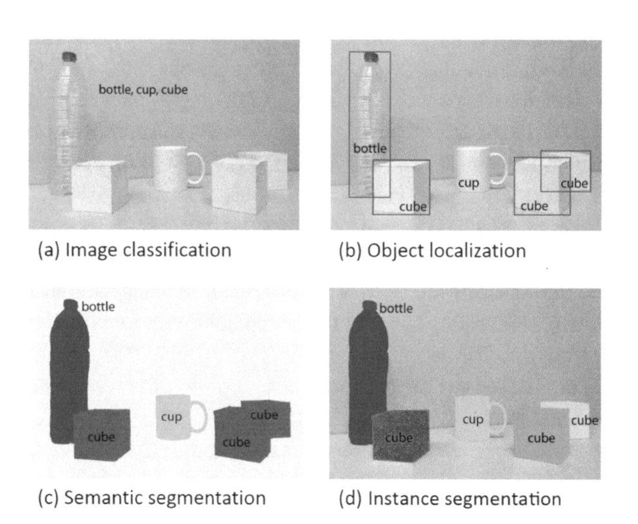

(a) Image classification (b) Object localization

(c) Semantic segmentation (d) Instance segmentation

Fig. 1. Evolution of object recognition or scene understanding from coarse-grained to fine-grained inference: classification, detection or localization, semantic segmentation, and instance segmentation.

Additional enhancements are possible, including part-based segmentation and instance segmentation (distinct labels for various instances of the same class) (low-level decomposition of already segmented classes into their components). The aforementioned progression is seen in Fig. 1.

2.1 Segmentation Methods

Deep learning methods have consistently outperformed other high-level computer vision techniques, particularly supervised methods like Convolutional Neural Networks (CNNs) for object or image detection [24–26]. This has inspired researchers to investigate the potential of such networks for pixel-level labeling issues like semantic segmentation. The main benefit of these deep learning techniques, which gives them an edge over conventional methods, is the ability to learn appropriate feature representations for the task at hand, such as pixel labeling on a specific dataset, in an end-to-end manner as opposed to using hand-crafted features that require domain expertise, effort, and frequently too much fine-tuning to make them work on a specific scenario.

A model must learn from a small number of labeled instances and a large number of unlabeled examples when using semi-supervised learning. Then, using fresh samples, this model needs to make predictions. The most efficient semi-supervised learning techniques include generative models like the Generative Adversarial Network (GAN) and Variational Autoencoder [24, 25]. Using a combination of tagged and unlabeled data, Hung et al. [25] trained the DNNs to improve segmentation masks using a fully convolutional discriminator. Mittal et al. [26] suggested a dual branch GAN based technique for semi-supervised semantic segmentation as a solution to the issue of inaccurate boundary detection and incorrect class assignment of large regions. They also suggested a technique for using unlabeled images to generate pseudo labels and then using these for network training to improve performance. On the PASCAL VOC dataset, their approach achieved 75.6% mIOU (mean intersection over union). Laine et al. [27] used an ensemble of many models trained using various regularizes and augmentation procedures in semi-supervised conditions to produce consensus predictions of unknown labels. Using only 500 labelled samples, they were able to reduce the classification error from 18 to 5.12% when using their strategy to the SVHN dataset. In order to reduce the variance in predictions of multiple passes of a data sample through the network during the training phase, Sajjadi et al. [28] gradient's descent-optimized unsupervised loss function was proposed. This function utilized randomized data transformation and augmentation and improved generalization during inference. They were able to achieve an error rate of 0.27% when their method was applied to the MNIST dataset using just 100 labelled samples. Tarvainen et al. [29] suggested averaging the model weights over various training phases when using the trained model's final weights to increase model resilience. They lowered the error rate to 4.18% when their approach was applied to the SVHN dataset using this technique and only 500 tagged samples. In their system for skin lesion segmentation, Li et al. [30] combined labeled and unlabeled data with a self-assembling strategy that encouraged the network to produce consistent predictions for the same input under various regularizations. They created a new performance benchmark (75.3% mIOU) on the international skin imaging collaboration (ISIC) dataset with just 300 labelled samples. Perone et al. [31] segmented MRI images using the Mean Teacher technique, which was first put forth in [32]. They achieved a mIOU of 55.5%, which is comparable to the 53.6% attained using supervised learning. French et al. showed that consistency regularization is a practical method of semi-supervised segmentation when the appropriate source of augmentation is applied, producing cutting-edge results. Additionally, their method was much simpler to use and deploy than GAN-style training.

There are three types of weakly supervised learning: incomplete supervision, where only a portion of the training data is labeled, inexact supervision, where only coarse-grained labels are assigned; and inaccurate supervision, where the assigned labels are not always precise. Using only the image-level labels provided during training, Pinheiro et al. [33] established a weakly supervised framework in which they created pixel-level labels of objects in images. They trained their CNN to focus on the pixels that were most important for classifying the image, and then they employed various smoothing priors to expand the application of the CNN to segmentation. They were successful in achieving benchmark performance (weakly supervised segmentation) using this method on the PAS-CAL VOC dataset. Which elements of an image cause neurons to fire are determined by the CAM (channel activation maps) family of algorithms [2, 34, 35]. Singh et al. [36] blocked off a section of the images at random to force their network to locate more than one discriminative region of an object in order to improve localization performance. Wei et al. [37] presented an adversarial erasing strategy to mine several discriminative object regions in order to address segmentation issues brought on by insufficient supervision. On the PASCALVOC dataset, they were able to get 55.7% mIoU by using this approach. Attention maps have been used in other research in an effort to enhance segmentation outcomes [38, 39]. To create segmentation masks, Huang et al. [40] introduced the deep seed region expanding algorithm. They suggested training a semantic segmentation network first using discriminative areas, then pixel-level supervision should be gradually increased through seeded region growth [41]. Their system attained a 66% mIOU using this technique on the PASCAL VOC dataset. It has also been researched to identify segmentation zones using box annotations that contain objects [42, 43].

Table 1. Summary of state-of-the-art algorithms in domain of weakly and semi supervised frameworks for image segmentation.

Authors	Year	Algorithm	Framework	Description
Singh, K. K. & Lee, Y. [36]	2017	Hide & Seek	Weakly-supervised	Trained the network by forcing it to locate more than one discriminative region of an object by blocking out the portions of an image at random
Huang, Z. et al. [40]	2018	Region Growing	Weakly-supervised	In contrast to conventional segmentation algorithms that use static labels, the authors here used seeded region growth algorithm to generate new labels during each training cycle

(continued)

Table 1. (*continued*)

Authors	Year	Algorithm	Framework	Description
Dai, J., K. He, & Sun [42]	2016	BoxSup	Weakly-supervised	For network training, bounding box annotations were utilized to produce candidate masks using unsupervised region proposal methods, these candidate masks improve with each iteration, providing more and more valuable information for CNN training
Wei, Y. et al. [37]	2017	Adversarial Erasing	Weakly-supervised	Used adversarial erasing approach for progressively mining the discriminative object regions during classification, then used these mined regions to generate complete dense objects
Pinheiro, P.O. & Collobert, R. [33]	2015	Overfeat + Pixel-wise Segmentation	Weakly-supervised	Firstly, they generated pixel-level labels of objects in images using only image-level labels provided during training, then used different smoothing priors on pixels, that played crucial role in image classification, to generate segmentation masks
Li, X. et al. [30]	2018	Self-Ensembling Model	Semi-supervised	The algorithm enabled the network-in-training to provide consistent predictions for the same input under different regularizations the network was optimized via weighted combination of supervised loss (labeled data) and a regularization loss (labeled + unlabeled data)

(*continued*)

Table 1. *(continued)*

Authors	Year	Algorithm	Framework	Description
Mittal, S. et al. [26]	2019	Dual-branch GAN	Semi-supervised	The authors generated pseudo labels from unlabeled images, which were then used to train the network, final segmentation masks were refined by MLMT (multi-label mean teacher) sub-network to improve performance
Hung, W.-C. et al. [44]	2018	Fully Convolutional Discriminator	Semi-supervised	The authors utilized a discriminator network to train a CNN for semantic segmentation task using both labelled and unlabeled images, the network was optimized by coupling adversarial and standard cross entropy loss

Semi-supervised learning that employments a combination of labeled and unlabeled tests can improve a model's execution. The generation of a segmentation label for a plant, even for a single image, is a time-consuming job. When very little data is available for training, weakly supervised algorithms are available, though these are difficult to implement, train, validate, and quantify. Moreover, semi- and weakly-supervised algorithms have complex model structures and frequently use pre-trained models to increase performance.

2.2 Segmentation Datasets

Data is arguably one of the most crucial components of any machine learning system. This significance is heightened even further when working with deep networks. For any segmentation system based on deep learning methods, collecting sufficient data into a dataset is essential. Gathering and constructing an appropriate dataset, which must have a scale large enough and represent the use case of the system accurately, needs time, domain expertise to select relevant information, and infrastructure to capture that data and transform it to a representation that the system can properly understand and learn. This task, despite the simplicity of its formulation in comparison with sophisticated neural network architecture definitions, is one of the hardest problems to solve in this context. Because of that, the most sensible approach usually means using an existing standard dataset which is representative enough for the domain of the problem. In fact, many datasets are part of a challenge that reserves some data for a competition in which many methods are tested, generating a fair ranking of methods according to their actual performance without any kind of data cherry-picking. This approach has another benefit for the community. Standardized datasets enable fair comparisons between systems. The most often used large-scale datasets for semantic segmentation are listed below.

Cityscapes [48] is a large database that focuses on the semantic interpretation of urban street scenes. It provides 30 classes separated into eight categories, as well as semantic, instance-wise, and dense pixel annotations (flat surfaces, humans, vehicles, constructions, objects, nature, sky, and void). The collection contains around 5000 highly tagged photographs and 20,000 coarsely annotated images. Data was collected over several months, throughout the day, in 50 cities with acceptable weather. It was originally shot as video; thus, the frames were hand-picked to include a range of dynamic things, scenario sets, and backdrops.

For real-time implementation, autonomous vehicles set onboard three types of sensors to get the awareness of their surroundings, mainly they are cameras that capture scenes and provide the semantics. The distance and velocities of other vehicles are established from LiDAR and Radar respectively. Datasets for driving scene understanding are satisfying for the research and experiments but they are far from being enough for full perception and a safe ride. KITTI [9] dataset which is based on LiDAR and Cityscapes on semantic image annotations has pioneered this area. Subsequent datasets have essentially as purpose to increase the scale [10], the number of tasks [11], and diversity [12].

However, these vehicles' AI systems still suffer from insufficient volume of data annotation to feed them, as the Deep learning methods are only efficient when the annotation is precise and huge in amount, thus helping the AI to accurately interpret the scenes in their field. Particularly, the need for an adequate estimation of the range or class to which a pixel belongs. Otherwise, a car cannot identify the physical object from a painted one on the street. Arguably, the solution to such an issue is to feed the system and train deep learning algorithms with more data by providing a huge dataset that contains thousands, or even better millions of images annotated.

Table 2. Popular large-scale segmentation datasets.

Datasets	Classes	Year	Purpose	Description
PASCAL VOC [45]	21	2012	Generic	PASCAL Visual Object Classes (VOC) this challenge consists of a ground-truth annotated dataset of images and five different competitions: classification, detection, segmentation, action classification, and person layout.

(continued)

Table 2. (*continued*)

Datasets	Classes	Year	Purpose	Description
Cityscapes [48]	8	2015	Urban	Cityscapes: is a large-scale database which focuses on semantic understanding of urban street scenes. It provides semantic, instance-wise, and dense pixel annotations for 30 classes grouped into 8 categories (flat surfaces, humans, vehicles, constructions, objects, nature, sky, and void).
KITTI-Layout [49]	3	2012	Urban/Driving	KITTI: is one of the most popular datasets for use in mobile robotics and autonomous driving. It consists of hours of traffic scenarios recorded with a variety of sensor modalities, including high resolution RGB, grayscale stereo cameras, and a 3D laser scanner.
Adobe's Portrait Segmentation [50]	2	2016	Portrait	Adobe's Portrait Segmentation: this is a dataset of 800 _ 600 pixels portrait images collected from Flickr, mainly captured with mobile front facing cameras.
MINC [51]	23	2015	Materials	Materials in Context (MINC): this work is a dataset for patch material classification and full scene material segmentation.
Stanford 2D-3D-S [54]	13	2017	Indoor	Stanford 2D-3D-S: is a multi-modal and largescale indoor spaces dataset extending the Stanford 3D Semantic Parsing work.

(*continued*)

Table 2. *(continued)*

Datasets	Classes	Year	Purpose	Description
SYNTHIA [47]	11	2016	Urban (Driving)	Synthetic Collection of Imagery and Annotations (SYNTHIA): is a large-scale collection of photorealistic renderings of a virtual city, semantically segmented, whose purpose is scene understanding in the context of driving or urban scenarios.
ShapeNet Part [53]	16	2016	Object/Part	ShapeNet Part: is a subset of the ShapeNet repository which focuses on fine-grained 3D object segmentation.

Nowadays, semantic segmentation annotation has been considered to be more complex than simple 2D bounding boxes[13–15], and numerous automotive corporates have already a massive dataset e.g., Tesla Autopilot's proprietary solution and dataset [16, 17] were obtained from 8 surround cameras and powerful vision processing provides 360 degrees of visibility at up to 250 m of range. (12 ultrasonic sensors and forward-facing radar can see vehicles through adverse conditions), and Cambridge-driving Labeled Video Database (CamVid) [18, 19] is also famous for its use of the traffic cam videos, and the extensively used dataset is Cityscapes [20, 21] which contains various stereo video sequences recorded in 50 cities and captures rich urban street scenes. Figure 1 shows examples of Cityscapes dataset annotation types. Yet we still lack the necessary volume of data with precise individual objects annotation.

The design of convolutional neural networks has boosted the semantic segmentation, mostly using fully convolutional architectures, the current state-of-the-art includes e.g., ANN with pyramid pooling modules, HRNet, and DeepLabv3+, we consider also CCNet with attention mechanisms and DANet to be a promising area to be investigated.

3 Experimental Details

Datasets. We evaluate our work on the pioneer dataset Cityscapes, is a large-scale database containing high-quality pixel-level annotations for 5000 images and 20000 coarse annotated ones. It focuses on the semantic understanding of urban street scenes. Although the scenes were captured in 50 different cities for several months which increases the scope of possibilities to study and compare results within the same dataset, some researches have shown concern about how accurate this dataset is when used for training and testing semantic segmentation methods on adverse visual conditions, as Cityscapes has a major drawback by including only images of daytimes, and normal-good weather conditions. However, Cityscapes still dominate the practical applications

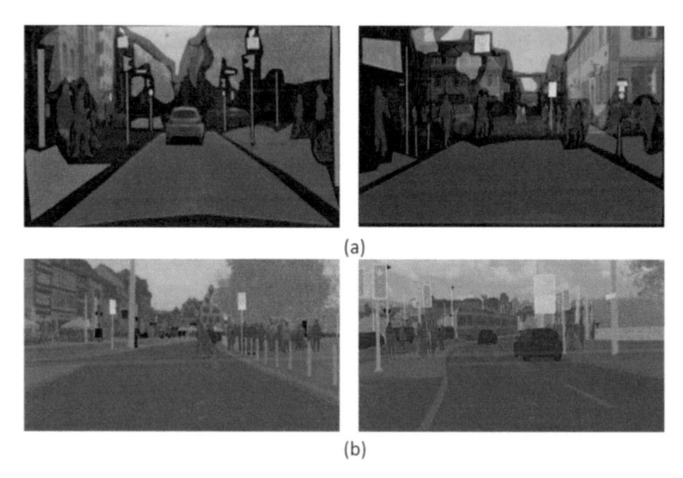

Fig. 2. Cityscapes dataset type of annotations examples (a) with coarse annotations, (b) with fine annotations.

and have demonstrated high performance compared to other datasets, as it distinguishes the data into 30 classes gathered into 8 categories: flat surfaces, humans, vehicles, constructions, objects, nature, sky, and void (Fig. 3).

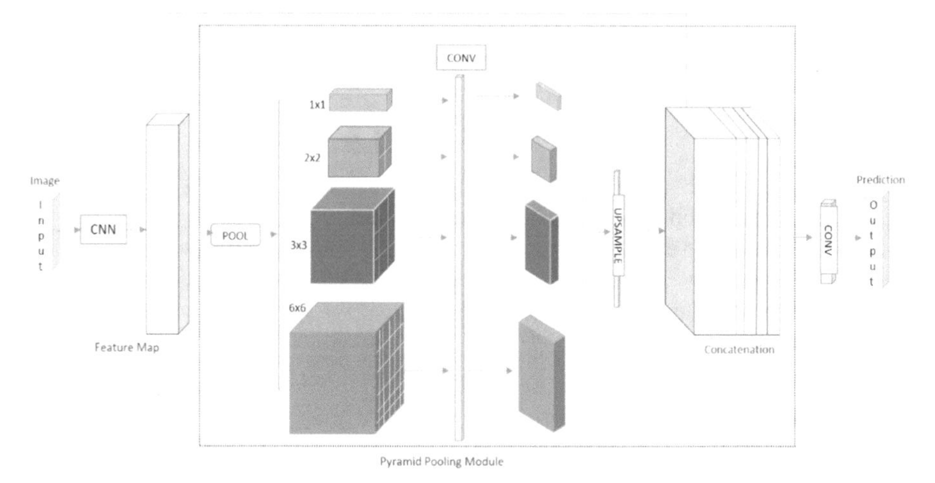

Fig. 3. Overview of Pyramid Scene Parsing Network (PSPNet). 1. The CNN to get the feature map of the last convolutional layer, 2. a pyramid parsing module to harvest different sub-region representations, then the upsampling and concatenation layers to form the final feature representation (local and global context information), 3. The final feature representation is fed into a convolution layer to get the final per-pixel prediction. (Adapted from [22]).

Deep Learning Model. We investigate the deep learning model, the Pyramid Scene Parsing Network (PSPNet) of real-time semantic segmentation for both fine and ground truth annotations, based on its performance in terms of accuracy in Cityscapes. The PSPNet architecture is illustrated in Fig. 2. Our focus is to extract features from the scenes and compare the training and the prediction results. This model functions efficiently in terms of computational cost, memory usage, and power. It is known to be effective to produce good quality results on the scene parsing task, and providing a superior framework for pixel-level prediction more reliable, thanks to its state-of-art pyramid pooling module.

Metrics. We provide an evaluation of the model based on one of the most extensively used measures for assessing the effectiveness of semantic segmentation algorithms mean Intersection over Union (mIoU). In addition, there are other means to evaluate the efficiency of a semantic image segmentation model, in other words, to state precisely the processing time, and the computational and memory load of the models. Thus, effectiveness and efficiency (frames per second (fps), inference time, and memory usage) are known to be the two main evaluate criteria most used.

Intersection over Union (IoU) is mathematically calculated by the intersection of the predicted segmentation map and the ground truth, divided by the area of union between the predicted segmentation map and the ground truth defined by Eq. (1), and the metric used for this work is *mean Intersection over Union* (mIoU) which is presented in Eq. (2) and defined as the average IoU over all classes.

$$IoU = \frac{\sum_{i=0}^{k} p_{ii}}{\sum_{i=0}^{k}\sum_{j=0}^{k} p_{ij} + \sum_{i=0}^{k}\sum_{j=0}^{k} p_{ji} - \sum_{i=0}^{k} p_{ii}} \tag{1}$$

$$mIoU = \frac{1}{k+1}\sum_{i=0}^{k} \frac{p_{ii}}{\sum_{j=0}^{k} p_{ij} + \sum_{j=0}^{k} p_{ji} - p_{ii}} \tag{2}$$

Such as p_{ij} the number of pixels of class i predicted/presumed as belonging to class j, for k + 1 classes (+1 class corresponds to background).

We have pretrained the PSPNet network on the Cityscapes dataset and then applied it to these data subsets to measure the image prediction accuracy, using mean Intersection over Union (mIoU) and compare them for fine and coarse ground truth annotations for the corresponding cities.

4 Results

Besides the fact that developing precise and high-quality datasets is an essential key to an efficient understanding of the scenes based only on images processing is a labor-intensive and time-consuming task, the complexity increases depending on the application, and datasets for autonomous cars like Cityscapes are challenging in terms of time and cost because they need to be designed and primarily fine tunned before real tasks especially to improve the major classes prediction e.g. road, cars, person and traffic (signs, lights).

We narrow our results to two cities per annotation type because of the dataset size. However, these results are optimistic and able to be projected to other urban cities and regions to give us every confidence in its reliability. Furthermore, our model was trained with fine annotated images of Cityscapes by the PSPNet network, Table 1 summarizes the results obtained using the mIoU metric to evaluate the effectiveness of the proposal (Table 3).

Table 3. The effectiveness of the semantic segmentation model PSPNet using mIoU metric.

		Road	Person	Car	Traffic light	Traffic sign	Other classes
Fine GT	Münster	98.6	52.1	89.8	41.5	68.7	54.8
	Düsseldorf	71.2	37.8	81.3	12.4	58.6	43.1
Coarse GT	Nuremberg	99.7	81.3	95.2	60.0	84.1	67.2
	Dortmund	99.8	72.1	97.8	68.3	86.2	71.0

Generally, the mean intersection over union (mIoU) shows how accurate is the semantic image segmentation, the following results have shown that coarse GT is a promising aspect to be explored for fine-tuning the pre-trained models. For instance, mIoU values of road and car have outperformed the fine GT. Thus, the results in Table 2. Have demonstrated positive prospects for the coarse GT as we have found that the accuracy (mIoU) with coarse GT outperforms the fine GT (Table 4).

Table 4. The mean per-class accuracy (mIoU) values for studied cities.

	road	person	car	Traffic light	Traffic sign
Fine GT	89.2	50.1	85.4	34.0	63.5
Coarse GT	99.6	79.2	96.3	64.1	82.9

The fine GT is the focus of most major applications because the coarse GT is frequently underrated due to the coarse annotated images being less detailed and alleged to be unreliable to be applied in a practical task that requires precise estimation of the object shape for the perception and the control components. This point can be argued against if it is assumed that object localization for autonomous cars can profit optimally from the coarse GT rather than being ignored, and be employed in tasks that lean toward localization that may be more crucial than acquiring the object's precise shape.

5 Conclusion

Our proposal is to exploit the coarse annotated images for the benefit of the models' training and prediction besides the focus on the fine ground truth. Although the fewer details we still obtain better results for the identification and localization of the objects using Cityscapes urban scenes and even outperform fine annotated images for some classes. The results prove the effectiveness of our proposal. This work is a first step in allowing us to choose the proper annotation type to improve our perception system for urban driving tasks, however, we are inspired to demonstrate the advantage of using coarse GT to train other models for future work and evaluate their performance compared to the fine GT.

Acknowledgment. W. Jebrane acknowledges support from the "Center National for Scientific and Technical Research" CNRST, Morocco.

References

1. Garcia-Garcia, A., Orts-Escolano, S., Oprea, S., Villena-Martinez, V., Garcia-Rodriguez, J.A.: Review on deep learning techniques applied to semantic segmentation, pp. 1–23 (2017)
2. Garcia-Garcia, A., et al.: A survey on deep learning techniques for image and video semantic segmentation. Appl. Soft Comput. J. **70**, 41–65 (2018)
3. Janai, J., Güney, F., Behl, A., Geiger, A.: computer vision for autonomous vehicles: problems, datasets and state of the art. Found. Trends® Comput. Graph. Vis. **12**, 1–308 (2020)
4. Kuo, W., Angelova, A., Malik, J. Lin, T.Y.: ShapeMask: Learning to segment novel objects by refining shape priors. In: Proceedings IEEE International Conference Computer Vision 2019, pp. 9206–9215, October 2019
5. Papadeas, I., Tsochatzidis, L., Amanatiadis, A., Pratikakis, I.: Real-time semantic image segmentation with deep learning for autonomous driving: a survey. Appl. Sci. **11** (2021)
6. Hua, B.S., et al.: SceneNN: a scene meshes dataset with annotations. In: Proceedings - 2016 4th International Conference 3D Vision, 3DV 2016, pp. 92–101 (2016)
7. Chen, P., et al.: Object localization under single coarse point supervision, vol. 2, pp. 4868–4877
8. Jing, L., Chen, Y., Tian, Y.: Coarse-to-fine semantic segmentation from image-level labels. IEEE Trans. Image Process. **29**, 225–236 (2020)
9. Geiger, A., Lenz, P., Urtasun, R.: Are we ready for autonomous driving? The KITTI vision benchmark suite. In: Proceedings IEEE Computer Society Conference Computer Vision Pattern Recognit, pp. 3354–3361 (2012)
10. Huang, X., et al.: The ApolloScape open dataset for autonomous driving and its application. IEEE Trans. Pattern Anal. Mach. Intell. **42**, 2702–2719 (2020)
11. Yu, F. et al.: BDD100K: a diverse driving dataset for heterogeneous multitask learning. Proc. In: IEEE Computer Society Conference Computer Vision Pattern Recognit, pp. 2633–2642 (2020)
12. Shelhamer, E., Long, J., Darrell, T.: Fully convolutional networks for semantic segmentation. IEEE Trans. Pattern Anal. Mach. Intell. **39**, 640–651 (2017)
13. Weng, W., Zhu, X.: UNet: convolutional networks for biomedical image segmentation. IEEE Access **9**, 16591–16603 (2021)
14. Neuhold, G., Ollmann, T., Bulo, S.R., Kontschieder, P.: The mapillary vistas dataset for semantic understanding of street scenes. In: Proceedings IEEE International Conference Computer Vision 2017, pp. 5000–5009, -October 2017
15. Richter, S.R., Vineet, V., Roth, S., Koltun, V.: Playing for data: ground truth from computer games. In: Leibe, B., Matas, J., Sebe, N., Welling, M. (eds.) ECCV 2016. LNCS, vol. 9906, pp. 102–118. Springer, Cham (2016). https://doi.org/10.1007/978-3-319-46475-6_7
16. Tesla. Autopilot: Full Self-Driving Hardware on All Cars. Tesla Motors 1 (2017). https://www.tesla.com/autopilot
17. Ingle, S., Phute, M.: Tesla autopilot : semi autonomous driving, an uptick for future autonomy. Int. Res. J. Eng. Technol. **3**, 369–372 (2016)

18. Brostow, G.J., Fauqueur, J., Cipolla, R.: Semantic object classes in video: a high-definition ground truth database. Pattern Recognit. Lett. **30**, 88–97 (2009)
19. Brostow, G., Shotton, J., Fauqueur, J., Cipolla, R.: Segmentation and recognition using SfM Point Clouds. Eccv, pp. 1–15 (2008)
20. Ramos, S., Rehfeld, T., Enzweiler, M., Benenson, R., Roth, S.: The cityscapes dataset for semantic urban scene understanding (2016)
21. Liu, X., et al.: Importance-Aware Semantic Segmentation in Self-Driving with Discrete Wasserstein Training (2017)
22. Zhao, H., Shi, J., Qi, X., Wang, X., Jia, J.: Pyramid scene parsing network. In: Proceedings 30th IEEE Conference Computer Vision Pattern Recognition, CVPR 2017, pp. 6230–6239, January 2017
23. Abdar, M., et al.: A review of uncertainty quantification in deep learning: techniques, applications and challenges. Inf. Fus. **76**, 243–297 (2021)
24. Kingma, D.P., et al.: Semi-supervised learning with deep generative models. In: Advances Neural Information Processing Systems, vol. 27 (2014)
25. Hung, W.-C., et al.: Adversarial learning for semi-supervised semantic segmentation. arXiv preprint: https://arxiv.org/abs/1802.07934 (2018)
26. Mittal, S., Tatarchenko, M., Brox, T.: Semi-supervised semantic segmentation with high-and low-level consistency. IEEE Trans. Pat. Anal. Mach. Intell. (2019)
27. Laine, S., Aila, T.: Temporal ensembling for semi-supervised learning. In: International Conference on Learning Representations (ICLR) (2017)
28. Sajjadi, M., Javanmardi, M., Tasdizen, T.: Regularization with stochastic transformations and perturbations for deep semisupervised learning. In: Advances Neural Information Systems (2016)
29. Tarvainen, A., Valpola, H.: Mean teachers are better role models: weight-averaged consistency targets improve semi-supervised deep learning results (2017)
30. Li, X., et al.: Semi-supervised skin lesion segmentation via transformation consistent self-ensembling model. arXiv preprint: https://arxiv.org/abs/1808.03887 (2018)
31. Perone, C.S., Cohen-Adad, J.: Deep semi-supervised segmentation with weight-averaged consistency targets. In: Deep Learning in Medical Image Analysis and Multimodal Learning for Clinical Decision Support (2018)
32. French, G., et al.: Semi-supervised semantic segmentation needs strong, varied perturbations. arXiv preprint: https://arxiv.org/abs/1906.01916 (2019)
33. Pinheiro, P.O., Collobert, R.: From image-level to pixel-level labeling with convolutional networks. In: Proceedings of the IEEE Conference on Computer Vision and Pattern Recognition (2015)
34. Selvaraju, R.R., et al.: Grad-cam: visual explanations from deep networks via gradient-based localization. In: Proceedings of the IEEE International Conference on Computer Vision (2017)
35. Zhou, B., et al.: Learning deep features for discriminative localization. In: Proceedings of the IEEE Conference on Computer Vision and Pattern Recognition (2016)
36. Singh, K.K., Lee, Y.J.: Hide-and-seek: forcing a network to be meticulous for weakly-supervised object and action localization. In: 2017 IEEE International Conference on Computer Vision (ICCV). IEEE (2017)
37. Wei, Y., et al.: Object region mining with adversarial erasing: a simple classification to semantic segmentation approach. In: Proceedings of the IEEE Conference on Computer Vision and Pattern Recognition (2017)
38. Li, K., et al.: Tell me where to look: guided attention inference network. In: Proceedings of the IEEE Conference on Computer Vision and Pattern Recognition (2018)
39. Zhang, X., et al.: Adversarial complementary learning for weakly supervised object localization. In: Proceedings of the IEEE Conference on Computer Vision and Pattern Recognition (2018)

40. Huang, Z., et al.: Weakly-supervised semantic segmentation network with deep seeded region growing. In: Proceedings of the IEEE Conference on Computer Vision and Pattern Recognition (2018)
41. Adams, R., Bischof, L.: Seeded region growing. IEEE Trans. Patt. Anal. Mach. **16**, 641–647 (1994)
42. Dai, J., He, K., Sun, J.: Boxsup: exploiting bounding boxes to supervise convolutional networks for semantic segmentation. In: Proceedings of the IEEE International Conference on Computer Vision (2015)
43. Khoreva, A., et al.: Simple does it: weakly supervised instance and semantic segmentation. In: Proceedings of the IEEE Conference on Computer Vision and Pattern Recognition (2017)
44. Everingham, M., Eslami, S.M.A., Van Gool, L., Williams, C.K.I., Winn, J., Zisserman, A.: The pascal visual object classes challenge: a retrospective. Int. J. Comput. Vision **111**(1), 98–136 (2015)
45. Mottaghi, R., et al.: The role of context for object detection and semantic segmentation in the wild. In: IEEE Conference on Computer Vision and Pattern Recognition (CVPR) (2014)
46. Chen, X., Mottaghi, R., Liu, X., Fidler, S., Urtasun, R., Yuille, A.: Detect what you can: Detecting and representing objects using holistic models and body parts. In: IEEE Conference (2017)
47. Ros, G., Sellart, L., Materzynska, J., Vazquez, D., Lopez, A.M.: The Synthia dataset: a large collection of synthetic images for semantic segmentation of urban scenes. In: Proceedings of the IEEE Conference on Computer Vision and Pattern Recognition (2016)
48. Cordts, M., et al.: The cityscapes dataset. In: CVPR Workshop on the Future of Datasets in Vision (2015)
49. Ros, G., Alvarez, J.M.: Unsupervised image transformation for outdoor semantic labelling. In: Intelligent Vehicles Symposium (IV). IEEE (2015)
50. Shen, X., et al.: Automatic portrait segmentation for image stylization. In: Computer Graphics Forum, vol. 35, no. 2. Wiley Online Library (2016)
51. Bell, S., Upchurch, P., Snavely, N., Bala, K., Material recognition in the wild with the materials in context database. In: Proceedings of the IEEE Conference on Computer Vision and Pattern Recognition (2015)
52. Lai, K., Bo, L., Ren, X., Fox, D.: A large-scale hierarchical multi-view RGB-D object dataset. In: Robotics and Automation (ICRA), 2011 IEEE International Conference on, pp. 1817–1824. IEEE (2011)
53. Yi, L., et al.: A scalable active framework for region annotation in 3D shape collections. SIGGRAPH Asia (2016)
54. Armeni, I., Sax, S., Zamir, A.R., Savarese, S.: Joint 2D-3D semantic data for indoor scene understanding. ArXiv e-prints, Febraury 2017
55. Hackel, T., Wegner, J.D., Schindler, K.: Contour detection in unstructured 3D point clouds. In: Proceedings of the IEEE Conference on Computer Vision and Pattern Recognition, pp. 1610–1618 (2016)

Fine-Tuning Pre-trained Vision Transformer Model for Anomaly Detection in Video Sequences

Abdelhafid Berroukham[1](✉), Khalid Housni[1], and Mohammed Lahraichi[2]

[1] L@RI Laboratory, MISC Team, Faculty of Sciences, Ibn Tofail University, Kenitra, Morocco
a.berroukham@gmail.com, housni.khalid@uit.ac.ma
[2] CRMEF, Casablanca, Morocco

Abstract. Detecting anomalous in video sequences is one of the most popular computer vision topics. It is considered a challenging task in video analysis due to its definition, which is subjective or context-dependent. Various deep learning models such as convolutional neural networks (CNNs) have been previously utilized for this purpose. This paper proposes a novel solution based on the state-of-the-art deep learning models called Vision Transformer, since it is a trendy topic nowadays and it is performance. We are going to fine-tune a pre-trained Vision Transformer model on the UCSD dataset, which enables the automatic classification of video frames (abnormal and normal objects). The evaluation of this model shows that it achieves a good Accuracy score.

Keywords: Video processing · Anomaly detection · Vision transformer · Deep Learning

1 Introduction

Anomaly detection and localization is a challenging task in video analysis due to the definition of "anomaly" that can have some degree of ambiguity within context. Visual behaviors are complicated and diverse in an unrestricted world, video quality, shadows, occlusion, illumination, moving camera, and complex backgrounds are challenges to overcome. In general, an event is considered an "anomaly" if it occurs infrequently or unexpectedly [1, 2].

Anomaly detection in the video is the task of recognizing frames from a video sequence that reflect occurrences that differ significantly from the normal, Identifying unusual incidents, such as fires, traffic accidents, or stampedes [3].

Anomaly detection is a growing field of research in itself. Several methods have been proposed to overcome this challenge but all of them have their limits namely the majority of existing methods rely on the presence of a labeled dataset including a set of normal events [3]. This assumption limits their domain of application, as it means that the system can't continuously be re-trained without human involvement.

Various approaches have been proposed in the literature for the detection of anomalies in video sequences. These approaches can be grouped in four categories: reconstruction

M. Lazaar et al. (Eds.): BDIoT 2022, LNNS 625, pp. 279–289, 2023.
https://doi.org/10.1007/978-3-031-28387-1_24

error [4, 5], future frame prediction [6], classifiers [2] and scoring [3]. The reconstruction error is one of the most used approaches for solving the anomaly detection problem. The basic assumption of using the reconstruction error is to train an auto-encoder on the normal data, and it is expected to produce higher reconstruction error for the abnormal inputs than the normal ones, which is used as a reference for identifying anomalies. The proposed work in [7] handled the anomaly detection problem by modelling of normal and abnormal event using regression model and density–based clustering to detect the abnormal event at clips level. As a result, the anomaly score is produced at the clip level along the video frame sequences and the anomaly is identified based on density of each clip. The proposed work in [8] used Generative Adversarial Nets (GANs) [9], which are trained to learn an internal representation of scene normality using normal frames and related optical-flow images. Since GANs has only been trained on normal data, it cannot generate abnormal events. Therefore, local difference between the real and generated images is used to detect possible abnormalities during testing time. Another work [10] based on generative adversarial networks (GANs) [9] and performs transfer learning algorithms on pre-trained convolutional neural network (CNN) to address Anomaly detection problem, this method transfers the knowledge of a pre-trained CNN (VGG16) to its discriminator CNN to solve this problem. The aim of video anomaly detection methods is to determine whether or not the current frame of a given video contains an anomaly.

Methods based on machine learning (ML) are ideally suited for such challenges, especially deep learning based methods. However, deep learning techniques require for a huge dataset sample, in order to acquire an effective performance measure for detecting anomalous events inside video frames.

Although Convolutional Neural Networks (CNNs) [11] are still the most popular building blocks in computer vision, recently a novel architecture called Vision Transformer (ViT) [12] has been developed. The Transformer neural network architecture was originally designed to analyze sequential data in Natural Language Processing (NLP) [13]. Moreover, recent research has been carried out on using transformer encoders for encoding images. Vision Transformers (ViT) is a deep learning neural network architecture that has been showing encouraging results in Computer Vision tasks such as image classification [14] and object detection [15]. Therefore, Transformers are quickly finding their place in computer vision applications.

In this work, we are going to apply Vision Transformer on the anomaly detection task, by classifying the frames, and detecting which frame contains the anomaly.

Finally, recent researches prove that ViT models are specifically good at classifying [14, 16], these successful results, can push Vision Transformer network architecture to be an important tool for many computer vision tasks.

The rest of this paper is organized as follows: the second section presents our proposed model structure and the used datasets, in the third section, we provide the experiment results and discussion. Finally, the concluding part.

2 Model and Data

In this section, the anomaly detection dataset is presented. Additionally, we discuss the vision transformer model that was fine-tuned on this dataset.

2.1 Proposed Model

In this work, we used Vision Transformer inspired from [12] which we adapted to our problem (anomaly detection in video sequences) as our pre-trained model. This choice is due to its recent success. Recently, it becomes the state-of-the-art of Deep Learning model that is used in image classification [12, 14, 16]. Our fine-tuned model is illustrated in Fig. 1.

LSTM [17] has mostly been replaced by Transformer-based models, which have been shown to be superior in quality for many sequence-to-sequence challenges. Transformers rely entirely on attention mechanisms to increase its speed by being parallelizable. Transformers are generally used to achieve parallelism and reduce memory complexity. For example, in comparison to RNN, the issue with RNN is the long-term dependencies that cause vanishing and/or exploding gradient problems and in addition, it cannot perform parallel calculations. Later, to tackle the RNN problems, LSTM introduced the idea of gated read, write, and forget operations. But, because of the dependency on learnable weights, it has a lot of parameters and requires a large amount of memory and the parallelism problem is still present. Transformers are different from the first two ones. They work based on the attention mechanism for long sentences. The idea of parallel computation is theoretically realized by this method since it avoids recursion. This will decrease the training period and improve performance.

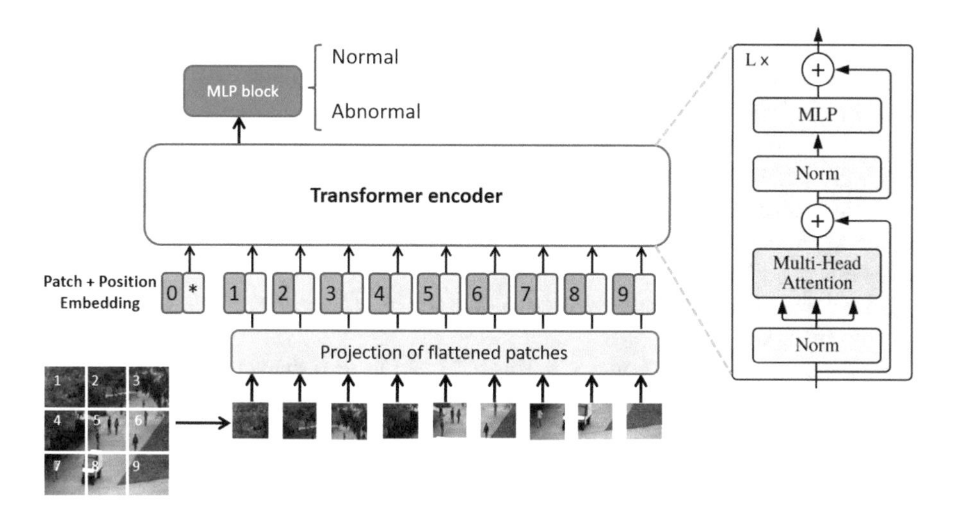

Fig. 1. Vision transformer used in our implementation

The concept of Vision Transformer (ViT) is an extension of the original concept of Transformer [13]. The Vision Transformer (ViT) [12] is essentially BERT [18] which is a stack of transformer encoders as a language model generator but applied to images. It achieves excellent results compared to state-of-the-art convolutional neural networks. In order to handle images in the model, each image is split into a sequence of flattened 2D patches (typically of resolution 16×16), called visual tokens. The visual tokens are embedded into a set of encoded vectors of fixed dimension (768). Then, a learnable embedding ($z_0^0 = X_{class}$) was prepended to the sequence of embedded patches in order to classify images. Next, The position of each patch in the image is embedded along with the encoded vector and fed into the transformer encoder network [13].

As shown in the right side of the ViT model overview figure, The transformer encoder is typically comprised of alternating layers of Multi-headed self-attention (MSA) [13] and Multi-layer perceptron (MLP) blocks. MLP contains two fully connected layers with a GELU function. Layer normalization (LN) [19] is applied before every block along with residual connections (Eq. 2, 3). The components of every block of the transformer encoder are illustrated in Fig. 2. Transformer encoders are stacked on top of each other (typically 12 Layers) to build the overall ViT architecture. Since ViT's primary goal is image classification, which does not require decoding the encoded image representation, a transformer decoder is not used.

$$z_0 = [x_{class}; x_p^1E; x_p^2E; \cdots ; x_p^NE] + E_{pos} \quad \text{such as} \quad E \in R^{(p^2.C)\times D}, E_{pos} \in R^{(N+1)\times D} \tag{1}$$

$$z_t' = MSA(LA(z_{t-1})) + z_{t-1} \quad \text{such as} \quad t = 1 \ldots L \tag{2}$$

$$z_t = MLP(LA(z_t')) + z_t' \quad \text{such as} \quad t = 1 \ldots L \tag{3}$$

$$y = LN(Z_L^0) \tag{4}$$

The self-attention/Multi-head self-attention (MSA) mechanism is the heart of every Transformer architecture, including the ViT architecture. The self-attention mechanism is a type of attention mechanism, which allows every element of a sequence to interact with every others and find out who they should pay more attention to.

Self-attention allows capturing 'long-term' information and dependencies between sequence elements. Self-attention is a weighted combination of all other patches embedding.

These attention maps help network to focus on most important regions in the image such as object(s).

In this paper, a video frame $x \in R^{H\times W\times C}$ has been reshaped into a sequence of flattened patches encoded and fed to the transformer encoder as illustrated in Fig. 1. (H, W) is original frame's resolution; C is the number of channels. The resulting number of the extracted patches is $N = (H \times W)/P^2$, where $P = 16$ is the resolution of each patch. The flattened patches are used through all of the transformer's layers after having been mapped.

Each flattened patch x in the created sequence of patches was mapped to a latent vector with hidden size $D = 768$, Then, A learnable embedding (z_0^0 = Xclass) prepend to the sequence of the embedded patches. Furthermore, a position embeddings Epos belonged to $R^{(N+1)\times D}$ are added to the patch embedding to hold the order of each patches that the transformer encoder receives as input (Eq. 1) [12]. The learnable class X_{class} that is the output of the model (z_L^0) serves as the frame's representation y. The frame representation is then associated with a classifier (Eq. 4), to classify the frame in the suited class.

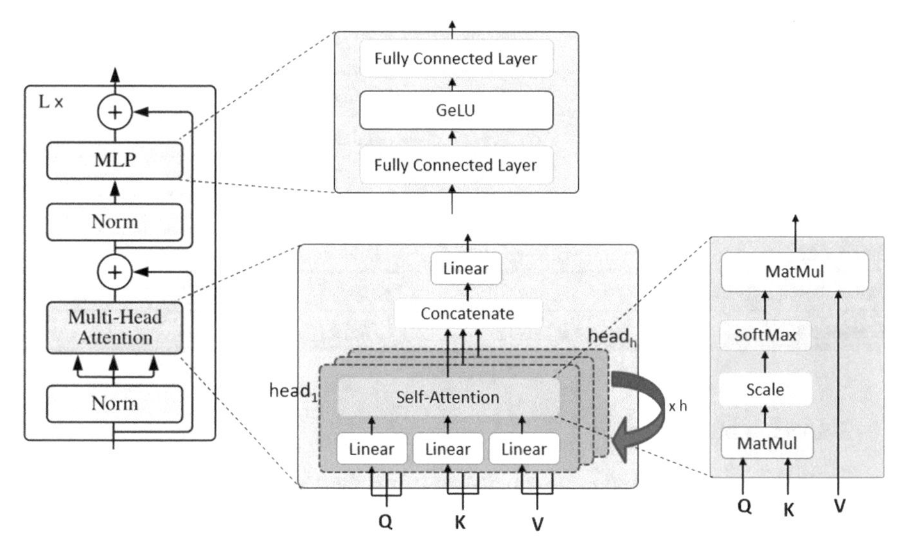

Fig. 2. Vision Transformer Encoder Layers

In short, the frame gets turned into patches and then projected to an embedding. The embedding then travels through the various layers and blocks of the model and the suitable class is returned.

The ViT-B16 contain approximately 86M parameters, and due to large computational needed to train the ViT model from scratch, where it could be trained using a standard cloud TPUv3 with 8 cores which cost 8$/hour in approximately 30 days [12]. Therefore, we try to leverage the pre-trained vision transformer model that already trained with a large dataset (ImageNet).

MLP Head is the output layer of the architecture; it converts the learned features of an input to a class output. Since we are working on frame classification. MLP head of the pre-trained model has out features of a thousand classes, we change that to our number of classes which is 2 (Normal, Abnormal), and we keep all the base layers because we don't want to update them during training when we are just using a feature extractor model. Rather than update the pre-trained weight during training on our custom dataset, we want to leverage them.

2.2 Hyper-parameter Setting

We train our Fine-tuned vision transformer on the Anomaly dataset UCSD for 5 epochs, In each epoch, the batch size is set to 8. We use Adam [20] optimizer for optimization and the learning rate is set to 2e−5 eps, Number of encoder layers L = 12.

30% of each class in the training set was chosen for validation in the experiments. The frames used in the dataset have been resized to H = 224, W = 224.

The vision transformer was fine-tuned using the weights of the pre-trained model on ImageNet-21K dataset [21]. The Pytorch framework was used for training, validation, and testing on an NVIDIA Tesla T4 GPU. Table 1 summarizes the hyperparameters used with our ViT model.

Table 1. Hyperparameter Configuration

Configuration	Value
Optimizer	Adam
Learning rate	2e−5
Epoch	10
Batch size	8
Batch normalization	True
Activation	GELU

2.3 Experimentation

As illustrated in Fig. 2, two components make up the vision transformer Encoder: a Self-Attention and MLP block. In order to understand better how the model trained to detect the anomalous frame. We illustrated the markers detected by ViT using the model's attention map.

The self-attention score of the model can be used to visualize the input frame. As we can see in Fig. 3, it displays samples of a few frames and their attention maps. The attention map for the Vision Transformer model demonstrated that the ViT-B16 is capable of quickly detecting the most crucial elements in a scene, which represent the anomaly object in our case.

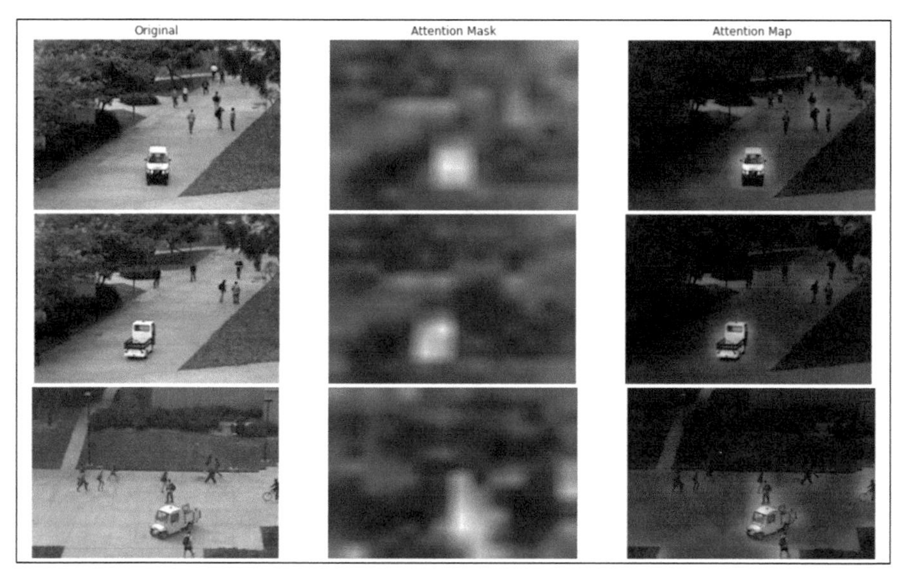

Fig. 3. Yellow color highlight important areas detected by the ViT model on the UCSD frames by using the attention weights from transformer block

Figure 4 shows the loss and accuracy curve that we got during the training and validation stage, which the model start from the loss value 0.72 in the first epoch and it reduced to a value of 0.02. Regarding the accuracy, the model achieved a good value, which is 0.99.

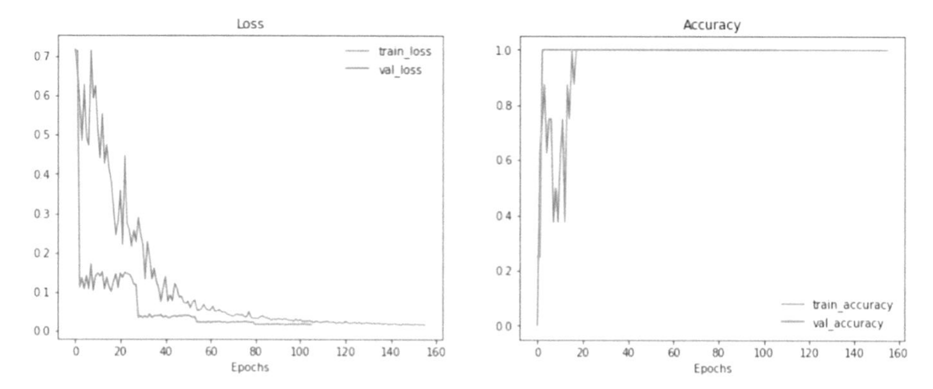

Fig. 4. Loss and Accuracy curve for both the training and validation steps

The objective in this paper is to test the performance of the ViT model for the classification task of anomaly.

2.4 Dataset

In this work, we use the UCSD dataset [22], the UCSD Pedestrian dataset includes two subsets: UCSD Peds1 dataset and UCSD Peds2 dataset. The dataset is split into training and testing data. There are no anomalies in the training data, it is all normal events and contains only pedestrians; however, there is at least one anomaly in every testing clip, the anomalous events are either: the movement of non-pedestrian entities through walkways or anomalous pedestrian motion. Common anomalies include bikes, skaters, small cars, and people walking across a walkway or in the grass, in certain frames, the anomalies appear in multiple locations. The difference between the two subsets is their frame size and the angle of the camera.

UCSD Pedestrian 1: This dataset has 34 video sequences for training, and 16 video sequences for testing in which one or more anomalies are present in some of the frames, pixel-level binary masks are given to a subset of 10 clips in the testing set to identify regions having anomalous events, each clip contains about 200 frames. There are about 3,400 anomalous and 5,500 normal frames, with resolution 238 $\times$ 158 pixels. In this dataset, the camera is placed at a high altitude.

Some examples of normal and abnormal images are shown in Fig. 5, all frames are resized to 224 $\times$ 224 pixels.

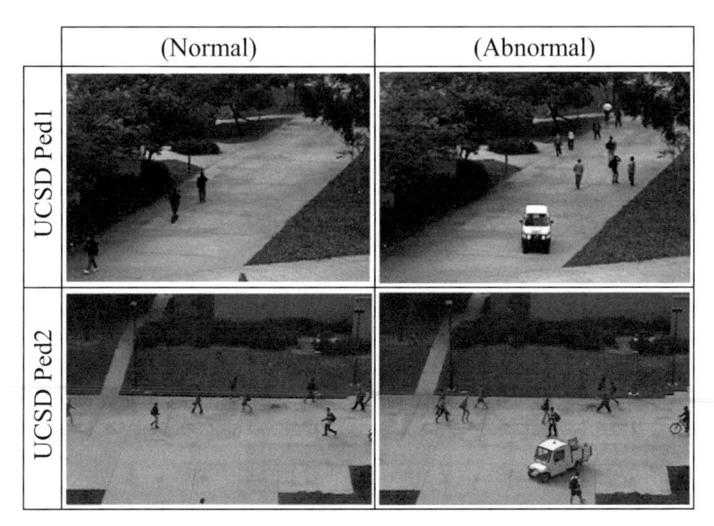

Fig. 5. Sample frames from the UCSD dataset; the left column shows normal pedestrian behavior, and the right shows the anomaly behavior in the scene

3 Results and Discussion

To evaluate our fine-tuned vision Transformer model, we used the accuracy, precision, recall, F1-score, and confusion matrix as criteria. Where true positive (TP), true negative (TN), false positive (FP), and false negative (FN) are used in the definition of these four criteria in Table 2.

Table 2. Metrics and Scoring

Criterion	Definition
Accuracy	$\frac{TP+TN}{TP+TN+FP+FN}$
Recall	$\frac{TP}{TP+FN}$
Precision	$\frac{TP}{TP+FP}$
F1-Score	$\frac{2xTP}{2xTP+FP+FN}$

From the confusion matrix that illustrates in Fig. 6 we see that 59% of the abnormal frames from the validation part of the dataset and 38% of the normal frames are correctly classified into the correct categories. This means that the effectiveness of the model in the classification task is very high. And for the accuracy metric which is the most fundamental evaluation criteria, we see in Fig. 7 that the model achieves 0.97 in both normal and abnormal categories. And no frame was reported incorrectly as abnormal.

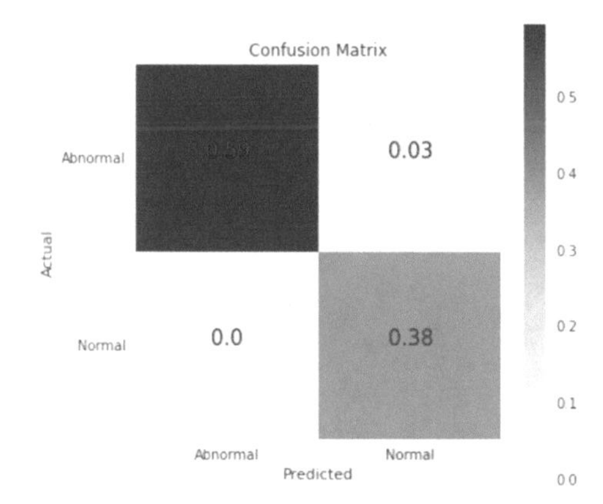

Fig. 6. Confusion matrix of the model

	precision	recall	f1-score
Abnormal	1.00	0.95	0.97
Normal	0.92	1.00	0.96
accuracy			0.97

Fig. 7. Evaluation metrics of the model

Figure 8 shows the results of anomaly detection in some frames, and as we can see, the model is able to detect if the frame presents the anomaly or not. Many objects that present the anomaly in frames have been detected such as cars, motorcycles, and pedestrian skis.

In this first work, we worked on the classification of frames as normal or abnormal, as opposed to object detection to keep the model complexity moderate. In the future, we will focus to localize the anomaly object in the frame.

Fig. 8. Samples of the normal and abnormal frame classification.

4 Conclusion

In this paper, we have presented our approach based on fine-tuning Vision transformer networks to classify frames into two categories, normal, and abnormal events. The result obtained shows its effectiveness in anomaly detection task. In future work, we want to make a model able to localize the object that presents the anomaly in the sequence.

References

1. Popoola, O.P., Wang, K.: Video-based abnormal human behavior recognition—a review. IEEE Trans. Syst. Man Cybern. C **42**(6), 865–878 (2012). https://doi.org/10.1109/TSMCC.2011.2178594
2. Sabokrou, M., Fayyaz, M., Fathy, M., Moayed, Z., Klette, R.: Deep-anomaly: Fully convolutional neural network for fast anomaly detection in crowded scenes. Comput. Vis. Image Underst. **172**, 88–97 (2018). https://doi.org/10.1016/j.cviu.2018.02.006
3. Pang, G., Yan, C., Shen, C., Hengel, A.V.D., Bai, X.: Self-trained deep ordinal regression for end-to-end video anomaly detection. In: 2020 IEEE/CVF Conference on Computer Vision and Pattern Recognition (CVPR), Seattle, WA, USA, June 2020, pp. 12170–12179 (2020). https://doi.org/10.1109/CVPR42600.2020.01219
4. Hasan, M., Choi, J., Neumann, J., Roy-Chowdhury, A.K., Davis, L.S.: Learning Temporal Regularity in Video Sequences, April 2016. arXiv:1604.04574 [cs]. http://arxiv.org/abs/1604.04574. Accessed 04 July 2021

5. Bidirectional Convolutional LSTM Autoencoder for Risk Detection. IJATCSE **9**(5), 8585–8589 (2020). https://doi.org/10.30534/ijatcse/2020/241952020
6. Liu, W., Luo, W., Lian, D., Gao, S.: Future frame prediction for anomaly detection–a new baseline. In: 2018 IEEE/CVF Conference on Computer Vision and Pattern Recognition, Salt Lake City, UT, June 2018, pp. 6536–6545 (2018). https://doi.org/10.1109/CVPR.2018.00684
7. Mahmood, S.A., Abid, A.M., Lafta, S.H.: Anomaly event detection and localization of video clips using global and local outliers. IJEECS **24**(2), 1063 (2021). https://doi.org/10.11591/ijeecs.v24.i2.pp1063-1073
8. Ravanbakhsh, M., Nabi, M., Sangineto, E., Marcenaro, L., Regazzoni, C., Sebe, N.: Abnormal Event Detection in Videos using Generative Adversarial Nets, August 2017. arXiv:1708.09644 [cs]. http://arxiv.org/abs/1708.09644. Accessed 25 July 2021
9. Goodfellow, I., et al.: Generative Adversarial Nets, p. 9
10. Atghaei, A., Ziaeinejad, S., Rahmati, M.: Abnormal Event Detection in Urban Surveillance Videos Using GAN and Transfer Learning , November 2020. arXiv:2011.09619 [cs]. http://arxiv.org/abs/2011.09619. Accessed 17 May 2021
11. O'Shea, K., Nash, R.: An Introduction to Convolutional Neural Networks, 02 December 2015. arXiv. http://arxiv.org/abs/1511.08458. Accessed 22 May 2022
12. Dosovitskiy, A., et al.: An Image is Worth 16x16 Words: Transformers for Image Recognition at Scale, 03 June 2021. arXiv. http://arxiv.org/abs/2010.11929. Accessed 22 May 2022
13. Vaswani, A., et al.: Attention Is All You Need, December 2017. arXiv:1706.03762 [cs]. http://arxiv.org/abs/1706.03762. Accessed 24 July 2021
14. Chen, H., et al.: GasHis-transformer: a multi-scale visual transformer approach for gastric histopathology image classification, 17 February 2022. arXiv. http://arxiv.org/abs/2104.14528. Accessed 01 June 2022
15. Carion, N., Massa, F., Synnaeve, G., Usunier, N., Kirillov, A., Zagoruyko, S.: End-to-End Object Detection with Transformers, 28 May 2020. arXiv. http://arxiv.org/abs/2005.12872. Accessed 18 May 2022
16. Lin, J.Y.-Y., Liao, S.-M., Huang, H.-J., Kuo, W.-T., Ou, O.H.-M.: Galaxy Morphological Classification with Efficient Vision Transformer, 03 February 2022. arXiv. http://arxiv.org/abs/2110.01024. Accessed 30 June 2022
17. Hochreiter, S., Schmidhuber, J.: Long short-term memory. Neural Comput. **9**(8), 1735–1780 (1997). https://doi.org/10.1162/neco.1997.9.8.1735
18. Devlin, J., Chang, M.-W., Lee, K., Toutanova, K.: BERT: Pre-training of Deep Bidirectional Transformers for Language Understanding, 24 May 2019. arXiv. http://arxiv.org/abs/1810.04805. Accessed 22 May 2022
19. Ba, J.L., Kiros, J.R., Hinton, G.E.: Layer Normalization, 21 July 2016. arXiv. http://arxiv.org/abs/1607.06450. Accessed 02 June 2022
20. Kingma, D.P., Ba, J.: Adam: A Method for Stochastic Optimization, 29 January 2017. arXiv. http://arxiv.org/abs/1412.6980. Accessed 12 June 2022
21. Krizhevsky, A., Sutskever, I., Hinton, G.E.: ImageNet classification with deep convolutional neural networks. Commun. ACM **60**(6), 84–90 (2017). https://doi.org/10.1145/3065386
22. Mahadevan, V., Li, W., Bhalodia, V., Vasconcelos, N.: Anomaly detection in crowded scenes. In: 2010 IEEE Computer Society Conference on Computer Vision and Pattern Recognition, San Francisco, CA, USA, June 2010, pp. 1975–1981 (2010). https://doi.org/10.1109/CVPR.2010.5539872

An Efficient Real-Time Moroccan Automatic License Plate Recognition System Based on the YOLO Object Detector

Zainab Ouardirhi[1,2](✉), Sidi Ahmed Mahmoudi[2], Mostapha Zbakh[1], Mohamed El Ghmary[3], Mohammed Benjelloun[2], Hamd Ait Abdelali[4], and Hatim Derrouz[4]

[1] Communication Networks Department, ENSIAS, Mohammed V University in Rabat, Rabat, Morocco

[2] Computer and Management Engineering Department, Polytechnic Faculty in Mons (FPMS-UMONS), Mons, Belgium

zainab_ouardirhi@um5.ac.ma

[3] Department of Computer Science, FSDM, Sidi Mohamed Ben Abdellah University, Fez, Morocco

[4] Microelectronics Embedded System Department, Moroccan Foundation for Advanced Science Innovation and Research, Rabat, Morocco

Abstract. In this paper, we adopt a powerful deep neural network (DNN) to tackle the problems of detecting car license plates (LPs) and recognizing Arabic letters in natural scene images. Our Automatic License Plate Recognition (ALPR) system is built using state-of-the-art methodologies and techniques to provide the optimal speed/accuracy trade-off at each level. Images of vehicles and Moroccan and European LPs were used to train the neural networks. The system was able to correctly detect and recognize in the test set all the characters existing in a LP with an accuracy of 91.11% and 98.89% taking into account that only one character is misread. Using a high-end GPU, our system likewise produced outstanding real-time execution results.

Keywords: Deep Learning · Image processing · CNN · YOLO · Zone of interest · Detection · Recognition · Voting system · License Plates

1 Introduction

Modern civilization has seen a huge growth in intelligent systems (IS) for the purpose of processing information. The combination of internet of things (IoT) and artificial intelligence (AI) technologies has allowed us to create intelligent information processing systems at an industrial level. However one of the difficult parts of developing and implementing such industrial IS is data collection.

M. Lazaar et al. (Eds.): BDIoT 2022, LNNS 625, pp. 290–302, 2023.
https://doi.org/10.1007/978-3-031-28387-1_25

Vehicles have now become one of the most frequently regarded as conceptual resources in information systems, as a result of the widespread adoption of information technology in various parts of modern life. Information from reality must be transformed into autonomous information systems. This can be done by human agents or particular intelligence equipment. Among these intelligent devices, we must mention Automatic License Plate Recognition (ALPR) system.

ALPR technology is growing in use, especially in Surveillance and Intelligent transportation systems, with relevant applications in several sectors such as command force, road safety, parking management [1], traffic flow control and toll collection [2].

Current ALPR systems achieve exemplary performance in controlled environments; however, performance is degraded when processing complex scenes. These devices can deal with a variety of constraints and uncontrollable conditions, such as particular detectors or viewing angles, uneven illumination, adequate lighting requirements, image blurring, capture in a predetermined zone, occlusions, and so on. Deep Learning (DL) methods have emerged as a useful element in the present sector. Indeed, with the advent of DL techniques, the accuracy of many pattern recognition tasks has improved.

In computer vision, many methods of achieving peak performance explore a kind of Convolutional Neural Network (CNN). The problem with these techniques is that they typically require tons of data to be trained from scratch, although most of the existing methods explore Transfer Learning to alleviate this problem. In addition, the more data you have, the deeper your network can be, allowing increasingly complex patterns to be recognized.

In this paper, we propose an end-to-end, efficient and layout-independent ALPR system using a YOLO-based model at all stages. In order to obtain the best speed-accuracy trade-off, the entire system was trained on a total of 25 510 annotated images that consists of vehicles and Moroccan and European LPs. We will start by reviewing related studies along with their limitations in Sect. 2. Presenting the proposed system in Sect. 3. Finally, before moving on to the conclusions, describe the experimental setup in Sect. 4.

2 Related Work

2.1 ALPR Systems

This section explores recent ALPR research that employs deep learning models. We'll go over the relevant literature from prior work in the areas of license plate detection (LPD) and license plate recognition (LPR), as well as their limitations.

The authors of [3] have suggested a real-time, end-to-end ALPR system based on deep convolutional neural networks (CNNs). In order to get better results, the method seeks to identify the front and back views of automobiles and LPs that are functioning in a cascaded mode. After that, characters within a cropped LP region are detected and recognized.

In [4], the authors suggested an ALPR system that was focused on locating and reading LPs in challenging situations. For LP detection and character segmentation, a CNN-based model called You Only Look Once version 3 (YOLOv3) was used, and a Convolutional Recurrent Neural Network (CRNN) [5] for character recognition (CR).

A study made in [6] presented a new system for the recognition of multinational LPs with a three-layer architecture, including LPD, Unified Character Recognition (UCR), and Multinational License Plate Layout Detection. In this approach, the authors used a more simplified YOLOv3 architecture for LPD and another YOLOv3-SPP with Spatial Pyramid Pooling block for CR stages.

Authors in [7] proposed a combination of YOLO and sliding-window processes. In their method, each character of Taiwan's vehicle LPs is detected by the sliding-window, the YOLO framework then identifies each class of object, plate, digit, and letter in the window.

In [8], authors used specific CNNs for each ALPR stage. The models used are: Fast-YOLO [9], YOLOv2 and CR-NET [10], an architecture inspired by Fast-YOLO for character segmentation and recognition.

It was proposed in [11] a novel LP detection and character recognition algorithm based on a combined feature extraction model and Backpropagation Neural Network (BPNN) which is adaptablein weak illumination and complicated backgrounds.

The authors of [12] have developed an ALPR system that can be used in a mobile application to recognize Egyptian LPs. The algorithm aims to first apply preprocessing on the acquired image, character detection using segmentation, and then character recognition by extracting features using Speeded Up Robust Features (SURF) [13].

In [14], the authors have proposed a method for Chinese vehicle LPR, which employs YOLOv2 detector for LPD and a CRNN for CR stages. Their recognition architecture consists of a CNN for context feature extraction, and a two-layered Gated Recurrent Unit (GRU) [15] for feature sequences decode.

In [16], authors used a cascade structure to read the LP. The model first detects the character region using a CNN classifier, employed in a sliding window fashion across the entire image, to generate bounding boxes independently at each scale using the run Length Smoothing Algorithm (RLSA) and Connected Component Analysis (CCA). Then, the generated boxes are filtered by geometric constraints and refined by the edge feature of LP. And finally, another CNN classifier was used to verify the remaining bounding box.

Authors of [17] used a cascaded framework composed of a fast region proposal network and a R-CNN network to extract LP. The model aims first to generate the LP candidates using a light-weight RPN network. Then extracts the Region of Interest (ROIs) from the original image using the sampler. And finally, using the R-CNN network, the model classifies the candidate plate and regresses four corners of the LP.

The authors in [18] developed a lightweight ALPRmodel that may be implemented entirely on embedded devices such as the Raspberry Pi3. To achieve the lowest memory consumption for the detection stage, they employed a mix of a MobileNet [19] feature extractor with fewer parameters and a Single Shot Detection model. They also used LPRNet20 for character recognition, which is a powerful but computationally economical network.

In [20], the authors developed a CNN-based method called MD-YOLO, inspired by the YOLO framework, to realize multi-directional car LPD. They proposed the angle deviation penalty factor (ADPF) to approximate the intersection ratio between predicted value and tag value. And they chose leaky and identity functions as activation functions rather than ReLU function in order to identify negative rotation angle values. A prepositive CNN attention model called ALMD-YOLO was employed prior to the implementation of MD-YOLO considering that the LP is usually very small.

2.2 Limitations

Most of the related work in the ALPR context that we discussed Sect. 2 has common limitations, as for instance the previous works that are dealing with multinational LPs were trained and tested on datasets from various countries, however, Moroccan LPs were never included. Certain previous works used many techniques, which results in an increase in the execution time of the ALPR system. Some of them also faced on the training faze the limitation of having smaller memory and computational power, and lack of training dataset.

3 Proposed System

In this section, we describe our proposed methodology to detect and recognize LPs. The input to the system is a video of a moving vehicle and the output is the textual form of that vehicle's LP content, as shown in Fig. 1.

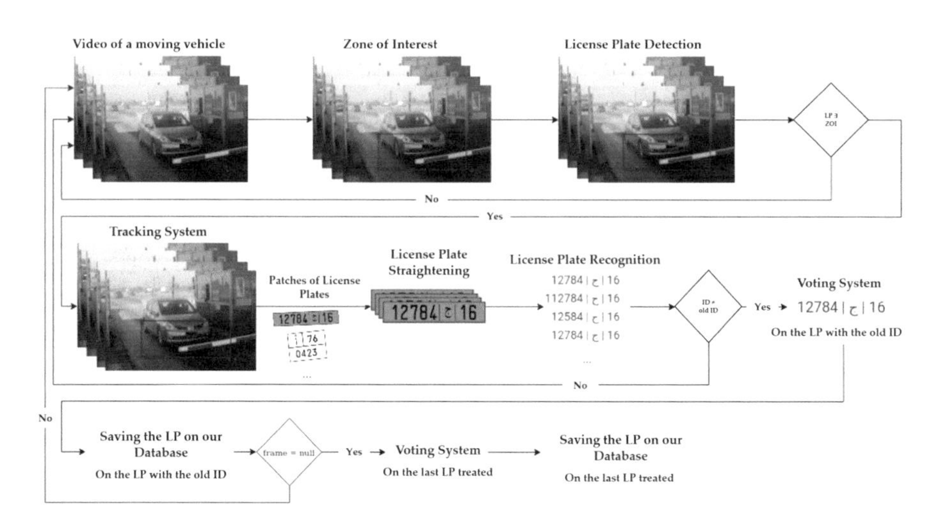

Fig. 1. Our proposed ALPR system block diagram: Following the acquisition of the input video, the system first detects the license plate when it appears in the manually defined zone of interest. For the period that this license plate is in this zone, it is tracked, extracted, and straightened in order to best recognize its content. Following the recognition phase, the system then employs a voting system that filters the most frequent license plate content in each image and saves it in our database for later use.

3.1 Zone of Interest

We created a Zone of Interest (ZOI) to allow the system to focus on one LP at a time, obtain a single good detection result for that LP, and only apply the recognition phase to plates with clear and easily recognizable characters. The next steps will be followed until the LP is validated to exist in that ZOI, as shown in Fig. 2.

Fig. 2. Before and after accessing the Zone of Interest.

3.2 License Plate Detection

Since LPs in our case tend to appear in scenes in small sizes, we devoted our research to find a model capable of detecting and recognizing an object of any size in an image. YOLOv3 [21] (Fig. 3), most well and reliable version of the YOLO model created by Joseph Redmon and his collaborators, is the most common deep object detector in practical applications. Indeed, it can detect

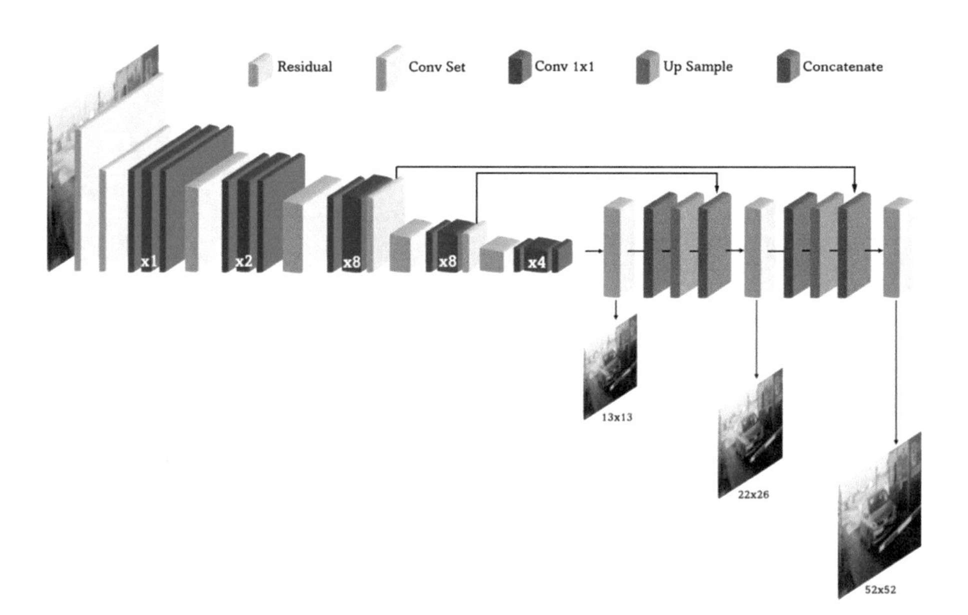

Fig. 3. YOLOv3 network architecture for detecting license plates.

small and overlapping compact objects. This YOLO version is slightly bigger than YOLOv2 [22], but it is more accurate and can detect faster.

Ultimately, we employed YOLOv3 network to detect LPs in order to achieve a decent balance between the accuracy rate and the execution duration. On a powerful computer, this network was established (and trained) to accomplish real-time detection. Darknet-53 is used by YOLOv3 as a feature extractor. It is still more efficient than ResNet-101 or ResNet-152 [23] and far more potent than the Darknet-19 used by YOLOv2.

Therefore, to use YOLOv3, we need to change the last number of layers taking into account the number of classes and the number of filters, we set them respectively to 18 and 1 where the class in our case is the LP (the last 3 layers of the network are shown in Table 1). The number of filters was calculated based on the Eq (1), where B is the predicting boxes and C is the number of classes we want the model to detect.

$$N\_Filters = (B \times (5 + C)) \tag{1}$$

Table 1. The last three layers of YOLOv3's feature extractor (Darknet-53) for LPD phase

Layer	Type	Filters	Size	Input	Output
...	...	...	...	...	...
103	conv	128	$1 \times 1/1$	$52 \times 52 \times 256$	$52 \times 52 \times 128$
104	conv	256	$3 \times 3/1$	$52 \times 52 \times 128$	$52 \times 52 \times 256$
105	conv	**18**	$1 \times 1/1$	$52 \times 52 \times 256$	$52 \times 52 \times 18$
106	detection				

3.3 Tracking System

Many computer vision and pattern recognition applications, including surveillance, vehicle navigation, and autonomous robot navigation, depend on object detection and tracking, also known as Object Tracking. Any tracking technique actually needs an object detection mechanism, whether it is in each frame or when the item first shows in the movie.

Additionally, object detection is technically unable to discriminate between several frames of the same object in an image. This has a negative impact on LP recognition and ultimately results in duplicate data. In order to overcome this problem, we will be tracking our LPs, so that each one of them becomes unique, as shown in Fig. 4.

The challenge of roughly tracking an object's trajectory in the picture plane as it travels across a scene is known as object tracking. Its goal is to find an object's position in each frame of the video in order to construct the path for that object over time.

The Tracking system we used, works well with occlusion, it employs a Hungarian algorithm [24] that can determine if an object in the current frame is

Fig. 4. Before and after applying the tracking system.

identical to one in the previous frame. It will be used for id attribution and association. The system makes also use of a Kalman Filter [25], which is a method for predicting future locations based on present ones. It may also estimate current position more accurately than the sensor or algorithm. It will be used to have better association.

3.4 License Plate Straightening Algorithm

Our LP does not automatically straighten after the detection and tracking phases, which makes the recognition phase ineffective. As shown in Fig. 5, we developed a license plate straightening (LPSt) algorithm to enhance the visibility of our discovered LP.

The algorithm works as follows: first, we obtain the image of the LP that needs to be straightened; next, we use the Canny Edge Detector to identify the edges of each object in the image; next, we use the Hough Line Transform to draw the longest lines, which in this case are the LP's edges; and finally, we rotate the resulting image in relation to the angle determined by two points that exist on the lines that we chose in the previous step.

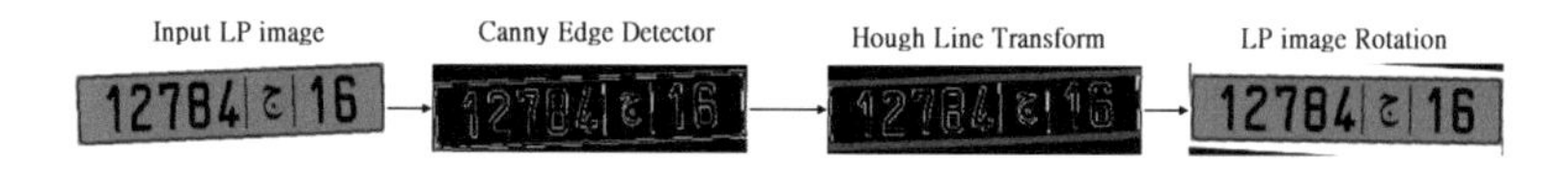

Fig. 5. Steps we followed to straighten the detected LP.

3.5 License Plate Recognition

We used YOLOv3 for the character recognition phase since as we explained in Sect. 3.2 it can detect small objects and accept input images of various sizes. Similarly to the LPD phase, we set the number of filters to 237 and the number of classes to 74, which represent the number of Arabic and French letters, separators and numbers.

3.6 Voting System

Each time an LP enters the ZOI, the system detects it and recognizes its content, as discussed in the preceding subsections. Assuming that this process is applied to the same LP multiple times and that inaccurate recognition results are occasionally obtained, our system saves the position and content of each LP that appears at the level of each frame of the input video, so that it can later select the most frequent LP content in each image and save it in our database for future use. As illustrated in Fig. 6, not only is content redundancy at the database level eliminated, but our results are also more efficient and accurate.

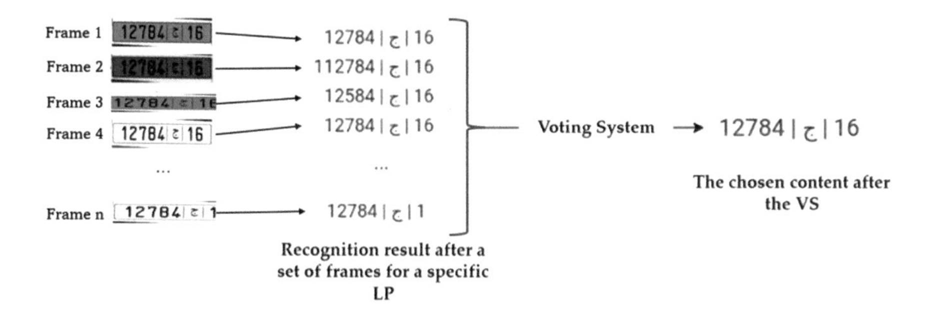

Fig. 6. Before and after the license plate voting system.

4 Experimental Results

In this section, we conduct the experiments to verify the effectiveness of the proposed system. Using a confidential database given by the MAScIR foundation, we trained our model with a Batch size of 64 and a Subdivision of 16 using 7 500 full resolution and high definition 1600×1200 size images of different types of vehicles for LPD, and 9 250 manually annotated Moroccan LPs and 8 760 manually annotated European LPs 400×100 size images for LPR, on an NVIDIA GTX 1050 Ti GPU (16 GB of memory and 960 CUDA10.2 cores), a machine with an Intel 6core processor and a frequency of 3.5 GHz per core under the Linux environment. We used 70% of our annotated data set for training and the rest for validation.

4.1 License Plate Detection

In order to know the performance of the YOLOv3 license plate detection model, we measured its relevance by calculating the precision, the recall and then conclude the F1 score using a confusion matrix which is built by putting respectively on the rows and columns of benchmark data which in our case are 1400 positive images containing vehicles, and 1400 negative images containing random objects, which were randomly selected from the internet.

According to the test performed on the system using sets of positive and negative images, we managed to obtain the confusion matrix presented in Table 2. Where images containing LPs and the system correctly detecting them are true positives (tp), images containing LPs and the system incorrectly detecting them are false positives (fp), images without LPs and the system did not detect

Table 2. Confusion matrix of the LPD phase

	Positive condition	Negative condition
Positive predicted	**tp** = 1247	**fp** = 53
Negative predicted	**fn** = 100	**tn** = 1400

anything are true negatives (*tn*), and images that don't contain LPs but the system has detected objects are false negatives (*fn*), same for images that do contain LPs but the system has not detected objects.

Table 3. Evaluation of the LPD phase.

Method	**Metric**		
	Accuracy	**Recall**	**F1 Score**
YOLOv3 [21]	66%	91%	76%
YOLO-LPD	95%	93%	94%

Based on the confusion matrix in Table 2, we can extract the Table 3 that concludes that our LPD system exhibits a very high level of accuracy. We were able to outperform the original YOLO model [21] with the help of our enhanced version (YOLO-LPD). In less than 680 ms and with a 94% F1 score, the YOLO-LPD network was able to detect one class in a 1600×1200 image.

4.2 License Plate Recognition

In order to know the performance of this model in this phase we tested our LPR on 2443 images of Moroccan and European LPs, and we obtained the results presented in the Table 4.

Table 4. Evaluation of the LPR phase.

Total number of images	Entirely correct	$\geq$1 CR error
2443	2336 (**95.62%**)	107 (**4.38%**)

This table shows that the model recognized 95.62% of the images well, which means our license plate recognition algorithm is very accurate.

4.3 Quantitative Comparison

In this section we report a comparison in order to show the influence of the proposed strategies on the accuracy rates of our proposed ALPR system. The results from testing the model without any techniques (YOLO-no LPSt and VS),

Table 5. The influence of the proposed strategies on the accuracy rates of our MALPR system.

Method	Metric		
	Accuracy	Recall	F1 Score
YOLO-no LPSt and VS [21]	67%	94%	78%
YOLO-LPSt	82%	97%	89%
YOLO-VS	89%	**98%**	93%
MALPR	**97%**	**98%**	**97%**

with the LP straining technique (YOLO-LPSt), with the voting system (YOLO-VS), and finally with all the approaches combined, which represents our final Moroccan Automatic License Plates Recognition (MALPR) system, are shown in Table 5.

This table makes the effects of the strategies we suggested very evident, and it shows how much more accurate our suggested MALPR system is. This just helps to illustrate how well our model handles the challenge of multi-directional car license plate detection and recognition as well as the removal of redundant data.

4.4 System Assessment

To evaluate the entire system, we calculated the accuracy for the two levels: the LPD level and the LPR level. For the first level, we calculated the percentage of the LPs detected, then for the second level, we calculated the percentage of LPs when all characters were misclassified, the percentage of LPs when a single character was misclassified, the percentage of LPs when only two characters were misclassified, and the percentage of LPs when more than two characters were misclassified.

We tested the system on 457 positive images containing vehicles. Table 6 shows the result of calculating metrics for the license plate detection part: recall and accuracy. The accuracy calculation results for the license plate recognition part are summarized in Table 7.

Table 6. Final output evaluation. LPD phase test results.

Number of positive images (A)	Number of LPs correctly detected (B)	False positives (C)	Recall (Bx100/A)	Precision (Bx100)/(B+C)
457	**450**	**7**	**98.47%**	**98.47%**

Table 7. Final output evaluation. LPR phase. Accuracy considering: fully corrected LPs, at least 1 wrong characters, at least 2 wrong characters and more then 2 wrong characters.

LP detected	Good recognition	1 character badly detected	2 characters badly detected	>2 characters badly detected
450	**410 (91.11%)**	**35 (7.78%)**	**5 (1.11%)**	**0 (0%)**

Conclusion

This paper presented an end-to-end ALPR system based on Deep Learning. The main objective for building such a system was to meet the needs based on content indexing, content analysis of images and videos, and recognition of LPs.

The development of the proposed system was mainly based on the use of the YOLO model on the two main subtasks of the system, namely the detection and recognition of the license plate.

We have performed several tips and techniques on the system to improve recognition results, such as defining an ZOI to speed up the process by working on one LP at a time, applying a few modifications on the detected LP to work with clear and easy to recognize characters, and finally run a voting system on each LP, mainly to avoid redundancies when adding it to our database.

This strategy was essential to obtain exceptional results because we avoided errors of often misclassified characters and also a remarkable number of predicted characters. This was proven by the tests we performed in the last section on our dataset. Our system was able to achieve a detection rate of 98.47% and a complete recognition rate of 91.11%. These results are satisfactory for some real-world automatic license plate recognition applications.

As future work, we intend to explore new CNN architectures to further optimize (in terms of speed and accuracy) the ALPR system; We also want to include vehicle manufacturer and model recognition in the system pipeline so that our dataset provides such information; Make the system capable of recognizing multinational LPs by annotating multinational LPs and characters; Improve the results obtained by analyzing cases of failure; Finally, it would be interesting to integrate this license plate recognition system into a real-time on-board system.

References

1. Sirithinaphong, T., Chamnongthai, K.: The recognition of car license plate for automatic parking system. In: ISSPA 1999. Proceedings of the Fifth International Symposium on Signal Processing and its Applications (IEEE Cat. No. 99EX359), vol. 1, pp. 455–457. IEEE (1999)
2. Davies, P., Emmott, N., Ayland, N.: License plate recognition technology for toll violation enforcement. In: IEE Colloquium on Image Analysis for Transport Applications, p. 7-1. IET (1990)

3. Silva, S.M., Jung, C.R.: Real-time license plate detection and recognition using deep convolutional neural networks. J. Vis. Commun. Image Represent. **71**, 102773 (2020)
4. Riaz, W., Azeem, A., Chenqiang, G., Yuxi, Z., Khalid, W., et al.: YOLO based recognition method for automatic license plate recognition. In: 2020 IEEE International Conference on Advances in Electrical Engineering and Computer Applications (AEECA), pp. 87–90. IEEE (2020)
5. Shi, B., Bai, X., Yao, C.: An end-to-end trainable neural network for image-based sequence recognition and its application to scene text recognition. IEEE Trans. Pattern Anal. Mach. Intell. **39**(11), 2298–2304 (2016)
6. Henry, C., Ahn, S.Y., Lee, S.-W.: Multinational license plate recognition using generalized character sequence detection. IEEE Access **8**, 35185–35199 (2020)
7. Chen, R.-C., et al.: Automatic license plate recognition via sliding-window darknet-YOLO deep learning. Image Vis. Comput. **87**, 47–56 (2019)
8. Laroca, R., et al.: A robust real-time automatic license plate recognition based on the yolo detector. In: 2018 International Joint Conference on Neural Networks (IJCNN), pp. 1–10. IEEE (2018)
9. Shafiee, M.J., Chywl, B., Li, F., Wong, A.: Fast YOLO: a fast you only look once system for real-time embedded object detection in video. arXiv preprint arXiv:1709.05943 (2017)
10. Liu, W., Zhang, C., Lin, G., Liu, F.: CRNet: cross-reference networks for few-shot segmentation. In: Proceedings of the IEEE/CVF Conference on Computer Vision and Pattern Recognition, pp. 4165–4173 (2020)
11. Xie, F., Zhang, M., Zhao, J., Yang, J., Liu, Y., Yuan, X.: A robust license plate detection and character recognition algorithm based on a combined feature extraction model and BPNN. J. Adv. Transp. **2018** (2018)
12. Nosseir, A., Roshdy, R.: Extraction of egyptian license plate numbers and characters using surf and cross correlation. In: Proceedings of the 7th International Conference on Software and Information Engineering, pp. 48–55 (2018)
13. Bay, H., Ess, A., Tuytelaars, T., Van Gool, L.: Speeded-up robust features (SURF). Comput. Vis. Image Underst. **110**(3), 346–359 (2008)
14. Sun, H., Fu, M., Abdussalam, A., Huang, Z., Sun, S., Wang, W.: License plate detection and recognition based on the YOLO detector and CRNN-12. In: Sun, S. (ed.) ICSINC 2018. LNEE, vol. 494, pp. 66–74. Springer, Singapore (2019). https://doi.org/10.1007/978-981-13-1733-0_9
15. Dey, R., Salem, F.M.: Gate-variants of gated recurrent unit (GRU) neural networks. In: 2017 IEEE 60th International Midwest Symposium on Circuits and Systems (MWSCAS), pp. 1597–1600. IEEE (2017)
16. Li, H., Wang, P., You, M., Shen, C.: Reading car license plates using deep neural networks. Image Vis. Comput. **72**, 14–23 (2018)
17. Dong, M., He, D., Luo, C., Liu, D., Zeng, W.: A CNN-based approach for automatic license plate recognition in the wild. In: BMVC (2017)
18. Alborzi, Y., Mehraban, T.S., Khoramdel, J., Ardekany, A.N.: Robust real time lightweight automatic license plate recognition system for Iranian license plates. In: 2019 7th International Conference on Robotics and Mechatronics (ICRoM), pp. 352–356. IEEE (2019)
19. Howard, A.G., et al.: MobileNets: efficient convolutional neural networks for mobile vision applications. arXiv preprint arXiv:1704.04861 (2017)
20. Xie, L., Ahmad, T., Jin, L., Liu, Y., Zhang, S.: A new CNN-based method for multi-directional car license plate detection. IEEE Trans. Intell. Transp. Syst. **19**(2), 507–517 (2018)

21. Redmon, J., Farhadi, A.: YOLOv3: an incremental improvement. arXiv preprint arXiv:1804.02767 (2018)
22. Redmon, J., Farhadi, A.: YOLO9000: better, faster, stronger. In: Proceedings of the IEEE Conference on Computer Vision and Pattern Recognition, pp. 7263–7271 (2017)
23. He, K., Zhang, X., Ren, S., Sun, J.: Deep residual learning for image recognition. In: Proceedings of the IEEE Conference on Computer Vision and Pattern Recognition, pp. 770–778 (2016)
24. Mills-Tettey, G.A., Stentz, A., Dias, M.B.: The dynamic Hungarian algorithm for the assignment problem with changing costs. Robotics Institute, Pittsburgh, PA, Technical report CMU-RI-TR-07-27 (2007)
25. Montella, C.: The Kalman filter and related algorithms. Alındığı tarih, vol. 30, p. 2020 (2011). https://www.researchgate.net/

Evaluation of Selected Machine Learning Models and Features for Electrical Consumption Prediction in Educational Institutions

Houda Daki(✉), Basma Saad, Asmaa El Hannani, and Hassan Ouahmane

Laboratory of Information Technologies, National School of Applied Sciences, University of Chouaib Doukkali, Route d'Azemmour, Nationale No 1, ElHaouzia, 24002 El Jadida, Morocco
{daki.h,saad.b,elhannani.a,ouahmane.h}@ucd.ac.ma

Abstract. Recently, predictive analytic is making a very good contribution to reliable power supply. It provides advanced techniques to process, interpret and analyze big energy data and make it more valuable. In this article, we have presented a benchmark of the most used features and forecast models to predict the electrical energy consumption of educational institutions. This study uses ASHRAE data set, that contains information about energy types based on historic usage rates and observed weather for many buildings type. The proposed system analyzes only electricity meter reading for college university of educational buildings. The objective of this work is to evaluate the prediction performance of each forecasting model using different type of features (weather and occupancy) and different regressors in order to investigate the impact of input variables.

Keywords: Big data · Machine learning · Smart grid · Electrical consumption forecasting

1 Introduction

1.1 Motivation

Over the last years, electrical consumption has exponentially increased in many sectors due to the new behaviors and policies adopted by energy users. Buildings sector which includes residential and commercial structures has known a peak demand that causes a great number of challenges related to energy management. Many sectors contribute to this high electricity consumption, but the building sector is the most concerned. Universities are among institutions with high electrical usage of the total amount of the entire institutions in building sector. Especially, due to the new practices in teaching and research, as well as the increase of the number of students and faculties. In order to support

M. Lazaar et al. (Eds.): BDIoT 2022, LNNS 625, pp. 303–315, 2023.
https://doi.org/10.1007/978-3-031-28387-1_26

this continuous demand, energy supplies try to better manage their resources, re-dimension the grid and control energy supply and demand.

In the same context, many educational buildings start to change their energy policy by developing a strategy to encourage local production and increase the use of renewable energy. Recently, a great number of universities in all the world try to become green schools and implement smart grid technologies to cover their electrical needs. But, the majority face the challenge of how to manage the over-production of electricity, because in same cases they can't inject the surplus on the national electrical grid neither store it using storage devices which are expensive or limited. Indeed, electricity is difficult to store because the existing storage devices are either poorly performing, expensive or limited by various constraints. For this reason, it is necessary that the production of the energy be permanently, in real time, equal to the consumption. Concretely, this production-consumption balance requires in particular the coupling of electricity consumption forecasting models with renewable energy production forecasting models. The purpose of this paper is to evaluate and compare the effectiveness of features and machine learning models in a unified framework for electrical consumption forecasting in an educational institution. Forecasting electrical consumption in educational institutions is an open problem and several systems have been proposed in the literature and much research has been done to identify the effective features and the accurate models [4,8,12,17].

1.2 Contribution

Most of studies either do not integrate all features and target only meteorological data or do not compare all predictive machine learning models as most approaches reported in the literature in Sect. 2 that work only with weather data (humidity, temperature, wind speed ...). Our work propose the use of both weather and the occupancy data (course duration, course type, fields, individuals...) to evaluate their effective impact on building electrical consumption. Our study focus on black box models, in order to describe the forecasting process. Contrarily to several studies, which compare prediction models for a stable system in terms of data volume, our system offers a benchmark of all the prediction models most used in the field for a case with an evolving quantity of data in the future.

In this paper we perform an extensive evaluation of features and models to predict electrical energy consumption for educational institutions. Firstly, we are interested in finding the most effective model(s). Secondly, we wanted to explore new features and select the most promising ones for this kind of forecasting. The rest of the paper is organized as follows. After an overview of related works in Sect. 2, we describe our proposed approach including the materials and methods in Sect. 3. In Sect. 4 we present the evaluation of the different features and models with a detailed discussion of the achieved results. Finally, in Sect. 5 we give some concluding remarks and future directions of this work.

2 Literature Review

2.1 Data Selection and Preparation

Data plays a very important role to improve prediction accuracy, because without the relevant data with high impact on buildings energy consumption the proposed solution would have no purpose. So, the system must collect the adequate data that really affect the energy use. Thus, the first step focuses on getting the most relevant information and eliminating irrelevant ones. For concreteness, we reviewed some researches interested to forecasting electrical consumption for education institutions to highlight the more relevant data to collect in this context [4,6,8,13,14,17,20]. According the these reviews, we conclude that the majority of these studies use meteorological data especially temperature, humidity and wind speed. But, few works use occupancy data to forecast electrical consumption. As a results, we found that there are two important kinds of data used to predict electrical consumption in educational institutions: occupancy data and meteorological data.

2.2 Model Training

After collecting the right data, we should train forecasting models over it, and provide an algorithm that can reason over and learn from this data. Many studies have discussed various aspects of electrical energy prediction to identify the accurate models. Bourdeau et al. [5] made a review on data-driven building energy modeling techniques used in building energy consumption modeling and forecast studies. They reviewed research papers interested to supervised machine learning for building energy consumption modeling from 2007 to 2019. Bourdeau et al. found that Artificial Neural Network (ANN) are the most used in this context, followed by Support Vector Machine (SVM) then regression models. Wei et al. [25] also reviewed data-driven techniques in building energy analysis. Wei et al. found that ANN, SVM are the most accurate to predict energy consumption in buildings. In the same context, Amasyali et al. [5] proposed an overview of the most used algorithms in this kind of predictions. They concluded that an overall of 47% of the energy consumption prediction models are based on ANN, while 25% used SVM, 4% DT and 24% other statistical models. A great number of the reviewed works found that ANNs are the best models to predict energy consumption [2,3,7,10,13,19,24]. According to this review, we ran our system over the most used models in this field including: MLP, SVM, DT random forest (RF) and gradient boosting (GB).

2.3 Model Evaluation

Evaluation metrics help to validate forecasting performances of data-driven algorithms. In fact, validation step is used in order to verify the quality of the models and features. Bourdeau et al. [5] proposed an overview study of the most used metrics. They found that the Root Mean Square Error (RMSE), the coefficient of variation of RMSE (CV-RMSE) and the Mean Average Error (MAE) assessed in 47%, 38% and 36% respectively. On the other hand, the coefficient of determination (R2), the Mean Square Error (MSE), the Mean Relative Error (MRE), the Mean Bias Error (MBE) and the Normalized Mean Bias Error (NMBE) are used only for 27%, 16%, 9%, 2% and 4% respectively. Zhang et al. [22] showed that the CV-RMSE must be the first performance measure to be selected followed by other metrics. Zhang et al. consider the CV-RMSE the most important metric followed by the RMSE. But, if these two values were unavailable, then the Mean Absolute Percentage Error (MAPE) was selected. However, this order is not usually respected, Runge et al. [21] have done a review of the most used error metrics, and they found that MAPE is predominately (38%) used as the main performance measure within forecasting papers, with CV-RMSE and R2 accounting for 17–20% of the performance metrics applied. Based on these reviews, we choosed to evaluate our models using three metrics: RMSE, MAE, NRMSE and NMAE.

3 Materials and Methods

3.1 Environment Description

In our experiments we used the Kaggle environment, Kaggle Kernel is a free Jupyter notebook server, it allows machine learning operations on cloud computers instead of doing it on physical computer. The GPU provided by Kaggle is Nvidia Tesla P100 GPU with 16 GB memory, it can also integrate GPU, and it's especially useful because it provides a K80/P100/T4 GPU for free.

3.2 Data Description

ASHRAE organised the Great Energy Predictor III (GEPIII) contest on Kaggle [1], in late 2019, with a purpose of identifying statistical and machine learning techniques that provide the best short-term predictions of building energy consumption. The competitors were provided with over 20 million points of training data from 2,380 energy meters collected for 1,448 buildings of different type (education, entertainment, food sales and service, lodging/residential, manufacturing, office, others, parking, public services, religious worship, retail, services, technology/science, warehouse/storage, utility) and from 16 sites worldwide and four energy meter reading including: electricity, chilledwater, steam, hotwater. This competition's overall objective was to predict over 41 million private and public test data points. This data is an hourly meter readings of 1,000 buildings from several sites around the world over three-year. These data are hourly measurements taken from energy metering systems at locations that had data from January 1, 2016, to December 31, 2018. For reasons of consistency and to serve the objective of our researches we have decided to select only electricity meter values meter id = 0: electricity. Reading for college university of educational building in site 4 and for the building id 582 so we omitted all the data that describe buildings(building id, flour count, square feet, year built) since it not significant at all with one building. We have also aggregated the data to be in daily granularity instead of hourly. Table 1 describes the data set used in this work.

Table 1. ASHRAE data sets input variables

Data field	Description
Timestamp	
Time	Time of measurement formatted as yyy-mm-dd hh:mm:ss
Target	
Meter-reading	Hourly energy consumption in kWh.
Weather data	
Air-temperature	Air temperature in degrees Celsius
Cloud-coverage	Portion of the sky covered in clouds in Oktas
Dew-temperature	Dew temperature in degrees Celsius
Precip-depth-1-hr	Precipitations in Millimeters
Sea-level-pressure	Sea level pressure in Millibar
Wind-speed	Wind speed in Meters per second
Occupancy data	
Semester	The semester type (Summer, Spring, Fall)
Instruction	Binary variable, it takes 0 for instructional day else 1
Holidays	Type of the holidays(Academic and Administrative Holiday, Big holiday, Spring Recess, Reading/Review/Recitation Week, no holiday)
Formal classes	Binary variable, it takes 0 if there is formal classes in that day else 1
Final exam	Binary variable, it takes 0 if there is no exam in that day else 1
Events	Type of events (Cal Day, Commencement, Convocation, no events)
Summer-sessions	The number of summer sessions per day

3.3 Data Preparation

Filling in the Missing Values. To handle the missing values there is two ways: either drop or impute the missing data. Dropping the missing values is not recommended because it leads to a loss of information and it affects the prediction accuracy. Imputing the data are more suitable, which refers to replacing the missing entries with substituted values depending on the nature of the problem and data. Many ways are presented in the literature [11,27,29] to handle the missing entries including: the use of constant values or a dictionary of values, imputation using the statistics, the use of algorithms that support missing values Prediction (interpolation) or imputation using Deep Learning Library (datawig). In this work, we decided to use a statistic method called the mean imputation, that imputes missing values using the mean of non-missing cases. This method is suitable in the case of no more than 5% of the data are missing and is used only to replace missing numerical values. So, the mean imputation is a good choice to meet our system requirements and needs. Mainly, because other imputations like features modeling are expensive in terms of resources and maintenance.

Data Aggregation. To switch from hourly to daily data resolution, data aggregation approach should be taken: For precipitations data the sum of hourly precipitations. Temperature (dry and air temperature), pressure and wind speed variables were divided into 3 subcategories, that are minimum, maximum, and mean, each subcategory is represented as a column. For cloud coverage the mode value is taken. However, for electricity consumption, the accumulated values during the day was considered.

Splitting the Data Set. Splitting the data is essential for an unbiased evaluation of the prediction performance. In most cases, the data sets are divided into three subsets:

- Training set: is applied to train, or fit, the model.
- Validation set: is used for unbiased model evaluation during hyper parameter tuning.
- Test set: is needed for an unbiased evaluation of the final model. It's not used for fitting or validation.

In machine learning the data should be splinted in many parts, in general training, validation and test data. The data sets with high quantity is used for training. In practice, data might be split at an (80,10,10) or a (70, 20,10) ratio of training, testing and validation. The exact ratio depends on the data, but a 70-20-10 ratio is an optimal choice for small data sets. Therefore, our data sets were split in three groups: training, validation and test data. In particular, we used 70% of the data (2017 and 2018) for training. The year 2019 is reserved for test and validation, 10% for validation and for testing we used 20%.

3.4 Training Model

The proposed solution is based on Python libraries, which offers the most known machine learning and statistical algorithms. In this study we investigate and compare the effectiveness of the top five machine learning models in the field of electrical consumption forecasting for educational buildings as reported in literature (see Sect. 2): MLP [28], SVR [15], DT [18], RF [23], and GB [15].

Machine learning models require some parameters, which are the internal coefficients set to train or optimize the model on a training data set. These parameters impact the model performance, so they should be set carefully, especially when we don't really know their optimal values in advance.

Machine learning models have many hyper-parameters, the best combination of values of the hyper-parameters searching should be done in a multi-dimensional space. So, hyper-parameter tuning is required to find the right values of the parameters. Grid Search method is very helpful in this cases. It is a search space as a grid of hyper parameter values which evaluate every position in the grid. Grid search is great for spot-checking combinations that are known to perform well generally. It is very useful for discovering and getting hyper parameter combinations that we would not have guessed intuitively, although it often requires more time to execute. In our case, we choose the optimal parameters for each model by experimentation. To investigate the importance of each data type we decided to run models using three scenarios:

- Using only weather data
- Using only occupancy data
- Using both weather and occupancy data

And to verify the quality of the models (DT, RF, GBT, SVR and MLP), three different metrics were used: RMSE, MAE, NRMSE and NMAE. The confidence interval of each model in each experiment is taken into consideration.

- Mean Absolute Error (MAE) [26]: this model evaluation metric describes the average of the absolute values of all differences between the forecast and the real values expressing the same phenomenon.

$$MAE = \frac{\sum_{1}^{N} |Y_{real} - Y_{pred}|}{N} \tag{1}$$

  where N is the total number of data points.
- Normalized Mean Absolute Error (NMAE) [16]: this metric presents the MAE by the mean of the time series, we can interpret the result as a weighted Mean Absolute Percentage Error.

$$NMAE = \frac{MAE}{mean(Y_{real})} * 100 \tag{2}$$

- Root Mean Square Error (RMSE) [9]: presents the concentration of data around the line of the best fit. Root Mean Squared Error is usually used in regression analysis for numerical predictions because it's a good measure of accuracy in the case of comparing prediction errors of different models or model configurations for a particular variable.

$$RMSE = \sqrt{1/N \sum_{i=1}^{N} (Y_{real_i} - Y_{pred})^2} \tag{3}$$

– Normalized Root Mean Square Error (NRMSE) [16]: this metric presents the RMSE by the mean of the time series, we can interpret the result as a weighted Mean Square Percentage Error.

$$NRMSE = \frac{RMSE}{mean(Y_{real})} * 100 \tag{4}$$

4 Results and Discussion

4.1 Weather Data Experiments

In this experiment, we used only ASHRAE dialy weather data after being prepared. The grid search method is used to get the optimal model parameters for weather scenario.

Table 2 and Fig. 1 present the RMSE, MAE, NRMSE and NMAE for the different models for weather scenario. The results show that our best system configuration is using GB model based on MAE metric. But, looking more closely to Table 2, we notice that error metrics of all models are overlapping if we take into consideration the confidence interval. So, all the choose regressors are accurate for this kind of predictions.

4.2 Occupancy Data Experiments

As reported in Sect. 2 many studies have highlighted the significance of occupancy data in electricity consumption prediction. However, most of these works either do not integrate these two main data types or do not use detailed occupancy data. In this experiment, we used daily occupancy data to see the impact of occupancy on predictions.

Table 2 and Fig. 1 present the RMSE, NRMSE, MAE and NMAE for the different models. The results show that occupancy data has the same impact as weather data since the error metrics of the two scenarios are closer to each other. MLP model is not a good choice in this case. The amount of data used is not enough to train the model. The rest of regressors are outperforming MLP model and giving close results.

4.3 Weather and Occupancy Data Experiments

In this experiment, we choose to combine both weather and occupancy data to see if it will help models to make more accurate predictions.

Table 2 and Fig. 1 present the RMSE, NRMSE, MAE and NMAE for the different models. Learning models are relatively similar in term of efficiency, the mean MAE error was 14.68, 13.67 and 13.84 for respectively weather, occupancy and weather+ occupancy combinations. Weather and occupancy had both notable impact on predictions, that's why the inclusion of those categories couldn't achieves interesting accuracy.

Table 2. The RMSE (kWh), MAE (kWh), NMAE(%) and NRMSE(%) error metrics of the five models

Metric	DT	RF	GBT	SVR	MLP
Weather scenario					
RMSE	21,23 ± 1,67	18,88 ± 1,54	18,60 ± 1,51	18,49 ± 1,49	18,58 ± 1,47
MAE	16,27 ± 1,67	14,10 ± 1,54	**13,94** ± 1,51	13,96 ± 1,49	14,17 ± 1,47
NMAE	10,20 ± 1,05	8,84 ± 0,96	8,74 ± 0,95	8,76 ± 0,93	8,89 ± 0,92
NRMSE	13,32 ± 1,05	11,84±0,96	11,67± 0,95	11,60± 0,93	11,65± 0,92
Occupancy scenario					
RMSE	18,79 ± 1,48	17,93 ± 1,41	17,14 ± 1,37	18,31 ± 1,46	13,27 ± 1,49
MAE	14,41 ± 1,48	13,8 ± 1,41	**13,03** ± 1,37	13,89 ± 1,46	17,99 ± 1,49
NMAE	9,04± 0,93	8,66± 0,88	8,17± 0,86	8,71± 0,91	11,28± 0,93
NRMSE	11,79± 0,93	11,25± 0,88	10,75± 0,86	11,48± 0,91	8,32± 0,93
Weather and occupancy scenario					
RMSE	18,19 ± 1,48	18,50 ± 1,5	18,26 ± 1,48	18,49±1,49	13,27±1,5
MAE	**13,70** ± 1,48	13,85 ± 1,5	**13,70** ± 1,48	13,95 ± 1,49	17,99 ± 1,5
NMAE	8,59± 0,93	8,69± 0,94	8,59± 0,93	8,75± 0,93	11,28± 0,94
NRMSE	11,41± 0,93	11,60± 0,94	11,45± 0,93	11,60± 0,93	8,32± 0,94

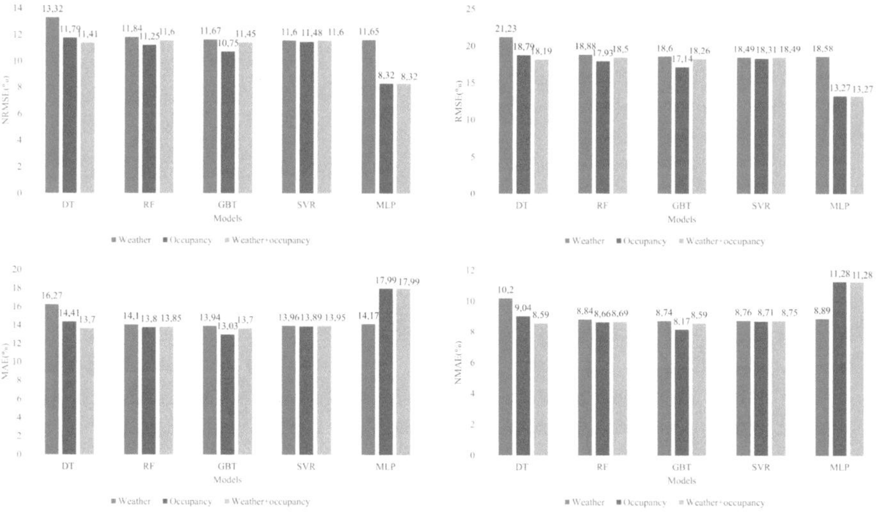

Fig. 1. Error metrics results for all scenarios

4.4 Weather and Occupancy Data Experiments with Features Selection

In this experiment we used feature selection approach for weather and occupancy scenario to select only the most significant features and removing the remaining

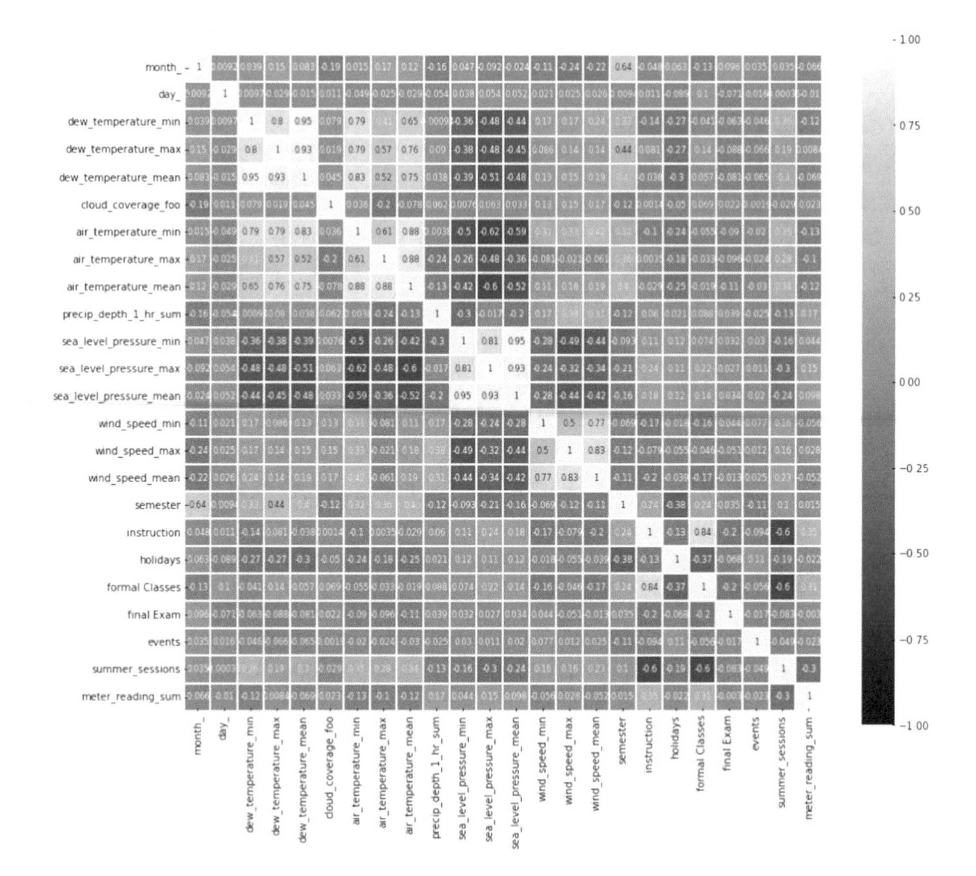

Fig. 2. Data correlation

ones from consideration. First of all, we ran the feature importance method, to have the score of each feature, the higher score represents the more important or relevant feature. For that reason, we used data correlation method which usually used to explain how one or more variables are related to each other. It plays a vital role in locating the important variables on which other variables depend. In general, the proper correlation analysis leads to better understanding of data. Figure 2 shows the correlation between all features. As we can see, the correlation coefficient varies from -0.62 to 1.

At this experience, we removed features within high correlation (correlation between input features), if two features have absolute correlation coefficient greater that 0.7, one of the features is removed. The kept feature is the one highly correlated with the target. As a results, the experiment ran only those features: month, day, cloud-coverage, air-temperature-min, air-temperature-max, precip-depth-1-hr-sum, sea-level-pressure-min, wind-speed-min, wind-speed-max, semester, instruction, holidays, final Exam, events, summer-sessions and meter-reading-sum'.

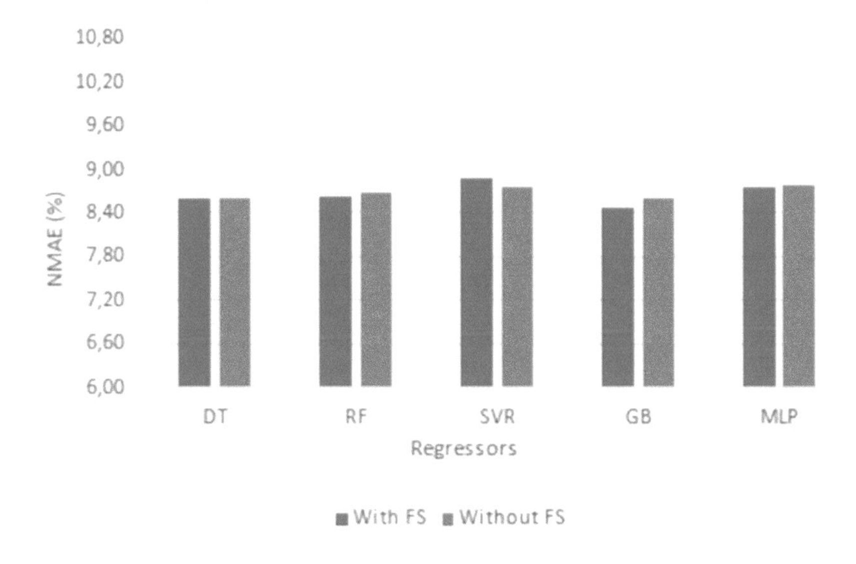

Fig. 3. Comparison of results with and without features selection approach

Results in Fig. 3 showed that features selection method is capable of retaining only relevant variables, so in all studied regressors, the NMAE of predictions before and after features selection is relatively similar. As we can see in Fig. 3, DT, RF and MLP keep the same behavior using feature selections. So, there is no added value with this approach. The only one that has been improved is the GB model. This result, shows that not the all used features are important, the ones choosed by feature selection are enough to get good performance.

5 Conclusion

In this paper, we have presented a comparative study of state-o-the-art forecasting models to predict daily electrical energy consumption in educational building. In particular, we were interested in both determine which model is more suitable for this kind of problem, and select the optimal features. In our experiments, We have proposed three scenarios (weather, occupancy and weather + occupancy) to select the best features and model. We have demonstrated that predictive models had closer efficiency by comparing four choose metrics as Fig. 1 presents. But, for a scalable solution the choose machine learning models are very accurate, which confirms literature review. MLP model needs more data to extract patterns. We conclude that, Weather and occupancy data had similar impact in electricity consumption prediction that's why the merge of those two categories doesn't achieve an impressive accuracy. We used also feature selection approach for weather and occupancy scenario to select only the most significant features. The experiment showed that features selection approach had no added value except retaining relevant variables, so before and after features selection is relatively similar.

References

1. ASHRAE - Great Energy Predictor III. https://www.kaggle.com/c/ashrae-energy-prediction. Accessed 22 Aug 2022
2. Ahmad, T., Chen, H., Guo, Y., Wang, J.: A comprehensive overview on the data-driven and large scale based approaches for forecasting of building energy demand: a review. Energy Buildings **165**, 301–320 (2018)
3. Ai, S., Chakravorty, A., Rong, C.: Household power demand prediction using evolutionary ensemble neural network pool with multiple network structures. Sensors **19**, 721 (2019)
4. Allab, Y., Pellegrino, M., Guo, X., Nefzaoui, E., Kindinis, A.: Energy and comfort assessment in educational building: Case study in a French university campus. Energy Buildings **143**, 202–219 (2017)
5. Amasyali, K., El-Gohary, N.M.: A review of data-driven building energy consumption prediction studies. Renew. Sustain. Energy Rev. **81**, 1192–1205 (2018)
6. Amber, K.P., et al.: Energy consumption forecasting for university sector buildings. Energies **10**, 1579 (2017)
7. Amber, K., Ahmad, R., Aslam, M., Kousar, A., Usman, M., Khan, M.S.: Intelligent techniques for forecasting electricity consumption of buildings. Energy **157**, 886–893 (2018)
8. Amber, K., Aslam, M., Hussain, S.: Electricity consumption forecasting models for administration buildings of the UK higher education sector. Energy Buildings **90**, 127–136 (2015)
9. Chai, T., Draxler, R.R.: Root mean square error (RMSE) or mean absolute error (MAE)?-arguments against avoiding RMSE in the literature. Geoscientific Model Develop. **7**, 1247–1250 (2014)
10. Chammas, M., Makhoul, A., Demerjian, J.: An efficient data model for energy prediction using wireless sensors. Comput. Electr. Eng. **76**, 249–257 (2019)
11. Che, Z., Purushotham, S., Cho, K., Sontag, D., Liu, Y.: Recurrent neural networks for multivariate time series with missing values. Sci. Rep. **8**, 1–12 (2018)
12. Daki, H., El Hannani, A., Ouahmane, H.: Big-data architecture for electrical consumption forecasting in educational institutions buildings. In: Proceedings of the 2nd International Conference on Networking, Information Systems & Security, p. 24. ACM (2019)
13. Grolinger, K., L'Heureux, A., Capretz, M.A., Seewald, L.: Energy forecasting for event venues: big data and prediction accuracy. Energy Buildings **112**, 222–233 (2016)
14. Hong, W.C., Li, M.W., Fan, G.F.: Short-Term Load Forecasting by Artificial Intelligent Technologies. MDPI, Basel (2019)
15. Kazemzadeh, M.R., Amjadian, A., Amraee, T.: A hybrid data mining driven algorithm for long term electric peak load and energy demand forecasting. Energy **204**, 117948 (2020)
16. Kolassa, S., Schütz, W., et al.: Advantages of the mad/mean ratio over the MAPE. Foresight Int. J. Appl. Forecast. **6**, 40–43 (2007)

17. Moon, J., Park, J., Hwang, E., Jun, S.: Forecasting power consumption for higher educational institutions based on machine learning. J. Supercomput. **74**, 3778–3800 (2018)
18. Priyam, A., Abhijeeta, G., Rathee, A., Srivastava, S.: Comparative analysis of decision tree classification algorithms. Int. J. Curr. Eng. Technol. **3**, 334–337 (2013)
19. Rahman, A., Srikumar, V., Smith, A.D.: Predicting electricity consumption for commercial and residential buildings using deep recurrent neural networks. Appl. Energy **212**, 372–385 (2018)
20. Ruiz, L.G.B., Rueda, R., Cuéllar, M.P., Pegalajar, M.: Energy consumption forecasting based on Elman neural networks with evolutive optimization. Expert Syst. Appl. **92**, 380–389 (2018)
21. Runge, J., Zmeureanu, R.: Forecasting energy use in buildings using artificial neural networks: a review. Energies **12**, 3254 (2019)
22. Sosnin, S., Vashurina, M., Withnall, M., Karpov, P., Fedorov, M., Tetko, I.V.: A survey of multi-task learning methods in chemoinformatics. Mol. Inf. **38**, 1800108 (2019)
23. Touw, W.G., et al.: Data mining in the life sciences with random forest: a walk in the park or lost in the jungle? Brief. Bioinform. **14**, 315–326 (2013)
24. Wahid, F., Ghazali, R., Shah, A.S., Fayaz, M.: Prediction of energy consumption in the buildings using multi-layer perceptron and random forest. Int. J. Appl. Sci. Technol. **101**, 13–22 (2017)
25. Wei, Y., et al.: A review of data-driven approaches for prediction and classification of building energy consumption. Renew. Sustain. Energy Rev. **82**, 1027–1047 (2018)
26. Willmott, C.J., Matsuura, K.: Advantages of the mean absolute error (MAE) over the root mean square error (RMSE) in assessing average model performance. Climate Res. **30**, 79–82 (2005)
27. Yadav, M.L., Roychoudhury, B.: Handling missing values: a study of popular imputation packages in R. Knowl. Based Syst. **160**, 104–118 (2018)
28. Yegnanarayana, B.: Artificial Neural Networks. PHI Learning Pvt. Ltd., Delhi (2009)
29. Zhang, S., Wu, X., Zhu, M.: Efficient missing data imputation for supervised learning. In: 9th IEEE International Conference on Cognitive Informatics, pp. 672–679. IEEE (2010)

Credit Card Fraud Detection Using SVM, Decision Tree and Random Forest Supervised Machine Learning Algorithms

Oussama Ndama(✉) and El Mokhtar En-Naimi(✉)

E-DSAI2S Research Team, Faculty of Sciences and Technologies, Abdelmalek Essaâdi University Tetouan, Tetouan, Morocco
oussama.ndama@gmail.com, en-naimi@uae.ac.ma

Abstract. Since the inception of e-commerce payment systems, there have always been those who look for novel ways to get unauthorized access to another person's assets.

Payment cards are easy to use because they only require sending a short number of characters to the bank in order to identify your account and authorize the transaction. Due to their simplicity, they are more exposed.

In this article, we'll talk about how to detect credit card fraud. Our vision is to develop a fraud detection system that uses machine learning techniques like decision trees, random forest and Support Vector Machine (SVM).

Keywords: Machine Learning · Credit card fraud detection · Random Forest · Decision Trees · Support Vector Machine

1 Introduction

Detecting credit card fraud entails finding fraudulent purchase attempts and rejecting them rather than processing the order. Many different tools and techniques are available for detecting fraud, and the majority of merchants use a combination of several of them.

Payment cards are simple to use since identifying your account and authorizing the transaction simply need sending a small amount of numbers to the bank. They are additionally exposed due of their simplicity.

The cost of credit card theft to the worldwide economy exceeds $24 billion annually [16], and the amount is rising.

The effects of fraud are more severe for smaller retailers, which is why it's crucial to have procedures and tools in place to identify fraud early on.

An approach called "Credit Card Fraud Detection using Machine Learning" requires a team of data scientists analyzing the data and creating a model that will be most efficient at detecting and blocking fraudulent transactions.

The data is then passed through a model that has been carefully trained to look for patterns and rules in order to categorize whether a transaction is legitimate or fraudulent.

M. Lazaar et al. (Eds.): BDIoT 2022, LNNS 625, pp. 316–327, 2023.
https://doi.org/10.1007/978-3-031-28387-1_27

2 Related Work

Numerous studies have been conducted in the area of detecting credit card fraud. The various research papers pertaining to credit card fraud detection are presented in this section. Furthermore, we place a lot of emphasis on the research that revealed fraud detection in the issue of class imbalance.

As a result, the primary approaches can be categories like DL, ML, for credit card fraud ensemble and feature ranking, as well as user authentication approaches [1, 2].

SVM is a useful supervised learning technique that can be used to solve classification and regression problems. The information in our dataset is classified using the best-fitting method determined by Support Vector Machine [3].

Decision Trees and Support Vector Machines were used in Sahin and Duman's [4] investigation into credit card fraud detection. The authors demonstrate that the suggested Decision Tree classifiers perform better in the identification of credit card fraud than SVM. SVM model detection accuracy is on par with Decision Tree models as training data scales, but it detects fewer frauds.

Quinlan [5] developed the Decision Tree technique, which can handle sequential data. The Decision Tree is a table of different tree appearances composed of internal, root, and leave nodes.

The Decision Tree was combined with Hunt's algorithm and Luhn's algorithm in a study by Save et al. [6] to identify fraudulent transactions. The shipping address and billing address of the legitimate user were verified by the paper. It is thought that these addresses must match in order for the transaction to be considered valid.

If not, the transaction is considered suspicious because a fake one is more likely to differ from the address of the legitimate user. The process of "Outlier detection" was described in the article, which came to the conclusion that the card validation was accurate and had few false alarms.

The Random Forest model is a classifier for aggregate data. Combining numerous decision tree classifiers results in the utilization of multiple trees. The basic goal of using many trees is to sufficiently train them so that each one contributes to the way a model is put together. The outcome would then be mixed after the tree had been built [7]. This model depends on a specific dataset using multiple decision trees.

Lakshmi et al. [8] examine the effectiveness of various ML algorithms, including Logistic Regression, Decision Tree, and Random Forest, for detecting credit card fraud. They made advantage of a well-known credit card transaction dataset from Kaggle, which consists of 284,808 credit card transactions from a set of data from European banks.

3 The State of the Art of Machine Learning Algorithms

In a broader sense, machine learning is a branch of artificial intelligence that gives systems the capacity to learn from experience automatically without human involvement and strives to make the most accurate predictions of future outcomes using a variety of algorithmic models. The traditional techniques to computing, where computers are deliberately designed to calculate or solve a problem, are quite different from machine learning. Machine learning is the study of the input data required to train a model, where the model discovers various patterns in the input data and makes use of that understanding to forecast unknowable outcomes.

Machine learning has a tremendously broad range of uses. It is utilized in many different applications, including the spam filter, face detection, autopilot, weather forecasting, stock market forecasting, medical diagnostics, and fraud detection, etc.

A machine learning method known as supervised learning is one in which the model being trained is given labels for both the input and the output. The supervised model extracts patterns from the input data by using the output and input labeled data for training. These uncovered patterns are applied to support next decisions.

Regression and classification are two further subcategories of supervised learning. The output variable in a classification issue is a category (e.g., fraud or genuine, rainy or sunny, etc.). The output variable in a regression issue is a real value (e.g., the price of a house, temperature, etc.).

There are typically three different classification methods: binary classification, multi-class classification (where there are more than two output labels), and multi-label classification (where the data samples are not mutually exclusive and each data sample is assigned a set of target labels). In this thesis, the output label is either normal or fraudulent, and the problem is one of binary classification.

On other hand we have unsupervised learning and the clustering is the process of grouping similar objects in data [10]. Various clustering techniques used to group the dataset produce various grouping results.

The K-means algorithm is an unsupervised technique. When there is no prior understanding of the specific class of observations in a dataset.

4 Comparative Approaches to Supervised Learning Algorithms

The different fields of machine learning could each handle a range of learning problems.

A credit card fraud detection solution is provided by the Machine Learning technique, such as random forest.

4.1 Random Forest

The random forest is the decision tree's ensemble. The supervised ML-based Decision Tree method is used to resolve regression and classification challenges. Root, decision, and leaf nodes are the three different sorts of nodes found in a Decision Tree. The algorithm's root node serves as its beginning point. In order to split the tree, a decision must be made at the decision node. A leaf node stands for a conclusion [11]. An ensemble of Decision Trees is used by the Random Forest technique to make predictions [12]. A decision is made in the Random Forest by a majority vote. The following is a mathematical definition of the Random Forest [13]:

A Random Forest is defined as $\{h(x, \Theta k), k = 1,\ldots\}$ where the $\{\Theta k\}$ is the number of independent, identically distributed trees that are allowed to vote on the input vector x. The prediction is the label with the most votes (Fig. 1).

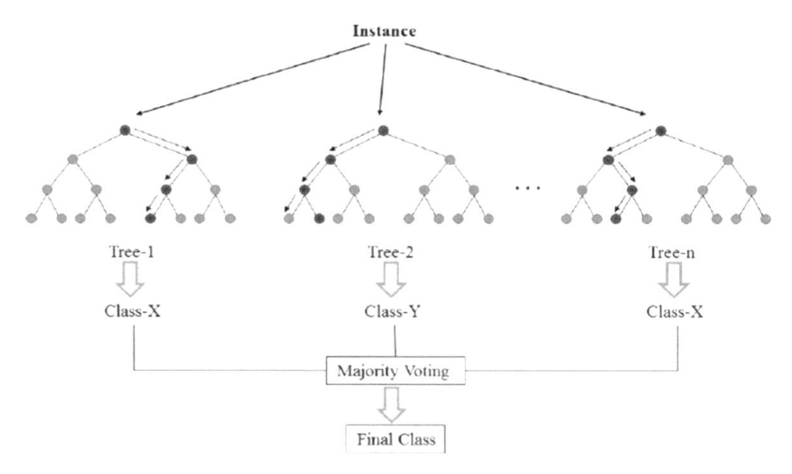

Fig. 1. Random Forest schema

Advantages:

- They are highly adaptable.
- Easy to implement, and simple to show and understand.

Disadvantages:

- The Prerequisites to check every condition one by one.

4.2 Random Forest vs Decision Tree

Understand how decision tree and random forest models differ from each other:

A decision tree model's accuracy increases when more splits are added to a given training dataset. Data can be readily overfit, so using the cross validation technique is advised.

The benefits of a straightforward decision tree model: simple to interpret because you can easily determine which variable and its value were utilized to split the data and predict the results [14].

A random forest looks like a dark box. You can specify the number of trees you want in your forest as well as the maximum number of attributes that can be used in each tree (n estimators). However, randomization cannot be controlled, including which data point is included in which tree and which characteristic is used for which tree. Accuracy increases with the number of trees added, but finally hits a fixed point. Random forest lessens the variance of error rather than minimizing the bias component. Unlike decision trees, it won't result in a severely biased model that overfits the data.

Random forest does not rely significantly on any particular set of features because features are selected at random during the training process. This differentiates random forests from bagging trees as a specific feature. Random forest always beats Decision Tree in terms of accuracy since it generalizes better and on validation datasets that have not yet been observed [14].

4.3 Support Vector Machine (SVM)

The machine learning algorithm known as a support vector machine (SVM) examines data for regression and classification purposes. The sorted data are output as a map by an SVM, with the margins between the two beings as far away as possible [15].

Due to the binary nature of SVM, the transactions are classified as either fraudulent or legitimate.

This enables us to recognize unusual user behavior, such as fraudulent user behavior. It employs regression analysis (Fig. 2).

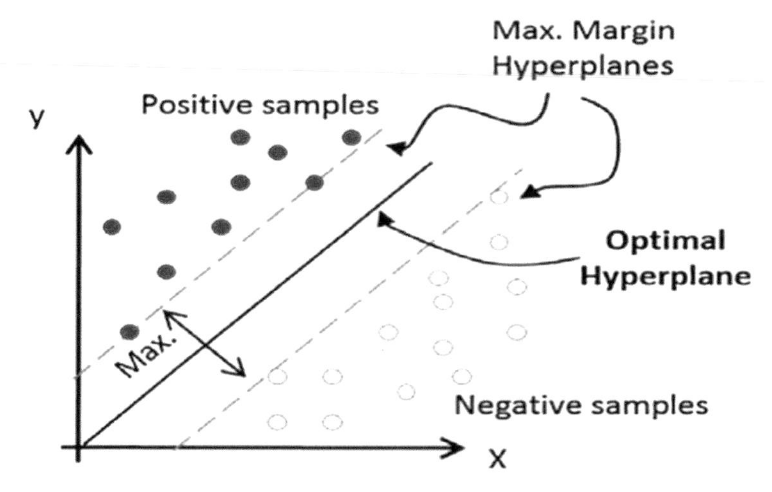

Fig. 2. Support Vector Machine Schema

Advantages:

- Since the optimality problem is curved by selecting an appropriate speculation grade, SVMs can be effective even when the preparation test contains some predisposition.

Disadvantages:

- Poor in process large dataset.
- Expensive and has low speed of detection.
- Medium accuracy.
- Lack of transparency of results.

5 Results

5.1 Environment

For this experiment we used python as coding language, Scikit-Learn, Pandas, Seaborn, Matplotlib and Numpy as a data science libraries, and the models are trained on a laptop equipped with an 11th generation Intel i5 11500H 2.90 GHZ (12 CPUs) processor and 16 GB RAM.

5.2 Data Description

The domain of machine learning highly depends on datasets. One of the most difficult responsibilities is data collection, especially if it involves financial issues like credit card fraud.

A dataset of 284,807 completely anonymous transactions is used in this project. Every transaction is classified as either fraudulent or not.

Take note that the dataset has very few instances of fraudulent transactions. The percentage of fraudulent card transactions is less than 0.18%.

The unbalanced class distribution can be visualized in a bar diagram given in Fig. 3 (Fig. 4).

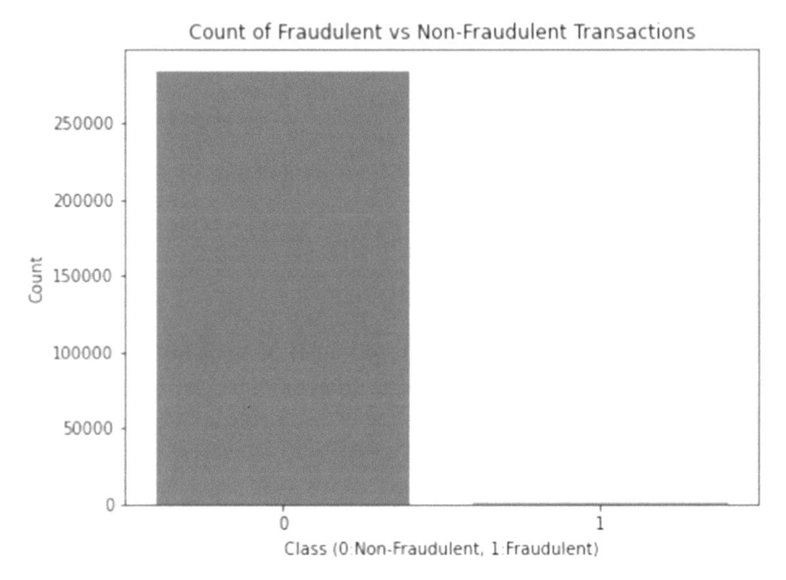

Fig. 3. Dataset distribution

```
print('Number of fraudulent transactions = %d or %d per 100,000 transactions in the dataset'
      %(len(data[data.Class==1]),len(data[data.Class==1])/len(data)*100000))
```

```
Number of fraudulent transactions = 492 or 172 per 100,000 transactions in the dataset
```

Fig. 4. Count of fraudulent transactions in the dataset

The dataset includes the numbers produced by the Principal Component Analysis (PCA) transformation. The original features have not been revealed, however, because of the confidentiality concerns. There are a total of 30 features, and 28 of them were generated through principal component analysis.

A dimensionality-reduction method called PCA condenses a large number of original variables into a smaller set of feature variables.

5.3 Data Standardization

Rescaling the features to give them the characteristics of a standard normal distribution with a mean of 0 and a standard deviation of 1 is known as "standardizing the features".

The features must be standardized before using machine learning techniques, which is a typical requirement for many machine learning models. The model's performance may be impacted if standardization is not done. We performed standardization on the 'Amount' feature using StandardScalar in the scikit-learn library (Fig. 5).

```
from sklearn.preprocessing import StandardScaler
data['normalizedAmount'] = StandardScaler().fit_transform(data['Amount'].values.reshape(-1,1))
data = data.drop(['Amount'],axis=1)
```

Fig. 5. Standardization on the amount of transaction

5.4 Confusion Metrics

A confusion metric that shows how well the model is predicted to suit the findings once connected to the earliest ones is a classification model visualization. The confusion metrics can be visualized using the association table as a heat map. Despite the fact that confusion metrics may be imagined using a variety of built-in ways, we can construct and visualize them depending on the score to enable better correlation.

The prediction results are presented using a confusion matrix to highlight true and false values:

TP: True Positive, which denotes the real data of the normal transactions are predicted as normal transactions.
TN: True Negative, which denotes normal transactions are predicted as fraud transactions.
FP: False Positive, which denotes fraud transactions are predicated as fraud transactions.
FN: False Negative. Fraud transactions are identified as normal transactions.

5.5 Roc Curve

An ROC curve (receiver operating characteristic curve) is a graph showing the performance of a classification model at all classification thresholds. This curve plots two parameters [17]:

- True Positive Rate
- False Positive Rate

True Positive Rate (TPR) is a synonym for recall and is therefore defined as follows:

$$TPR = \frac{TP}{TP + FN}$$

False Positive Rate (FPR) is defined as follows:

$$FPR = \frac{FP}{FP + TN}$$

AUC: Area Under the ROC Curve. AUC stands for "Area under the ROC Curve." That is, AUC measures the entire two-dimensional area underneath the entire ROC curve (think integral calculus) from (0,0) to (1,1) [17].

5.6 Support Vector Machine

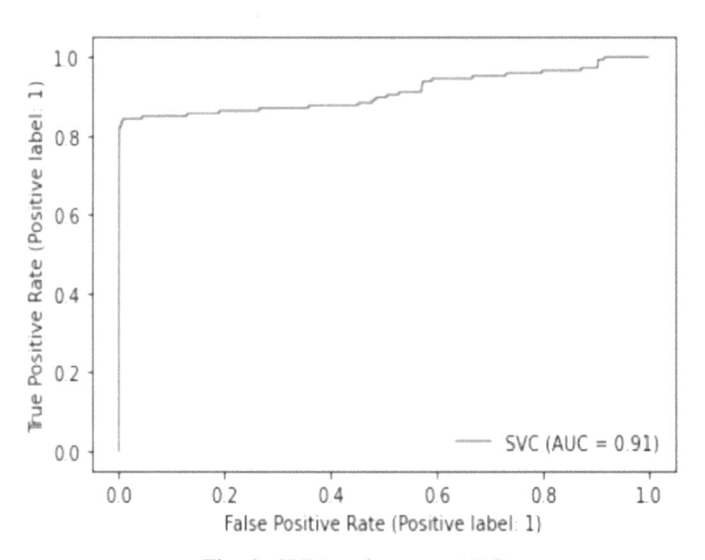

Fig. 6. SVM performance AUC

Random Forest:

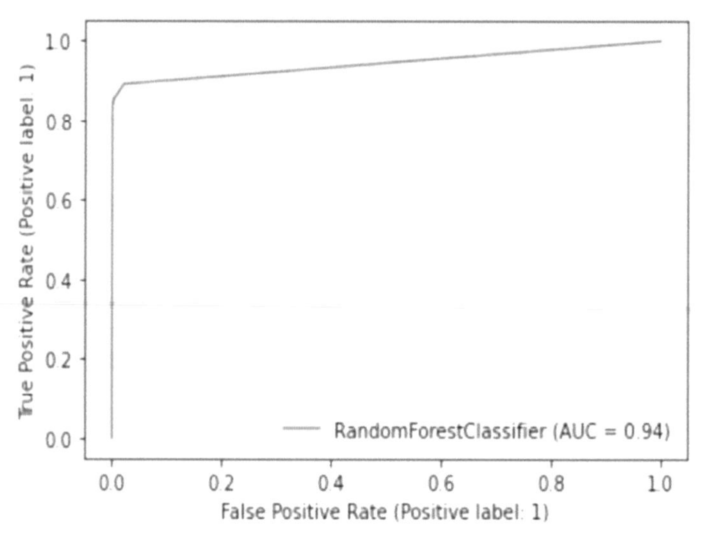

Fig. 7. RF performance AUC

Decision Tree:

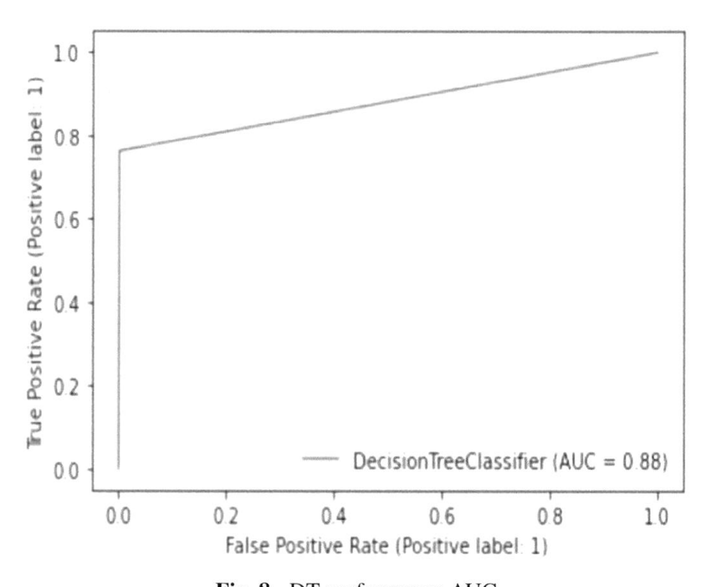

Fig. 8. DT performance AUC

The higher the AUC, the better the performance of the model at distinguishing between the positive and negative classes.

So we notice based on Figs. 6, 7, 8 that Random forest win with an AUC equal to 0.94.

6 Discussion

We construct tree different types of classification models during this step, Decision tree, Random forest and SVM. These are the models that are most frequently implemented to solve classification problems, regardless of the fact that we have access to a wide range of alternative models.

All of these models can be effectively created by utilizing the scikit-learn packages. The result of the implemented ML algorithms is shown in Fig. 9 and Fig. 10.

	Model	Accuracy	FalseNegRate	Recall	Precision	F1 Score
0	SVM	0.999368	0.215447	0.784553	0.839130	0.810924
1	RandomForest	0.999520	0.238095	0.761905	0.949153	0.845283
2	DecisionTree	0.999309	0.238095	0.761905	0.823529	0.791519

Fig. 9. Results on test dataset

	Model	Accuracy	FalseNegRate	Recall	Precision	F1 Score
0	SVM	0.999368	0.215447	0.784553	0.839130	0.810924
1	RandomForest	0.999856	0.071138	0.928862	0.987041	0.957068
2	DecisionTree	0.999793	0.071138	0.928862	0.950104	0.939363

Fig. 10. Results on the whole dataset

The training and testing phases of Decision Trees, Random Forest and SVM algorithms provide the core of the implementation for this study. These machine learning techniques may be trained and tested using structured based classification. The data cleansing procedure can be made even better and executed more quickly.

Given their overall f1 scores, the above tables clearly demonstrate that random forest outperformed the SVM and decision tree model on both test dataset and on the whole dataset.

This demonstrates the efficiency of ensemble techniques, which can boost performance even when there is a class imbalance problem.

But we still haven't reached the goal of detecting all fraudulent transactions.

7 Conclusion and Perspectives

Financial organizations are increasingly at risk from credit card fraud. Fraudsters frequently develop novel fraud techniques. A strong Machine learning model can manage the evolving fraud landscape. A fraud detection system's top aim is to accurately anticipate fraud situations while lowering the number of false-positive cases.

Machine Learning approaches function differently depending on the specific business scenario. Different Machine Learning approaches are mostly driven by the type of incoming data. The effectiveness of the model for identifying credit card fraud is heavily influenced by the quantity of data, volume of transactions, and correlation between the features.

Traditional algorithms cannot detect credit card fraud as effectively as machine learning techniques like Random Forest whose example is in this thesis.

The random forest model is the best algorithm when comparing all algorithms performances side by side, as shown in Fig. 9 and 10, with an accuracy of 99.95%, f1-score of 84.53% in test dataset and 99.98%, f1-score 95.07% over the whole dataset and a rate of false negative of only 7% comparing to 21% for SVM.

The algorithm receives much more samples from one class, leading it to be biased towards that class. It does not learn what makes the other class "different" and does not comprehend the underlying patterns that allow us to differentiate between classes.

So when a highly unbalanced class distribution is provide as input data, the predictive model is more likely to favor samples that are in the majority.

The models would score high on their loss functions just by predicting the majority class. In these instances, the Accuracy Paradox appears. So, accuracy does not holds good for imbalanced data.

In order to solve this issue, to improve the performance, and to reach the goal of 0% false negative (Fraudulent transactions classified as Non-Fraudulent) and 100% recall, several sampling techniques need to be applied in the future.

To conclude, future research might look at how deep learning algorithms and multiple sampling techniques could be applied to improve the outcomes shown in this paper and address the issue of unbalanced data.

References

1. Abakarim, Y., Lahby, M., Attioui, A.: An efficient real time model for credit card fraud detection based on deep learning. In: Proceedings of the 12th International Conference on Intelligent Systems: Theories and Applications, October 2018
2. Arora, V., Leekha, R.S., Lee, K., Kataria, A.: Facilitating user authorization from imbalanced data logs of credit cards using artificial intelligence. Mob. Inf. Syst. **2020**, 1–13 (2020)
3. Adepoju, O., Wosowei, J., Jaiman, H.: Comparative evaluation of credit card fraud detection using machine learning techniques. In: 2019 Global Conference for Advancement in Technology (GCAT). IEEE (2019)
4. Sahin, Y., Duman, E.: Detecting credit card fraud by decision trees and support vector machines. In: Proceedings of International Multi-Conference of Engineers and Computer Scientists (IMECS 2011), vol. 1, pp. 1–6, 16–18 March 2011
5. Quinlau, R.: Induction of decision trees. Mach. Learn. **1**(1), 1–106 (1986)

6. Save, P., Tiwarekar, P., Jain, K., Mahyavanshi, N.: A novel idea for credit card fraud detection using a decision tree. Int. J. Comput. Appl. **161**(13) (2017)
7. Xuan, S., Liu, G., Li, Z., Zheng, L., Wang, S., Jiang, C.: Random forest for credit card fraud detection. In: 2018 IEEE 15th International Conference on Networking, Sensing and Control (ICNSC), pp. 1–6 (2018)
8. Lakshmi, S., Kavilla, S.: Machine learning for credit card fraud detection system. Int. J. Appl. Eng. Res. **13**(24), 16819–16824 (2018)
9. Esakkiraj, S., Chidambaram, S.: A predictive approach for fraud detection using hidden Markov model. Int. J. Eng. Res. Technol. (IJERT). **2**(1), January- 2013 C
10. https://developers.google.com/machine-learning/clustering/algorithm/advantages-disadvantages
11. Liang, J., Qin, Z., Xiao, S., Ou, L., Lin, X.: Efficient and secure decision tree classification for cloud-assisted online diagnosis services. IEEE Trans. Dependable Secure Comput. (2019)
12. Ghiasi, M.M., Zendehboudi, S.: Application of decision tree-based ensemble learning in the classification of breast cancer. Comput. Biol. Med. (2021)
13. Breiman, L. Random forests. Mach. Learn. (2001)
14. https://www.analyticsvidhya.com/blog/2020/05/decision-tree-vs-random-forest-algorithm/
15. https://www.techopedia.com/definition/30364/support-vector-machine-svm
16. https://review42.com/resources/ecommerce-fraud-statistics/
17. https://developers.google.com/machine-learning/crash-course/classification/roc-and-auc

SNDAE: Self-Normalizing Deep AutoEncoder for Recommendation

Nouhaila Idrissi[1(✉)], Ahmed Zellou[2], and Zohra Bakkoury[1]

[1] AMIPS Research Team, EMI, Mohammed V University, Rabat, Morocco
nouhaila_idrissi@um5.ac.ma

[2] SPM Research Team, ENSIAS, Mohammed V University, Rabat, Morocco

Abstract. Users and consumers on the web are inundated with massive marketing and information notices while they crave directly personalized and precise content from online businesses. Bearing in mind the circumstances of the pandemic's impact that weighted in to help online sales growth, it is self-evident that companies should enhance employed recommendation engines. Engaging with relevant recommendations proves that users' needs are understood, and their preferences are genuinely catered for, leading to a superlative and satisfactory experience. This paper proposes a novel Self-Normalizing Deep AutoEncoder (SNDAE) to overcome sparsity issues of recommender systems. The proposed SNDAE first leverages a greedy layer-by-layer training procedure to learn a deep generative model for efficient feature representation. Furthermore, SNDAE exploits self-normalization properties, yielding enhanced convergence and improved regularization performance. Experimental results demonstrate that the proposed SNDAE outperforms state-of-the-art approaches in sparse rating prediction.

Keywords: Recommender systems · AutoEncoder · Sparsity · Collaborative filtering

1 Introduction

Since their appearance in the mid-1990s, Recommender systems (RSs) have grown commonplace on the Internet to sift through extensive data information and assist users in making decisions [1,2]. Such systems deliver personalized experiences tailored to users' needs while fulfilling potentially significant service demands. Nevertheless, the recommendation falls under impaired performance due to the challenging gap in modeling user-item interactions and inherent sparsity hurdles [2]. Matrix factorization (MF) was applied in RSs to deal with sparse and missing entries in rating matrices [3,4]. The MF-based approaches are significantly improved by employing Deep Learning (DL) models to overcome linear flaws of factorization techniques, specifically when dealing with heterogeneous and complex real-world data. DL-based recommendations have evolved to use Neural networks (NNs) for efficient feature extraction of both users and

M. Lazaar et al. (Eds.): BDIoT 2022, LNNS 625, pp. 328–338, 2023.
https://doi.org/10.1007/978-3-031-28387-1_28

items in Collaborative Filtering (CF) applications [5]. Such models perform the training of numerous layers and parameters, which can be a tedious task with high-time complexity and may get stuck at local minima [6]. We propose, in this paper, the Self-Normalizing Deep AutoEncoder (SNDAE) model for sparse rating prediction. The proposed SNDAE uses Deep Belief Networks (DBNs)' strong and complex feature extraction ability for dealing with typical NNs' issues. The DBN learns high-level hidden factors by employing a greedy layer-wise procedure [6] using stacked Restricted Boltzmann Machines (RBMs) [7], where each RBM's hidden layer acts as a visible layer for the following RBM. Furthermore, by introducing self-normalizing properties, we enhance the generalization and regularization performance of the proposed SNDAE.

The main contributions of this paper are summarized as follows:

- We propose a novel model named Self-Normalizing Deep AutoEncoder (SNDAE) that leverages efficient DBN's pre-training to initialize parameters of a Deep AutoEncoder for missing rating prediction.
- To further improve SNDAE's regularization and convergence, the Scaled Exponential Linear Unit (SELU) is employed to render self-normalization properties.
- We empirically show the impact of the DBN's unsupervised greedy pre-training on the performance of supervised classification.
- Finally, extensive experimental comparisons with six state-of-the-art baseline approaches prove the significance of the proposed SNDAE.

The remaining sections of the paper are organized as follows: Sect. 2 summarizes the related work. Section 3 presents a detailed description of the proposed SNDAE model. Experimental results and evaluations are discussed in Sect. 4. Finally, Sect. 5 concludes the paper and provides future directions.

2 Related Work

AutoEncoders (AEs) are used in recommendation systems for relevant latent feature learning. Scarce input entries of the user-item rating matrix are reconstructed for improved representation learning by modeling feature vectors of each instance with an AE and then decoding latent vectors to reconstruct the input. Several AE extensions are employed to learn deep generative models of sparse data [2]. Undercomplete AEs generate a compressed representation of user-item interactions with a hidden code layer including a smaller dimension than input entries [8], thus reducing dimensionality, and overcoming shortcomings of Principal Component Analysis (PCA), especially for non-linear manifold learning. Sparse AEs are also leveraged to build hybrid CF models to apprehend hidden factors from ratings [9]. Such approaches do not require a reduction in the dimension of feature layers. Otherwise, the input is partially corrupted with noise in denoising AE-based models [10]. These techniques are subsequently trained to map feature representations to recover the original input data. Variational AE-based approaches [11,12], on the other hand, offer a probability distribution for each latent feature. The bottleneck, which represents the compressed latent space, is used to generate an output reconstruction of sparse input data. Some AE-based hybrid approaches are proposed to learn hidden user-item interaction by incorporating available ratings with additional information to overcome sparsity limitations. Nevertheless, accentuating such linked representations by AE-based hybrid techniques for accurate missing rating prediction may lead to impaired recommendation performance resulting in overfitting due to irrelevant various and multimodal factors of users' and items' information [2].

3 Self-Normalizing Deep AutoEncoder (SNDAE)

Leveraging the unsupervised pre-training phase of DBN, we first construct a deep generative model for efficient feature learning. Such a greedy layer-wise training stage helps tackle gaps related to traditional deep neural networks' gradient descent training procedure that can swiftly get stuck with local minima [6]. The DBN is a deep model primarily built from stacked RBMs [6,7], which serve as feature detectors to reconstruct input entries probabilistically. The RBM can be represented as a stochastic two-layer neural network composed of a visible layer $\{v_0, ..., v_i, ..v_p\}$ and a hidden layer $\{h_0, ..., h_j, ..., h_q\}$ connected through weights W_{ij} and biases (b_i, c_j) in which the energy of connected visible units and hidden units is given by:

$$E(v,h) = \sum_{i}^{p}\sum_{j}^{q} W_{ij}v_i h_j - \sum_{i \in v} b_i v_i - \sum_{j \in h} c_j h_j \tag{1}$$

The probability distributions of binary visible and hidden units can be obtained from the energy as follows:

$$\begin{aligned} p(h) = \textstyle\sum_v p(v,h) = \frac{1}{\wp}\sum_v e^{-E(v,h)} \\ \wp = \textstyle\sum_v \sum_h e^{-E(v,h)} \end{aligned} \tag{2}$$

where $\wp$ is the partition constant used for normalization. Given v, for instance, the conditional distribution of a hidden configuration of a state h can be estimated by:

$$\begin{aligned} p(h \mid v) = \textstyle\prod_{j=1}^{q} p(h_j \mid v) \\ p(h_j = 1 \mid v) = \sigma(\textstyle\sum_{i=1}^{p} w_{ij} v_i + c_j) \end{aligned} \tag{3}$$

where σ is the activation function. Training such a stochastic network is done by maximizing the log probability between the target distribution $< v_i h_j >^0$ and the RBM's modeled distribution $< v_i h_j >'$. Taking the derivative of the negative log-likelihood and including the learning rate α allows updating the parameters of the network as follows:

$$\begin{aligned} \triangle W_{ij} = \alpha \frac{\partial p(v)}{\partial W_{ij}} \\ \triangle W_{ij} = \alpha(< v_i h_j >^0 - < v_i h_j >') \end{aligned} \tag{4}$$

Once the model is pre-trained, we use two symmetrical DBNs to form the encoder and the decoder of the SNDAE. Hence, instead of the traditional random weight

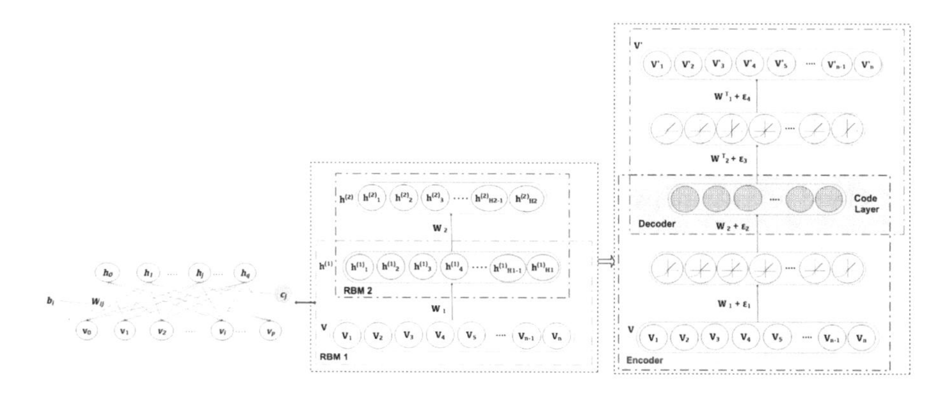

Fig. 1. The proposed Self-Normalizing Deep AutoEncoder

initialization, the contrastive divergence is utilized to learn the initial parameters of the deep AE [6,7]. Figure 1 depicts how the resulting weights from the DBN pre-training stage are unrolled and leveraged to initialize the SNDAE. After the greedy layer-by-layer training procedure, the model can be further finetuned with back-propagation [6], thus enhancing the generalization performance. Given the input vector V, an AE in the SNDAE first maps V_k to a hidden representation x_k through the encoding function $\beta_{\theta E}$ using the activation function δ:

$$x_k = \beta_{\theta E}(V_k) = \delta_\beta(W_E V_k + bias_E) \tag{5}$$

While the decoding function $\zeta_{\theta D}$ maps the hidden vector x to reconstructed input vector y as follows:

$$y_k = \zeta_{\theta D}(x_k) = \delta_\zeta(W_D x_k + bias_D) \tag{6}$$

The aim is to optimize the parameter set $\theta = \{W_E, bias_E, W_D, bias_D\}$ that represents the weight matrix and the bias of the encoder and decoder, respectively. Such optimization is performed by minimizing the reconstruction function ℓ referring to squared error loss between the input V and reconstruction vector y:

$$\begin{aligned} \ell(V, y) = \textstyle\sum_{i=1}^{n} \parallel V_k - y_k \parallel^2 \\ \textstyle\sum_{i=1}^{n} \ell(V_k, y_k) = \frac{1}{2n} \textstyle\sum_{i=1}^{n} (\zeta_{\theta D}(\beta_{\theta E}(V_k)) - V_k) \end{aligned} \tag{7}$$

where V_k and y_k are respectively the kth dimension of input V and reconstructed output vector y.

The Scaled Exponential Linear Unit (SELU) is employed as an activation function δ in hidden and output layers to render self-normalization [13], thereby enhancing the SNDAE's regularization and convergence.

$$\begin{aligned} \delta(a) = \kappa a \quad if\ a > 0 \\ \delta(a) = \kappa \eta e^a - \eta \quad if\ a \leq 0 \end{aligned} \tag{8}$$

The SELU is defined by Eq. 8. Here $\kappa \simeq 1.0507$ and $\eta \simeq 1.6733$. δ can get below 0, allowing fast network convergence.

As an optimization technique, Adaptive Moment Estimation (Adam) is leveraged to combine the advantages of AdaGrad and RMSProp [14] into effective learning with sparse gradients and non-stationary settings. Let g_t be the gradient distribution, the optimizer memorizes the exponentially decaying average of previous gradients to determine the momentum μ_t and also estimates the second uncentered variance of past squared gradients ν_t:

$$\begin{aligned} \mu_t &= \iota_1 \mu_{t-1} + (1 - \iota_1) g_t \\ \nu_t &= \iota_2 \nu_{t-1} + (1 - \iota_2) g_t^2 \end{aligned} \tag{9}$$

The weighting hyper-parameters ι_1 and ι_2 are set to 0.9 and 0.999, respectively, to ensure a slow variance estimate than the momentum. The first and second-moment estimators, after bias correction, are defined as follows:

$$\hat{\mu_t} = \frac{\mu_t}{1-\iota_1^t} \quad ; \quad \hat{\nu_t} = \frac{\nu_t}{1-\iota_2^t} \tag{10}$$

The parameters can be updated according to the rule ϕ by scaling the learning rate τ for each parameter, where the threshold $\epsilon = 1e^{-08}$:

$$\phi_t = \phi_{t+1} - \tau \frac{\hat{\mu_t}}{\sqrt{\hat{\nu_t}} + \epsilon'} \tag{11}$$

4 Experiments

Experiments are conducted to address the following two Research Questions:

RQ1. Does the unsupervised pre-training have an impact on the accuracy and performance of supervised classification?
RQ2. Compared with the state-of-the-art approaches, what is the performance of the proposed SNDAE model?

4.1 Experimental Setup

Datasets. Experimental evaluations are carried out using the MovieLens 100K (ML-100K) benchmark dataset. Statistics and the sparsity level are shown in Table 1. Where N_{Us}, N_{Is}, and N_{Rs} are, respectively, the number of users, items, and ratings.

Table 1. Description of Experimental Dataset

Datasets	N_{Us}	N_{Is}	N_{Rs}	Sparsity
MovieLens 100K	943	1 682	100 000	$(1 - \frac{N_{Rs}}{N_{Us} \times N_{Is}}) = 93.7\%$

Metrics. The Root Mean Square Error (RMSE) is used to evaluate the efficiency of the proposed SNDAE and all compared models in missing rating prediction. We utilize the metrics briefly presented in Table 2 to evaluate the classification accuracy. The improvement in precision generally comes at the expense of recall and vice versa. Therefore, the F1 score, considered a harmonic mean of the two metrics, reflects the classification quality according to the classes but does not consider the possible imbalance between them.

Table 2. Classification measures

Model	Measure	Formula
Precision	$\frac{TP}{TP+FP}$ or $\frac{TP}{Actual}$	The proportion of individuals, among all those for whom the classifier has predicted a particular class, who really belong to it.
Sensitivity (Recall)	$\frac{TP}{TP+FN}$ or $\frac{TP}{Predicted}$	Indicates the proportion of total efficient outcomes accurately classified by the model.
Specificity	$\frac{TN}{TN+FP}$	Evaluates the accuracy of the designation as positive class by dividing the true negatives by the number of all negatives.
Accuracy	$\frac{TP+TN}{TP+TN+FP+FN}$	The proportion of right predictions by the classifier out of the total number of individuals in the data set.
F1 Score	$2 \times \frac{Precision \times Recall}{Precision+Recall}$	Summarizes the recall and precision measurements for a class in a single indicator, by calculating their harmonic mean.

Baselines. The following six benchmark models are used for comparison:

- **SSAERec** [9] fuses stacked sparse AEs into MF using Singular Value Decomposition (SVD++) for rating prediction.
- **CAVAE** [11] integrates additional variational AE with probabilistic MF to learn latent vectors from side information.
- **VABMF** [12] is a Bayesian DL-based model that employs stochastic gradient variational Bayes for non-linear representation of user-item interactions.
- **ReDa** [15] fully exploits rating information to capture hidden representations of users and items using AEs without auxiliary data.
- **CAPR** [16] models latent feature representation for users and items with two different collaborative-based AEs and graph regularization.
- **Distrib_AE** [17] is a distributed deep AE with optimization that apprehends non-linear interactions between users and items for missing rating prediction.

Parameter Settings. Based on the validation results, we choose a momentum of 0.1, a batch size of 32, and a dropout of 0.5. We set the learning rate α as 0.0005. SNDAE can achieve better performance when the number of hidden features of the first and second hidden layers is set to 60 and 30, respectively.

4.2 Overall Performance

RQ1. The Impact of Unsupervised Pre-training: We investigate the influence of the unsupervised pre-training phase on supervised classification by empirically showing the impact of pre-trained weights and biases resulting from a DBN on training a Multilayer perceptron (MLP). We classify the $[N \times F]$ labels, which are the F hidden features that characterize the behaviors of N users obtained from the DBN. We compare the results of label classification obtained in three cases by using: (A) MLP with no pre-training: randomly initialized weights and biases. (B) MLP with pre-trained weights by the DBN. (C) MLP with pre-trained weights and biases resulted from the DBN model. The NN consists of two layers: a Tanh activation function in the first layer and a Sigmoid function in the second hidden layer. The learning rate used to train the model is 0.00001. Classification results of the three cases are described in Table 3, while losses in Mean Squared Error (MSE) are presented in Fig. 2.

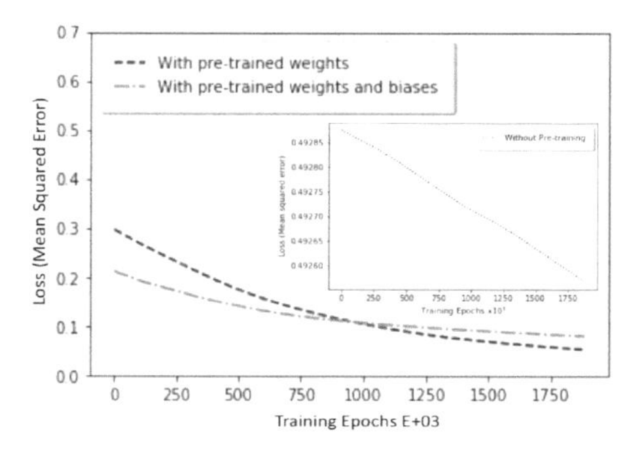

Fig. 2. MSE losses of MLP classifier of DBN labels with and without pre-trained weights and biases

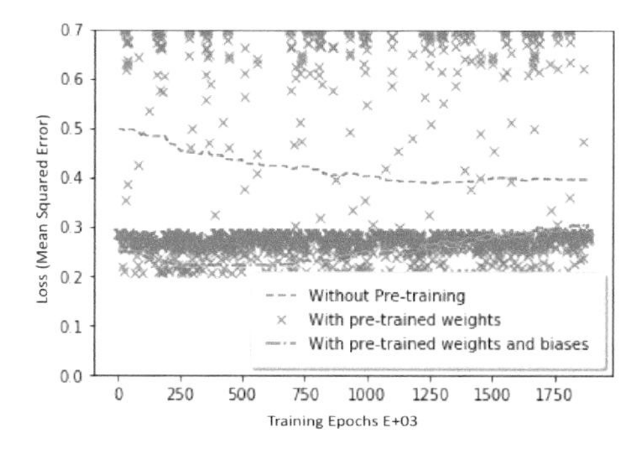

Fig. 3. MSE losses of MLP classifier of gender features with and without pre-trained weights and biases

We also examine the performance of the MLP classification of other users' features. We perform a user gender classification in three cases: (1) Classification with random weights and biases. (2) Classification using DBN pre-trained weights and random biases. (3) Classification using DBN pre-trained weights and biases. The MSE losses according to the training epochs are depicted in Fig. 3, while confusion matrices of the three cases are presented in Fig. 4.

Table 3. MLP Classification results of DBN labels

Metric	MLP without pre-training	MLP with pre-training
Precision	49.79	98.81
Recall	99.91	100
Specificity	0.11	7.74
Accuracy	49.79	98.81
F1 Score	66.46	99.4

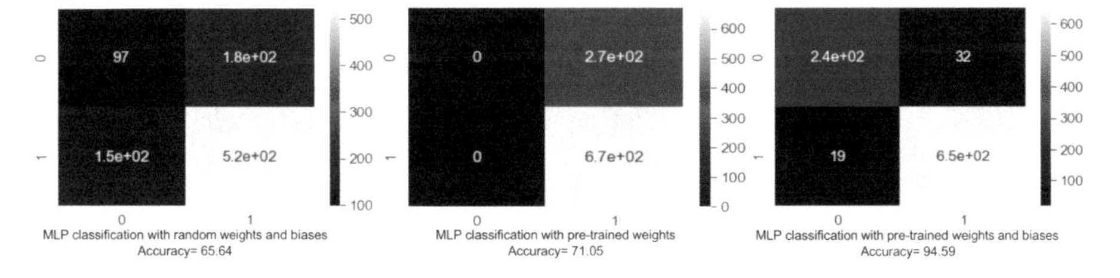

Fig. 4. Confusion matrices of MLP classification of users gender features with and without pre-trained weights and biases

Experimental findings prove the unsupervised pre-training's benefit in enhancing the accuracy and performance of supervised classification. We observe that leveraging the weights and biases from the DBN model leads to higher performance and accuracy when classifying the labels obtained from the DBN. Likewise, the classification using pre-trained weights and biases achieved the best accuracy when user gender labels were classified. We, therefore, conclude that there is a correlation between the features obtained by using the DBN model and the features of the same users.

RQ2. The Performance of the SNDAE Model: Fig. 5 summarises the RMSE values for the proposed SNDAE model and benchmark approaches using the MovieLens dataset. We can observe that the proposed model outperforms all the baseline methods. The improvement is 13.21%, compared with the best

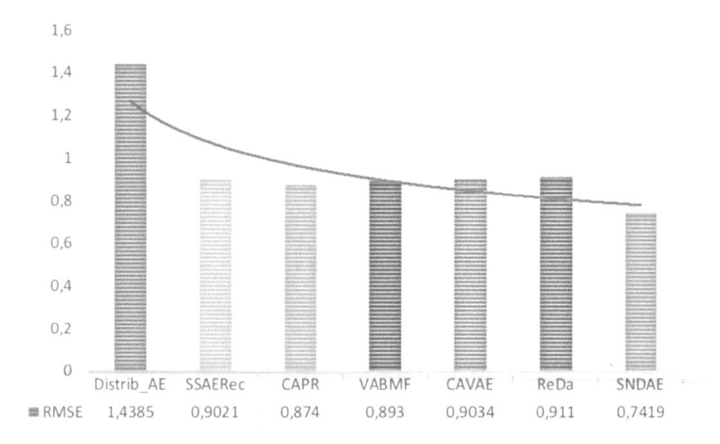

Fig. 5. RMSE results for the proposed SNDAE model and benchmark approaches

method among baseline approaches, namely, CAPR. Compared with the state-of-the-art techniques, for instance, Reda; CAPR uses two different AE for user-based and item-based latent feature representation. However, SNDAE is superior in rating accuracy and vastly outperforms MF-based approaches (CAVE, SSAEREC) that show close prediction results. One major drawback of CAVE is that it relies on side information as inputs to learn the latent factors, which may not be helpful in cases where additional data is unavailable. On the other hand, VABMF, which integrates a variational Bayes estimator, slightly outperforms SSAEREC, which employs AEs with traditional SVD++ MF for rating prediction, which may impair the model's effectiveness in capturing complex non-linear hidden vectors. Otherwise, Distrib_AE that uses three parallel AEs without pre-training for missing rating prediction demonstrates the worst performance. This further proves the effectiveness of Greedy layer-wise pre-training of the SNDAE, especially for reliable parameter initialization, thus resulting in enhanced performance by additionally leveraging self-normalization properties. The outcomes prove that the proposed SNDAE can efficiently overcome sparsity issues and improve recommendation prediction performance without relying on auxiliary information.

5 Conclusion

The proposed Self-Normalizing Deep AutoEncoder (SNDAE) takes advantage of DBN's pre-training to introduce deep generative learning for high-level latent feature representations for users and items for relevant missing rating prediction. Furthermore, the proposed SDNAE exploits self-normalization properties to overcome typical NNs' limitations and enhance convergence and regularization performance. Experimental evaluations and comparisons with baseline approaches demonstrate the effectiveness and accuracy of SNDAE recommendations, especially when dealing with sparsity limitations. As future work, we will investigate the impact of pre-training and self-normalization on other DL models such as Convolutional Neural Network (CNN) and combine the proposed SNDAE with hybrid models for further enhanced recommendation performance.

References

1. Zheng, Y., Wang, D.X.: A survey of recommender systems with multi-objective optimization. Neurocomputing **474**, 141–153 (2022)
2. Idrissi, N., Zellou, A.: A systematic literature review of sparsity issues in recommender systems. Soc. Netw. Anal. Min. **10**(1), 1–23 (2020). https://doi.org/10.1007/s13278-020-0626-2
3. Zheng, X., Ni, Z., Zhong, X., Luo, Y.: Kernelized deep learning for matrix factorization recommendation system using explicit and implicit information. IEEE Trans. Neural Netw. Learn. Syst., 1–12 (2022)
4. Idrissi, N., Zellou, A., Hourrane, O., Bakkoury, Z.: Addressing cold start challenges in recommender systems: towards a new hybrid approach. In: Proceedings of the International Conference on Smart Applications, Communications and Networking (SmartNets), pp. 1–6. IEEE (2019)
5. Idrissi, N., Zellou, A., Hourrane, O., Bakkoury, Z., Benlahmar, E.H.: A new hybrid-enhanced recommender system for mitigating cold start issues. In: Proceedings of the 2019 11th International Conference on Information Management and Engineering, pp. 10–14 (2019)
6. Hinton, G.E.: A practical guide to training restricted Boltzmann machines. In: Montavon, G., Orr, G.B., Müller, K.-R. (eds.) Neural Networks: Tricks of the Trade. LNCS, vol. 7700, pp. 599–619. Springer, Heidelberg (2012). https://doi.org/10.1007/978-3-642-35289-8_32
7. Idrissi, N., Hourrane, O., Zellou, A.: A restricted Boltzmann machine-based recommender system for alleviating sparsity issues. In: Proceedings of the 1st International Conference on Smart Systems and Data Science (ICSSD), pp. 1–5. IEEE (2019)
8. Vagliano, I., Galke, L., Scherp, A.: Recommendations for item set completion: on the semantics of item co-occurrence with data sparsity, input size, and input modalities. Inf. Retrieval J. **25**(3), 1–37 (2022). https://doi.org/10.1007/s10791-022-09408-9
9. Zhang, Y., Zhao, C., Chen, M., Yuan, M.: Integrating stacked sparse auto-encoder into matrix factorization for rating prediction. IEEE Access **9**, 17641–17648 (2021)
10. Abinaya, S., Kavitha Devi, M.K.: Trust-based context-aware collaborative filtering using denoising autoencoder. In: Ranganathan, G., Bestak, R., Palanisamy, R., Rocha, Á. (eds.) Pervasive Computing and Social Networking. Lecture Notes in Networks and Systems, vol. 317, pp. 35–49. Springer, Singapore (2022). https://doi.org/10.1007/978-981-16-5640-8_4
11. He, M., Meng, Q., Zhang, S.: Collaborative additional variational autoencoder for top-N recommender systems. IEEE Access. **7**, 5707–5713 (2019)
12. Aldhubri, A., Lasheng, Yu., Mohsen, F., Al-Qatf, M.: Variational autoencoder Bayesian matrix factorization (VABMF) for collaborative filtering. Appl. Intell. **51**(7), 5132–5145 (2021). https://doi.org/10.1007/s10489-020-02049-9

13. Klambauer, G., Unterthiner, T., Mayr, A., Hochreiter, S.: Self-normalizing neural networks. In: Advanced in Neural Information Processing System, vol. 30 (2017)
14. Kingma, D.P., Ba, J.: Adam: a method for stochastic optimization. In: International Conference on Learning Representations (2014)
15. Zhuang, F., Zhang, Z., Qian, M., Shi, C., Xie, X., He, Q.: Representation learning via dual-autoencoder for recommendation. Neural Netw. **90**, 83–89 (2017)
16. Zhu, Y., Wu, X., Qiang, J., Yuan, Y., Li, Y.: Representation learning with collaborative autoencoder for personalized recommendation. Expert Syst. Appl. **186**, 115825 (2021)
17. Ravi Kumar, R.R.S.: Distributed deep autoencoder for recommendation system. Turkish J. Comput. Math. Educ. (TURCOMAT) **12**(10), 3851–3855 (2021)

Intelligent Evaluation System Using NLP

Smail Admeur[1](✉), Outman Haddani[1], Souad Alaoui[2], Souad Amjad[1], and Hicham Attariuas[1]

[1] Abdelmalek Essaadi University, Tétouan, Morocco
s.admeur@uae.ac.ma
[2] Université Sidi Mohamed Ben Abdellah University, Fès, Morocco

Abstract. This article presents a comparative study of different evaluation systems. It does so with the proposal of a new architecture for the development of an autonomous generator of evaluation exercises that represents two main functionalities: automatic instantaneous evaluation and automatic improvement of the proposed questions.

Our proposed solution allows each author to build a model of exercises according to his own pedagogical choices. This model is framed by the domain knowledge that each author must have, which will then be automatically instantiated for the generation of several exercise variants.

The main feature of our generator is that it proposes different types of exercises. Each exercise is independent in its domain and adopted to the profile of each learner. This gives the possibility to use it in many disciplines. In addition, the use of domain knowledge reduces the workload of the author in defining the exercise templates and the associated corrections. Moreover, the automatic improvement of the proposed questions is done through the application of sentiment analysis techniques on the recorded answers.

Keywords: Online Learning · Knowledge management · Assessment systems · Deep learning · Sentiment analysis

1 Context of Work

Generally, the evaluation is part of the continuous improvement of our actions. It represents several advantages such as feedback improvement [1], following the traces, authentic evaluation, the analysis and correction of texts.

However, due to its various advantages, the use of evaluation automation has become very important. Thus, several researches have been made for the adaptation and improvement in different areas. For instance, Farida's researches [2] propose an approach called "ODALA+" that allows the generation of adapted exercises of the learner's profile and the results from the former assessments. In addition, LIN [3] also propose a model to evaluate the educational competitiveness of universities. However, in the work of [4], he shows that a static-dynamic evaluation index system can be used perfectly to measure the quality of the university education according to the Chinese model.

M. Lazaar et al. (Eds.): BDIoT 2022, LNNS 625, pp. 339–350, 2023.
https://doi.org/10.1007/978-3-031-28387-1_29

However, the automation of evaluation is highly demanded [5] in order to overcome the challenges of traditional testing. It allows the system to be more dynamic in a variety of ways according to each individual's progress. For example, the content can be designed to:

- Generate questions based on the previous answer(s).
- Propose sub-questions based on the original answer.
- Give advice or suggestions during the progress, on request or according to the given answer.
- Formulate questions or answers in form of illustrations to manipulate, animations, multimedia files, etc.

All of these features can be combined in various ways to collect data in order to present individualized and standardized tutoring [6].

The objective of this paper is to build a system that is capable of facilitating the generation of exercises with a punctual follow-up of each learner's course. We have proposed an architecture of a semi-automatic evaluation system based on an exercise generator. This allows the author to freely choose the model that will later be instantiated to automatically create various exercises linked to the same resource. The main purpose of an exercise generator is to facilitate the self-assessment process and the follow-up of individuals' progress in online learning environment in particular and in a knowledge management platform in general. It proposes to the learners several variations of exercises of the same content without having to go back to the resource's main author each time.

The idea of offering different exercises related to the same content is very relevant for the follow-up of each learner and for not influencing the evaluation by previous resolutions. Learners may fail the first few attempts to answer the exercises if the knowledge is not mastered. Therefore, it is possible that a learner may have to answer the same exercise several times before being successful. This is why we have chosen to use exercise generators that the author can easily use in any field.

In order for the teacher to be able to intervene in the exercises' choice, he should use a semi-automatic process of generating exercises. This principle facilitates the creation of several instances based on the exercise model that is defined only once at the beginning, by the main author of the resource. On one hand, our proposed system will save the author's time as well as facilitate the problems of evaluation. On the other hand, it will have a diversity of adaptable exercises for each student with an immediate overview of his level (Table 1).

Table 1. Comparative table of the different exercise generators.

	Generator	**Different Exercises**	**Edition**	**Multi domains**	**immediate analysis**	**Fast**
Manual	Moodle	-	X	X	-	-

(continued)

Table 1. (*continued*)

	Generator	Different Exercises	Edition	Multi domains	immediate analysis	Fast
	Hot Potatoes	-	X	X	-	-
Automatic	Aplusix	X	-	-	X	X
	Reading Tutor	X	-	-	X	X
Semi automatic	Ambre	X	-	-	X	X
	PepiGen	X	-	-	X	X
	Geppetop	X	-	X	-	X

This generic architecture (Fig. 1) of the exercise generators allows the distinction of four levels: the general level contains the knowledge independent of the domain one wishes to generate an exercise. For example, the knowledge allowing to write a statement in J2EE language. The domain level contains the knowledge specific to the application domain. For example, the knowledge of calculation. The generation level contains the processes specific to the creation of an exercise: the definition of constraints on an exercise pattern stored in an exercise structure, the instantiation of this structure to generate an exercise and its solution and then the layout to provide exercises with a homogeneous presentation. Finally, the exercise level contains the specific documents to the created exercise, including the exercise structure and its instantiation (an exercise statement and its correction).

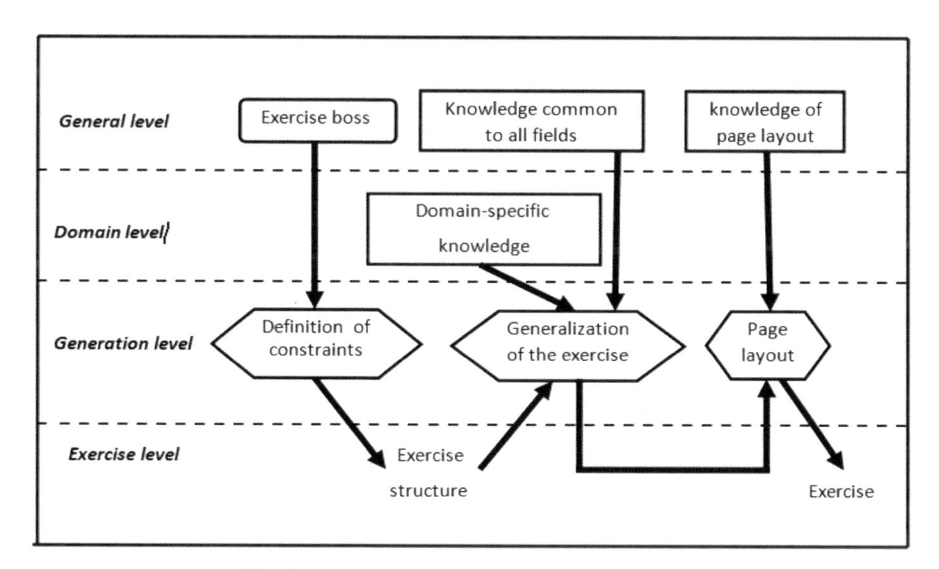

Fig. 1. Generic architecture of exercise generator

2 Text Mining: Sentiment Analysis

Sentiment analysis (sometimes called opinion mining) [11, 12] is a part of text mining that tries to define the opinions, feelings and attitudes presented in a text or a set of texts. It consists in building automatic tools able to extract subjective information from texts in natural language, in order to create structured and exploitable knowledge that can be used by a decision support system or a decision maker.

Generally, sentences are either objective or subjective [13]. When a sentence is objective, no other basic task is required. However, when a sentence is subjective, its polarities (positive, negative or neutral) must be estimated (Fig. 2).

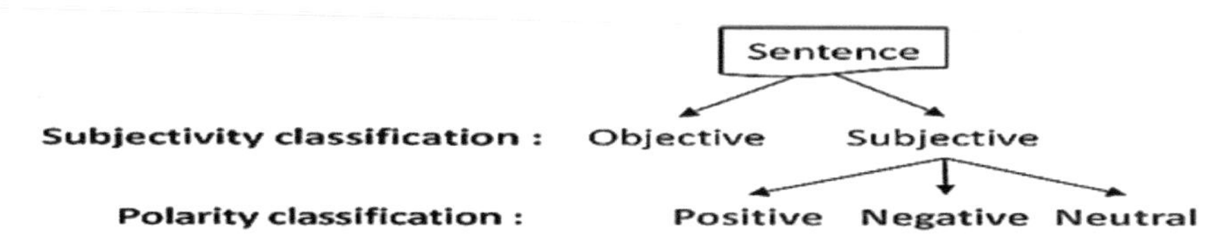

Fig. 2. Generic architecture of exercise generator.

It is important to note that Natural Language [14] indicates the differences between positive and negative emotions in a certain document. Yet, it does not identify specific positive and negative emotions. For instance, "angry" and "sad" are both considered negative emotions. However, when Natural Language analyzes text that is considered "angry" or text that is considered "sad", the response only indicates that the feeling in the text is negative, not that it is sad or angry.

Recently, the use of sentiment analysis has seen great growth in several areas. Through sentiment analysis, companies can learn about customers' opinions on their products or services in order to improve their productivities and increase revenues [15]. In the field of commerce, the decisions of the majority of customers are based on the opinions of other consumers. In fact, they are even willing to pay more for a product with a more favorable opinion than another. Moreover, in the political domain, sentiment analysis is strongly presented [16]. Politicians have followed the trend of sentiment analysis before declaring a new law. Moreover, we can cite the experience of "Google Play" [17], which uses sentiment analysis to classify applications according to the degree of user satisfaction, by extracting the overall attitude (positive or negative) on a given comment.

In conclusion, we consider that sentiment analysis can be used to improve evaluation systems in various fields of application.

3 Our Approach

We have chosen the semi-automatic generators, which combine the advantages of the other two categories. In order to take advantage of the automatic generators' functionalities and giving the author the editorial freedom on the created exercises at the same time, these generators propose to the author to define an exercise model which is then instantiated to produce to a large number of exercises [9]. Some distance learning platforms propose exercises involving variables, which are close to the notion of exercise model [7]. Our main expected goals for our proposed system are:

- Generation of various types of exercises.
- Customized exercises to the learner's profile.
- Allow the author to choose the content of the exercise.
- No technical skills required for the author.
- Easy adaptation of the system for all areas.
- Evaluation is automatic and fast (time saving).

In order to build our semi-automatic generator adapted to many domains and including other types of exercises, we have chosen to use the GEPPETO approach [6]. This approach allows the teacher to express constraints on the exercises to be generated. To do this, it is necessary to have a model of the types of exercises that can be generated, in order to know the type of constraints that the author should be allowed to express.

The GEPPETO approach (Fig. 3) proposes models and processes allowing to personalize the pedagogical activities proposed to the learners by respecting the pedagogical objectives of each personalization agent. It proposes a meta model which will guide the modeling of the activity to be personalized. Thus, the activity meta-model constrains the expression of the knowledge that an expert must provide in order to describe pedagogical activities. This knowledge, once provided by the expert, forms a model allowing its customization. From an activity model, it is possible to define constraints on the choice of activities. These constraints on activities are defined by each agent according to its personalization needs. These constraints are then interpreted by a system to personalize the pedagogical activities provided to the learners:

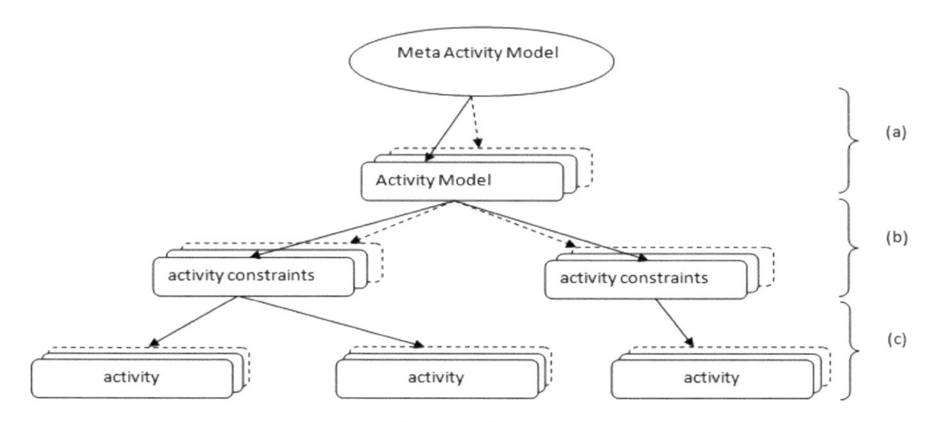

Fig. 3. The GEPPETO approach.

The processes of the GEPPETO approach can be broken down into three main phases. The first one (a) is called the initialization phase. It is performed only once. In this phase, an expert uses the activity meta model to define an activity model for each type of activity. In the second phase, an agent uses an activity model created by the expert in the previous phase to constrain the personalization of the activities proposed to the learners it manages. This process of defining constraints on activities is decomposed in two steps: the proposal of an interface adapted to the activity model allowing to limit the possibilities when defining constraints on activities, and the recording of these constraints. In the last phase (c), the defined constraints are then used to customize the pedagogical activities proposed to the learners. This activity generation process is also divided into two steps: the use of constraints to create an activity and its formatting.

Our proposal:

This part will be intoduced by the detailed presentation of our proposed system which is inspired by the GEPPETO approach and the work of Lefevre [18], which allows:

- Semi-automatic generation of various exercise variants.
- The production of personalized evaluation to the profile of each learner.
- Automatic diagnosis of answers.
- The proposed solutions are based on the knowledge of the field.
- Self-evaluation and improvement of the proposed questions.

Our system has two main actors namely the author of the resource and the user who will follow this resource (Fig. 4). Therefore, each actor will perform its own tasks and interact with the system in four steps (Fig. 5).

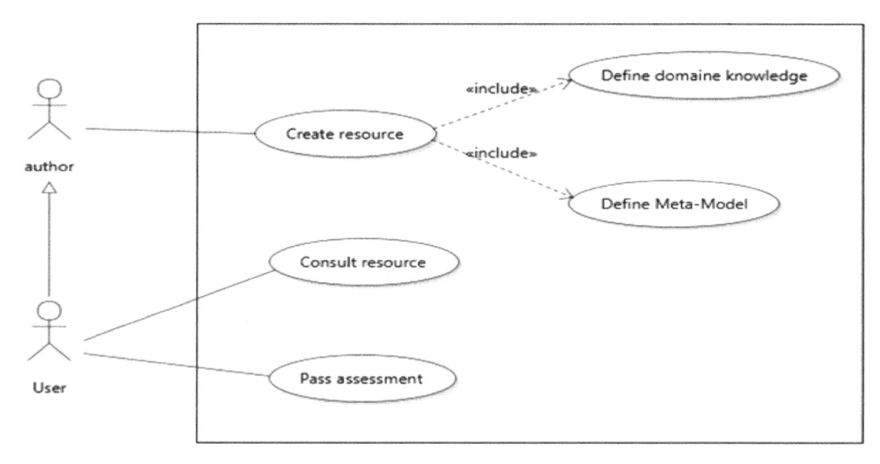

Fig. 4. Use case diagram.

Our system is composed of four steps (Fig. 5). In the first step, the process is triggered as soon as the author creates a resource, determines a set of domain knowledge and describes an exercise meta-model. In the second step, the system generates different exercises for a given resource based on the domain knowledge and the meta-model that has already been determined by the author in the previous step. This way, the system will propose several exercise models for the same resource. This is the strong point of our tool that allowed a free choice of exercise model to promote the learning plan. Then, in the third step, the system will analyze the recorded answers and display the results in a fast and automatic way. Finally, in the fourth step, our system will proceed to the improvement of the proposed questions, through the analysis of the feelings and the calculation of the score for each recorded answer taking into consideration that the good question is the one that collects a positive score.

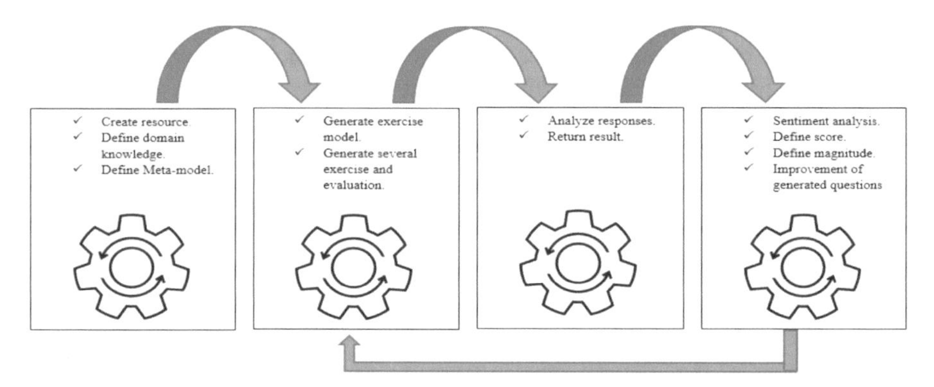

Fig. 5. The architecture of our system.

4 Experimentation and Implementation

To evaluate and validate our proposal, our system has been implemented in a knowledge management platform [19]. Thus, we have created five different courses that are made available to 9 students of the 2nd year of computer science at Higher Normal School of Tetouan (ENES) during a period of 5 weeks.

We have defined six indicators to analyze the result of our experimentation:

Nbr_Course: the number of courses followed.
Nbr_Exercise: the number of exercises treated.
Nbr_Exercise_V2: the number of exercises treated in several variations.
Freq_C: the frequency of the courses followed by an individual compared to the whole of the courses, we use the following Eq. (1).
Freq_E: the frequency of the exercises of an individual compared to the whole of the exercises generated, we use the following Eq. (2).
Avr_Score: the average sentiment score recorded on all the responses of each individual.

$$\text{Freq\_C} = \left(\sum(\text{User\_course})\right)/\left(\sum(\text{All\_cours})\right) \times 100 \tag{1}$$

Or:
$\sum$ User_course: Number of course visited by the user.
$\sum$ All_course: The sum of courses visited by all users.

$$\text{Freq\_E} = \left(\sum(\text{User\_exercise})\right)/\left(\sum \text{All\_Exercice}\right) \times 100 \tag{2}$$

Or:

$\sum$ User_Exercice: Number of exercise visited by the user.
$\sum$ All_Exercice: The sum of exercise visited by all users.

The Table 2 represents the results of our experimentation, which allowed us to extract some important findings. Indeed, we have discovered that there is a strong correlation between the number of courses followed and the number of exercises generated, especially with those in several versions (Fig. 6). We have taken the example of the student number 6 who uses the greatest number of exercises and who registers at the same time the best participation rate (Number of courses, questionnaires…).

Table 2. The results of our experimentation.

Learner	Nbr_ Cours	Nbr_Exercise	Nbr_Exercise_V2	Freq_C	Freq_Q	Avr_Score
1	3	6	2	12%	12%	0.6
2	2	3	1	8%	6%	0.3
3	5	11	4	19%	21%	0.8
4	1	1	1	4%	2%	0.1
5	4	7	3	15%	13%	0.6
6	5	12	5	19%	23%	0.9
7	2	3	1	8%	6%	0.4
8	1	1	1	4%	2%	−0.8
9	3	8	2	12%	15%	0.5

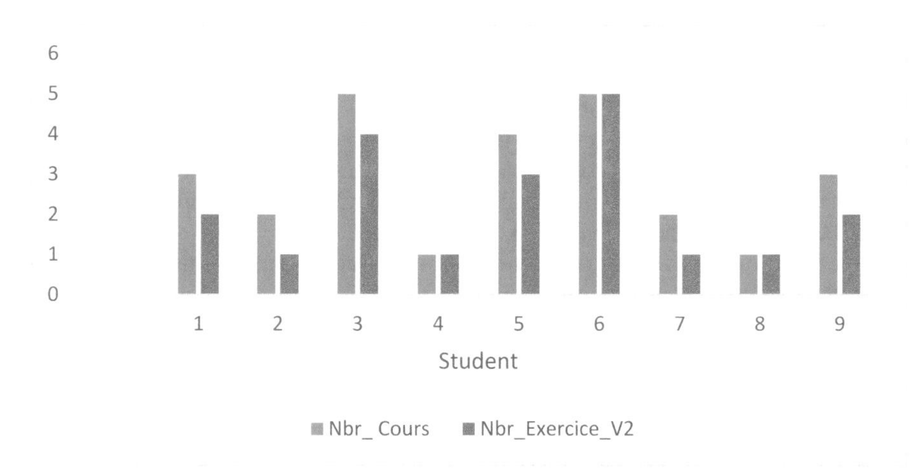

Fig. 6. Comparison of Nbr_Cours AND Nbr_Exercise_v2.

Additionally, we have found that the feelings and satisfaction of individuals also influence their participation frequencies (Fig. 7), taking the example of two students number 8 and 3 who successively record positive and negative scores, from which we see that the feelings score directly influences the learning activity in both directions.

Thus, we can confirm that our proposal is validated. In fact, the automation adaptation and improvement of the evaluation system improve the quality of teaching and the participation rate of individuals. Furthermore, we can confirm that sentiment analysis can be exploited to develop the involvement and satisfaction of individuals during the learning process.

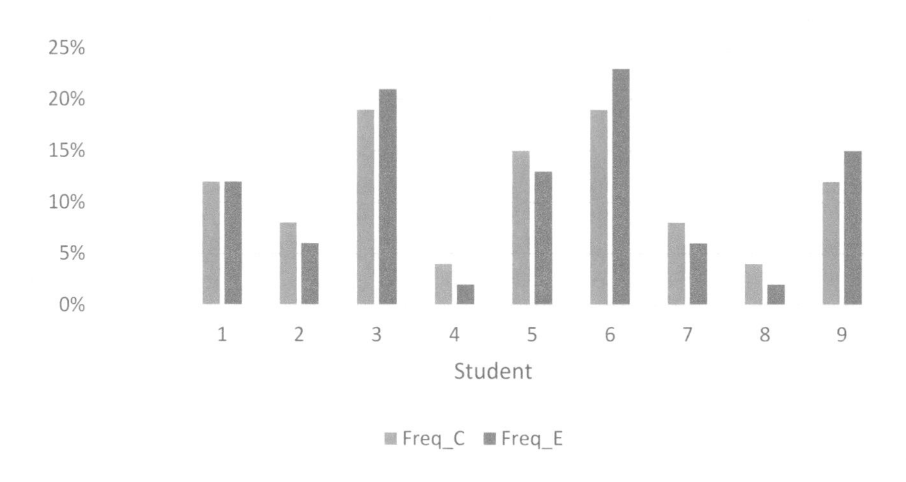

Fig. 7. Comparison of Freq_C AND Freq_E.

Figure 8 shows the evaluation part in the platform, where learners can consult their results and easily choose the name of the course and the evaluation model that will follow.

Fig. 8. Assessment choices.

Figure 9 shows an example of an assessment for the ANDROID course of the type OPEN QUESTION, where the learner can navigate between questions and pages interact and answer freely.

Fig. 9. Evaluation passage.

Figure 10 shows an example of the implementation used for sentiment analysis. We used "Cloud Natural Language API" to exploit the learners' responses. This is a very powerful API offered partially by Google for the scientific research domain, which allows to determine the global score of the feelings found in a given text, which is between -1.0 (negative) and 1.0 (positive).


```
{
  "documentSentiment": {
    "score": 0.3,
    "magnitude": 3.8
  },
  "language": "FR",
   "sentences": [
    {
      "text": {
        "content": "Il y a vingt ans,
        Notre Faculté des sciences de TETOUAN a engendrée
        une nouvelle vision d'éducation,conçue dans la
        liberté et vouée à la proposition que tous les
        personnes ont le droit à la recherche scientifique.",
        "beginOffset": 0
      },
      "sentiment": {
        "magnitude": 0.7,
        "score": 0.7
      }
    },
   ...
}
```


Fig. 10. Sentiment analysis example.

5 Conclusion

We have presented in this paper an intelligent assessment system. This later allows to generate and automatically improve various assessment tests adapted to the learner's profile, which can also be adapted and used in many domains (Knowledge Management, companies, University).

The result of our experimentation is very acceptable. It allowed us to confirm our proposals, and that the adaptation of the evaluation systems favors the loyalty of the learners and increases the rate of their participation in the learning process. In addition, the feelings of individuals directly influence their learning activity.

Our proposed system allows a significant optimization of time in terms of evaluations. Thus, this work has allowed us to discover new research areas, namely the exploitation of our proposal with machine learning for the prediction of suitable courses and the participation rate of new learners.

References

1. Alexopoulou, A., Batsou, A., Drigas, A.S.: Effectiveness of assessment, diagnostic and intervention ICT tools for children and adolescents with ADHD. Int. J. Recent Contrib. Eng. Sci. & IT (iJES) **7**(3), 51–63 (2019). https://doi.org/10.3991/ijes.v7i3.11178
2. Farida, B.D., Malik, S.M., Catherine, C., Pierre Jean, C.: Adaptive exercises generation using an automated evaluation and a domain ontology: the ODALA+ approach. Int. J. Emerg. Technol. Learn. (iJET) **6**(2), 4–10 (2011). https://doi.org/10.3991/ijet.v6i2.1562
3. Lin, L.: An evaluation system and its model for educational competitiveness of universities. Int. J. Emerg. Technol. Learn. (iJET) **15**(11), 188–201 (2020). https://doi.org/10.3991/ijet.v15i11.14521
4. Liu, J., Wang, Y.: Static and dynamic evaluations of college teaching quality. Int. J. Emerg. Technol. Learn. (iJET) **17**(02), 114–127 (2022). https://doi.org/10.3991/ijet.v17i02.29005
5. Damaj, I., Zaher, A., Yousafzai, J.: Assessment and evaluation framework with successful application in ABET accreditation. Int. J. Eng. Pedagogy (iJEP) **7**(3), 73–91 (2017). https://doi.org/10.3991/ijep.v7i3.7262
6. Haddani, O., Amjad, S., Dahmani, A.: Personalized recommendation of human resources based on preferences and personality types a collaborative filtering-based approach. In: Third International Conference on Systems of Collaboration (SysCo), Casablanca, pp. 1–6 (2016). https://doi.org/10.1109/SYSCO.2016.7831333
7. Auzende, O., Giroire, H., Calvez, F.: Propositions d'extensions à IMS-QTI 2.1 pour l'expression de contraintes sur les variables d'exercices mathématiques Extensions à QTI 2.1 pour l'expression de contraint es (2007)
8. Daouas, T., Jmal, J.: Prise en compte des aspects de l'autoregulation dans des exercices interactifs en ligne. Int. J. Inf. Sci. Decis. Making Inf. Savoirs, Décisions & Médiations. ISDM32 (2008)
9. Duclosson, N., Jean-Daubias, S., Riot, S.: AMBRE-enseignant: un module partenaire de l'enseignant pour créer des problèmes. Revue: Environnements Informatiques pour l'Apprentissage Humain, Montpellier, pp. 353–358 (2005)
10. Stéphanie, J., Marie, L., Nathalie, G.: Adapte, un outil générique pour proposer des activités pédagogiques personnalisées. Workshop Prise en Compte de l'Utilisateur dans les Systèmes d'Information, INFORSID 2009, Toulouse, France. Revue scientifique, pp. 51–62, 26 (2009)
11. Amazon, C.: Analyser des sentiments dans du texte (2019). https://aws.amazon.com/fr/getting-started/hands-on/analyze-sentiment-comprehend/. consulté le Aug 2019
12. Rakotomalala, R.: Introduction au Textmining (2017). http://eric.univlyon2.fr/~ricco/cours/cours_text_mining.html. consulté le Sept 2020
13. Hadji, M.: Analyse des sentiments: Généralités (2019). https://medium.com/@mehdihadji/analyse-des. consulté le May 2021
14. https://cloud.google.com/natural-language. consulté le Oct 2021
15. Tseng, K.-K., Lin, R.-Y., Zhou, H., Kurniajaya, K.J., Li, Q.: Price prediction of e-commerce products through Internet sentiment analysis. Electron. Commer. Res. **18**(1), 65–88 (2017). https://doi.org/10.1007/s10660-017-9272-9
16. Aniruddha Prabhu, B.P., Ashwini, B.P., Anwar Khan, T., Das, A.: Predicting election result with sentimental analysis using twitter data for candidate selection. In: Saini, H.S., Sayal, R., Govardhan, A., Buyya, R. (eds.) Innovations in Computer Science and Engineering. LNNS, vol. 74, pp. 49–55. Springer, Singapore (2019). https://doi.org/10.1007/978-981-13-7082-3_7

17. https://support.google.com/googleplay/android-developer. consulté le Mar 2020
18. Lefevre, M., Guin, N., Cablé, B., Buffa, B.: ASKER: un outil auteur pour la création d'exercices d'auto-évaluation (2015)
19. Haddani, O., Amjad, S., Jellouli, I.: IMS-LIP-KM: extension of IMS-LIP standard for modeling a new user profile. Int. J. Comput. Appl. **177**, 18–27 (2020). https://doi.org/10.5120/ijca2020919959

Automatic License Plate Recognition with YOLOv5 and Faster-RCNN

Mohamed El Ghmary[1(✉)], Younes Ouassine[2], and Ali Ouacha[2]

[1] Department of Computer Science, FSDM, Sidi Mohamed Ben Abdellah University, Fez, Morocco
mohamed.elghmary@usmba.ac.ma

[2] Department of Computer Science, Faculty of Science, Mohammed V University in Rabat, Rabat, Morocco
{younes.ouassine,a.ouacha}@um5r.ac.ma

Abstract. Artificial intelligence is a field that can help us, humans, to do things that are a bit difficult. In some of the fields of artificial intelligence, we find image processing. One of the most interesting applications of image processing is vehicle license plate recognition, or what is known as automatic license plate recognition (ALPR). Most of these applications use optical character recognition (OCR) methods. The Automatic License Plate Recognition system is not only focused on parking lots but can be used in all those facilities necessary to control, monitor, and have a record of all vehicles that pass through certain access. Example: private garages of companies, shopping centers, tolls, hospitals, etc. In this paper, we will explain how we applied two object detection algorithms to build a system capable of recognizing and extracting Moroccan license plates from images. We used transfer learning on YOLOv5 and Faster-RCNN and trained them both on 724 images of Moroccan registered vehicles to obtain a system that can support a parking system.

Keywords: Plate localization · characters detection · YOLOv5 · Faster-RCNN

1 Introduction

Automatic license plate recognition is a method used in traffic management, digital security monitoring, vehicle recognition, and parking management in large cities. This task is a complex problem due to many factors, such as poor lighting conditions, blurred images, Arabic characters of license plate numbers in the Moroccan case, and variability of plates. The ALPR system must function properly under natural conditions, meaning that it must be able to perform in a wide variety of environments while maintaining a high level of accuracy. ALPR systems generally involve two steps: License plate detection (LP) and character recognition (CR). In our approach, we decided to build two models separately instead of using directly libraries like easyOCR or Tesseract because of their weaknesses in the treatment of the variance of the Moroccan license plates, among the weaknesses, we find that the Moroccan license plates contain foreign numbers but the letters sometimes we find them foreign and sometimes Arabic. The first model has been

M. Lazaar et al. (Eds.): BDIoT 2022, LNNS 625, pp. 351–361, 2023.
https://doi.org/10.1007/978-3-031-28387-1_30

trained to detect the license plate which will then be cropped from the original image, and then be passed into the second model which has been trained to detect the characters. The main objective of this work is to compare the Faster-RCNN and YOLOv5 algorithms for each step of our ALPR system on Moroccan plates.

The work done in our research is divided into five steps. The first one consists in discussing the articles that deal with the topics related to our work, then we move to the second one, which is a paragraph that contains the data sources used in the training of the two architectures YOLOv5 and Faster-RCNN. For these two last ones, we have detailed their architectures and functioning in the following part. Concerning the last part made to discuss the results obtained for each architecture.

2 Related Works

In this part of our paper, we will discuss some related work for license plate recognition systems and object detection with Faster-RCNN and YOLOv5.

2.1 License Plate Recognition System

A license plate Detection System is a system to recognize each plate on a vehicle and then read all the characters on the plate. Automation of plate reading will facilitate many traditional jobs that are distracted by human error. Several papers have discussed this with a tendency to use two different neural networks for efficiency and accuracy problems. In the paper [1], the License Plate Recognition System uses a concept called the character position method which checks whether a given candidate region is a number plate or not. In the paper [2] the writers used the contours detection methods for plate license extraction and for character candidate classification they used The KNN classifier.

In [3] and [4], both researchers used a projection profile method for their license plate character segmentation process. Two projections for their license plate character segmentation process which is horizontal projection and vertical projection respectively. In [3], the segmentation process was performed using a method called horizontal license plate projection. Basically, it is a process of finding the maximum peak or also known as the space between characters by iteratively projecting the graph vertically. A vertical projection method was also used in [4] for a license plate character segmentation process on a cell phone. In this method, the vertical axis is scanned for a black dot in the images to detect the character.

2.2 Object Detection with Faster-RCNN and YOLOv5

Using Fast-RCNN for real-time object detection is possible if the region proposals are already pre-computed. This task can take from about 0.2 s to one or two seconds for an image depending on the method.

YOLOv5 is one of the best models available for object detection today. The advantage of this deep neural network is that it is very easy to retrain the network on your own custom data set and is available in four models, namely s, m, l, and x, each offering different accuracy and detection performance. This task can take from about 0.2 s to one or two seconds for an image depending on the method. This task can take very little time to detect for an image less than 0.05 s for the model YOLOv5 version s.

3 Datasets

We used a dataset containing images of vehicles registered in Morocco collected by the Laboratory of Modelling, Simulation, and Data Analysis (MSDA) of the Polytechnic University Mohammed 6. The total data we used to train the license plate detection model was 724. A total of 580 images were used as train data and 144 images as test/validation data. We used the same dataset for the plate character recognition model. Since we did not have enough data, we used pre-trained models on the COCO dataset. COCO Is a free object recognition database used to train machine learning programs published by Microsoft. This database includes hundreds of thousands of images with millions of objects already labelled for training (Fig. 1).

Fig. 1. Example of dataset

The following two figures show statistics of some examples of annotation: the Fig. 2 is for the annotation of plates and the Fig. 3 for the notation of characters.

	Class	x min	y min	x max - x min	y max - ymin	Image name	Width	Height
0	LP	322	162	139	40	20200614_162449b.jpg	720	960
1	LP	262	178	168	32	20200614_162450b.jpg	720	960
2	LP	299	128	85	30	20200614_162501b.jpg	720	960

Fig. 2. Examples of statistics for plate annotation

4936	7	105	19	36	73	20201106_224622.jpg	503	138
4937	9	142	21	36	74	20201106_224622.jpg	503	138
4938	3	181	25	37	72	20201106_224622.jpg	503	138
4939	h	311	47	50	46	20201106_224622.jpg	503	138
4940	6	433	41	39	72	20201106_224622.jpg	503	138

Fig. 3. Examples of statistics for character annotation

4 Architectures

This part will focus on the description of the architectures used in our work.

4.1 Faster-RCNN

In this section we describe Faster-RCNN architecture design in detail (Fig. 4).

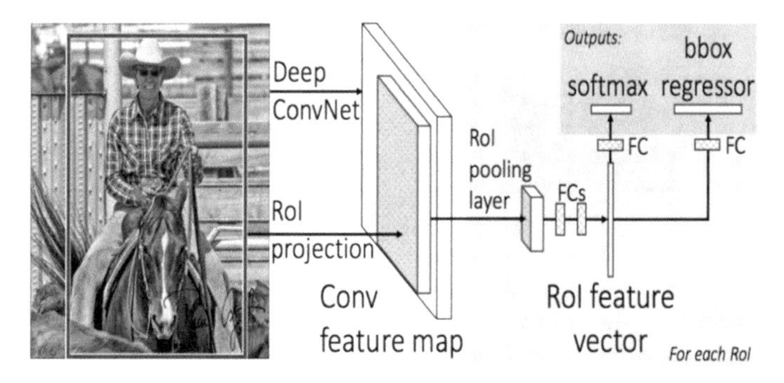

Fig. 4. Faster-RCNN architecture

- In this approach, the input image is passed through a convolutional neural network (CNN) to obtain a map of the features of the objects in the image. This part of the Faster R-CNN architecture is called the "backbone" network.
- This feature map is then used by a Region Proposal Network (RPN) to generate region proposals (bounding boxes that contain the relevant objects in the image) using anchors (fixed-size reference boxes placed uniformly in the original image to detect objects). These regions are then filtered by NMS (Non-Maximum Suppression). NMS is a method that allows you to screen the proposed regions and choose only those that are interesting.
- The feature map extracted by CNN and the relevant object bounding boxes are used to generate a new feature map by pooling the regions of interest (RoI) determined in the second step.

- The clustered regions then pass through fully connected layers for the prediction of object field coordinates and output classes. This part of the Faster R-CNN architecture is called the header network.

4.2 YOLOv5

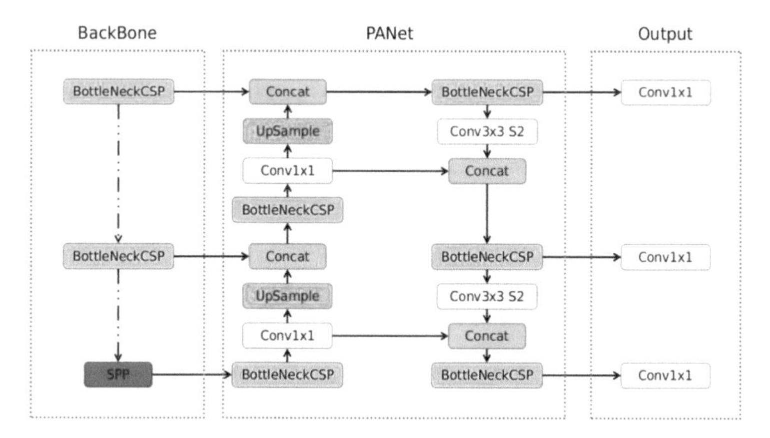

Fig. 5. YOLOv5 architecture

YOLOv5 is, like many neural networks dedicated to image processing, based on layers of convolutions. The architecture of YOLOv5 can be separated into three parts, as shown in Fig. 5.

These three parts can be assimilated into a set of layers, composing a functional unit. Each of these parts corresponds to a role, established by this set of layers, which were "pieces" of networks already existing in the literature and which were simply slightly modified.

- Input: This box in Fig. 5 does not correspond to a part of the network as presented just above. It simply corresponds to the data taken as input by the network, i.e., an image, a batch of images, or even a video.
- Backbone: The first part of the network is called the backbone. It allows the network to be optimized by improving the gradient descent. It is this part of the network that allows a very short inference time, and thus the use of the network in real-time. It also serves as the basis for feature extraction for the network.
- Neck: The Model Neck is used to create feature pyramids. Feature pyramids help models generalize successfully when scaling objects. They make it easier to identify the same object in different sizes and scales. Feature pyramids are very useful in helping models run efficiently on unseen data. Other models, such as FPN, BiFPN, and PANet, use different kinds of feature pyramid approaches. PANet is used as a move in YOLOv5 to obtain feature pyramids it corresponds to an improvement for YOLO, which allows to obtain better predictions.

- Head: The Head model is primarily responsible for the final detection step. It uses anchor boxes to construct the final output vectors with class probabilities, objectivity scores, and bounding boxes.

5 Training and Inference

Both object detection models were pre-trained on the COCO dataset, because we didn't have enough data, so it was plausible to take advantage of transfer learning of models that were trained on such a rich dataset. The training of the models was carried out on a workstation with 48 processor cores, 128 Gb of RAM, and a GeForce RTX 3090 24 GiB graphics card.

5.1 Training and Evaluation of Plate Detection Model

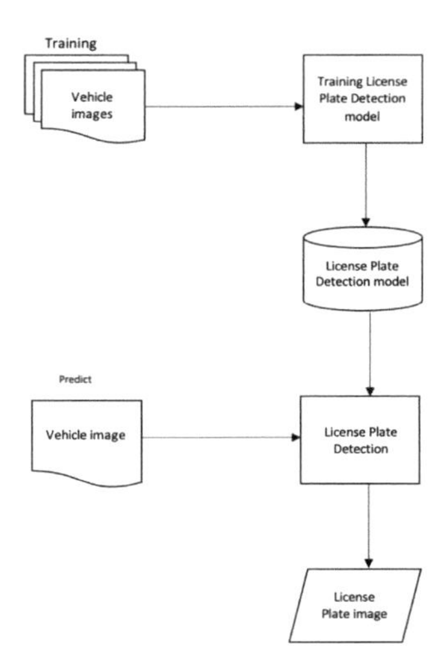

Fig. 6. Flow training of plate detection model

For the License Plate Detection model, we trained the YOLOv5 and Faster-RCNN by using the transfer learning method. We trained the models to detect license plate objects in the image. We trained the model on 580 vehicles with registration plate images and then validated the trained model in 144 images (Fig. 6).

5.2 Training and Evaluation of Characters Detection Model

For the License Plate Detection model, we trained the YOLOv5 and Faster-RCNN by using the transfer learning method. We trained the models to detect license plate objects in the image. We trained the model on 580 license plate images with registration plate images and then validated the trained model in 144 license plate images (Fig. 7).

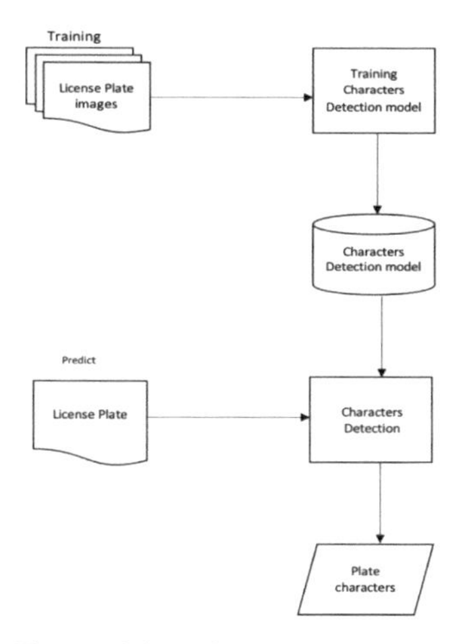

Fig. 7. Flow training of character detection model

5.3 Inference

As shown in Fig. 8, we use the License Plate Detection model and the character Detection model that was previously trained in our License Plate Recognition System as follows:

1. First, the system will receive a car image as an input.
2. Then, the image will be processed by the License Plate Detection model. The model detected and cropped the plate number object from the original image.
3. The cropped image will be processed by the Character Detection model. The model detects characters in the cropped image and returned a sequence of digit numbers (Figs. 9 and 10).

Fig. 8. Inference flow

6 Results and Analysis

For the last part of our work, we will discuss the detailed results for each architecture and each part of the system.

Fig. 9. Example of our detection system using Faster-RCNN

Fig. 10. Example of our detection system using YOLOv5

6.1 License Plate Detection

It can be seen from the first line in Table 1 that using YOLOv5 and from the second line in the same table that using Faster-RCNN.

- Precision: our models trained to detect vehicle plates has the ability to make True Positive (TP) comparisons with the number of positive predicted data of 89.50% for the YOLOv5 model and 96.36% for the Faster-RCNN.
- inference time: For the YOLOv5 model has the ability to detect a vehicle plate on 8 ms, on the other side, the Faster-RCNN model does the same detection in 100 ms.
- model size: The checkpoints generate after training the YOLOv5 model takes 14.8 MiB from storage memory and the checkpoints generate after training the Faster-RCNN model takes 330 MiB.

Table 1. Experiment result of License Plate Detection

Detector	Precision	Recall	Inference time	Model size
YOLOv5	89.50%	90.6%	8 ms	14.8 MiB
Faster-RCNN	96.63%	94.4%	100 ms	330 MiB

6.2 Character Detection

It can be seen from the first line in Table 2 that using YOLOv5 and from the second line in the same table that using Faster-RCNN.

- Precision: our models trained to detect plate characters has the ability to make True Positive (TP) comparisons with the number of positive predicted data of 97.50% for the YOLOv5 model and 95.89% for the Faster-RCNN.
- inference time: For the YOLOv5 model has the ability to detect a vehicle plate at 9 ms, on the other side, the Faster-RCNN model does the same detection at 42 ms.
- model size: The checkpoints generate after training the YOLOv5 model takes 14.9 MiB from storage memory and the checkpoints generate after training the Faster-RCNN model takes 166 MiB.

Table 2. Experiment result of Plate Characters Detection

Detector	Precision	Recall	Inference time	Model size
YOLOv5	97.50%	95.1%	9 ms	14.9 MiB
Faster-RCNN	95.89%	94.8%	42 ms	166 MiB

Mathematical formulas to calculate the precision and recall are:

$$\text{Precision} = \text{True Positive}/(\text{True Positive} + \text{False Positive})$$

$$\text{Recall} = \text{True Positive}/(\text{True Positive} + \text{False Negative})$$

The biggest differences between the two systems are the size of the model and the inference time, since our system is designed to be installed in service cameras, so it must be a model of small size and must be able to do its job in the shortest possible time. Therefore, the system based on the YOLOv5 architecture is the best one to do the job of character plate detection.

7 Conclusion

In this paper, we create two systems to help the automatization of a Moroccan parking system. Our license plate recognition system consists of two parts, each part is equipped with an object detection model.

The first system composes of two object detection models for Object License Plate Detection, we used transfer learning on YOLOv5 to build the Object License Plate Detector. Our model achieved a precision of 89.50% with a detection time of 8 ms, In the Characters Detection model we also used transfer learning on YOLOv5 and we get a precision of 97.50% with a detection time of 9 ms. The second system composes of two object detection models for Object License Plate Detection. Our model achieved a precision of 96.63% with a detection time of 100 ms, for the Characters Detection we get a precision of 95.89% with a detection time of 42 ms.

For future work may involve improving the models to detect license plates well, for example by using data augmentation to have more images of the plates or taking new images and annotating them to have new images and plates that do not exist in the data set. The future work is not only the improvement it can be the deployment of models using different methods for example use onnx, torchserve, etc.

References

1. Singh, B., Kaur, M., Singh, D., Singh, G.: Automatic number plate recognition system by character position method. Int. J. Comput. Vis. Robot. **6**(1–2), 94–112 (2016)
2. Vanshika, R.: Automatic number plate recognition. In: International Conference on Innovative Computing & Communication (ICICC) (2021)
3. Mutholib, A., Gunawan, T.S., Kartiwi, M.: Design and implementation of automatic number plate recognition on android platform. In: 2012 International Conference on Computer and Communication Engineering (ICCCE), pp. 540–543. IEEE (2012)
4. Wang, C.M., Su, C.Y.: Fast license plate location and recognition using wavelet transform in android. In: 2012 7th IEEE Conference on Industrial Electronics and Applications (ICIEA), pp. 1035–1038. IEEE (2012)
5. Ravirathinam, P., Patawari, A.: Automatic license plate recognition for Indian roads using faster-RCNN. In: 2019 11th International Conference on Advanced Computing (ICoAC), pp. 275–281. IEEE (2019)
6. Ap, N.P., Vigneshwaran, T., Arappradhan, M.S., Madhanraj, R.: Automatic number plate detection in vehicles using faster R-CNN. In: 2020 International Conference on System, Computation, Automation and Networking (ICSCAN), pp. 1–6. IEEE (2020)
7. Raj, S., Gupta, Y., Malhotra, R.: License plate recognition system using Yolov5 and CNN. In: 2022 8th International Conference on Advanced Computing and Communication Systems (ICACCS), vol. 1, pp. 372–377. IEEE (2022)
8. Jiang, H., Learned-Miller, E.: Face detection with the faster R-CNN. In: 2017 12th IEEE International Conference on Automatic Face & Gesture Recognition (FG 2017), pp. 650–657. IEEE (2017)
9. Amatya, M.L., Sarvanan, K.N.: The state of the art–Vehicle Number Plate Identification–a complete Survey. Int. Res. J. Eng. Technol. **4**(2), 785–792 (2017)
10. Fan, Q., Brown, L., Smith, J.: A closer look at Faster R-CNN for vehicle detection. In: 2016 IEEE Intelligent Vehicles Symposium (IV), pp. 124–129. IEEE (2016)
11. Zhang, L., Fang, Y., Sheng, J., Zhang, Y., Shi, J.: Research and implementation of database operation recognition based on YOLO v5 algorithm. In: 2021 International Conference on Computer Information Science and Artificial Intelligence (CISAI), pp. 367–372. IEEE (2021)
12. Sung, J.Y., Yu, S.B.: Real-time automatic license plate recognition system using YOLOv4. In: 2020 IEEE International Conference on Consumer Electronics-Asia (ICCE-Asia), pp. 1–3. IEEE (2020)

13. Chandra, A., Stefanus, R.: An end-to-end optical character recognition pipeline for Indonesian identity card. In: 2021 9th International Conference on Information and Communication Technology (ICoICT), pp. 307–312. IEEE (2021)
14. Tra, H.T.H., Trung, H.D., Trung, N.H.: YOLOv5 based deep convolutional neural networks for vehicle recognition in smart university campus. In: Abraham, A., et al. (eds.) HIS 2021. LNNS, vol. 420, pp. 3–12. Springer, Cham (2022). https://doi.org/10.1007/978-3-030-96305-7_1
15. Nguyen, N.V., Rigaud, C., Burie, J.C.: Comic characters detection using deep learning. In: 2017 14th IAPR International Conference on Document Analysis and Recognition (ICDAR), vol. 3, pp. 41–46. IEEE (2017)
16. Wu, T.H., Wang, T.W., Liu, Y.Q.: Real-time vehicle and distance detection based on improved Yolo v5 network. In: 2021 3rd World Symposium on Artificial Intelligence (WSAI), pp. 24–28. IEEE (2021)
17. Nepal, U., Eslamiat, H.: Comparing YOLOv3, YOLOv4 and YOLOv5 for autonomous landing spot detection in faulty UAVs. Sensors **22**(2), 464 (2022)
18. Peiyuan, J., Daji, E., Fangyao, L., Ying, C., Bo, M.: A review of Yolo algorithm developments. In: The 8th International Conference on Information Technology and Quantitative Management (ITQM 2020 & 2021): Developing Global Digital Economy after COVID-19 (2022)

The Perception of the Combination of Simulations and Laboratory Experiments by Moroccan Students

Mohammed Chekour[1](✉), Yassine Zaoui Seghroucheni[2], Mouenis Anouar Tadlaoui[3], Younes Hamzaoui[1], and Abdelaziz Bouchaib[1]

[1] Ibn Tofail University, Kenitra, Morocco
mohamed.chekour@uit.ac.ma
[2] University Mohammed V of Rabat, Rabat, Morocco
[3] Abdelmalek Essaadi University, Tetouan, Morocco

Abstract. All over the world, secondary school students are losing interest in the physical sciences. The causes of this problem are diverse. To increase the motivation of learners, pedagogues recommend the integration of practical work in an effective way in the teaching process of experimental disciplines such as physical sciences. However, this work does not fill the gap between theory and practice. Simulation is of great interest in electricity, as it allows for the simplification of the real systems under study. Also, simulation can serve as a link between theoretical concepts and laboratory work. In this paper, we have proposed an approach based on the combination of simulation (virtual experiment) and laboratory work (real experiment) to teach electrical concepts. This approach is used to teach concepts related to capacitor operations. The proposed simulation allows students to change the parameters of the experiment and explore the phenomenon of capacitor charging. To measure the degree of user satisfaction with this approach, a questionnaire was administered to eighty students from a randomly selected high school. The results obtained show a favorable perception among the students of the positive effects of the combination of virtual and real experiences of electrical phenomena in the acquisition of electrical concepts.

Keywords: Physics · Simulation · Laboratory Work · Electrical Concepts · Capacitor Charging Phenomenon · Motivation · PSPICE Simulator

1 Introduction

Skills related to the experimental method and critical thinking are paramount to effective teaching of physics [1]. With technological advances, there are new options for introducing these skills using simulators [2]. Certainly, students in the digital age are growing up with electronic devices and can navigate the virtual world. Hands-on activities can be a novelty compared to simulations [3]. Moreover, several research studies show that the use of simulations offers interesting opportunities to exercise higher order skills such as reflective thinking, abstraction, and advanced problem-solving skills [4]. In addition,

M. Lazaar et al. (Eds.): BDIoT 2022, LNNS 625, pp. 362–375, 2023.
https://doi.org/10.1007/978-3-031-28387-1_31

recent research results show that hands-on work and simulations support students' learning of electrical circuits in a similar way [5]. Other studies indicate that the combination of computer simulation and traditional teaching has a positive effect on the teaching of physics and especially electricity [6, 7].

In this article, we propose an approach based on the combination of virtual experiments using simulation and real experiments using laboratory work. This approach aims at making the learning of electrical concepts more sustainable by linking the theoretical and practical knowledge of the learners. The simulation part of this approach gives learners the opportunity to be autonomous and free in their investigation and construction of scientific knowledge. Moreover, the simulations overcome the weak points of laboratory work: Learners are strictly guided, the fragility of the laboratory equipment, and the risks of mishandling this equipment. Therefore, the added value of our approach lies in the fact that real and virtual experiments are complementary and allow to reach the finalities of the experimental method based on the formulation of hypotheses, proposal of experiments and analysis of results. To achieve the objective of this research, we simulated the phenomenon of capacitor charging via the PSPICE simulator. The choice of this phenomenon is not arbitrary. This phenomenon is quite complex and difficult to assimilate by most high school students. The result of the simulation is followed by the real realization of the experiment using the laboratory equipment. To measure the degree of user satisfaction with this approach, a questionnaire was administered to eighty students. The main question of this research is: "What impact would the use of the approach based on the combination of simulation and laboratory work have on the acquisition of concepts in electricity?

The rest of this paper is organized as follows: the next section presents the theoretical framework in which this work is embedded. Section 3 is dedicated to the methodology of our research. Section 4 is dedicated to the presentation of the scenario of our approach based on the combination of virtual and real experiments. In Sect. 5, we present the results of the questionnaire which aims at measuring the degree of satisfaction of our approach. The last section concludes the paper.

2 Theoretical Framework

2.1 The Benefits of the Inquiry-Based Educational Approach in Sciences

Many countries want to increase enrollment in science courses to meet an expected demand for more scientists [1]. Learners around the world are losing interest in science [8–10]. The causes of this problem are diverse [11]. The abstract nature of some of the concepts taught is responsible for the lack of motivation of students towards science education [12]. Indeed, one of the most important factors that negatively influence students' attitudes towards science is the curriculum, the syllabus, and the professional practices of teachers [13]. Thus, educationalists emphasize the importance of teaching methods in improving the learning of experimental sciences and especially the learning of electricity [14]. Investigative learning places learners at the center of the learning process. They act as scientists and discover the phenomena studied by applying the experimental method based on generating hypotheses, designing experiments and interpreting results [15]. Indeed, inquiry-based learning is an approach that enhances science education by engaging students in real-life investigations. This pedagogical approach increases students' motivation and prepares them to become active constructors of their own knowledge [16]. The inquiry-based approach is recommended by several researchers [5–7, 17]. However, this approach is little used in physical science classes and is often used in an ineffective way. This is due to several factors, such as the problem of using scientific materials that allow students to conduct investigations in an independent manner [18], the difficulty of incorporating abstract concepts into the investigation [16], the difficulty of achieving the actual objectives of the laboratory work [18] and the absence of cognitive bridges between theoretical and practical knowledge [19]. Also, improving students' critical thinking skills is among the most important objectives of science education [20]. However, physics teachers spend more time communicating information in class than doing experimental activities [21]. The teaching of these sciences, which is intended to be experimental, is in most cases theoretical without making strong links between theory and practice [22]. According to an exploratory study, most of these teachers carry out less than 50% of the experiments programmed in the textbook [23]. According to the same research, the lack of scientific equipment needed to carry out practical work is the source of this problem. To overcome this problem, computer simulation can replace some of the real experiments to make the learning of electricity concepts more interactive [24]. Also, simulation helps students bridge the gap between theory and reality, in the case of electrical circuits. It is a source of constructive feedback, helping students to identify and correct their misconceptions [25].

2.2 The Added Value of Simulation in the Educational Context

Simulation is an explanatory means to define a system, an analysis vector to determine results, a design evaluator to analyze and evaluate proposed solutions [26]. In scientific disciplines, the computer was used very early because of the computational capabilities it offered. This speed provides a new tool for testing working hypotheses [7]. By simulating the results of a theory, one can quickly assess its validity and make much faster progress in the process of theory development [27]. This approach consists of making progress in the understanding of a real phenomenon by proposing a numerical model that can be compared and evaluated with the results of laboratory work [18]. Moreover, simulation allows the real systems studied to be simplified [28]. It presents itself as a "unique" didactic tool to overcome the problems caused by experiments that require long, dangerous, or expensive manipulations [18]. In the field of education, simulation allows virtual experiments to be carried out, giving students the opportunity to interact with the simulation software with total freedom, which is not always possible in laboratory work [29]. Also, simulations can be used as a complementary tool to the laboratory. Indeed, the combination of virtual experiments (via the simulation) and real experiments (via the laboratory equipment) saves time by reducing the duration of the laboratory session [30].

2.3 Common Point and Differences Between Simulations and Laboratory Experiments

Several researchers discuss the similarities and differences between real (using laboratory equipment) and virtual (using simulation software) experiments [31, 32]. Gilbert and Troitzsch state that simulation questions the model and not the phenomenon itself [33]. However, the material aspect, which is an important element for the external validation of experiments, is absent in simulation [34]. Certainly, the distinction between real and virtual experience must be based on the difference between internal and external validity. When formulating a hypothesis about a physical phenomenon, the internal validity of the virtual experiment (simulation) is simply the first step. The second step is to compare the internal results of the simulation with the external results of the experiments. The external validity of the experimental results is crucial to confirm or reject the original hypothesis. Thus, Winsberg [35] states that despite the differences between real and virtual experiments can make inferences about the world. Robin Millar and colleagues [36] point out that the discrepancy between teachers' and learners' goals influences the effectiveness of experimental activities. In principle, teachers' goals are inspired by the curriculum. After deciding on these objectives, the teacher designs the practical task. When this task is implemented, the teacher is focused on the process of investigation, while the learners are more concerned with giving a "right answer". To address this problem, physical science teachers are encouraged to promote autonomy and empower the learner in the experimentation phase [37]. However, poor choices by learners during the experimental phase can lead to multiple dangers. Certainly, the integration of simulation in the teaching of physical sciences can avoid such problems. Moreover, simulation can be more beneficial than laboratory experiments when unobservable or difficult-to-observe phenomena are addressed [18].

In short, several studies indicate that combining simulation with experiment-based teaching can have a beneficial effect on the learning process [5]. However, this requires a reform of pedagogical approaches to take advantage of the benefits of simulation in an optimal way. Furthermore, the integration of simulation into physical science teaching attracts the attention of students [38]. Students, by succeeding in their own simulations, acquire not only knowledge about the phenomenon represented by the simulation but also additional motivation.

3 Methodology

3.1 General Context

Our research is intended for Moroccan high school students in the final year of the scientific section. To achieve the objectives of this research, a questionnaire was administered to eighty students at the public high school Abou Bakr Essediq. These students used the simulation and laboratory experiments in the teaching/learning process of electricity concepts for 3 months. The implementation of the study lasted 24 h and was spread over 12 weeks (2 h per week) in the multimedia (ICT) room of Abou Bakr Essediq High School.

3.2 Participants

The study population consisted of 212 secondary school students in an urban high school in Tetouan (Morocco), of whom 80 (37.7%) participated in the study. The high school was selected in a random way. Also, two classes were randomly selected to serve as the sample for the present study. The students of these two classes used the PSPICE simulation software (virtual experiments) in the teaching/learning process of electricity concepts before moving on to laboratory experiments (real experiments).

3.3 Instruments and Procedures

The questionnaire aims at measuring the satisfaction of the students who benefited from the teaching based on the combination of virtual experiments and real experiments of the capacitor charging phenomenon. The virtual experiments were carried out by the PSPICE simulation software. The combination of these two types of experiments aims at bridging the gap between the theoretical and practical knowledge of the learners. The questionnaire is organized around four questions that allow the students to give their opinions about this new approach. The questionnaire is individual. It takes place in the Abou Bakr Eddedik high school, after the session of carrying out the virtual experiment of the capacitor charging phenomenon.

3.4 Data Analysis

The data of our research was constituted from an individual questionnaire allowing Moroccan high school students to express their opinions concerning the effect of the combination of virtual and real experiments on the acquisition of electrical concepts. The questionnaire addresses the following issues:

- The PSPICE simulator is easy to use.
- The PSPICE simulator promotes autonomy.
- Simulation is an appropriate tool for the acquisition of the capacitor charging phenomenon.
- The combination of PSPICE simulator and laboratory work makes it easier to learn electrical phenomena.

These questions were developed based on the literature review [18, 21, 29, 30, 39, 40] and on open interviews with physical science teachers about improving the academic performance of Moroccan high school students in physics.

The answers to the questions in our questionnaire are of the 4-point Lickert scale type ("strongly agree", "slightly agree", "slightly disagree" and "strongly disagree").

After the collection of the completed questionnaires, the data was coded and compiled in an Excel database. The collection of data required the authorization of the headmaster of the Abou Bakr Essediq high school as well as the regional directorate of the Ministry of Education in Tetouan. The teachers and students interviewed were informed of the objectives and the course of the present research.

3.5 Example of Simulation: Capacitor Charging Phenomenon

The set-up (see the following Fig. 1) consists of a resistor (R = 10 KΩ), a capacitor (C = 10 nF) and a DC voltage source (6 V). To calculate the voltage across the source and the capacitor, two voltmeters are placed.

Fig. 1. Diagram of the capacitor charging phenomenon

The green curve is the voltage across the voltage source and the red curve is the voltage across the capacitor (see Fig. 2).

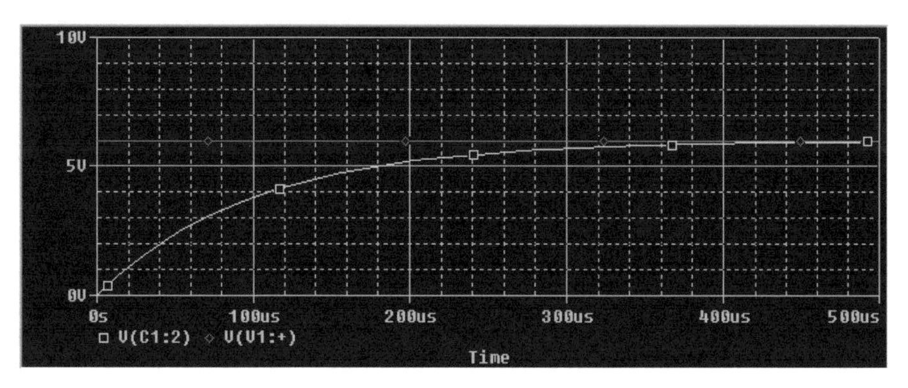

Fig. 2. Result of the simulation

4 Scenario for Combining Simulations with Laboratory Work

The cornerstone of this article is based on the scenario shown in Fig. 3. At the beginning of the session, the teacher orients the scientific debate in the classroom towards the pedagogical objective to be achieved. He facilitates this debate to detect a scientific problem appropriate to the curriculum to be taught. Once the scientific problem is carefully identified, the teacher invites his students to propose working hypotheses (provisional solutions to the problem to be solved). Then, the teacher encourages the students to propose and then carry out the simulation (virtual experiment) using the PSPICE simulator. In case the results of the simulation allow to solve the problem detected at the beginning of the session, the teacher invites the students to carry out the same experiment but this time with the real scientific material of the laboratory (real experiment). In short, the approach used in this work is based on this scenario that combines real and virtual experiments in the acquisition of concepts related to the phenomenon of capacitor charging.

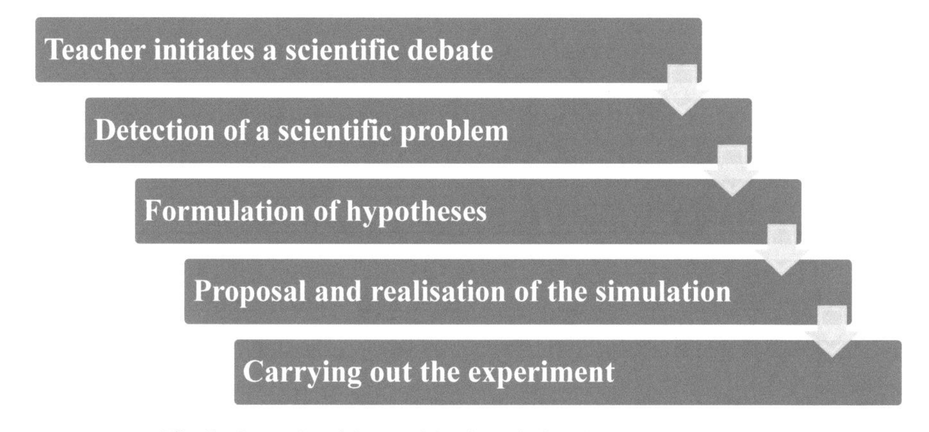

Fig. 3. Scenario of the combination of virtual and real experience

5 Results

The objective of the simulation of the capacitor charging phenomenon is to facilitate the cognitive task of Moroccan high school students in the process of acquiring concepts in electricity. It also aims at providing more freedom to the learners during their learning process to establish cognitive bridges between theoretical and practical knowledge. To measure the degree of user satisfaction with our approach based on the combination of virtual and real experiences, the following axes were proposed:

- The PSPICE simulator is easy to use
- The PSPICE simulator promotes autonomy
- The simulation is a suitable tool for the acquisition of the capacitor charging phenomenon

- The combination of PSPICE simulator and laboratory work makes it easier to learn electrical phenomena

5.1 The PSPICE Simulator is Easy to Operate

The choice of the PSPICE simulator is not arbitrary. It is recommended by many educators for teaching electricity. However, it is interesting to check whether this tool is appropriate for Moroccan high school students. For this reason, we have programmed these five questions: The PSPICE simulator is easy to operate:

- At the level of installation
- At the level insertion of electrical components
- At the level of drawing the diagrams
- At the level of simulation parameterization
- At the level of visualisation of the results

The following Fig. 4 summarizes the results of the Moroccan high school students' answers.

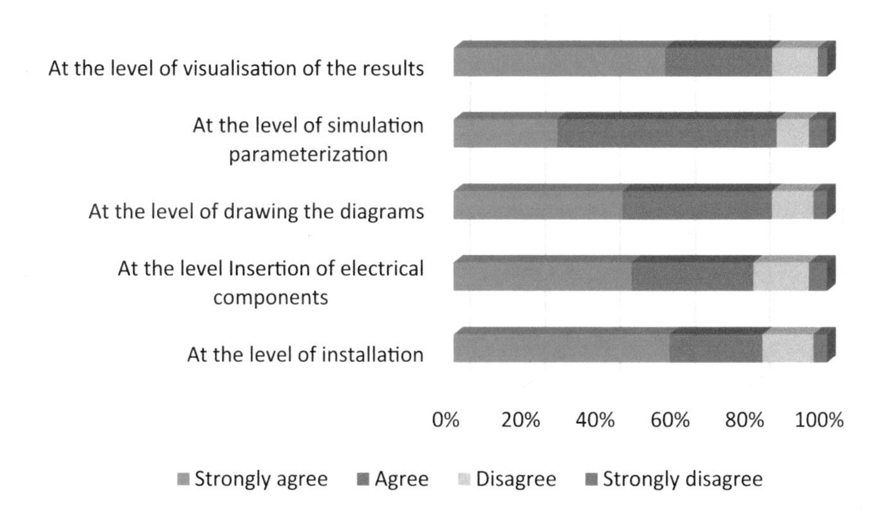

Fig. 4. The PSPICE simulator is easy to operate

The results show that the majority of students (more than 80%) find that PSPICE is easy to exploit in terms of installation, at the level of the insertion of electrical components, in terms of drawing of the diagrams, in terms of configuration of the simulation and at the level of visualization of the results of the simulations.

5.2 The PSPICE Simulator Promotes Autonomy

To measure the degree of satisfaction of students at the level of self-regulation of assimilated knowledge and at the level of autonomy, we have programmed these three questions: The PSPICE simulator promotes autonomy:

- In terms of extracurricular activities
- In terms of the investigation approach
- In terms of knowledge self-regulation

The following Fig. 5 summarizes the results of the Moroccan high school students' answers.

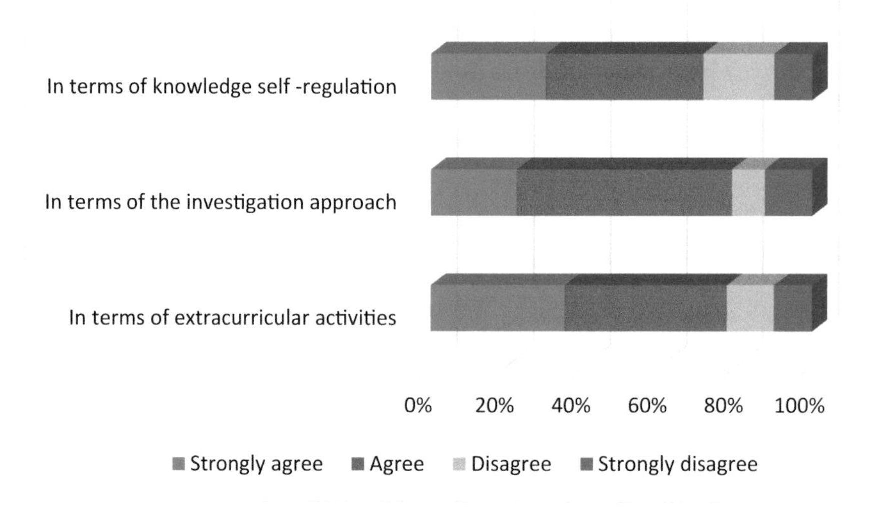

Fig. 5. The PSPICE simulator promotes autonomy

The results show that more than 65% of the students find that PSPICE can be used as a tool for self-regulation of knowledge and learner autonomy in class and in extracurricular activities.

5.3 Simulation is an Appropriate Tool for Learning About Capacitor Charging

To measure the degree of satisfaction of the students in using the PSPICE simulator to study the phenomenon of capacitor charging, we programmed the following question: Is simulation an appropriate tool for the acquisition of the capacitor charging phenomenon?

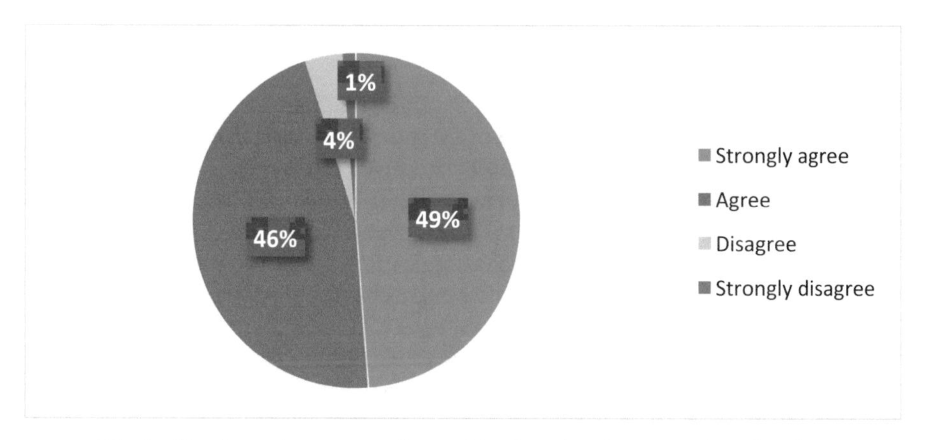

Fig. 6. Simulation is an appropriate tool for learning about capacitor charging

According to the same questionnaire, 95% of the students surveyed found that the integration of PSPICE helped to assimilate the concepts of the capacitor charging phenomenon (see Fig. 6).

5.4 The Combination of the PSPICE Simulator and the Laboratory Work Makes It Easier to Learn Electrical Phenomena

To measure the students' perception of the use of our approach to learn electrical concepts, we programmed the following question: The combination of real and virtual experiments makes it easier to learn electrical phenomena.

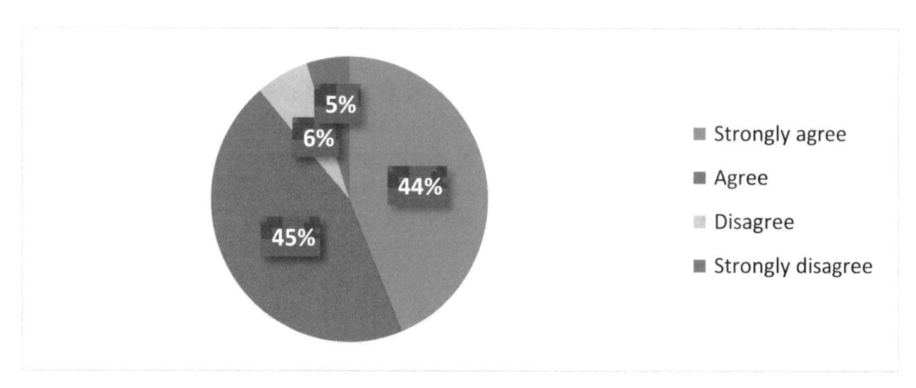

Fig. 7. The combination of the PSPICE simulator and the laboratory work makes it easier to learn electrical phenomena

In the last question of our questionnaire, the students' answers show that there is a positive perception of the combination of the PSPICE simulator and the laboratory work. Moreover, they consider that this combination makes it easier to learn about electrical phenomena (see Fig. 7).

6 Discussion

The results of the questionnaire show that the PSPICE simulator is easy to operate for most of the students surveyed and most of these students think that PSPICE promotes autonomy in them. Also, 95% of the students surveyed felt that the simulation was able to reduce some of the difficulties related to the acquisition of the concepts of the capacitor charging phenomenon. Furthermore, the students interviewed expressed a positive perception of our approach based on the combination of real and virtual experiments. This is in harmony with the results of other research conducted in at least similar contexts [21, 25, 29, 30, 40]. Furthermore, during the facilitation and supervision of the workshops that incorporate our approach, we found that the students were remarkably motivated and that our approach enabled these learners to construct their own knowledge by modifying the parameters of the simulation software and immediately visualizing the results of their modifications. However, the results obtained in this research are based on declarative data. The students interviewed express their opinions about the effect of the integration of our approach on the acquisition of electrical concepts. These opinions may not reflect reality. In this situation, it will be important to consolidate the results obtained through research that aims to study the behavior of Moroccan high school students in a more profound way when combining real and virtual experiences.

In short, virtual, and real experiences can be complementary in the learning process of physical concepts. Indeed, it will be beneficial to apply the experimental method using simulation before moving on to real experiments. In this way, learners can be given more freedom in the investigation process. This represents an opportunity for learners to build cognitive bridges between theoretical and practical knowledge. As a result, learners are motivated, active, and responsible for their learning.

7 Conclusion

Laboratory experiments and computer simulations can achieve similar goals, such as increasing students' interest in science and raising their understanding of electrical concepts. While the learners are guided in the laboratory work because of several constraints, on the contrary, they are totally free when using the simulation software. This represents an opportunity for them to be free in the experimental method and especially in the verification of hypotheses with the help of virtual experiments. On the other hand, the lack of experimental activities is the main cause of the introduction of misrepresentations in learners [41]. The latter is due to the lack of physical facilities in the institutions [23]. In this situation, simulations will be an alternative to perform virtually inaccessible experiments. Also, the results obtained in this study show that high school students are motivated to integrate our approach based on the combination of real and virtual experiments in the teaching/learning process of electricity concepts and they consider that the same approach was able to reduce some of the difficulties in acquiring the concepts related to the capacitor charging phenomenon. Despite the various advantages of simulation, it cannot replace experiments. Moreover, it can lead to erroneous behavior associated with the physical model used, which aims to simplify the phenomenon studied. To avoid confusing the learner, the model should be neither too small nor too complex [42]. Simulations should be introduced at the right time in the course, using the right pedagogical strategy and with very precise pedagogical objectives. However, there is a risk that users of simulators may confuse a real phenomenon with its representation in simulation. To avoid this risk, some researchers insist that it is necessary to clearly separate reality, simulation, and theory [43].

References

1. Villeret, O.: Les obstacles à la mise en place d'une démarche d'investigation problématisante par des enseignants débutants de sciences physiques: identification et travail en formation. PhD Thesis, Nantes (2018). https://www.theses.fr/2018NANT2005
2. Falloon, G.: Using simulations to teach young students science concepts: an experiential learning theoretical analysis. Comput. Educ. **135**, 138–159 (2019). https://doi.org/10.1016/j.compedu.2019.03.001
3. Al-Baadani, A.A., Abbas, M.: The impact of coronavirus (COVID19) pandemic on higher education institutions (HEIS) in Yemen: challenges and recommendations for the future. Eur. J. Educ. Stud. **7**(7) (2020). https://www.oapub.org/edu/index.php/ejes/article/view/3152
4. Fessakis, G., Gouli, E., Mavroudi, E.: Problem solving by 5–6 years old kindergarten children in a computer programming environment: a case study. Comput. Educ. **63**, 87–97 (2013). https://doi.org/10.1016/j.compedu.2012.11.016
5. Leung, P.K., Cheng, M.M.: Practical work or simulations? Voices of millennial digital natives. J. Educ. Technol. Syst. **50**(1), 48–72 (2021). https://doi.org/10.1177/00472395211018967
6. Zacharia, Z.: Beliefs, attitudes, and intentions of science teachers regarding the educational use of computer simulations and inquiry-based experiments in physics. J. Res. Sci. Teach. **40**(8), 792–823 (2003). https://doi.org/10.1002/tea.10112
7. Chekour, M., Laafou, M., Janati-Idrissi, R.: What are the adequate pedagogical approaches for teaching scientific disciplines? Physics as a case study. J. Educ. Soc. Res. **8**(2), 141–148 (2018). https://doi.org/10.2478/jesr-2018-0025
8. Andersen, L., Chen, J.A.: Do high-ability students disidentify with science? A descriptive study of US ninth graders in 2009. Sci. Educ. **100**(1), 57–77 (2016). https://doi.org/10.1002/sce.21197
9. Regan, E., DeWitt, J.: Attitudes, interest and factors influencing STEM enrolment behaviour: an overview of relevant literature. In: Henriksen, E., Dillon, J., Ryder, J. (eds.) Understanding Student Participation and Choice in Science and Technology Education, pp. 63–88. Springer, Dordrecht (2015). https://doi.org/10.1007/978-94-007-7793-4_5
10. Sheldrake, R., Mujtaba, T., Reiss, M.J: Students' changing attitudes and aspirations towards physics during secondary school. Res. Sci. Educ. **49**(6), 1809-1834 (2019). https://doi.org/10.1007/s11165-017-9676-5
11. Reiss, M.J.: Students' attitudes towards science: a long-term perspective. Can. J. Math Sci. Technol. Educ. **4**(1), 97–109 (2004). https://doi.org/10.1080/14926150409556599
12. Gagić, Z.Z., Skuban, S.J., Radulović, B.N., Stojanović, M.M., Gajić, O.: The implementation of mind maps in teaching physics: educational efficiency and students' involvement. J. Balt. Sci. Educ. **18**(1), 117–131 (2019). https://doi.org/10.33225/jbse/19.18.117
13. Donnelly, J.F., Jenkins, E.W.: Science Education Policy, Professionalism and Change. Prabhat Prakashan (2000)
14. Ornek, F.: Models in science education: applications of models in learning and teaching science. Int. J. Environ. Sci. Educ. **3**(2), 35–45 (2008). https://files.eric.ed.gov/fulltext/EJ894843.pdf

15. de Jong, T.: Scaffolding inquiry learning: how much intelligence is needed and by whom? In: AIED, p. 4 (2005)
16. Kubieck, J.: Inquiry-based learning, the nature of science, and computer technology: new possibilities in science education. Can. J. Learn. Technol. Rev. Can. L'apprentissage Technol. **31**(1) (2005). https://www.learntechlib.org/p/42864/
17. Hudha, M.N., Batlolona, J.R.: How are the physics critical thinking skills of the students taught by using inquiry-discovery through empirical and theorethical overview? Eurasia J. Math. Sci. Technol. Educ. **14**(2), 691–697 (2017). https://doi.org/10.12973/ejmste/80632
18. Chekour, M.: Contribution à l'amélioration d'acquisition de concepts en électricité chez les lycéens marocains. Université Abdelamalek Essaadi (2019). https://doi.org/10.13140/RG.2.2.19550.36160
19. Pulgar, J.: Classroom creativity and students' social networks: theoretical and practical implications. Think. Ski. Creat. **42**, 100942 (2021). https://doi.org/10.1016/j.tsc.2021.100942
20. DiPasquale, J., Hunter, W.: Critical thinking in asynchronous online discussions: a systematic review. Can. J. Learn. Technol. Rev. Can. L'apprentissage Technol. **43**(2) (2017). https://www.learntechlib.org/p/183614/
21. Chekour, M.: Teaching electricity between pedagogy and technology. In: Personalization and Collaboration in Adaptive E-Learning, IGI Global, pp. 304–314 (2020). https://doi.org/10.4018/978-1-7998-1492-4.ch015
22. Cobbinah, C., Bayaga, A.: Physics content and pedagogical changes: ramification of theory and practice. EURASIA J. Math. Sci. Technol. Educ. **13**(6), 1633–1651 (2017). https://doi.org/10.12973/eurasia.2017.00689a
23. Chekour, M., Laafou, M., Janati-Idrissi, R.: Les facteurs influençant l'acquisition des concepts en électricité. Cas des lycéens marocains. Adjectif En Ligne (2015). http://www.adjectif.net/spip/spip.php?article354
24. Kortelainen, J., Mikkola, A.: Semantic data model in multibody system simulation. Proc. Inst. Mech. Eng. Part K J. Multi-Body Dyn. **224**(4), 341–352 (2010). https://doi.org/10.1243/14644193JMBD257
25. Ronen, M., Eliahu, M.: Simulation—a bridge between theory and reality: the case of electric circuits. J. Comput. Assist. Learn. **16**(1), 14–26 (2000). https://doi.org/10.1046/j.1365-2729.2000.00112.x
26. Law, A.M., Kelton, W.D., Kelton, W.D.: Simulation Modeling and Analysis. McGraw-Hill, New York (1991)
27. Kotiadis, K.: Using soft systems methodology to determine the simulation study objectives. J. Simul. **1**(3), 215–222 (2007). https://doi.org/10.1057/palgrave.jos.4250025
28. Lin, J., Lu, Q., Ding, X., Zhang, Z., Zhang, X., Sadayappan, P.: Gaining insights into multicore cache partitioning: bridging the gap between simulation and real systems. In: 2008 IEEE 14th International Symposium on High Performance Computer Architecture, pp. 367–378 (2008). https://doi.org/10.1109/HPCA.2008.4658653
29. Chekour, M.: The impact perception of the resonance phenomenon simulation on the learning of physics concepts. Phys. Educ. **53**(5), 055004 (2018). https://doi.org/10.1088/1361-6552/aac984
30. De Jong, T., Linn, M.C., Zacharia, Z.C.: Physical and virtual laboratories in science and engineering education. Science **340**(6130), 305–308 (2013). https://doi.org/10.1126/science.1230579
31. Makransky, G., Terkildsen, T.S., Mayer, R.E.: Adding immersive virtual reality to a science lab simulation causes more presence but less learning. Learn. Instr. **60**, 225–236 (2019). https://doi.org/10.1016/j.learninstruc.2017.12.007
32. Thees, M., Kapp, S., Strzys, M.P., Beil, F., Lukowicz, P., Kuhn, J.: Effects of augmented reality on learning and cognitive load in university physics laboratory courses. Comput. Hum. Behav. **108**, 106316 (2020). https://doi.org/10.1016/j.chb.2020.106316

33. Gilbert, N., Troitzsch, K.: Simulation for the Social Scientist. Open University Press, Buckingham, UK (1999)
34. Morgan, M.S., et al.: Experiments without material invention: model experiments, virtual experiments and virtually experiments. In: Radder, H. (eds.) The Philosophy of Scientific Experimentation, pp. 216–235. University of Pittsburgh Press, Pittsburgh, PA (2003). http://dare.uva.nl/record/1/183568
35. Winsberg, E.: Simulated experiments: methodology for a virtual world. Philos. Sci. **70**(1), 105–125 (2003). https://doi.org/10.1086/367872
36. Millar, R., Tiberghien, A., Le Maréchal, J.F.: Varieties of labwork: a way of profiling labwork tasks. In: Psillos, D., Niedderer, H. (eds.) Teaching and Learning in the Science Laboratory. Science & Technology Education Library, vol. 16, pp. 9–20. Springer, Dordrecht (2002). https://doi.org/10.1007/0-306-48196-0_3
37. Albe, V.: Enseignement médiatisé des travaux pratiques de physique en DEUG: compte rendu d'innovation. Didaskalia **15**(1), 159–166 (1999). http://documents.irevues.inist.fr/handle/2042/23879
38. Kranjc, T.: Simulations as a complement and a motivation element in the teaching of physics. Metod. Obz. Časopis Za Odgoj.-Obraz. Teor. Praksu **6**(12), 175–187 (2011). https://hrcak.srce.hr/71293
39. Chekour, M., Laafou, M., Janati-Idrissi, R.: Distance training for physics teachers in Pspice simulator. Mediterr. J. Soc. Sci. **6**(3), 232 (2015). https://doi.org/10.5901/mjss.2015.v6n3s1p232
40. Olympiou, G., Zacharia, Z.C.: Examining students' actions while experimenting with a blended combination of physical manipulatives and virtual manipulatives in physics. In: Mikropoulos, T. (eds.) Research on e-Learning and ICT in Education, pp. 257–278. Springer, Cham (2018). https://doi.org/10.1007/978-3-319-95059-4_16
41. Noupet Tatchou, G.: Conceptions d'élèves du secondaire sur le rôle de l'expérience en sciences-physiques: cas de quelques expériences de cours en électrocinétique. Mémoire de Diplôme d'Etudes Approfondies en Sciences de l'Education, ENS, Dakar (2004). http://www.fastef-portedu.ucad.sn/cesea/cuse/tatchou.pdf
42. Richoux, B., Salvetat, C., Beaufils, D.: Simulation numérique dans l'enseignement de la physique: enjeux, conditions. Bull. Union Phys. **842**, 497–521 (2002)
43. Droui, M., El Hajjami, A.: Simulations informatiques en enseignement des sciences: apports et limites. EpiNet 164 (2014). http://www.epi.asso.fr/revue/articles/a1404e.htm

Deep Learning Models for Medical Image Analysis in Smart Healthcare System: A Review

Souad Kamal[1], Mohamed Lazaar[1](✉), Mohammed Bennani Othmani[2], Farid Bourzgui[3], and Oussama Mahboub[4]

[1] ENSIAS, Mohammed V University, Rabat, Morocco
{Souad_kamal,Mohamed_Lazaar}@um5.ac.ma
[2] Laboratory of Medical Informatics, Faculty of Medicine and Pharmacy, Hassan II University, Casablanca, Morocco
[3] Faculty of Dental Medicine, Hassan II University, Casablanca, Morocco
[4] ENSA, Abdelmalek Essaadi University, Tetouan, Morocco
omahboub@uae.ac.ma

Abstract. The concept of Smart Healthcare has become a reality thanks to advances in information and communication technologies (ICT). The medical image analysis in smart healthcare system is the fastest emerging area in medical science field. In parallel the artificial intelligence has become an increasingly important aspect in this field. The major component of artificial intelligence (AI) is many machine learning (ML) models which apply in medical diagnosis and treatment with the advancement of healthcare system industry. Deep Learning has been used extensively to analyze the data generated by Internet of Medical Things (IOMT) devices in a smart healthcare by several scientific researchers. The aim of this paper is to discuss the use of deep learning techniques to analyze medical images in the smart healthcare systems. In the literatures studies, we did not focus on a single specific task, we talked about several issues in the field of health: A state of the art on the determination of the best deep learning models that have been used for medical image analysis, our research was also around the study of the use of deep learning in other diseases and more precisely in the analysis of dental images.

Keywords: Deep Learning · Smart Healthcare · Transfer Learning · Healthcare Decision Making · Computer Aided Diagnosis · Medical Image Analysis

1 Introduction

In recent years, scientific research gives great importance to artificial intelligence (AI), it has been one of the most popular fields in research, and plays a role in the daily lives of many people [1–4]. Artificial intelligence (AI) is the ability of machines to learn from data input, it aims to produce machines capable of simulating human intelligence and solving problems in a flexible way without the help of human beings from finding an optimal and adaptive approach as well as applying a set of theories and techniques [5]. The major component of artificial intelligence (AI) is machine learning (ML) which is

M. Lazaar et al. (Eds.): BDIoT 2022, LNNS 625, pp. 376–387, 2023.
https://doi.org/10.1007/978-3-031-28387-1_32

the science of getting a computer to act without programming. It is made to analyze the input data and process the accumulated experimental data from the use of computational methods and the training of these data. In fact, computers learn from input data and reinforce their properties by the errors they have made without specific programming or establishment of a mathematical model [5, 6]. Deep learning is a subset of machine learning that, in very simple terms, it is composed of algorithms that permit software to train itself to perform tasks, like speech and image recognition, by exposing multilayered neural networks to vast amounts of data [7]. In this context, the emergence of Deep Learning has inspired numerous innovative solutions for image analysis problems in interdisciplinary fields. For example, in image processing [8] and analysis. Also, it has provided significant results in autonomous driving [9]. Deep learning is also showing vast potential in particular in the healthcare domain. In the literature, deep learning studies applied to injury prediction have reported similar and sometimes superior performance to that of professional clinicians in some tasks [10, 11]. The application of deep learning-based medical image analysis has the potential to provide decision support to clinicians and improve the accuracy and efficiency of various diagnostic and treatment processes, thereby stimulating new research and development efforts in computer-aided diagnosis (CAD) [12]. For this reason, AI researchers have applied powerful deep learning algorithms called the convolutional neural network (CNN) to segment, classify objects and detect disease [13].

The field of deep learning is growing rapidly thanks to recent developments in convolutional neural network (CNN) architectures [14]. "Convolutional Neural Networks (CNNs) have become a popular choice for medical image analysis due to their rapid development. Clinical skin screenings, mammography, and eye exams for diabetic retinopathy have successfully used CNNs" [15].

In order to develop convolutional neural networks capable of performing medical image analysis tasks, these deep learning algorithms must be trained on large image datasets [16]. "However, a significant challenge with advancing these algorithms includes the limited availability of supervised histopathological datasets" [17]. For this reason, the procedure called transfer learning has been employed, which is a pre-trained model on a large data-set that can be used in similar tasks without retraining the all model from the beginning, we can only adapt the input and the output of the pre-trained model and it can give interesting results. It is a strategy in which features extracted by a convolutional neural network (CNN) from data are useful in solving a different but related task, involving new data, which usually turns out to be of a smaller population to train a convolutional neural network (CNN) from scratch [18]. And in the rest of this paper we will discuss his usefulness in healthcare and we will cite some papers that treat it in medical image analysis in the following sections.

Research Methodology. Our research was around the usefulness of deep learning in the field of healthcare as well as where is this application represented in the analysis of medical images? And what are the most commonly used deep learning methods in this context? First we took as a case study the COVID-19 disease, but we found a lot of researches that are similar in terms of the medical images analysis related to this disease, as this was represented in the COVID-19 detection by comparing the performance of some powerful pre-trained convolutional neural network models [19].

Then we took as another case study the analysis of dental medical images of which we found several tasks which were carried out among which detecting periodontal bone loss in panoramic dental X-rays, detecting and classifying teeth in dental panoramic radio-graphs, classifying panoramic radio-graph into the two genders, detecting the three classes ("cavity", "filling" and "implant") from the panoramic dental radio-graphs. The articles we have chosen are published in indexed journals and we have tried to search for all the articles that can cover the forms of dental medical images analysis between recent years from 2019 until 2022.

Our research questions were for what are the main problems we are facing while using Deep Learning models in medical image analysis? What can we glean from studies that compare CNN performance to that of clinicians?

The rest paper is structured as follow: in Sect. 2 we review the different deep learning approaches to some important work done by other healthcare researchers, as we have focused mainly on the field of dentistry, in Sect. 3 we compare the results of these studies via a comparative table. Thus, in the following we summarize our paper with a general analysis (Fig. 1).

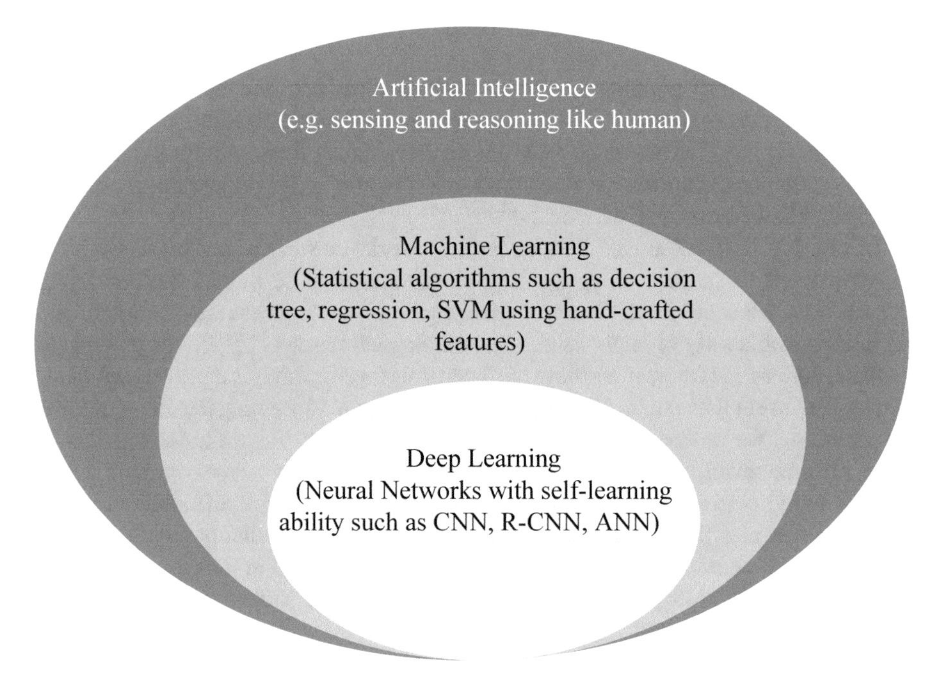

Fig. 1. The context of artificial intelligence, machine learning, and deep learning. SVM: Support Vector Machine. CNN: convolutional neural networks, R-CNN: recurrent CNN, ANN: artificial neural networks [20].

2 Literature Survey

In recent years, many scientific articles had the objective of studying and researching how to develop more and more CNN approaches in the field of medical image analysis. This section discusses recent advances and different techniques in this context.

In the pilot study by Schwendicke et al. [21], they aimed to detect carious lesions by applying deep convolutional neural networks (CNNs) on Near-Infrared-Light Transillumination (NILT) images. For this they used 226 NILT images and each image focused on one tooth (113 premolars, 113 molars), the images were segmented into 2 groups (decayed tooth, non-decayed tooth). For training they used two state-of-the-art CNNs (Resnet18 [22], Resnext50 [23]). To compare the performance of the two models the area-under-the-receiver-operating-characteristics-curve (AUC), sensitivity, specificity, and positive/negative predictive values (PPV/NPV) were used. The best model in terms of AUC was Resnext50, it reached up to 74%. The CNNs showed a satisfactory discriminatory ability to detect carious lesions, but there were limitations in this study, among them the authors consider that due to the complexity and non-linearity of the structure of the prediction models, these neural networks remain opaque and this has a negative impact on decision-making, also there was a limitation of the reliability of the examiners that it constitutes the test of reference. The fact of using images, each of which focuses on a single tooth, contributes to another limitation of this study.

Another paper by Schwendicke et al. [24] they carried out a review of the scope of dental imaging studies where the most commonly used type of image was panoramic radiographs, periapical radiographs and Cone Beam Computed Tomography (CBCT) [24]. Almost half of all studies reported using the transfer learning technique including pre-trained convolutional neural network (CNN) variants. When evaluating pre-trained CNNs, AlexNet [25] was leveraged the most, followed by VGG16 [26], ResNet [27], U-Net [28], GoogLeNet [29], and V-net [30] or DenseNet [31].

Since there was a diversity of studies, tasks and metrics used, this led to a variation in the performance of the models used.

The majority of the studies were tasked with tooth classification, with an average precision between 0.77 and 0.98 and detection of carious lesions, with an average precision between 0.82 and 0.89, on the other studies comparing the convolutional neural network (CNN) to clinicians, there were those who found that convolutional neural network (CNN) worked better than clinicians. On the other hand, others have found similar performance in convolutional neural network (CNN) and clinician outcomes, but there was just one study where convolutional neural network (CNN) performed worse than humans [24].

In the article by Kim et al. [32], the authors used the panoramic image which is a first intention examination, often necessary [33]. In this study, transfer learning was used through a method called DeNTNet [32] which detects periodontal bone loss in panoramic dental x-rays, and even gives tooth counts corresponding to the lesion. This study applies a comparison between the proposed network and the dental clinician's performance in detecting the lesion. The proposed model was able to achieve higher periodontal bone loss (PBL) detection performance than dental clinicians, since DeNTNet [32] had the F1 score of 0.75 on the test set, while the average performance of dental clinicians was 0.69. This study had a few limitations, among them which panoramic dental x-rays capture a large field of view, resulting in low resolution for each individual tooth. This hampered the overall sensitivity performance of dental clinicians and DeNTNet [32] for detecting periodontal bone loss was limited.

In the article by Muramatsu et al. [34], they proposed a CNN-based method which can be useful in automatic filing of dental charts for forensic identification and preprocessing step for computerized image analysis of dental disease [34]. First the authors perform the tooth detection task using bounding boxes, this procedure achieved 96.4% of sensitivity, using DetectNet [35]. Then, these bounding boxes were classified and to achieve this task they considered two levels of classification, the first was to classify the type of teeth are they incisors, canines, premolars or molars, the second level was to detect tooth conditions, including non-metallic restorations, partially restored or fully restored. Classification accuracies for tooth types was 93.2% and dental conditions was 98.0%, using the ResNet50 [36] architecture. This study had several limitations: one was the small dataset of panoramic images used, they didn't include the third molars also "it was difficult to distinguish between the teeth restored with metal and nonmetal so the use of tooth condition categories including a nonmetal restored group was another limitation" [34].

Staying in forensic medicine, in the article by Ataş [37], they consider that gender identification is the initial stage in human identification who have been exposed to fatal natural disasters or catastrophes [37]. Therefore, using a high processing speed, accuracy and stability methods which do not require any manual feature adjustment is a good way for gender prediction. They assessed a deep learning model called DenseNet121 [31] for binary classification. The Panoramic Dental Radiography (PDR) image were used to train the proposed model because it can cover the entire mandible and contribute positively to biometric identification for gender classification for gender classification, for that the female and male patient ratio in the dataset was 58% and 42% respectively. The model DenseNet121 [38] was compared to another deep transfer learning models VGG16 [26], ResNet50 [36] and EfficientNetB6 [38]. So the proposed model outperformed the other approaches by achieving a success rate of 97.25% in gender classification, also it was trained in a very short time compared to other models. Among the limitations that complicated the gender identification, were differences in terms of contrast, location and resolution parameters in the panoramic dental radiographs (PDR) images.

In the paper by Cantu et al. [11] they started their study with the hypothesis that the use of deep learning (DL) in the detection of proximal caries lesions is more precise than individual clinical dentists [11]. Indeed, they followed their literature in healthcare, and more precisely in the analysis of medical images using deep learning methods, in particular using convolutional neural networks (CNNs), for example detecting breast cancer in mammographies [39], skin cancer [40], or diabetic retinopathy [41]. in the field of dentistry, deep learning has been used in several detection issues, for example in detecting periodontal bone loss and caries lesions on peri-apical radiographs [24], and knowing that deep learning has been little applied for the analysis and detection of caries on bitewings, the authors decided to carry out this study in this context of detection by comparing the performance of convolutional neural networks with dental experts [11]. So to do this study, they used the architecture U-Net, first published by Ronneberger et al. (2015), which is more used for segmentation tasks in medical imaging [28] it is a fully convolutional neural network (CNN) and which is characterized by two parts, the encoder part and the decoder part, in this study EfficientNet-B5 [42] has been used as an encoder, this model has been widely used in classification tasks, for this study it is a binary classification, such that for each pixel in the processed image they classify it whether it belongs to a carious lesion, or the contrary. Finally the neural network reached an accuracy of 0.80, while the dental experts, their accuracy was less than 0.71 [11]. With that the authors endorsed their previously stated hypothesis. But there remains the big major problem for the neural network employed as well as the dentists is the generalizability [11].

The article by AL-Ghamdi et al. [15] is in the same context of applying deep learning (DL) in dental radiographs analysis, and this from the treatment of the problem of multiclassification on panoramic radiographs as the authors consider the three classes to be detected on X-ray images: "cavity", "filling" and "implant" [15]. They compared three convolutional neural network models, one is in the form of Neural Search Architecture Network (NASNet) [43], AlexNet [25] and convolutional neural network (CNN) model. Neural Search Architecture Network (NASNet) [43] was the best model with 96% accuracy [15].

3 Comparative Studies of Medical Image Analysis: Dentistry Case

We compare the results of the studies mentioned above via a comparative table, these studies are between 2019 and 2022, they always use the deep learning technique of which it is the transfer learning by the application the pre-trained convolutional neural network, the models of deep learning and the tasks as well as the evaluation metrics differ from one study to another (Table 1).

Table 1. State of the art techniques

Authors	Years	Techniques	Models	Tasks	Datasets	Results
Kim et al. [32]	2019	Pre-trained convolutional neural network (PT-CNN): Transfer learning	DeNTNet [32]	Detecting periodontal bone loss in panoramic dental X-rays, giving tooth counts corresponding to the lesion to reduce the workload involved in reporting tooth numbers and comparing detection precision with dental clinicians	12179 panoramic dental radiographs collected from Korea University of Anam Hospital after removing identifiable patient information	F1-score equal to 75%
Schwendicke et al. [21]	2020	Pre-trained convolutional neural network (PT-CNN): Transfer learning	Resnet18 [22], Resnext50 [23]	Detecting caries lesions in Near-Infrared-Light Transillumination (NILT) images	226 extracted posterior teeth (113 premolars and 113 molars) were obtained with informed consent under an ethics-approved protocol,	The best model was Resnext50 with respect to area-under-the-receiver-operating-characteristics-curve (AUC) metric, and it reached up to 74%
Muramatsu et al. [34]	2021	Pre-trained convolutional neural network (PT-CNN): Transfer learning	DetectNet [35] based on GoogLeNet [29] for detection task, ResNet50 [36] for classification task	Detecting and classifying teeth in dental panoramic radio-graphs for automatic structured filing of the dental charts	The dental panoramic radiographs were obtained at Asahi University Hospital as a part of routine examinations using the QRmaster-P (Takara Telesystems Corp., Osaka, Japan)	For detection task: 96.4 of sensitivity, for classification task accuracy: 93.2%

(continued)

Table 1. *(continued)*

Authors	Years	Techniques	Models	Tasks	Datasets	Results
Ataş [37]	2022	Pre-trained convolutional neural network (PT-CNN): Transfer learning	DenseNet121 [38], VGG16 [26], ResNet50 [36] and Efficient-NetB6 [38]	Classifying Panoramic Radiograph images into the two genders (male or female), and comparing the models to obtain the best	A dataset of 24000 PDR images was examined from patients aged between 18 and 77 who received dental treatment in Diya barik Oral and Dental Health Hospital Periodontology clinic between 2015 and 2020	The best model was DenseNet121 and it achieved a success rate of 97.25% in terms of accuracy, with over 8 M parameters, it reached 96.80%, 97.69%, 97.25% and 96.80 in terms of precision, recall, F1-Score and specificity successively
Cantu et al. [11]	2020	Pre-trained convolutional neural network (PT-CNN): Transfer learning	U-Net [28]	Classifying each pixel in the processed images whether it belongs to a carious lesion, or the contrary, comparing the result with the dental clinicians	Total of 3686 bitewings originating from routine care provided at a dental clinic	The U-Net model used reached 80% in terms of accuracy
AL-Ghamdi et al. [15]	2022	Pre-trained convolutional neural network (PT-CNN): Transfer learning	Neural Search Architecture Network (NASNet) [43], AlexNet [25] and Convolutional neural network (CNN) model	Detecting the three classes ("cavity", "filling" and "implant") from the panoramic dental radiographs, comparing the three models	The dental image dataset consisted of unidentified and anonymized panoramic dental X-ray images of a total of 116 patients, acquired from Noor Medical Imaging Center, Qom, Tran	The best model was Neural Search Architecture Network (NASNet) and it achieved 96.51% with data augmentation and 93.36% without data augmentation in terms of accuracy

4 Analysis and Discussion

We report in this work that the authors have solved several problems in the field of dentistry thanks to deep learning techniques, precisely the convolutional neural network from the integration of the transfer learning technique which refers to the use of pre-trained architectures so as not to waste time during training as there was in most studies a low availability of the dataset. They studied the detection of periodontal bone loss in dental panoramic radiographs, Detection and classification of teeth in dental panoramic radiographs for automatic structured filing of the dental charts, the classification of panoramic radiographs in both genders (male or female), the detection of carious lesions etc. Also they always compare the performance of their best transfer learning models with the results of expert clinicians, and most of the time the authors find that these models outperform these clinicians, which shows that deep learning has a positive impact on healthcare in terms of precision when analyzing medical images.

Except that there were many limitations, among them we found that in the article by Muramatsu et al. [34] there were several limitations: one was the small dataset of panoramic images used, they did not include third molars and "it was difficult to distinguish between teeth restored with metal and not metallic, so the use of tooth condition categories including a restored non-metallic group was another limitation" [34]. In the article by Ataş [37], among the limitations that complicated the gender identification, were differences in terms of contrast, location and resolution parameters in the panoramic dental radiographs (PDR) images. In the paper by Cantu et al. [11], the big major problem for the neural network employed as well as the dentists was the generalizability. In the article by AL-Ghamdi et al. [15] they deal with the problem of multiclassification on panoramic radiographs as the authors consider the three classes to be detected on X-ray images: "cavity", "filling" and "implant" [15], and as a future work, they propose that "multiclass classification can be done in large datasets with a greater number of classes. A new advanced deep learning based on hybrid deep learning can be implemented that can enhance the existing performance. Multiclass classification can be done in large datasets with a greater number of classes. A new advanced deep learning based on hybrid deep learning can be implemented that can enhance the existing performance" [15]. For studies that used panoramic dental x-ray (PDR) images in which panoramic dental x-rays capture a wide field of view, resulting in low resolution for each individual tooth. For instance in the article by Kim et al. [32] this hampered the overall sensitivity performance of dental clinicians and DeNTNet in detecting periodontal bone loss which was limited [32]. For this I propose to go deeper and make comparisons between imaging methods like bitewing and periapical radiographs which are the most suitable because they are easily acquired, cheap and provide high-resolution images [44–46], or cone beam CT (CBCT) or intraoral radiographs with panoramic radiography or cone beam CT (CBCT) or intraoral radiographs with panoramic radiography [34–37].

5 Conclusion and Perspectives

The present review shows that convolutional neural networks (CNNs) are increasingly employed for medical image analysis in research settings and it is important to improve their generalizability and especially that it was found in the literature that when comparing the performance of convolutional neural network (CNN) against an independent test group of medical professionals, most studies found that convolutional neural network (CNN) performed superior and sometimes similarly to clinicians. This has clinical significance is that convolutional networks (CNNs) can be used in diagnostic support smart healthcare systems, thereby assisting clinicians in more comprehensive, systematic, and faster analysis and documentation of medical images. Almost all the studies that have been reported in the literature use individually constructed convolutional networks (CNN) architectures, including variants of pre-trained models via the method called transfer learning. The models performances varied widely between studies, tasks and used metrics.

In dentistry the most commonly used image type were panoramic radiographs. Panoramic dental x-rays capture a wide field of view, resulting in low resolution for each individual tooth. For instance this hampered the overall sensitivity performance of dental clinicians and pre-trained models in detecting periodontal bone loss which was limited. So if we make comparisons between medical imaging methods in the evaluation of deep learning models in order to perform the analysis of medical images, it will be better. In our research, we seek to make a study that is characterized by originality in relation to the Moroccan population therefore these limitations and others that are stated previously in the analysis and discussion section will be developed in the future taking into consideration our purpose depending on the availability of sufficient Moroccan dental images to apply the deep learning models.

References

1. Thakur, A., Mishra, A.P., Panda, B., et al.: Application of artificial intelligence in pharmaceutical and biomedical studies. Curr. Pharm. Des. **26**(29), 3569–3578 (2020)
2. Dolci, R.: IoT solutions for precision farming and food manufacturing: artificial intelligence applications in digital food. In: 2017 IEEE 41st Annual Computer Software and Applications Conference (COMPSAC), pp. 384–385 (2017)
3. Wang, N., Liu, Y., Liu, Z., Huang, X.: Application of artificial intelligence and big data in modern financial management. In: 2020 International Conference on Artificial Intelligence and Education (ICAIE), pp. 85–87 (2020)
4. Yanhua, Z.: The application of artificial intelligence in foreign language teaching. In: 2020 International Conference on Artificial Intelligence and Education (ICAIE), pp. 40–42 (2020)
5. Ren, R., Luo, H., Su, C., et al.: Machine learning in dental, oral and craniofacial imaging: a review of recent progress. PeerJ **9**, e11451 (2021). https://doi.org/10.7717/peerj.11451
6. Erickson, B.J., Korfiatis, P., Akkus, Z., Kline, T.L.: Machine learning for medical imaging. Radiographics **37**, 505–515 (2017). https://doi.org/10.1148/rg.2017160130
7. Rasanayagam, G.: AI vs. machine learning vs. deep learning. In: Nerd Tech. https://medium.com/nerd-for-tech/ai-vs-machine-learning-vs-deep-learning-60b3d0611fe9 (2021). Accessed 3 Sep 2022
8. Nixon, M.S., Aguado, A.S.: 3 - image processing. In: Nixon, M.S., Aguado, A.S. (eds.) Feature Extraction and Image Processing for Computer Vision (Fourth Edition). Academic Press, pp. 83–139 (2020)
9. Gonsalves, T., Upadhyay, J.: Chapter eight - integrated deep learning for self-driving robotic cars. In: Shaw, R.N., Ghosh, A., Balas, V.E., Bianchini, M. (eds.) Artificial Intelligence for Future Generation Robotics. Elsevier, pp. 93–118 (2021)

10. Liu, X., Faes, L., Kale, A.U., et al.: A comparison of deep learning performance against health-care professionals in detecting diseases from medical imaging: a systematic review and meta-analysis. Lancet Digit. Health **1**, e271–e297 (2019). https://doi.org/10.1016/S2589-7500(19)30123-2
11. Cantu, A.G., Gehrung, S., Krois, J., et al.: Detecting caries lesions of different radiographic extension on bitewings using deep learning. J. Dent. **100**, 103425 (2020). https://doi.org/10.1016/j.jdent.2020.103425
12. Chan, H.-P., Samala, R.K., Hadjiiski, L.M., Zhou, C.: Deep learning in medical image analysis. In: Lee, G., Fujita, H. (eds.) Deep Learning in Medical Image Analysis. AEMB, vol. 1213, pp. 3–21. Springer, Cham (2020). https://doi.org/10.1007/978-3-030-33128-3_1
13. Alzubaidi, L., et al.: Review of deep learning: concepts, CNN architectures, challenges, applications, future directions. J. Big Data **8**(1), 1–74 (2021). https://doi.org/10.1186/s40537-021-00444-8
14. Wu, X., Sahoo, D., Hoi, S.C.H.: Recent advances in deep learning for object detection. Neurocomputing **396**, 39–64 (2020). https://doi.org/10.1016/j.neucom.2020.01.085
15. AL-Ghamdi, A.S.A.-M., Ragab, M., AlGhamdi, S.A., et al.: Detection of dental diseases through X-ray images using neural search architecture network. Comput. Intell. Neurosci. **2022**, e3500552 (2022). https://doi.org/10.1155/2022/3500552
16. Gu, J., Wang, Z., Kuen, J., et al.: Recent advances in convolutional neural networks. Pattern Recognit. **77**, 354–377 (2018). https://doi.org/10.1016/j.patcog.2017.10.013
17. Lagree, A., Mohebpour, M., Meti, N., et al.: A review and comparison of breast tumor cell nuclei segmentation performances using deep convolutional neural networks. Sci. Rep. **11**, 8025 (2021). https://doi.org/10.1038/s41598-021-87496-1
18. Weiss, K., Khoshgoftaar, T.M., Wang, D.: A survey of transfer learning. J. Big Data **3**(1), 1–40 (2016). https://doi.org/10.1186/s40537-016-0043-6
19. Apostolopoulos, I.D., Mpesiana, T.A.: Covid-19: automatic detection from X-ray images utilizing transfer learning with convolutional neural networks. Phys. Eng. Sci. Med. **43**(2), 635–640 (2020). https://doi.org/10.1007/s13246-020-00865-4
20. Akkus, Z., Aly, Y., Attia, I., et al.: Artificial intelligence (AI)-empowered echocardiography interpretation: a state-of-the-art review. J. Clin. Med. **10**, 1391 (2021). https://doi.org/10.3390/jcm10071391
21. Schwendicke, F., Elhennawy, K., Paris, S., et al.: Deep learning for caries lesion detection in near-infrared light transillumination images: a pilot study. J. Dent. **92**, 103260 (2020). https://doi.org/10.1016/j.jdent.2019.103260
22. Napoletano, P., Piccoli, F., Schettini, R.: Anomaly detection in nanofibrous materials by CNN-based self-similarity. Sensors **18**(1), 209 (2018). https://doi.org/10.3390/s18010209
23. Go, J.H., Jan, T., Mohanty, M., et al.: Visualization approach for malware classification with ResNeXt. In: 2020 IEEE Congress on Evolutionary Computation (CEC), pp. 1–7 (2020)
24. Schwendicke, F., Golla, T., Dreher, M., Krois, J.: Convolutional neural networks for dental image diagnostics: a scoping review. J. Dent. **91**, 103226 (2019). https://doi.org/10.1016/j.jdent.2019.103226
25. Team GL: AlexNet: the first CNN to win image net. In: Gt. Blog Free Resour. What Matters Shape Your Career (2020). https://www.mygreatlearning.com/blog/alexnet-the-first-cnn-to-win-image-net/. Accessed 20 Aug 2022
26. Simonyan, K., Zisserman, A.: Very deep convolutional networks for large-scale image recognition. arXiv preprint arXiv:1409.1556 (2015)
27. Szegedy, C., Ioffe, S., Vanhoucke, V., Alemi, A.: Inception-v4, inception-ResNet and the impact of residual connections on learning. In: Proceedings of the AAAI Conference on Artificial Intelligence, vol. 31, No. 1 (2016)

28. Ronneberger, O., Fischer, P., Brox, T.: U-Net: convolutional networks for biomedical image segmentation. In: Navab, N., Hornegger, J., Wells, W.M., Frangi, A.F. (eds.) MICCAI 2015. LNCS, vol. 9351, pp. 234–241. Springer, Cham (2015). https://doi.org/10.1007/978-3-319-24574-4_28
29. Szegedy, C., Liu, W., Jia, Y., et al.: Going deeper with convolutions. In: Proceedings of the IEEE Conference on Computer Vision and Pattern Recognition, pp. 1–9 (2014)
30. Milletari, F., Navab, N., Ahmadi, S.-A.: V-Net: fully convolutional neural networks for volumetric medical image segmentation. In: 2016 Fourth International Conference on 3D Vision (3DV). IEEE, Stanford, CA, USA, pp. 565–571 (2016)
31. Zhou, T., Ye, X., Lu, H., et al.: Dense convolutional network and its application in medical image analysis. BioMed Res. Int. **2022**, 1–22 (2022). https://doi.org/10.1155/2022/2384830
32. Kim, J., Lee, H.-S., Song, I.-S., Jung, K.-H.: DeNTNet: deep neural transfer network for the detection of periodontal bone loss using panoramic dental radiographs. Sci. Rep. **9**, 17615 (2019). https://doi.org/10.1038/s41598-019-53758-2
33. de Oliveira Capote, T.S., de Almeida Gonçalves, M., Gonçalves, A., Gonçalves, M.: Panoramic radiography—diagnosis of relevant structures that might compromise oral and general health of the patient. In: Emerging Trends in Oral Health Sciences and Dentistry. IntechOpen (2015)
34. Muramatsu, C., et al.: Tooth detection and classification on panoramic radiographs for automatic dental chart filing: improved classification by multi-sized input data. Oral Radiol. **37**(1), 13–19 (2020). https://doi.org/10.1007/s11282-019-00418-w
35. Riquelme, D., Akhloufi, M.: Deep learning for lung cancer nodules detection and classification in CT scans. AI **1**, 28–67 (2020). https://doi.org/10.3390/ai1010003
36. Boesch, G.: Deep residual networks (ResNet, ResNet50) - 2022 guide (2022). https://viso.ai/deep-learning/resnet-residual-neural-network/. Accessed 20 Aug 2022
37. Atas, I.: Human gender prediction based on deep transfer learning from panoramic radiograph images. arXiv preprint arXiv:2205.09850 (2022)
38. Rahman, A.I., Bhuiyan, S., Reza, Z.H., Zaheen, J., Khan, T.A.N.: Detection of intracranial hemorrhage on CT scan images using convolutional neural network. Doctoral dissertation, Brac University (2021)
39. Becker, A.S., Marcon, M., Ghafoor, S., et al.: Deep learning in mammography: diagnostic accuracy of a multipurpose image analysis software in the detection of breast cancer. Invest. Radiol. **52**, 434–440 (2017). https://doi.org/10.1097/RLI.0000000000000358
40. Esteva, A., Kuprel, B., Novoa, R.A., et al.: Dermatologist-level classification of skin cancer with deep neural networks. Nature **542**, 115–118 (2017). https://doi.org/10.1038/nature21056
41. Gulshan, V., Peng, L., Coram, M., et al.: Development and validation of a deep learning algorithm for detection of diabetic retinopathy in retinal fundus photographs. JAMA **316**, 2402–2410 (2016). https://doi.org/10.1001/jama.2016.17216
42. Agarwal, V.: Complete architectural details of all EfficientNet models. In: Medium (2020). https://towardsdatascience.com/complete-architectural-details-of-all-efficientnet-models-5fd5b736142. Accessed 21 Aug 2022
43. Tsang, S.-H.: Review: NASNet — neural architecture search network (image classification). In: Medium (2021). https://sh-tsang.medium.com/review-nasnet-neural-architecture-search-network-image-classification-23139ea0425d. Accessed 21 Aug 2022
44. Jeffcoat, M.K.: Current concepts in periodontal disease testing. J. Am. Dent. Assoc. (1939) **125**(8), 1071–1078 (1994). https://doi.org/10.14219/jada.archive.1994.0136
45. Reddy, M.S.: Radiographic methods in the evaluation of periodontal therapy. J. Periodontol. **63**, 1078–1084 (1992). https://doi.org/10.1902/jop.1992.63.12s.1078
46. de Faria Vasconcelos, K., Evangelista, K., Rodrigues, C., et al.: Detection of periodontal bone loss using cone beam CT and intraoral radiography. Dentomaxillofacial Radiol. **41**, 64–69 (2012). https://doi.org/10.1259/dmfr/13676777

Application of Artificial Intelligence in the Supply Chain: A Systematic Literature Review

Mohamed Kriouich[1], Hicham Sarir[2], and Oussama Mahboub[1](✉)

[1] Advanced Science and Technology Team, ENSA Tetouan, Abdelmalek Esaadi University, Tetouan, Morocco
mohamed.kriouich@etu.uae.ac.ma, omahboub@uae.ac.ma
[2] Information System and Software Engineering, ENSA Tetouan, Abdelmalek Esaadi University, Tetouan, Morocco
hsarrir@uae.ac.ma

Abstract. This research presents a systematic literature review of using artificial intelligence (AI) in the supply chain (SC). We examine the works done so far in relation to AI research, which still needs to be investigated in light of the potential benefits of implementing AI in supply chain management. Dynamic supply chain processes require artificial intelligence techniques capable of dealing with their increasing complexity. Artificial intelligence techniques such as machine learning, Deep learning, natural language processing, expert systems, knowledge representation, Fuzzy logic and models, social intelligence, Artificial neural networks, Support vector machines, Gaussian models, robotics, Ant colony, and computer vision can transform and optimize the supply chain. In order to classify the study material, we looked at 50 research papers from the Scopus database that were published between 2002 and 2022 and assigned each one a grade based on two fundamental structural factors are as follows: AI techniques and sector of application.

Keywords: Artificial Intelligence · Supply chain · Systematic literature review · Taxonomy of artificial intelligence

1 Introduction

In an era where risks and epidemics such as Covid-19 are increasing, and in addition to disasters that have occasionally swayed, this has made today's supply chains very different from just a few years ago, and they continue to evolve especially in the last two years [1]. SCM is defined as The management of a network of relationships within a company and between interdependent organizations and business units made up of material suppliers, purchasing, production facilities, logistics, marketing, and related sys- -tems that enable the forward and reverse flow of goods, services, money, and information from the original producer to the final consumer with the advantages of adding value, maximizing profitability through efficiency, and achieving customer satisfaction [2].

M. Lazaar et al. (Eds.): BDIoT 2022, LNNS 625, pp. 388–401, 2023.
https://doi.org/10.1007/978-3-031-28387-1_33

Uncertainty about demand, increased competitiveness, as well as higher supply risks, compelled their optimization. The goal of research in this period is to explore and improve ways to manage information and data with a view to making timely decisions. One heavily adopted method is AI techniques because they clearly affect SC in all its fields (e.g. Inventory, Logistics, transportation), AI applications are highly present in previous research [3]. AI has seen a wider spread in academic discourse and has contributed to many areas: Like commercial research, artificial intelligence has become present in more comprehensive research [4–6]. SC is an area that benefits heavily from artificial intelligence techniques, given the interest of practitioners and researchers by this area [7–9]; Nevertheless, SC still needs more and more research and studies to make more use of the applications of artificial intelligence, with many researchers stating the need for these studies [10, 11]. Our research presents a recent study of papers published in recent years involving the application of artificial intelligence techniques in the supply chain. Through conducting a systematic review, we address the knowledge gap and offer a vision for AI applications in the supply chain, by answering the following research questions:

- Q1: What Areas for SC are a fertile field of AI applications?
- Q2: What taxonomy of artificial intelligence techniques exist in this study?

The rest of this article is divided into the following sections: The methodology adopted in Sect. 2, the results and analysis Sect. 3, a discussion is presented in Sect. 4, and a conclusion is provided in Sect. 5.

2 Methods

We used a defined approach for data collection, selecting analytical tools, and responding to research questions based on the goals of the study. We undertook a thorough literature review to achieve this goal (SLR). Reviews of narrative literature that are not systematic tend to be biased and frequently lack rigor, whereas systematic reviews employ a systematic technique to synthesize the evidence on research topics with an extensive and thorough plan of study [1, 12, 13].

2.1 Material Collection

The data for this research were collected using Scopus database. Although the two databases Scopus and the Web of Science are the most used, the Scopus database covers 60% more than the Web of Science Database [14]. The Scopus database has been demonstrated to cover a variety of scientific, technological, medical, artistic, and social science topics in past research comparing literature reviews across several literature databases [15]. To ensure high quality, Scopus offers comprehensive data about each document, along with full profiles of the author and their institution. We used diaries and essays that were only authored in English for this investigation. In order to create a thorough and accurate systematic assessment of the literature, we downloaded a database and used the terms «Artificial Intelligence» and «Supply Chain ", which were chosen with great precision. These phrases allow us to narrow the scope of the search. We located 1573 papers using a preliminary search of Scopus database. Following that, we carefully sorted the papers, keeping only those that included material of the highest caliber and were pertinent to the topic. After then, we stopped searching for contributions such as book chapters, conference papers, editorials, conference proceedings, book series, books, and trade journals. Only publications with the designation "ABS rating 3 journals" were chosen since this level contains highly cited publications that produce well-structured original studies and adhere to strict criteria of scientific accuracy. Additionally, it is the research field's most trustworthy setup that supports our study's objectives. According to [16] it is also the most widely adopted ranking in research [1]. Because of their publication of research with a low standard of quality and rigor, journals with an "ABS rating 2 journal" and an "ABS ranking 1 journal" were avoided (Fig. 1).

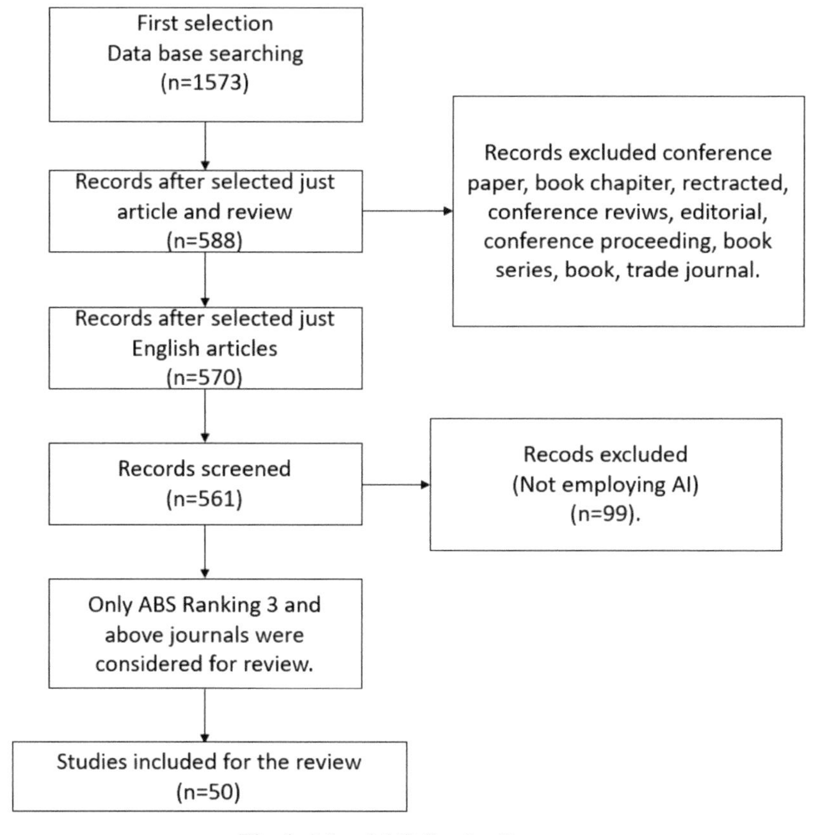

Fig. 1. Material Collection Process.

2.2 Review Classification Framework

In the literature, many review classification frameworks have been used to extract the data required. In this work, we used two basic structural dimensions: area of application, and artificial intelligence techniques.

AI Techniques

During this study, we evaluated all the selected papers, the aim being to extract all the artificial intelligence techniques used by each author in his research, as we were able to extract all the techniques found in the papers. This process has contributed to our increased understanding in terms of the extent to which AI is involved in solving supply chain problems. This data can be utilized to create a relationship between particular problem categories and the algorithms that are most effective for solving them.

Field of Application

We categorized the articles according to the matter of the sector where AI was used to solve the problems of SC. The field of the application shows both the sectors or areas that are adopting AI for supply chains and those that still don't have a clear idea of how AI will fit into their supply chain. It also allowed us to identify which industries were still a long way from transforming their supply chains with AI and, as a result, required more focus and study to help them do so. This factor can aid businesses in comprehending the significance of and possibilities for supply chain transformation by AI.

3 Results and Classification

After the literature review, we got these results: (i) AI techniques species utilized in the field of SC, (ii) Applications of AI in many sectors of industry (Appendix 1).

3.1 AI Techniques Employed in the Field of SC

Based on the selected papers, 30 AI algorithms or techniques are highlighted. The first technique used is Genetics Algorithm (15 papers), followed by Particle swarm intelligence (3 papers). The other techniques were used in two or one paper. In this study GA technique dominated in AI application; however other papers used new AI techniques, like Fuzzy wavelet neural networks, C4.5 algorithm, Kalman filter, Clustering, and Graph Neural Networks. AI techniques are applied in many supply-chain fields such as logistics, production, and others (Table 1).

Table 1. Technique/algorithms

Technique/algorithms	Study	Total
Genetic Algorithms	[1–14]	15
Particle swarm intelligence	[15–17]	3

(continued)

Table 1. (*continued*)

Technique/algorithms	Study	Total
SMART algorithm	[18, 19]	2
Nearest neighbor	[20, 21]	2
Artificial immune system algorithm	[16, 22]	2
Artificial Neural networks	[10, 23]	2
Neural networks	[24]	1
Hybrid Technique	[25]	1
Ant colony optimization	[26]	1
Graph Neural Networks	[27]	1
Fuzzy Wavelet Neural Networks	[20]	1
Dimensionality Reduction	[28]	1
C4.5 algorithm	[19]	1
Kalman filter	[17]	1
Clustering	[18]	1
Support vector machine	[29]	1
Bayesian networks	[23]	1
Particle swarm Optimization	[17]	1
Regression trees	[30]	1
Random forest	[30]	1
Logistic regression	[29]	1
Decision tree	[31]	1
Bioinspired algorithms	[25]	1
Multi-objective artificial bee colony	[32]	1
Artificial WD	[33]	1
Intelligent optimization algorithms	[34]	1
Association rule mining	[35]	1
Adaptive tabu search (ATS) algorithm	[36]	1
Intelligent agent	[37]	1
Artificial bee colony	[2, 15]	1

3.2 AI Applications Across the Different Industries

The use of artificial intelligence in the supply chain has led to its development and improvement. AI techniques have been applied in many industries. Our review shows which industries have taken an interest in artificial intelligence to become more predictive. AI techniques have improved these industries by improving decision-making. AI technologies enhance decision support by being proactive and predictive. Many industrial fields, such as the automotive, food, manufacturing, and remanufacturing ones were some of the sectors where AI techniques have been applied. AI techniques have also been used in other industries such as the healthcare industry, retail industry, oil industry, green supply chain, and aerospace (Appendix 1).

4 Discussion

We conducted a combined systematic literature review for this study in an effort to analyze the research activity from 2002 to 2022 relevant to AI applications in SC. Additionally, a detailed classification of the papers is provided in this study. These components worked well together to provide answers to our study questions.

Through our paper review, our analysis revealed that AI is applied in various sub-areas of supply chains, such as demand forecasting, supplier selection, supply chain network design, supply chain risk management, inventory management, as well as other areas [1]. The diversity of artificial intelligence techniques provides for the development of sub-fields, by improving the decision-making process by analyzing data, forecasting, and prediction.

4.1 Q1: What Areas of SC are Fertile Fields for AI Applications?

Application of AI in Demand Forecasting
In recent years, SCM has had several problems forecasting demand, such as distorting demand. The treatment of this problem under SCM has made it increasingly deformed. The use of AI techniques has been shown to produce significant results. For instance, the inventory replenishment system and prediction model reduce inventory levels while raising customer service standards. As a result, inventory performance improved [59]. In this field, machine learning was the most AI technique. The use of AI in the demand forecasting process results in fewer product shortages, fewer excess stocks, and thus less product loss, which contributes to waste reduction and environmental preservation.

Application of AI in Management and Resilience
Supply chain is exposed to risk due to its complexity and global nature of operations. To protect strong supply chain systems, preventive measures are required. Artificial intelligence techniques are increasingly being used to reduce SC risk, resulting in better supply chain management. To avoid danger SCRM can be tracked in either an interactive or proactive strategy, where the former is applied after the risk occurs and the latter identifies and evaluates risks before they occur. Because delayed risk responses have been shown to be highly harmful, researchers are focusing more on proactive strategies. Proactive strategies are characterized by being able to accurately anticipate risks. This proactive predictive ability can be obtained thanks to artificial intelligence [39].

Application of AI in Transportation
Several issues plague transportation network design, including vehicle steering and

scheduling issues, charging standardization issues, multimodal communication issues, and other issues such as road network design, traffic, and parking space utilization.

Application of AI in Supplier Selection

Recent research confirms that artificial intelligence techniques outperform traditional methods of delivering supplier performance and identifying the best suppliers. Assessing and selecting suppliers is a complex process involving several standards. [60] used a genetic algorithm to develop a new smart model for evaluating and prioritizing supplier performance. Proposing an intelligent approach to forecast categorization suppliers' effectiveness in the cosmetics sector [61]. A framework for handling supplier selection that employs mysterious logic and neural networks [62]. This technique has been used successfully in supplier evaluation and selection.

Application of AI in Inventory Management

Among the many responsibilities of inventory management are forecasting the volume of orders, assuring product availability, and guaranteeing timely responsiveness. This can be adjusted by artificial intelligence systems, which automatically track sales and store inventory data in real-time, allowing us to track demand and avoid overstocking or reduction. ANN, which is known for its ability to handle data more accurately, is one of the most commonly used technologies in this field. ANN technique is used to explore the feasibility and importance of ADP algorithms to improve inventory decisions [63]. Because random multifactory and multi-product inventories are so complicated, there aren't many research papers that can help producers make better judgments. Producing actual proof that the vendor-managed inventory supply chain might experience considerable cost gains from the use of the genetic algorithm-based Decision Support System [19]. This technique has demonstrated its superiority over more traditional methods.

Application of AI in Purchasing

For businesses to maintain a competitive advantage, they must predict and respond to market developments [59]. Busch (2016) highlights collaborative procurement platforms to take advantage of speed, creativity, innovation, and intelligence [60]. AI helps buyers tackle complex problems that were previously unsolvable. Organizations must consider the future role of procurement given the favorable effects of digitization on procurement [60]. In addition to assisting the procurement manager in a variety of strategic and tactical procurement decisions, an agent-based procurement system can take the position of a human decision-maker [18].

4.2 Q2: What Taxonomy of Artificial Intelligence Techniques Exist in this Study?

We extracted many AI techniques from the literature, when we investigated these techniques, we were able to classify many of them. We discovered that the majority of these technologies are related to ML, which is divided into several techniques and algorithms, which in turn are divided into other algorithms. Algorithms and techniques extracted from articles belonging to ML are as follows: Genetic Algorithms, support vector machines, Clustering, Decision Tree Algorithms, FWNN, Bayesian, Dimensionality Reduction, Association rule mining and Neural Networks. There are three classifications of machine learning: supervised learning, unsupervised learning, and reinforcement learning [52]. Supervised learning includes the following techniques: Decision Tree Algorithms, Regression, Ensemble, support vector machines [53], and Association rule mining. Unsupervised learning involves the following techniques: FWNN [54], Genetic Algorithms and Clustering. Reinforcement learning includes the following techniques: Neural networks, Deep Learning and Bayesian. The C4.5 algorithm is a kind of Decision Tree Algorithms. Bayesian algorithms include Bayesian network techniques. Logistic regression techniques are a kind of regression Algorithms. Random Forest techniques is a kind of Ensemble Algorithm [53], Look Fig. 2.

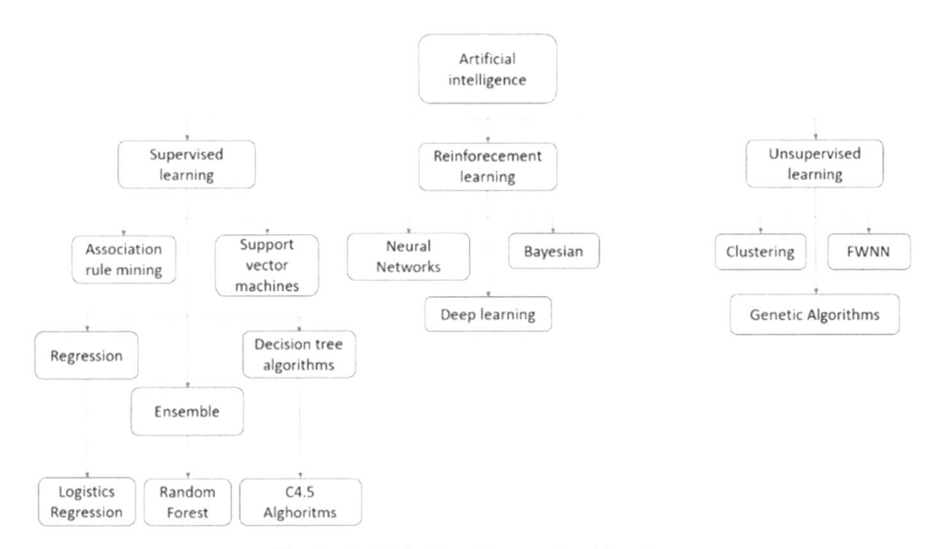

Fig. 2. Artificial Intelligence classification.

5 Conclusion

This study gives an overview of AI-supply research evolution and tendency. According to our analysis results, the number of publications related to the relationship between AI and supply chain management is in an increase, and more attention should be paid to it. Moreover, this increase also applies to working more in supply chain and supply chain management. The main objective of our contributions is to update the reviews that have been done before in the literature and propose a classification of the actual research articles database. The purpose of this classification is to give an overview of the actual state of literature and to get an understanding of the importance of AI in the supply chain. Furthermore, this study can be a guideline to AI supply chain academics and practitioners in their future works.

Appendix 1

Author(s)	AI Techniques	Field of Application
Tirkolaee et al. (2022)	A hybrid technique	Simulation-based
Eluubek kyzy et al. (2021)	Ant colony optimization	Agricultural supply chain system
Riahi et al. (2021)	AI algorithms review	Review
Kosasih et al. (2021)	Graph Neural Networks	Automotive
Belhadi et al. (2021)	Fuzzy Wavelet Neural Networks	Simulation-based
Hewitt et al. (2020)	Mixed integer linear programming	Simulation-based
Badakhshan et al. (2020)	Genetic algorithms	Simulation-based
Brevik et al. (2020)	Genetic algorithms	Broiler production supply chain
Thürer et al. (2020)	AI algorithms review	Agri-food
Flores and Villalobos (2020)	Dimensionality Reduction	Agriculture
Priore et al. (2019)	C4.5 algorithm	Wholesaling
Baryannis et al. (2019)	SVM/ANN/Bayesian networks	Review
Sharahi and KhaliliDamghani (2019)	Genetic algorithms	Natural gas industry
Hong et al. (2018)	Particle swarm Optimization	Liquor industry
Villegas and Pedregal (2018)	Kalman filter	Grocery retail
Papagiannidis et al. (2018)	Clustering	Urbanism
Dellino et al. (2018)	Genetic algorithm	Food
Simchi and Wu (2018)	Regression trees/random forest	Online retail
Ma et al. (2018)	Logistic regression/support vector Machine/decision tree	Manufacturing
Wanke et al. (2017)	Genetic algorithm/ANN	Retail
Xu et al. (2017)	AI algorithms review	Simulation-based
Chang et al. (2017)	Genetic algorithm	Food
Zhang et al. (2017)	Bioinspired algorithm	Simulation-based
Kumar et al. (2017)	Artificial immune systems/particle swarm optimization	Manufacturing
Zhang et al. (2016)	Multi-objective artificial bee colony	Simulation-based

(*continued*)

(*continued*)

Author(s)	AI Techniques	Field of Application
Moncayo and Mastrocinque (2016)	Artificial WD	Simulation-based
Dev et al. (2016)	Decision tree	Simulation-based
Nikolopoulos et al. (2016)	Nearest neighbor	Automotive
Zimmer et al. (2016)	AI algorithms review	Review
Borade and Sweeney (2015)	Genetic algorithm	Food manufacturing
Guo et al. (2015)	Intelligent optimization algorithms	Clothing manufacturing
Rodger (2014)	Genetic algorithm	Aerospace industry
Mascle and Gosse (2014)	AI algorithms review	Food manufacturing
Holimchayachotikul et al. (2014)	Particle swarm intelligence	
Manzini et al. (2014)	Nearest neighbor	Distribution
Roozbeh et al. (2014)	Genetic algorithm	Simulation-based
Ting et al. (2014)	Association rule mining	Red wine industry
Miao et al. (2014)	Adaptive tabu search (ATS)	Simulation-based
Latha et al. (2013)	Genetic algorithm	Pump manufacturing
Moon et al. (2012)	Genetic algorithm	Simulation-based
Kadadevaramath et al. (2012)	Particle swarm intelligence	Simulation-based
Marchetta et al. (2012)	Intelligent agent	Automotive
Kumar et al. (2011)	Artificial immune system	Simulation-based
Warren 2010	Ant colony	Simulation-based
Li et al. (2010)	Neural networks	Food
Qu et al. (2010)	Genetic algorithm	Simulation-based
Kumar et al. (2010)	Genetic algorithm/Particle swarm	Simulation-based
O'Donnell et al. (2009)	Genetic algorithm	Simulation-based
Giannoccaro and Pontrandolfo (2002)	SMART algorithm	Simulation-based
Pontrandolfo et al. (2002)	SMART algorithm	Simulation-based

References

1. Riahi, Y., Saikouk, T., Gunasekaran, A., Badraoui, I.: Artificial intelligence applications in supply chain: a descriptive bibliometric analysis and future research directions. Expert Syst. Appl. **173** (2021)

2. Stock, J.R., Boyer, S.L.: Developing a consensus definition of supply chain management: a qualitative study. Int. J. Phys. Distrib. Logist. Manag. **39**, 690–711 (2009). https://doi.org/10.1108/09600030910996323
3. Huin, S.F., Luong, L.H.S., Abhary, K.: Knowledge-based tool for planning of enterprise resources in ASEAN SMEs (2003)
4. Canhoto, A.I., Clear, F.: Artificial intelligence and machine learning as business tools: a framework for diagnosing value destruction potential. Bus. Horiz. **63**, 183–193 (2020). https://doi.org/10.1016/j.bushor.2019.11.003
5. Dirican, C.: The impacts of robotics, artificial intelligence on business and economics. Procedia Soc. Behav. Sci. **195**, 564–573 (2015). https://doi.org/10.1016/j.sbspro.2015.06.134
6. Soni, N., Sharma, E.K., Singh, N., Kapoor, A.: Artificial intelligence in business: from research and innovation to market deployment. Procedia Comput. Sci. **167**, 2200–2210 (2020)
7. Jarrahi, M.H.: Artificial intelligence and the future of work: human-AI symbiosis in organizational decision making. Bus. Horiz. **61**, 577–586 (2018). https://doi.org/10.1016/j.bushor.2018.03.007
8. Kaplan, A., Haenlein, M.: Rulers of the world, unite! the challenges and opportunities of artificial intelligence. Bus. Horiz. **63**, 37–50 (2020). https://doi.org/10.1016/j.bushor.2019.09.003
9. Nishant, R., Kennedy, M., Corbett, J.: Artificial intelligence for sustainability: challenges, opportunities, and a research agenda. Int. J. Inf. Manag. **53**, 102104 (2020). https://doi.org/10.1016/j.ijinfomgt.2020.102104
10. Min, H.: Artificial intelligence in supply chain management: theory and applications. Int. J. Log. Res. Appl. **13**, 13–39 (2010). https://doi.org/10.1080/13675560902736537
11. Dubey, R., et al.: Big data analytics and artificial intelligence pathway to operational performance under the effects of entrepreneurial orientation and environmental dynamism: a study of manufacturing organisations. Int. J. Prod. Econ. **226**, 107599 (2020). https://doi.org/10.1016/j.ijpe.2019.107599
12. Phulwani, P.R., Kumar, D., Goyal, P.: A systematic literature review and bibliometric analysis of recycling behavior. J. Glob. Mark. **33**, 354–376 (2020). https://doi.org/10.1080/08911762.2020.1765444
13. Tawfik, G.M.: A step by step guide for conducting a systematic review and meta-analysis with simulation data. Trop. Med. Health **47** (2019)
14. Zhao, D., Strotmann, A.: Analysis and visualization of citation networks
15. Fahimnia, B., Sarkis, J., Davarzani, H.: Green supply chain management: a review and bibliometric analysis. Int. J. Prod. Econ. **162**, 101–114 (2015)
16. Skjølsvik, T., Pemer, F., Løwendahl, B.R.: Strategic management of professional service firms: reviewing ABS journals and identifying key research themes. J. Prof. Organ. **4**, 203–239 (2017). https://doi.org/10.1093/jpo/jox005
17. Chang, Y., Erera, A.L., White, C.C.: Risk Assessment of deliberate contamination of food production facilities. IEEE Trans. Syst. Man Cybern. Syst. **47**, 381–393 (2017). https://doi.org/10.1109/TSMC.2015.2500822
18. Dellino, G., Laudadio, T., Mari, R., Mastronardi, N., Meloni, C.: A reliable decision support system for fresh food supply chain management. Int. J. Prod. Res. **56**, 1458–1485 (2018). https://doi.org/10.1080/00207543.2017.1367106
19. Borade, A.B., Sweeney, E.: Decision support system for vendor managed inventory supply chain: a case study. Int. J. Prod. Res. **53**, 4789–4818 (2015). https://doi.org/10.1080/00207543.2014.993047
20. Latha Shankar, B., Basavarajappa, S., Kadadevaramath, R.S., Chen, J.C.H.: A bi-objective optimization of supply chain design and distribution operations using non-dominated sorting algorithm: a case study. Expert Syst. Appl. **40**, 5730–5739 (2013). https://doi.org/10.1016/j.eswa.2013.03.047

21. Latha Shankar, B., Basavarajappa, S., Chen, J.C.H., Kadadevaramath, R.S.: Location and allocation decisions for multi-echelon supply chain network - a multi-objective evolutionary approach. Expert Syst. Appl. **40**, 551–562 (2013). https://doi.org/10.1016/j.eswa.2012.07.065
22. Rodger, J.A.: Application of a fuzzy feasibility Bayesian probabilistic estimation of supply chain backorder aging, unfilled backorders, and customer wait time using stochastic simulation with Markov blankets. Expert Syst. Appl. **41**, 7005–7022 (2014). https://doi.org/10.1016/j.eswa.2014.05.012
23. Wanke, P., Alvarenga, H., Correa, H., Hadi-Vencheh, A., Azad, M.A.K.: Fuzzy inference systems and inventory allocation decisions: exploring the impact of priority rules on total costs and service levels. Expert Syst. Appl. **85**, 182–193 (2017). https://doi.org/10.1016/j.eswa.2017.05.043
24. O'Donnell, T., Humphreys, P., McIvor, R., Maguire, L.: Reducing the negative effects of sales promotions in supply chains using genetic algorithms. Expert Syst. Appl. **36**, 7827–7837 (2009). https://doi.org/10.1016/j.eswa.2008.11.034
25. Kumar, S.K., Tiwari, M.K., Babiceanu, R.F.: Minimisation of supply chain cost with embedded risk using computational intelligence approaches. Int. J. Prod. Res. **48**, 3717–3739 (2010). https://doi.org/10.1080/00207540902893425
26. Qu, T., Huang, G.Q., Zhang, Y., Dai, Q.Y.: A generic analytical target cascading optimization system for decentralized supply chain configuration over supply chain grid. Int. J. Prod. Econ. **127**, 262–277 (2010). https://doi.org/10.1016/j.ijpe.2009.08.008
27. Moon, I., Lee, J.H., Seong, J.: Vehicle routing problem with time windows considering overtime and outsourcing vehicles. Expert Syst. Appl. **39**, 13202–13213 (2012). https://doi.org/10.1016/j.eswa.2012.05.081
28. Roozbeh Nia, A., Hemmati Far, M., Akhavan Niaki, S.T.: A fuzzy vendor managed inventory of multi-item economic order quantity model under shortage: an ant colony optimization algorithm. Int. J. Prod. Econ. **155**, 259–271 (2014). https://doi.org/10.1016/j.ijpe.2013.07.017
29. Sharahi, J., Khalili-Damghani, K.: Fuzzy type-II De-Novo programming for resource allocation and target setting in network data envelopment analysis: a natural gas supply chain. Expert Syst. Appl. **117**, 312–329 (2019). https://doi.org/10.1016/j.eswa.2018.09.046
30. Brevik, E., Lauen, A., Rolke, M.C.B., Fagerholt, K., Hansen, J.R.: Optimisation of the broiler production supply chain. Int. J. Prod. Res. **58**, 5218–5237 (2020). https://doi.org/10.1080/00207543.2020.1713415
31. Badakhshan, E., Humphreys, P., Maguire, L., McIvor, R.: Using simulation-based system dynamics and genetic algorithms to reduce the cash flow bullwhip in the supply chain. Int. J. Prod. Res. **58**, 5253–5279 (2020). https://doi.org/10.1080/00207543.2020.1715505
32. Kumar, V.N.S.A., Kumar, V., Brady, M., Garza-Reyes, J.A., Simpson, M.: Resolving forward-reverse logistics multi-period model using evolutionary algorithms. Int. J. Prod. Econ. **183**, 458–469 (2017). https://doi.org/10.1016/j.ijpe.2016.04.026
33. Hong, J., Diabat, A., Panicker, V.V., Rajagopalan, S.: A two-stage supply chain problem with fixed costs: an ant colony optimization approach. Int. J. Prod. Econ. **204**, 214–226 (2018). https://doi.org/10.1016/j.ijpe.2018.07.019
34. Pontrandolfo, P., Gosavi, A., Okogbaa, O.G., Das, T.K.: Global supply chain management: a reinforcement learning approach. Int. J. Prod. Res. **40**, 1299–1317 (2002). https://doi.org/10.1080/00207540110118640
35. Giannoccaro, I., Pontrandolfo, P.: Inventory management in supply chains: a reinforcement learning approach
36. Manzini, R., Accorsi, R., Bortolini, M.: Operational planning models for distribution networks. Int. J. Prod. Res. **52**, 89–116 (2014). https://doi.org/10.1080/00207543.2013.828168

37. Nikolopoulos, K.I., Babai, M.Z., Bozos, K.: Forecasting supply chain sporadic demand with nearest neighbor approaches. Int. J. Prod. Econ. **177**, 139–148 (2016). https://doi.org/10.1016/j.ijpe.2016.04.013
38. Kumar, V., Mishra, N., Chan, F.T.S., Verma, A.: Managing warehousing in an agile supply chain environment: an F-AIS algorithm based approach. Int. J. Prod. Res. **49**, 6407–6426 (2011). https://doi.org/10.1080/00207543.2010.528057
39. Baryannis, G., Validi, S., Dani, S., Antoniou, G.: Supply chain risk management and artificial intelligence: state of the art and future research directions. Int. J. Prod. Res. **57**, 2179–2202 (2019)
40. Li, Y., Kramer, M.R., Beulens, A.J.M., van der Vorst, J.G.A.J.: A framework for early warning and proactive control systems in food supply chain networks. Comput. Ind. **61**, 852–862 (2010). https://doi.org/10.1016/j.compind.2010.07.010
41. Eluubekkyzy, I., Song, H., Vajdi, A., Wang, Y., Zhou, J.: Blockchain for consortium: a practical paradigm in agricultural supply chain system. Expert Syst. Appl. **184** (2021). https://doi.org/10.1016/j.eswa.2021.115425
42. Kosasih, E.E., Brintrup, A.: A machine learning approach for predicting hidden links in supply chain with graph neural networks. Int. J. Prod. Res. **60**, 5380–5393 (2021). https://doi.org/10.1080/00207543.2021.1956697
43. Belhadi, A., Kamble, S., Fosso Wamba, S., Queiroz, M.M.: Building supply-chain resilience: an artificial intelligence-based technique and decision-making framework. Int. J. Prod. Res. (2021). https://doi.org/10.1080/00207543.2021.1950935
44. Flores, H., Villalobos, J.R.: A stochastic planning framework for the discovery of complementary, agricultural systems. Eur. J. Oper. Res. **280**, 707–729 (2020). https://doi.org/10.1016/j.ejor.2019.07.053
45. Priore, P., Ponte, B., Rosillo, R., de la Fuente, D.: Applying machine learning to the dynamic selection of replenishment policies in fast-changing supply chain environments. Int. J. Prod. Res. **57**, 3663–3677 (2019). https://doi.org/10.1080/00207543.2018.1552369
46. Villegas, M.A., Pedregal, D.J.: Supply chain decision support systems based on a novel hierarchical forecasting approach. Decis. Support Syst. **114**, 29–36 (2018). https://doi.org/10.1016/j.dss.2018.08.003
47. Papagiannidis, S., See-To, E.W.K., Assimakopoulos, D.G., Yang, Y.: Identifying industrial clusters with a novel big-data methodology: are SIC codes (not) fit for purpose in the Internet age? Comput. Oper. Res. **98**, 355–366 (2018). https://doi.org/10.1016/j.cor.2017.06.010
48. Ma, H., Wang, Y., Wang, K.: Automatic detection of false positive RFID readings using machine learning algorithms. Expert Syst. Appl. **91**, 442–451 (2018). https://doi.org/10.1016/j.eswa.2017.09.021
49. Simchi-Levi, D., Wu, M.X.: Powering retailers' digitization through analytics and automation. Int. J. Prod. Res. **56**, 809–816 (2018). https://doi.org/10.1080/00207543.2017.1404161
50. Dev, N.K., Shankar, R., Gunasekaran, A., Thakur, L.S.: A hybrid adaptive decision system for supply chain reconfiguration. Int. J. Prod. Res. **54**, 7100–7114 (2016). https://doi.org/10.1080/00207543.2015.1134842
51. Zhang, X., et al.: An intelligent physarum solver for supply chain network design under profit maximization and oligopolistic competition. Int. J. Prod. Res. **55**, 244–263 (2017). https://doi.org/10.1080/00207543.2016.1203075
52. Zhang, S., Lee, C.K.M., Wu, K., Choy, K.L.: Multi-objective optimization for sustainable supply chain network design considering multiple distribution channels. Expert Syst. Appl. **65**, 87–99 (2016). https://doi.org/10.1016/j.eswa.2016.08.037
53. Moncayo-Martínez, L.A., Mastrocinque, E.: A multi-objective intelligent water drop algorithm to minimise cost of goods sold and time to market in logistics networks. Expert Syst. Appl. **64**, 455–466 (2016). https://doi.org/10.1016/j.eswa.2016.08.003

54. Guo, Z.X., Ngai, E.W.T., Yang, C., Liang, X.: An RFID-based intelligent decision support system architecture for production monitoring and scheduling in a distributed manufacturing environment. Int. J. Prod. Econ. **159**, 16–28 (2015). https://doi.org/10.1016/j.ijpe.2014.09.004
55. Ting, S.L., Tse, Y.K., Ho, G.T.S., Chung, S.H., Pang, G.: Mining logistics data to assure the quality in a sustainable food supply chain: a case in the red wine industry. Int. J. Prod. Econ. **152**, 200–209 (2014). https://doi.org/10.1016/j.ijpe.2013.12.010
56. Miao, Z., Cai, S., Xu, D.: Applying an adaptive tabu search algorithm to optimize truck-dock assignment in the crossdock management system. Expert Syst. Appl. **41**, 16–22 (2014). https://doi.org/10.1016/j.eswa.2013.07.007
57. Marchetta, M.G., Mayer, F., Forradellas, R.Q.: A reference framework following a proactive approach for Product Lifecycle Management. Comput. Ind. **62**, 672–683 (2011). https://doi.org/10.1016/j.compind.2011.04.004
58. Xu, J., Ding, C.: A class of chance constrained multiobjective linear programming with birandom coefficients and its application to vendors selection. Int. J. Prod. Econ. **131**, 709–720 (2011). https://doi.org/10.1016/j.ijpe.2011.02.020
59. Bala, P.K.: Improving inventory performance with clustering based demand forecasts. J. Model. Manag. **7**, 23–37 (2012). https://doi.org/10.1108/17465661211208794
60. Fallahpour, A., Wong, K.Y., Olugu, E.U., Musa, S.N.: A predictive integrated genetic-based model for supplier evaluation and selection. Int. J. Fuzzy Syst. **19**(4), 1041–1057 (2017). https://doi.org/10.1007/s40815-017-0324-z
61. Vahdani, B., Iranmanesh, S.H., Mousavi, S.M., Abdollahzade, M.: A locally linear neuro-fuzzy model for supplier selection in cosmetics industry. Appl. Math. Model. **36**, 4714–4727 (2012). https://doi.org/10.1016/j.apm.2011.12.006
62. Lau, H.C.W., Hui, I.K., Chan, F.T.S., Wong, C.W.Y.: Expert Systems (2002)
63. Çimen, M., Kirkbride, C.: Approximate dynamic programming algorithms for multidimensional flexible production-inventory problems. Int. J. Prod. Res. **55**, 2034–2050 (2017). https://doi.org/10.1080/00207543.2016.1264643
64. Majeed Alneamy, J.S., Hameed Alnaish, Z.A., Mohd Hashim, S.Z., Hamed Alnaish, R.A.: Utilizing hybrid functional fuzzy wavelet neural networks with a teaching learning-based optimization algorithm for medical disease diagnosis. Comput. Biol. Med. **112** (2019). https://doi.org/10.1016/j.compbiomed.2019.103348

Innovative Practices and the Contribution of Digital Technology to the University

Rabha Kissani[1](✉) and Abdelghani Es-Sarghini[2]

[1] Ibn Tofail University, ESEF, Kenitra, Morocco
rabha.kissani@uit.ac.ma
[2] Mohamed V University, Rabat, Morocco
abdelghani.essarghini@um5r.ac.ma

Abstract. This article questions the innovative pedagogical practices and the contribution of the digital technologies during the "pedagogical continuity" phase. It aims to clarify certain aspects of remote learning at the Ibn Tofail university. This non programmed learning method aims to find an effective response to the urgent situation related to the spread of the pandemic of "Covid-19".

To collect the necessary data, we used two types of online questionnaires. The first one was for professors in order to examine the way to manage remote courses (adaptation of courses - satisfaction - achievement of learning objectives - student participation). The second questionnaire was addressed to students, which was interesting not only to raise the satisfaction criteria and the constraints encountered but also to relieve their capacity to learn independently.

Based on research work related to the role of digital technology in the act of learning, mainly remote learning, this study has made it possible to identify relevant information that would promote the improvement of courses via the Moodle platform. It highlighted the effectiveness of the digital resources used to ensure pedagogical continuity and in particular the ability of students to overcome the physical absence of the professor and to learn independently.

Keywords: E-learning · Pedagogical Innovation · Autonomy in Learning · Platforms · Innovative Action

1 Introduction

The Covid-19 pandemic is undoubtedly the first truly global crisis of the 21st century, as no one was prepared for it: all sectors of life were affected, including the teaching and education sectors. So, we had to improvise, act, continue working, teach, and monitor students, but how?

The magnitude of the coronavirus pandemic has pushed educational decision-makers worldwide to opt for distance learning. We point out that several institutions, mainly universities, already had platforms; we mention the case of the Ibn Tofail University of Kenitra, where we conducted a survey on professors and students to study the innovative pedagogical practices adopted to ensure pedagogical continuity and to see the

M. Lazaar et al. (Eds.): BDIoT 2022, LNNS 625, pp. 402–413, 2023.
https://doi.org/10.1007/978-3-031-28387-1_34

involvement of students in the process of their autonomous learning accompanied by their tutors.

Professors sometimes have had to take the initiative to keep in touch with their students via the social networks (Facebook, WhatsApp), or the institutional platforms already designed but less exploited. In this context, Ibn Tofail University, as well as several other universities are experimenting distance learning to ensure "pedagogical continuity" via the Moodle platform. Indeed, to innovate and refine the quality of its educational system, the Ibn Tofail University has, since 2007, a digital working environment. The Moodle platform (Modular Object-Oriented Dynamic Learning Environment) that offers various learning activities such as video conferences, forums, discussions (...). However, despite the implementation of this digital space and its educational tools, professors were mostly providing face-to-face courses except some initiatives of introducing students to distance learning practices via Moodle. So, how do professors and students use this digital space? What are their perceptions of this new form of learning? How do they conceive the act of remote learning? And How do distance or hybrid synchronous or asynchronous teaching forms promote student's autonomy?

In this article, we shed the lights on innovative practices at Ibn Tofail University in the digital era and the coronavirus pandemic in order to see the contribution of e-learning during this health crisis. Firstly, we will recall the theoretical background of this research work. Then, we will check the contribution of the Moodle platform to innovative teaching practices. In addition to that, we will analyze the evolution of the postures of both professors and students during remote learning. And we will discuss the results of this research work that will allow us to answer our main problematic question.

2 Theoretical Framework

Technological advance has now reached all sectors of human life. The emergence of Internet and new devices such as laptops, and tablets in our daily life has generated new uses; even the educational sector has not escaped the trend. Recently, the use of information and communication technologies in education has become a common objective for several countries: western countries have already started the digital transition of their educational systems, particularly regarding distance learning. Indeed, educational staff understood the necessity to adapt to this new digital environment, and also to adjust to a new type of learner: a modern learner, who studies, works and communicates through new technologies. Therefore, the use of these new techno in different disciplines is at the heart of the debate: between absolute rejection and a total approval. Their backgrounds are completely different leading to various opinions about the modalities of introducing distance learning. In this context, several theoretical and empirical studies have been carried out exposing the contribution of these technological tools for the learning of languages [1]. The aim is to respond to multiple learning difficulties from primary school to university. Other research works (Peraya, 1996; Peraya & Campion, 2008), have tried to provide an appropriate response to these different opinions by showing the contributions of ICTE (Information and Communication Technologies for Education) in the context of language learning. These tools can transform pedagogy by providing professors and learners with new learning environments that can meet the requirements. In this matter, several platforms and websites offer learning opportunities through courses, activities, as well as complementary pedagogical support tools [2, 3].

According to (Boéchat-Heer, 2011; CTIE, 2006, 2007; Karsenti & Larose, 2005), the integration of information and communication technologies (ICT) into teaching practice is a complex process that requires, in addition to pedagogical and didactic skills in ICT, the adoption of innovative postures inherent to the professional situations experienced [4]. The report of the OECD Education Committee (1998) cited in Karsenti, Savoie-Zajc, and Larose (2000) states that "it is not enough to graft the use of computer tools onto existing pedagogies; it is preferable to adapt teaching to the new possibilities and advantages that are available" (p. 91) [5]. Similarly, Perrenoud (1998) and Peraya (2002), quoted in Karsenti, Peraya, and Viens (2002), emphasize that: The effort to integrate ICTE would only be of interest in so far as the technologies allow either the instructor to improve his or her pedagogy, or the learner to establish a better relationship with knowledge. The integration of ICTE is thus the ideal opportunity to rethink pedagogy, and the conception of the school, both from the point of view of teaching and learning [6]. In fact, during distance learning (DL), professors enter a pedagogical relationship with their students at a distance: they use work strategies and techniques (homework, questioning, self-assessment, collaborative work) to "tame the distance", according to the expression of Jacquinot [7].

A context that encourages pedagogical innovation (modes of learning, evaluation, support, and reinforcement of learning…), research conducted in this framework, among others (Perrenoud, 2003), (Béchard and Pelletier, 2001), see that this action can promote certain changes at the pedagogical level as long as it seeks to improve learning substantially [8]. Thus, innovation is conditioned by values that determine the action is implemented: it is a question of adherence to values that the actors must share which is linked to worldviews and broader strategies. It is connected to risk and uncertainty and generates fears among the actors. So, it's a question of managing uncertainty. Therefore, change in practices must be seen as a process and not as a simple political or administrative decision. A change is accomplished in stages where actors develop the necessary knowledge to set up new practices by implementing and adapting already known methods (project-based and problem-based learning, reverse classes, collaborative research, digital tools,…). Indeed, it is not enough to introduce a digital tool in a classroom and claim innovation. It must be perceived as a change in the link of the three poles of the pedagogical triangle including attitudes, connections, practices, evaluation devices and so on, with a view to success and efficiency, and considering existing values.

3 Methodology

In this research, we focus on the description of the variables that are involved in remote learning at Ibn Tofail University during Covid-19. We presented and discussed the percentages obtained for each of their modalities. Two types of online questionnaires (made using Google Forms online application) were used. The first questionnaire was for professors at Ibn Tofail University examining the management of remote courses (course adaptation – satisfaction – achievement of learning objectives – student participation). This questionnaire was sent to 100 professors using the Moodle platform via their institutional emails, we received 90 responses. The target in the second questionnaire was students who used Moodle. Among 220 students, we received 121 responses. This questionnaire aims to identify student's opinion towards the use of Moodle (Initiation to the use of Moodle, the different tools that were used, the most or least relevant tools, the satisfaction criteria, and the constraints encountered as well as their ability to learn independently. The questionnaire was anonymous; we explained that the answers are interesting if they provide important indications regarding the use of Moodle during "Covid-19" pandemic and that it would verify the role of full remote learning in ensuring educational continuity while recognizing the role of the two protagonists of the pedagogical act "the professors" and the "students". We drew inspiration from various research works and more particularly from the works carried out by (Albero, 2003, p. 12; Nissen, 2007, p. 142) [19]. In addition, we specify that the nature of the scope of this survey allowed us to analyze both quantitatively and qualitatively, which gives more interest to this work. Data analysis was conducted using Google Forms.

4 Results

4.1 Moodle Platform: Usability and Functionality

One of the major challenges of education at university in the digital age and the coronavirus pandemic "Covid-19" is the suggestion of a pedagogical model appropriate to this situation. Starting an innovative approach to its educational system, Ibn Tofail University had already designed its digital space, offering several categories of tools. For example, communication tools allow exchanging and collaborating between different users in asynchronous and synchronous modes (Chat). There are also collaborative production tools (homework submissions, feedbacks,). However, this platform was only used on a large scale when all the university's departments were required to move from face-to-face to distance learning, this is what we noticed when we analyzed our survey results: 65.5% of the professors who responded to the questionnaire said that they had never adopted distance learning before the health crisis of "Covid-19" (Fig. 1).

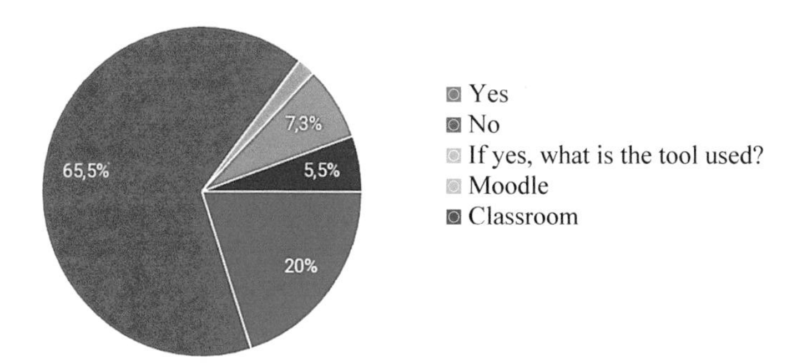

Fig. 1. Frequency of DL (Distance learning) before the health crisis related to the spread of the "Covid-19".

This statement raises questions in the university educational sphere, given that the Moodle platform was designed in 2007. However, we note a positive involvement of both students and professors. Even though 41.5% of professors said that they didn't receive any training and they managed on their own, they were actually able to monitor students. We will come back to this point later, but first we'll see how students were able to learn remotely and in what way?

According to survey results, most students prefer learning via Moodle, with a value of 7.8% (Fig. 2).

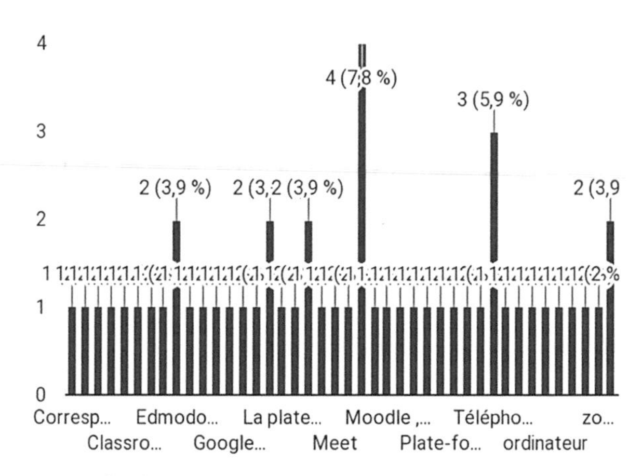

Fig. 2. Methods used for DL (distance learning).

According to the comments drawn from a few examples of the 90 responses received regarding the use of Moodle, a set of factors contributed to a relevant use of this digital space (Initiation to the use of Moodle as soon as they register at the university, ownership of an institutional account, help provided by the professor and by the class leader, availability of a video illustrating the steps to follow to connect to Moodle, and familiarity with social networks…). Here are the responses and extracts representing most students (see Fig. 3).

Student 1: *I have been using it since 1st semester.*
Student 2: *The professor showed us how to access our institutional account.*
Student 3: *Everything is clear and easy especially when the person is familiar with social networks.*
Student 4: *The steps are easy, and I accessed my institutional account without any problems.*
Student 5: *It's very simple.*
Student 6: *Just follow the steps.*
Student 7: *I just followed the steps suggested in the video and was able to access the platform easily.*
Student 8: *I clicked Moodle and I put in French studies and then I put in my account, and I logged in.*
Student 9: *The class leader helped us.*
Student 10: *Because it is accessed regularly.*
Student 11: *It's easy to access because as soon as you go to the faculty website, you see the Moodle icon, and then everything becomes clear.*

Fig. 3. Extracts from students' responses on the use of Moodle.

In addition to that, the appropriation of technological tools by students has made it easier for them to connect to their institutional space and to follow the courses given by professors remotely [10]. 47.9% of the students questioned said they had already studied by distance. They also connect to the Internet via their laptop (52.1%). Regarding activities most presented by professors in the distance courses, homework submission ranks first with a percentage of 70%, followed by video conferencing (37%), and chat with a rate of 24.1% (see Fig. 4).

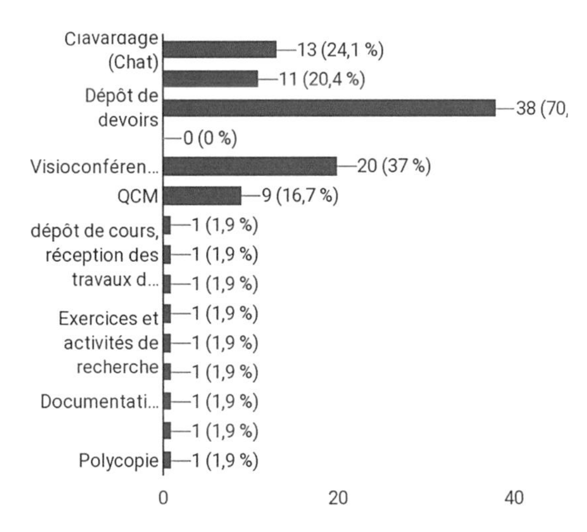

Fig. 4. Activities used in distance learning courses

These different tools and digital supports of the Moodle platform allow activities' diversification in each training which brings a didactic diversity. The chat as a tool for exchange and interaction with professors and between learners [8] is among the three most used activities in distance learning with a rate of 24.1%. Besides, the homework tool is widely used by professors, recording a rate of 70%. The production can be written or oral. It is evaluated by professors and followed by written feedback to students. In

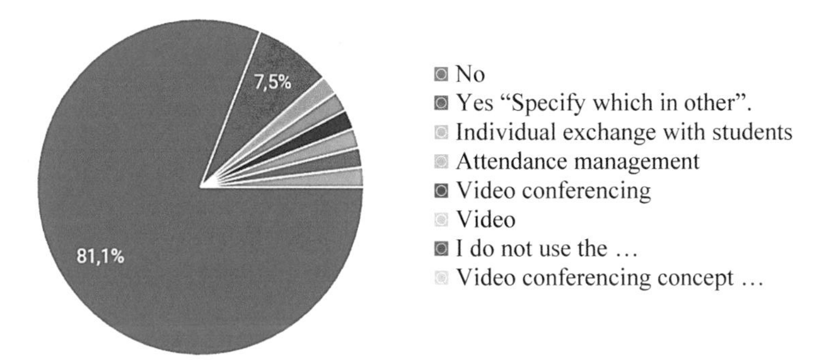

Fig. 5. Professors' satisfaction with activities presented by Moodle.

addition to that, 81.1% of surveyed professors were not satisfied with the digital resources and activities available in Moodle (see Fig. 5).

This digital work environment was applied during the suspension of face-to-face courses, following the coronavirus pandemic. According to professors, it is a space that offers various digital supports (texts, images, audios, and videos) to students.

4.2 Moodle for Distance Learning Continuity

Even though Internet and new technologies has become unavoidable in student's daily life. It changed their personal habits and their involvement in online learning. Although several platforms have been designed, particularly Moodle at Ibn Tofail University, to respond to classes' overcrowding and to present to students a solution adequate to their needs (digital supports). Distance learning has not been fully integrated into the Moroccan educational system. However, the result of our survey highlights encouraging rates (see Fig. 6) of students' adherence to distance learning during the period of "Covid-19" (Table 1).

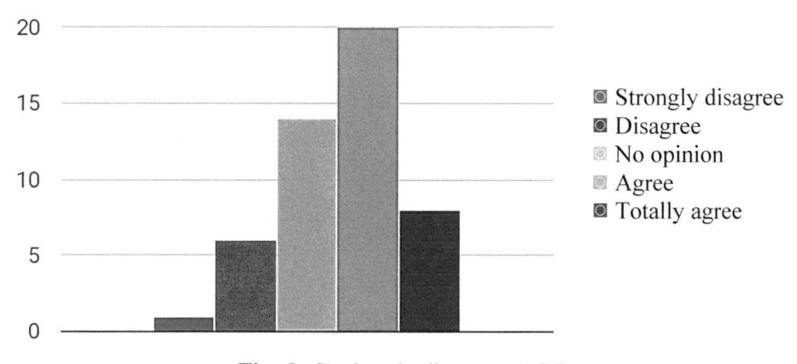

Fig. 6. Students' adherence to DL.

Table. 1. Professor's opinion on student adherence to DL.

Professors' opinions on student adherence to DL	Strongly disagree	Disagree	No opinion	Agree	Totally agree
The value recorded out of 20	1	6	14	20	8

In addition to that, students were able to take benefits of the online courses although this mode of teaching is marked by the separation in time and space of the learner and the tutor.

Let us remember that the application of distance learning has prevented students from missing a year and has pushed political and pedagogical decision-makers to take a close interest about it as a response to mass education. A recent study in France, carried out by the CNAM (Centre National des Arts et Métiers), underlines the importance of this teaching/learning method. It allows, on one hand, to reduce trips for both students and professors in order to gain time particularly if they live far from university. And, on the other hand, to manage classes at the time level that became more and more crowded. However, it should be noted that DL requires, in addition to the appropriation of technological tools, the ability of students to overcome the physical absence of the professor and to be able to learn independently.

4.3 Autonomy in Distance Learning

The acquisition of skills that promote autonomy in learning is an issue that rises both in the classroom and online. In the case of physical absence of the professor, taking a course partially or entirely remotely requires a great deal of student's autonomy but also the expertise of professors in terms of design, production, dissemination, supervision, evaluation, and reflexivity. However, the emergency related to "Covid-19" turned the act of teaching and training towards distance learning in an unprepared way at the macro or micro level which may impact the experience of distance learning for students [13].

The survey results of professors and students showed that their appropriation of digital tools and familiarity with working on Moodle before the lockdown made the transition to distance learning less difficult for them. Thus, most professors agreed that distance learning was able to develop students' autonomy (see Fig. 7).

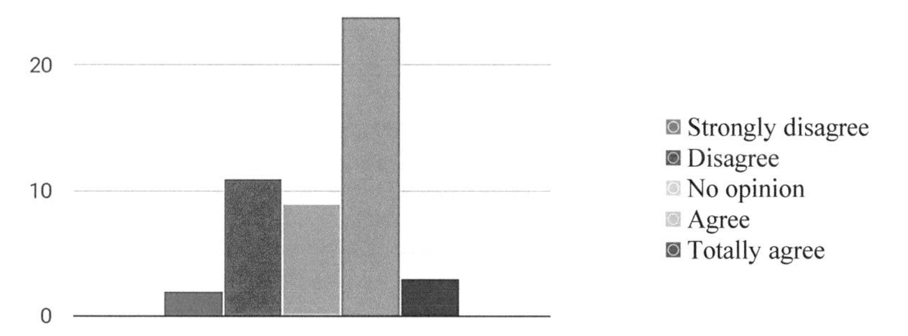

Fig. 7. Professors' views on student autonomy development.

Indeed, one of the major challenges of distance education is to prepare students to work independently. Therefore, it is important to identify their learner profiles on a cognitive and socio-emotional level, anticipate the possible difficulties, and provide them with resources to make them more autonomous and perseverant. The significant outcome of this field survey is the increased participation of most students in their education and training by initiating research parallel to various courses they took online (Fig. 8).

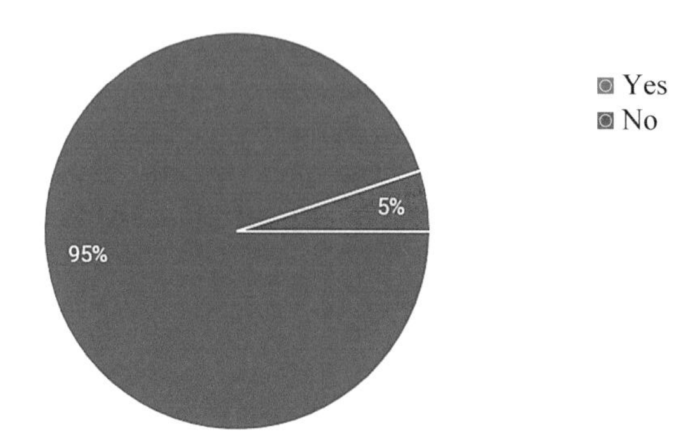

Fig. 8. Students' participation in their education and training.

In this context, students take part of the methods implemented to achieve the learning activities by planning their time, choosing adequate digital supports, and especially maintaining their motivation. It is a question of (see Fig. 9) growing their knowledge, completing their know-how to understand a course by using multiple supports (video sequences, files, images, etc.) so that they can build themself their knowledge [18].

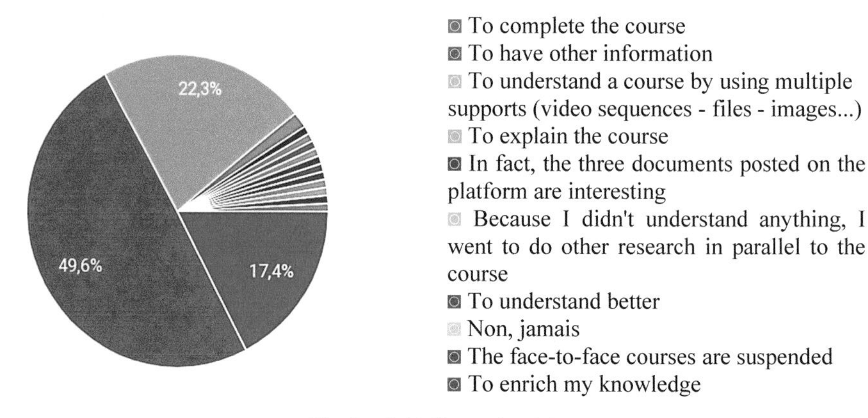

Fig. 9. Self-directed training.

5 Discussion

According to our study, student's adherence to distance learning has ensured pedagogical continuity as well as the supervision of students' work including final projects and assignments. This result confirms the findings of Jacquinot (1993) about the accessibility [7] of this mode of teaching/learning, characterized by synchronous (chat) and asynchronous (PDF, Word documents, videos, etc.) teaching/learning situations, which has solved the spatial and temporal constraints that have blocked access to face-to-face classes. Indeed, students appreciated the access to this digital work environment. And so, these results align with (Raby, Karsenti, Meunier & Villeneuve, 2011), especially to access information (grades, schedules, course documents, forum,...) [9]. Thus, DL made it easier for them to learn instantly, facilitating the integration of scientific knowledge (Pépin, 1994) and the transfer of knowledge [12]. Our study also confirms also that appropriation is a crucial phase in the use and integration of platforms and ICT in general [11]. Furthermore, educational leaders constrained to move from face-to-face to distance learning, noticed that students were autonomous and could follow courses and build their disciplinary skills without difficulty. Although the concept of autonomy is still debated from a theoretical and practical perspective (Garrison & Baynton, 1987; Hostler, 1986), it remains central to the philosophy of adult education and a key concept in distance learning [14]. According to (Moore, 1997, p. 22): "...an autonomous learner can formulate his or her learning need in terms of goals and specific objectives and set, more or less explicitly, criteria for achievement. In this process, he or she gathers the information he or she wants, practices the skills, works to solve the problem, and achieves the desired goals..." [15]. Indeed, the professors' survey results align with Blin (2010) and Deschênes (1991), who consider that autonomous students should not only learn but also gradually take control of their learning process, from preparation to evaluation through execution [16, 17]. Therefore, this research study shed the lights on the importance of remote learning in the improvement of students' autonomy.

6 Conclusion

This study raises the urgency to review the role of distance learning in student's education, particularly in the context of the Moroccan university. It is a question for educational decision-makers to recognize the place of a pedagogical paradigm of empowerment as a solution to DL. This research also shed light on the role of institutional platforms in maintaining distance learning courses when the physical presence of professors and students is suspended due to the spread of "Covid-19".

As far as pedagogical continuity is concerned, it has been found that distance learning, although programmed without rigorous prior planning, has allowed professors to guarantee the training and follow-up of their students. Its implementation seems to be limited by the academic deadlines of an institutional context based largely on face-to-face teaching. On the other hand, the apparition of MOOCs (Massive Open Online Courses) and their success in terms of flexibility, especially during a health crisis, has led educational institutions to question the offer to be made to learners in the third millennium.

The results of this research work show that professors could monitor their students through Moodle platform even though they received no training and managed on their own. Most professors used the platform even if they were unsatisfied with its digital activities. Submitting written or oral assignments were their most used activity, in addition to video conferencing and chat. Our results also show that the appreciation of the digital environment and the appropriation of the technological tools by the students allowed them to learn easily by distance using the Moodle platform. Thus, the significant student adherence to distance learning during the period of containment related to the spread of "Covid-19" ensured "pedagogical continuity". Furthermore, our research highlights that distance learning could develop students' autonomy, meaning that students could take control of their motivation and learning process by using the different resources offered by their professors on the platform.

References

1. Mangenot, F.: Réseau Internet et apprentissage du français. Études Linguist. Appliquée Rev. Didactologie Lang.-Cult. **110**, 205–214 (1998). https://edutice.archives-ouvertes.fr/edutice-00000230/document
2. Peraya, D.: Educational mediated communication, distance learning, and communication technologies. J. Res. Educ. Media India Counc. Res. Educ. Media ICREM **III**(2), 11–24 (1996). https://tecfa.unige.ch/tecfa/publicat/peraya-papers/icrem/india.html
3. Peraya, D., Campion, B.: Introduction d'un changement d'environnement virtuel de travail dans un cours de second cycle: contribution à l'étude des dispositifs hybrides. Rev. Int. Technol. En Pédagogie Univ. Int. J. Technol. High. Educ. **5**(1), 29–44 (2008). https://dial.uclouvain.be/downloader/downloader.php?pid=boreal:69371&datastream=PDF_01
4. Karsenti, T., Larose, F.: Les TIC au cœur des pédagogies universitaires. Presses de l'Université du Québec, Canada (2001). https://extranet.puq.ca/media/produits/documents/53_9782760516717.pdf
5. Komis, V., Depover, C., Karsenti, T., Tselios, N., Filippidi, A.: Comprendre l'usage des plateformes d'enseignement et les outils Web 2.0 dans des contextes universitaires de formation hybride: aspects méthodologiques. Form. Prof. **21**(2), 48–64 (2013). https://doi.org/10.18162/fp.2013.34
6. Peraya, D., Viens, J., Karsenti, T.: Introduction: formation des enseignants à l'intégration pédagogique des TIC: Esquisse historique des fondements, des recherches et des pratiques. Rev. Sci. L'éducation **28**(2), 243–264 (2002). https://doi.org/10.7202/007353ar
7. Jacquinot, G.: Apprivoiser la distance et supprimer l'absence ? ou les défis de la formation à distance. Rev. Fr. Pédagogie 55–67 (1993). https://www.persee.fr/doc/rfp_0556-7807_1993_num_102_1_1305
8. Soughati, N., Kissani, R.: L'enseignement à distance au département de Langue et de Littérature Françaises, Le clavardage, un outil d'apprentissage synchrone médiatisé et multidimensionnel. En Ligne
9. Raby, C., Karsenti, T., Meunier, H., Villeneuve, S.: Usage des TIC en pédagogie universitaire: point de vue des étudiants. Revue internationale des technologies en pédagogie universitaire. Int. J. Technol. High. Educ. **8**(3), 6–19 (2011). https://doi.org/10.7202/1006396ar
10. Proulx, S.: Usages de l'Internet: la «pensée réseaux» et l'appropriation d'une culture numérique. In: Guichard, D.E. (ed.) Comprendre les usages de l'Internet, Paris, France, pp. 139–145 (2001). http://sergeproulx.uqam.ca
11. Jouët, J.: Retour critique sur la sociologie des usages. Réseaux **18**(100), 487–521 (2000). https://doi.org/10.3406/reso.2000.2235
12. Pépin, Y.: Savoirs pratiques et savoirs scolaires: une représentation constructiviste de l'éducation. Rev. Sci. L'éducation **20**(1), 63–85 (1994). https://doi.org/10.7202/031701ar
13. La conception de cours à distance. Le tableau, vol. 2, no. 1 (2013). https://pedagogie.uquebec.ca/sites/default/files/documents/numeros-tableau/letableau-v2-n1-2013.pdf
14. Garrison, D.R., Baynton, M.: Beyond independence in distance education: the concept of control. Am. J. Distance Educ. **3**, 3–15 (1987). https://doi.org/10.1080/08923648709526593
15. Moore, M.: Theory of transactional distance. In: Keegan, D. (ed.) Theoretical Principles of Distance Education, pp. 22–38. Routledge, New York (1997). http://www.c3l.uni-oldenburg.de/cde/found/moore93.pdf

16. Blin, F.: Concevoir des cybertâches pour l'autonomie de l'apprenant: vers une activité théorique modèle pédagogique (2010)
17. Deschênes, A.-J.: Autonomie et enseignement à distance. Rev. Can. Pour Létude Léducation Adultes **5**(1), 32–54 (1991). https://cjsae.library.dal.ca/index.php/cjsae/article/download/2295/2007
18. Peterson, P.L.: Making learning meaningful: lessons from research on cognition and instruction. Educ. Psychol. **23**(4), 365–373 (1988). https://doi.org/10.1207/s15326985ep2304_4
19. Albero, B.: Self-training in open and distance learning systems: instrumenting the development of autonomy in learning. In: Saleh, I., Lepage, D., Bouyahi, S. (eds.) ICT at the Heart of Higher Education. Proceedings of the Study Day of 12 November 2002, pp. 139–159 (2003)

PKN: A Hybrid Intrusion Detection System Installed on Fog Architecture

Ait Moulay Rachid[1](✉) and El Ghazi Abdellatif[2]

[1] Algebra and Functional Analysis Group, Mohamed V University, Rabat, Morocco
rachid_aitmoulay@um5.ac.ma

[2] TIC Lab, Information and Communication Technology Laboratory, Rabat, Morocco
abdellatif.elghazi@uir.ac.ma

Abstract. The Internet of things (IoT) is a technology aiming to connect sensors and actuators to the Internet by sharing and transferring data transparently and seamlessly without human intervention. While the usage of IoT increases every day, its protection against the insider and outsider threats becomes challenging using conventional security measures. This work suggests a hybrid intrusion detection installed on the fog architecture to give near real-time responses to attacks with minimum resource consumption. This Ids is composed of Convolutional Neural Network (CNN) for feature extraction, PCA and LDA to minimize the features created by the CNN. Lastly, we will employ KNN for multiclass classification to categorize attacks and give insights to users about the ongoing attacks on IoT networks.

Keywords: Intrusion detection · Cloud Computing · Machine Learning · Fog Computing

1 Introduction

Today they are over 14 billion Internet of Things (IoT) devices available and ready for use by customers, ranging from home appliances, baby monitors, routers, and smartwatches, transmitting a large amount of data and communicating with each other with minimal human intervention [1, 2]. These connected devices can evaluate data and adjust its behaviour by analysing information using statistics and applying machine learning (ML), Artificial Intelligence (AI), as well Deep Learning (DL) algorithms to produce desirable results in near real-time [3–5]. IoT devices lack processing power, memory, and storage. They are usually cheap and market-oriented with little care for security, making them vulnerable to attackers eager to exploit network vulnerabilities to cripple, spy, or use them as an entry for attacks like Distributed Denial of service (DDOS) or Malware. Furthermore, as the number of IoT devices available on the market is exponentially expanding with an annual growth of 28.7% between 2018 and 2025, protection and security become more urgent with each passing day as they lack identity and device management, authentication, authorisation, and efficient protection mechanism [6, 7].

M. Lazaar et al. (Eds.): BDIoT 2022, LNNS 625, pp. 414–427, 2023.
https://doi.org/10.1007/978-3-031-28387-1_35

To provide an efficient and real-time experience to IoT device users, Cisco has coined a new architecture called Fog Computing, an addition and a replacement for cloud computing that provides data processing and computing power close to the devices creating or collecting the information. This architecture addresses network congestion issues and latency-sensitive service problems by decreasing communication between IoT devices and cloud servers and reducing latency. After being processed by fog devices like routers, switches, microcontrollers, or raspberry pies, heavy analytics and data storage are performed on the cloud servers later in the day.

By placing ML and AI close to sensors and actuators, fog computing can help thwart large-scale attacks on the IoT environment [8]. In addition to architecture, multiple security measures should be applied to develop security mechanisms against IoT attacks. IoT security strategy must aim to defend three pillars that constitute the functioning core of connected devices and their applications [7] as follows:

- Confidentiality: is essential to guarantee data privacy and protect against unauthorised parties' eavesdropping. It is important to ensure that the data is readable only by the proposed destination.
- Integrity: ensures that the data has been received without modification or alteration. It guarantees the information has not been tampered with on its way to remote systems and data centers from heterogeneous networks where IoT devices like sensors and actuators are installed.
- Availability: is a driving key factor for IoT devices to guarantee the availability of collected information and applications to the users or other IoT devices. It must provide safe data access without any delay and at any place and time.

Ids Applied methodologies allow us to separate it into three categories a: anomaly detection, b: signature detection, c: hybdrid detection.

Anomaly-based detection is a system that learns from a model of normal network traffic by applying algorithms and heuristics to find network anomalies. It analyzes network traffic to distinguish between normal and network traffic deviation (attack). It can identify exploits that have never been seen before (zero-day attacks), however, it frequently produces false positives and negatives. Signature based detection is a system that It examines patterns of known vulnerabilities and attack signatures. It consistently maintains its collection of knowledge regarding hackers' latest attack techniques. One of their strengths is that they don't produce a lot of false positives and negatives, but their main shortcoming is that they can't identify zero-day attacks that haven't been taught yet in the learning phase. Hybrid detection it is an effective system that combines anomaly and signature Ids features. This combination of two intrusion detection systems enables the identification of zero-day attacks and ensures a greater detection rate with minimal false positives and negatives [9–12].

To resolve the many challenges above, we propose a Pkn Ids that operates on the fog layer, using a Raspberry Pie 4, it grants advanced computing and processing power to protect IoT devices, also provides near real-time response and lower latency which improves IoT security when facing large scale attacks. Pkn Ids will be trained offline on a cloud machine to ease the training, especially because of the usage of CNN that can take time to extract important features, while we employ PCA and LDA to accelerate the training of the KNN model. The PKN Ids will update its knowledge base regularly from the cloud server to avoid having an obsolete database and to detect threats rapidly before it spreads in the network, and to identify the right attack category to help decision maker take adequate and immediate action to safeguard the network.

Our goal is to create a fast response with low processing time Ids using CNN for key attribute identification [26], PCA and LDA for dimensionality reduction to ease processing and minimize resource consumption, and the K-Nearest Neighbour algorithm (KNN) for multiclass classification. We test this pipeline on an imbalanced dataset like the NSL KDD to confirm the model's generalisation and adaptability to new attacks.

1.1 Motivation

The proposed PKN Ids will adhere to the following conditions:

- The hybdird Ids will apply KNN algorithm to detect with reasonable accuracy common attacks.
- PKN Ids will have a high packet processing rate to avoid large packet queue waiting to be processed by the Ids.
- We will use PCA and LDA algorithms for feature selection, to choose the best attributes for the PKN Ids.
- We will train a multiclass model using NSL-KDD dataset.

The rest of the paper is organized as follows Sect. 2 presents the recent works related to Ids, Sect. 3 details the composition of the proposed architecture and algorithms, Sect. 4 Experimentations and Sect. 5 will conclude our paper.

2 Related Works

Kitsune Ids is a Plug and Play and efficient online Network Intrusion Detection System (NIDS) that watches for network abnormalities by monitoring traffic and applying statistics using an ensemble of auto-encoders that operates with a small memory footprint and a low computational complexity.

The NIDS is composed of 4 modules. The first is a packet capturer that intercepts traffic passing through the internal and external networks, and the second is a packet parser that gets meta information from the raw packets captured in the previous module using tools like tshark or capture++.

The third module is a feature extractor that extracts n features from arrived packets to create instances that the anomaly detector will use. A fourth module, a feature mapper is used to create vector v with a maximum number of features m from instances x with n features ($n \geq m$) by calculating a function f using agglomerative hierarchical clustering on incrementally summary data. Then the last and final module is the anomaly detector that uses an ensemble of auto-encoders by calculating the squared error of the vector v to compute the sensitivity s of abnormality in the training mode, which will be used after in the execution mode.

The Kitsune NIDS is well-performed when tested on a raspberry pie, it can process over 1000 packets per second. The TPR (true positive rates) for MIRAI, SSDP flood, and fuzzing attacks attained 99%. Still, they didn't reach the same results for other attacks like MITM and syn floods and Os scanning with a receiver operating characteristics (AUC) value of 58%, 94%, and 95% [13].

Another Ids called Bootstrap Your Own Latent (BYOL), a self-supervised intrusion detection system, is trained on unlabeled data to find helpful information and significant features in network traffic. To choose features that would improve model accuracy and eliminate those that would penalize the BYOL model, they added BOTNet with a multi-head self-attention technique. Additionally, they used data augmentation strategies to raise the training data quality and increase model generalization to derive valuable representations.

By performing the first operation V [horizontal flip, vertical flip, random crop], and a second one is designated as V' by applying [vertical flip, random shuffle (fisher Yates shuffle), random crop]. Then, to improve the feature representation's ability to distinguish between data perturbations resulting from the information augmentation strategy, they employed V as an input for the target network and V' as an input for an online network in the BYOL model.

They added a modified search harmony algorithm (MHSA) and a self-attention mechanism to summarize helpful information in the low resolution and abstract features to increase the model's accuracy.

The proposed model attained excellent results close to the conventional supervised models, with an accuracy of 92.67%, 99.25%, 96.7%, and 97.55% for NSL KDD, KDD CUP 99, CIC IDS2017, and CIDDS_001 respectively [14].

A lightweight Ids for edge devices using SVM was proposed to detect only DOS attacks. They used the transmission rate of the packet field to train their model because they remarked this attribute was increasing or decreasing depending on the attack's type or the execution stage (exfiltration of data, malware update).

From this attribute, they derived three features composed of mean, maximum, and median to avoid under-fitting. They also created multiple feature sets to test the performance of the lightweight SVM with three different kernels: Linear kernel, Polynomials, and radial basis function. They conducted numerous experiments to choose better parameters for the SVM based Ids, and they have found out that the lightweight linear SVM is much better than the other two kernels. Then they compared the performance of this Ids using accuracy and CPU time with different lightweight algorithms like a Genetical based SVM [15] and A-IDS [16], and wfs-IDS [17]. Their Ids outperformed the other three algorithms in accuracy and CPU time, which confirmed its lightweight property because the CPU time is less than the mentioned three algorithms; thus, it will not consume more energy and resources for an accuracy of 98.3%. The problem with their Ids is that it has not been used on other attacks like remote to local or unauthorized access on the edge node, which limited their Ids to detect only DOS, so it can't be generalized to other attack types [18].

An adaptive Ids using Artificial Neural Network (ANN) is implemented to protect fog nodes and enhance IoT security. Depending on the threat level, it can measure threats and self-protect against attacks by closing connections or asking for authentication. They created a risk management unit that utilises the output calculated by the Artificial Neural Network model to evaluate the threat activity level by verifying the interval of output τ. This unit can monitor logs to check the authenticity and the periodicity of the actions raised to the risk management unit to measure its threat levels. They trained three models depending on the resources they were trying to monitor. These resources include memory availability, buffer consumption and CPU usage.

The architecture of ANN is composed of 10 neurons in a hidden layer, an activation function using a symmetric sigmoid function, two delays unit and one linear output function. They used the levenberg Marquardt back propagation algorithm for the algorithm enhancement, which efficiently showed its ability to distinguish between normal and abnormal activity. The Framework can protect against DOS, flooding with accuracy that can reach 97% and precision of 98.4% and recall of 98.9%, with little overhead to the fog node, which can be categorized as lightweight because it did not stress the fog node resources [19].

The authors [20] used the DNN-KNN algorithm operating on the fog to maximise the detection rate, which implemented a binary classification model. In the first step, they used DNN to classify events into malignant or non-malignant, it is composed of one input layer and two hidden layers, and each of them has the same number of neurons. In the hidden Layer, they used hyperbolic tangent as an activation function, for the output layer, they implemented two neurons, one neuron for the malignant activity and the other one for the normal behaviour, they used the softmax activation function in the Output Layer. If one of the neurons does not achieve a defined limit to conclude the classification of the activity as normal or malicious, the suspicious activity will be sent to a feature reduction module implementing the Information Gain algorithm, which selects the best features for classification. These features are redirected to the KNN algorithm to be finally classified. The result from the k nearest neighbour is considered final. The DNN-KNN algorithm showed an accuracy rate of 99.77% and a recall rate of 99.76% for the NSL-KDD dataset. When used on the CICDS2017 dataset, it showed a higher accuracy and recall rate, attaining 99.85% and 99.87% compared to other implementations using the same datasets.

The authors suggested a generic lightweight intrusion detection system using fuzzy logic and Fog based approach called GLSF IOT, adhering to the principle of not trusting network traffic and running with minimal memory and processing overhead. This Ids is divided into three layers edge device layer, fog layer and cloud layer to achieve near real-time processing and a high detection rate of attacks.

The authors proposed a trust management mechanism installed on the cloud to deal with behavior irregularities in the edge and fog layers to identify authentic fog nodes that can execute anomaly detection to discern insider and outsider threats.

GLSF IOT executes a fuzzy logic algorithm to figure out the most reliable fog nodes by grouping them into two categories trusted and untrusted ones and running three types of fuzzy logic algorithms to calculate the confidence threshold of fog and edge nodes. The first parameter is rank consistency, where a fuzzy logic algorithm computes the deviation sum value of each fog and edge node by an anomaly detector. The second parameter is the energy consumption analysis of edge devices. The last criterion is the change in control packet where trusted fog nodes verify specific packet characteristics coming from edge nodes, where the fuzzy rule compares normal node traffic attributes with the intercepted packet length and packet send number (because malicious nodes communicate regularly with command-and-control servers) to perceive malicious nodes, who deviated from a calculated threshold limit of average transmitted packet and size. This trust node mechanism regularly updates the list of malicious fog nodes and edge nodes to be excluded from the network to keep only the reliable fog nodes that can scan the network for malicious edge nodes.

The authors tested this Ids on a simulated environment using Contiki hybrid simulator on four attack types like DDOS, BlackHole, Selective Forward, Collusion which attained a high true positive rate of 97.5%, 96%, 94%, and 91% respectively. In addition this Ids is lightweight and can detect zero-day attacks in near real-time by applying a multi-layer security trust mechanism [21].

3 Proposed Method

3.1 Convolution Neural Networks

Our proposed architecture is composed of Convolutional Neural Network (CNN), a deep learning technique inspired by the animal visual cortex composition and is akin to the connectivity pattern of Neurons in the Human Brain. This architecture can learn instantly and seamlessly from low to high-level spatial hierarchy patterns. It can perform image classification, pattern recognition, object recognition, natural language processing, classification, segmentation, and feature extraction [27, 28].

In our case we will use CNN for complex pattern and feature extraction, it will automatically detect abstract representations without manual feature engineering and human intervention. First we will transform NSL-KDD 41 features into 121 after applying one-hot encoding to features like protocol types: UDP, TCP, and ICMP, that are changed to three-dimensional vectors (1, 0, 0), (0, 1, 0), (0, 0, 1), as well as flag type feature that is transformed to nine-dimensional vector and finally 67-dimensional vector for the service type attribute.

The 121 features are converted to 11×11 pixels grayscale image, and then we resize and center the image using zero padding technique to achieve 48×48 pixels image with a centered 11×11 pixels image inside it (NSL KDD).

Our CNN will use 48 × 48 pixels as an input for the first Convolutional layer. We implemented 44 convolution filters, also 5 kernel size, and applied Relu activation function to this convolutional layer. We then employed Batch normalization to speed up learning, converge faster, and improve accuracy [31], afterwards we added max pooling filter to select the feature based on the maximum value of a region to minimize the number of parameters and spatial dimension to avoid overfitting and accelerate learning [28]. This pooling layer is composed of 2 filters and a stride of 2.

We added another convolutional layer with 84 filters, 5 kernel size and Relu activation function, followed by Batch normalization and max pooling layer with 2 filters and a stride of 2. The last convolution layer comprises 124 filters, 3 kernel size and Relu activation function and then max pooling layer like the previous two.

At the end of the CNN, we add three fully connected layers to perform feature extraction of important representation of the NSL KDD dataset to create 625 × 1 vector dimensions, which then is used as an input for the Principal Component Analysis (PCA) for feature reduction.

3.2 Data Set Description

The NSL KDD dataset is an improved version compared to the KDD 99 dataset, it allows to create an efficient Ids that are not biased to the frequent records or duplicates, which can make it incapable of learning uncommon attacks like R2L, U2R. Also it provides a test set that does not contain duplicate records, which allows assessing the performance of the models by eliminating redundant attacks that falsely result in higher detection rates [30]. NSL KDD dataset comprises 4 anomaly types: Dos, Probe, U2R, R2L, and normal behavior.

To train our Pkn Ids we will use KDDTrain+.txt file, which is composed of 125973 records, and to test our solution we will employ KDDTest+.txt file with 22544 rows (18% of the training set) to confirm our result generalizability. Figure 1, 2 show NSL KDD training and test set different label distribution.

3.3 Data Augmentation

Due to the imbalanced characteristics of the NSL KDD dataset, where U2R and R2L are poorly represented in the training set, we use Adaptive Synthetic Sampling Approach (ADASYN) for the imbalanced dataset to generate a varied number of samples depending on an estimate of the local distribution of the class to be oversampled. This method employs a weighted distribution for specific minority class examples based on their learning difficulty to create more synthetic data generated for minority class examples that are more difficult to learn than minority class examples that are simpler to learn [23, 29]. After applying ADASYN oversampling techniques to our NSL KDD dataset, we reached 336654 records. Table 1 shows training dataset repartitions before and after oversampling.

3.4 Data Normalization

To prepare the NSL KDD dataset we will adopt min-max normalization to improve accuracy and converge faster to the global minimum by having equal distribution and limiting the range of different columns without losing essential details and features.

Min-max normalization is defined by:

$$x_{norm} = \frac{x - x_{min}}{x_{max} - x_{min}} \tag{1}$$

where x_{min} is the minimum value of a feature, x_{max} is the maximum value of the same feature, x is the original feature we are trying to rescale, and x_{norm} is the result of the feature scaling (data normalization).

3.5 Principal Component Analysis

Principal Component Analysis (PCA) is an exploratory data analysis technique used to minimize dataset dimensions. Its goal is to increase dataset interpretability and reduce memory usage and processing time.

The algorithm searches for a new dimension that finds new uncorrelated variables minimizing information loss, and augmenting features variance. To reduce data dimension, PCA tries to solve the eigenvalue/eigenvector problem with the following equation:

$$\mathrm{Ax} = \alpha \mathrm{x} \tag{2}$$

where A is the covariance matrix, x is eigenvector unit, and α is the eigenvalue. In our case we will apply PCA to minimize the dimension of the features extracted (625 attributes) from our lightweight CNN [24].

3.6 Linear Discriminant Analysis

Linear Discriminant Analysis (LDA) is used as a dimensionality reduction algorithm. It reduces the number of features from original to G. the algorithm projects the training dataset onto a lower-dimensional vector space to have a new feature space that can achieve maximum separation and discrimination between classes, this new calculated feature will be the new training dataset.

Table 1. NSL KDD over sampling with ADASYN

Dataset	Number of Normal	Number of Attack	Total	Type
Train	67343	58630	125937	Original
Balanced train	67343	269311	336654	Fixed

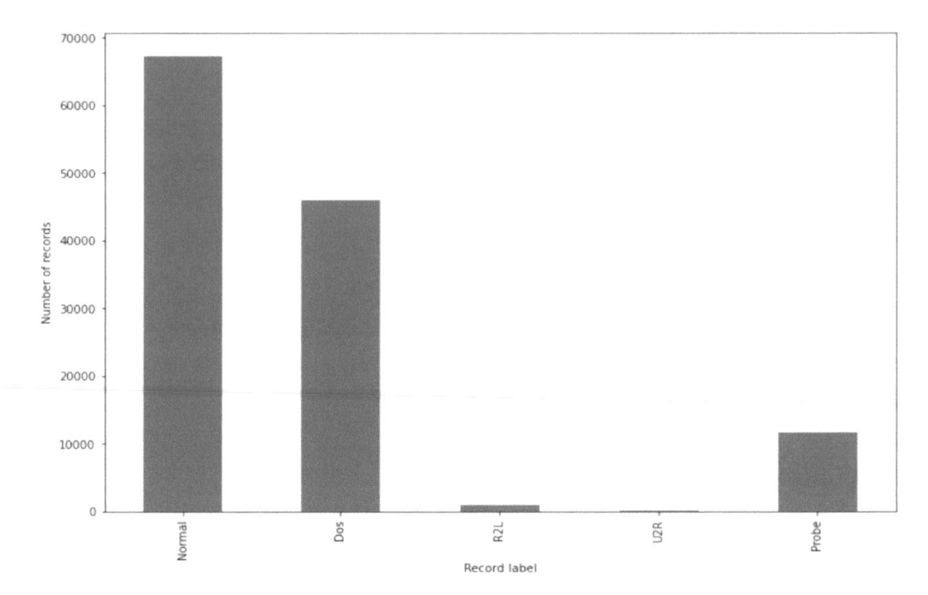

Fig. 1. Number of instances in NSL KDD training set

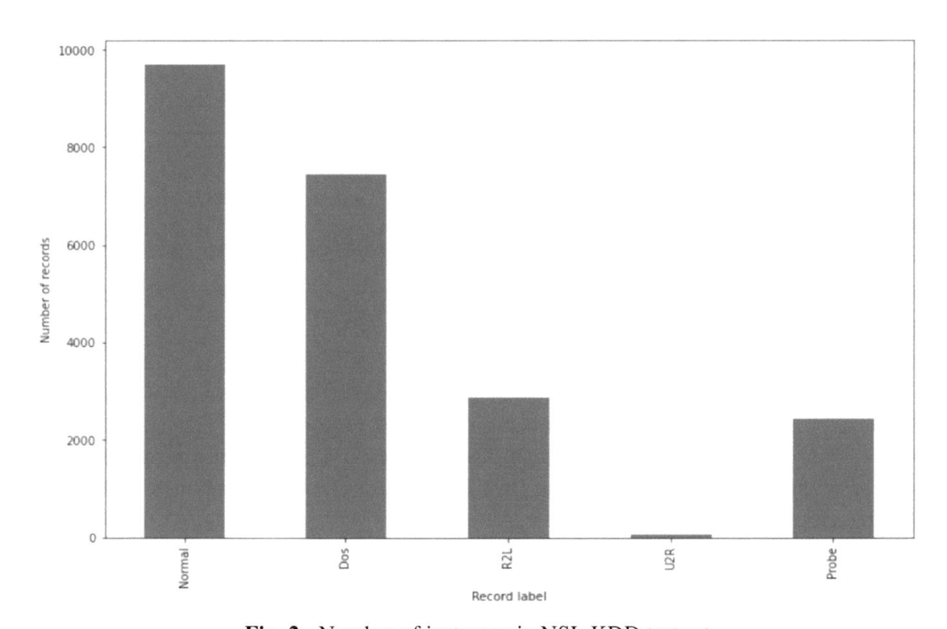

Fig. 2. Number of instances in NSL KDD test set

This algorithm will reduce the features coming from the PCA to create new attributes that will be used by KNN signature-based Ids to compare with its signature to find similarities with the attack database.

To select features and project it to the new projection space, LDA calculate the variance of the projected classes:

$$s_1^2 = \sum\nolimits_{x_i \in C_1} (a_i - u_1)^2 \tag{3}$$

$$s_2^2 = \sum\nolimits_{x_i \in C_1} (a_i - u_2)^2 \tag{4}$$

where s_i^2 is the variance and a_i, u_i are respectively the features and the means. The objective of the LDA algorithm is to find the best features that maximizes the means and minimizes the variance by applying the following formulation:

$$max_{v:||v||=1} \frac{(u_1 - u_2)^2}{s_1^2 + s_2^2} \tag{5}$$

3.7 K-Nearest Neighbour

It is a supervised learning algorithm, it calculates similarity to associate new data points to the existing one if they are similar, it's a straightforward algorithm to implement depending on the value of neighbours k and the distance function used like Manhattan or Euclidian distance. The performance of this algorithm degrades on large datasets because it must compute the distance between a point and each existing one in its database to find similarity (Brute Force KNN), also it does not scale well on high dimensional datasets [24, 25].

We used both PCA and LDA to enhance KNN because after PCA minimizes the dimension of the properties generated by the CNN dataset. LDA will be applied to further reduce the dimension of the attributes produced by PCA, as well to maximize the accuracy and precision rate. To calculate the distance between two points, we will use the Euclidian distance as follows:

$$\text{Euclidean Distance(p, q)} = \sqrt{\sum\nolimits_{i=1}^{n} (p_i - q_i)^2} \tag{6}$$

where p and q are two distinct vectors of n dimensions.

4 Experimentation

To evaluate our Pkn Ids accuracy and precision we will train our model using google colab Pro [32], with 2GHZ CPU, 32 GB memory and NVIDIA P100. The experiments are conducted using Keras package [33], Panda library, Scikits Learn (Sklearn), and Python 3 Language.

We tested our architecture on the NSL KDD dataset and compared it with a simple PCA LDA KNN architecture and a two-step ensemble approach [34].

Our proposed PKN Ids transforms the 121 features of NSL KDD into 625 key attributes, and then these features are reduced to 117 using PCA, after that these properties are transformed and reduced into 4 important key attributes by LDA. After that we employed KNN for multiclass approach. We will opt for K = 9 to classify different attacks on the NSL KDD dataset. Figure 3 shows Pkn Ids processing steps.

Our proposed architecture achieved 75.2% accuracy, showing an increased precision and accuracy when compared to the simple PCA, LDA and KNN, due to the usage of CNN to increase features contributing and ameliorating classification.

The other architecture used for comparison implements PCA to minimize 121 features of the dataset into 100 features, and then we applied LDA to reduce the features from 100 to two. After that we used Knn and choose k = 10 neighbours to train our model to discover normal, and abnormal behaviour.

This architecture attained about 68% of accuracy, it showed low performance on some attacks like R2L and U2R because of the dataset is unbalanced, and these attack types are not well represented in the training dataset. The third approach using two-step ensemble learning achieved 92.7% accuracy with precision rate of 84.62%, 97.29%, 99.91%, 99.93%, and 99.76% of U2R, R2L, Probe, Dos attacks and normal behaviour respectively, while in our case we achieved 38%, 92.19%, 68%, 93.47%, 67.75% of U2R, R2L, Probe, Dos attacks and normal behaviour respectively.

When inspecting the accuracy and precision rate we can see there is a slight improvement when employing CNN about 7.2% enhancement. While we see a large difference when we compare our architecture the ensemble learning architecture, because of the lost details when applying PCA and LDA feature reduction when training our model. But theoretically we can process a large amount of dataset and have near realtime response because we minimized NSL KDD features from 121 to 4 and equivalent of 35.25 times data optimization.

The Fig. 4, 5 shows a confusion matrix of the PKN Ids, and the PCA+LDA+KNN Ids where:

- 0: represents normal behaviour.
- 1: represents DOS attacks.
- 2: represents R2L attacks (Remote to local).
- 3: represents U2R (User to root).
- 4: represents Probe.

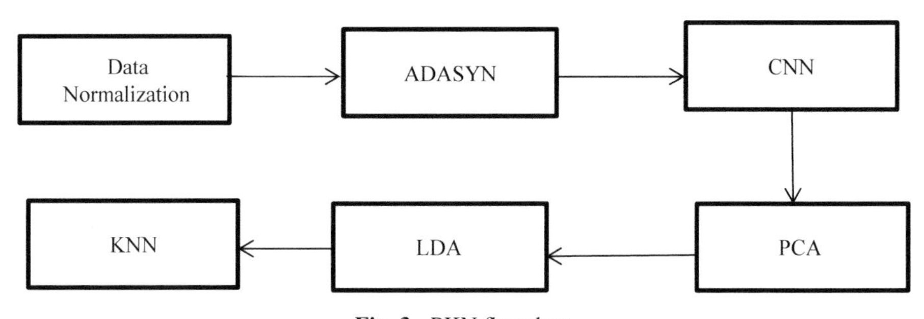

Fig. 3. PKN flowchart

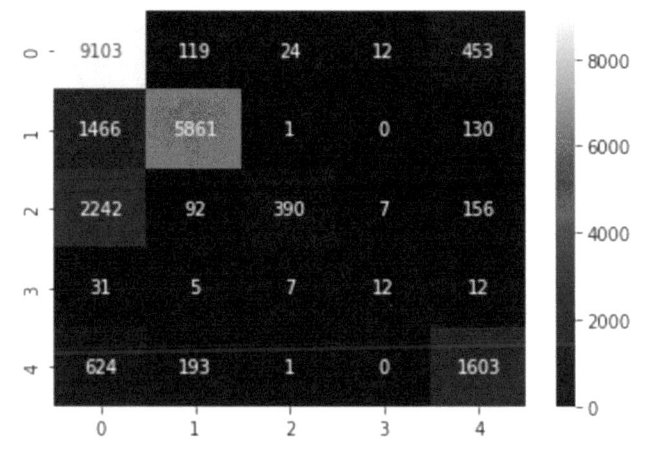

Fig. 4. PKN Confusion Matrix

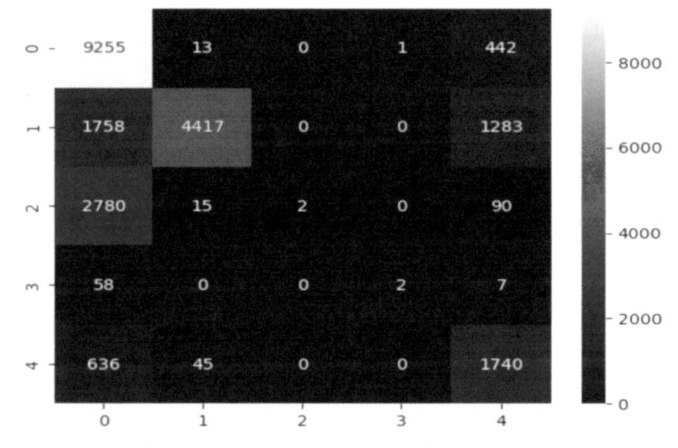

Fig. 5. PCA LDA KNN Confusion Matrix

5 Conclusion

The proposed Pkn Ids was evaluated on the NSL KDD dataset which showed a slight improvement when CNN was involved to extract important features compared to only PCA+LDA+KNN. But it showed low accuracy and precision rate when compared to other state of the art Ids, because of information loss due to employing multiple feature reduction techniques like PCA and LDA.

In the future we plan to test our Pkn Ids on other datasets like CICIDS2017 and install it on raspberry pie to test our solution on real environment to measure energy and resources consumption to confirm its lightweight capability.

References

1. Atzori, L., Iera, A., Morabito, G.: Internet of things a survey. Comput. Netw. **54**(15), 2787–2805 (2010)
2. Biljana, L., Risteska, S., Kire, V.: A review of internet of things for smart home: challenges and solutions. J. Clean. Prod. **140**(3), 1454–1464 (2017)
3. Noura, M., Atiquzzaman, M., Gaedke, M.: Interoperability in internet of things infrastructure: classification, challenges, and future work. In: Lin, Y.-B., Deng, D.-J., You, I., Lin, C.-C. (eds.) IoTaaS 2017. LNICSSITE, vol. 246, pp. 11–18. Springer, Cham (2018). https://doi.org/10.1007/978-3-030-00410-1_2
4. Al-Garadi, M., Amr, M., Al-Ali, A., Du, X., Ali, I., Guizani, M.: A survey of machine and deep learning methods for internet of things (IoT) security. IEEE Commun. Surv. Tutor. **22**(3), 1646–1685 (2020)
5. Hussain, F., Hussain, R., Hassan, S.A., Hossain, E.: Machine learning in IoT security: current solutions and future challenges. IEEE Commun. Surv. Tutor. **22**(3), 1686–1721 (2020)
6. Kolias, C., Kambourakis, G., Stavrou, A., Voas, J.: DDoS in the IoT: mirai and other botnets. Computer **50**, 80–84 (2017)
7. Muhammad, F., Anjum, W., Mazhar, K.S.: A critical analysis on the security concerns of internet of things (IoT). Int. J. Comput. Appl. **111**, 1–6 (2015)
8. Bonomi, F., Milito, R., Zhu, J., Addepalli, S.: Fog computing and its role in the internet of things. In: Proceedings of the First Edition of the MCC Workshop on Mobile Cloud Computing, pp. 13–16. ACM (2012)
9. Granjal, J., Monteiro, E., Silva, J.S.: Security for the internet of things: a survey of existing protocols and open research issues. IEEE Commun. Surv. Tutor. **17**, 1294–1312 (2015)
10. Ammar, M., Russello, G., Crispo, B.: Internet of things: a survey on the security of IoT frameworks. J. Inf. Secur. Appl. **38**, 8–27 (2018)
11. Stellios, I., Kotzanikolaou, P., Psarakis, M., Alcaraz, C., Lopez, J.: A survey of IoT-enabled cyberattacks: assessing attack paths to critical infrastructures and services. IEEE Commun. Surv. Tutor. **20**, 3453–3495 (2018)
12. Buczak, A.L., Guven, E.: A survey of data mining and machine learning methods for cyber security intrusion detection. IEEE Commun. Surv. Tutor. **18**, 1153–1176 (2015)
13. Mirsky, Y., Doitshman, T., Elovici, Y., Shabtai, A.: Kitsune: an ensemble of autoencoders for online network intrusion detection - arxiv.org (2018)
14. Wang, Z., Li, Z., Wang, J., Li, D.: Network intrusion detection model based on improved BYOL self-supervised learning. Secur. Commun. Netw. Hindawi **2021**, 23 (2021)
15. Tao, P., Sun, Z., Sun, Z.: An improved intrusion detection algorithm based on GA and SVM. IEEE Access **6**, 13624–13631 (2018)
16. Aljawarneh, S., Aldwairi, M., Bani Yassein, M.: Anomaly-based intrusion detection system through feature selection analysis and building hybrid efficient model. J. Comput. Sci. **25**, 152–160 (2018)
17. Li, Y., Wang, J.L., Lu, Z.H., Young, C.: Building lightweight intrusion detection system using wrapper-based feature selection mechanisms. Comput. Secur. **28**(6), 466–475 (2009)
18. Jan, S.U., Ahmed, S., Shakhov, V., Koo, I.: Towards a lightweight intrusion detection system for the internet of things. IEEE Access **7**, 42450–42471 (2019)
19. Pacheco, J., Benitez, V.H., Herrán, L.C., Satam, P.: Artificial neural networks-based intrusion detection system for internet of things fog nodes. IEEE Access **8**, 73907–73918 (2020)
20. Antonio de Souzaa, C., Westphall, C.B., Machado, R.B., Sobral, B.M., dos Santos Vieira, G.: Hybrid approach to intrusion detection in fog-based IoT environments. Comput. Netw. (2020)

21. Zahra, S.R., Chishti, M.A.: A generic and lightweight security mechanism for detecting malicious behavior in the uncertain internet of things using fuzzy logic and fog-based approach. Neural Comput. Appl. **34**, 6927–6952 (2021)
22. Su, J., Vargas, D.V., Prasad, S., Sgandurra, D., Feng, Y., Sakurai, K.: Lightweight classification of IoT malware based on image recognition, CoRR (2018)
23. Zhang, X., Ran, J., Mi, J.: An intrusion detection system based on convolutional neural network for imbalanced network traffic. In: IEEE 7th International Conference on Computer Science and Network Technology (ICCSNT) (2019)
24. Doshi, R., Apthorpe, N., Feamster, N.: Machine learning DDoS detection for consumer internet of things devices. In: 2018 IEEE Security and Privacy Workshops (SPW), pp. 29–35 (2018)
25. Deng, Z., Zhu, X., Cheng, D., Zong, M.: Efficient kNN classification algorithm for big data. Neurocomputing **195**, 143–148 (2016)
26. Manjunath, J., Meghana, R.K.: Feature extraction using convolution neural and deep learning. In: 2018 3rd IEEE International Conference on Recent Trends in Electronics, Information & Communication Technology (2018)
27. Jarrett, K., Kavukcuoglu, K., Ranzato, M.A., LeCun, Y.: What is the best multi-stage architecture for object recognition? In: IEEE 12th International Conference on Computer Vision (2009)
28. Krizhevsky, A., Sutskever, I., Hinton, G.E.: ImageNet classification with deep convolutional, neural networks. In: Advances in Neural Information Processing Systems, NeurIPS
29. He, H., Bai, Y., Garcia, E.A., Li, S.: ADASYN: adaptive synthetic sampling approach for imbalanced learning. In: IEEE International Joint Conference on Neural Networks (IEEE World Congress on Computational Intelligence), pp. 1322–1328 (2008)
30. https://www.unb.ca/cic/datasets/nsl.html
31. Ioffe, S., Szegedy, C.: Batch normalization: accelerating deep network training by reducing internal covariate shift. In: Proceedings of the 32nd International Conference on Machine Learning, vol. 37, pp. 448–456. PMLR (2015)
32. https://colab.research.google.com
33. https://keras.io/
34. Antonio de Souza, C., Westphall, C.B., Machado, R.B.: Two-step ensemble approach for intrusion detection and identification in IoT and fog computing environments. Comput. Electr. Eng. (2022)

The Emotional Intelligence of Moroccan Consumers: Educational Level as a Determining Factor

Mouad Ennakra[1(✉)], Moulay Smail Alaoui[1], and El Hassan El – Hassouny[2]

[1] Cognitive Sciences Laboratory, FLSH Dhar El Mahraz, USMBA, Fez, Morocco
ennakramouad@gmail.com

[2] ERIPDS, ENS, Abdelmalek Essaadi University, Tetouan, Morocco

Abstract. Scientific research on emotional intelligence supports that socio-demographic variables are a primary lead to know the degree of emotional intelligence of consumers. These variables can provide information about the intentions, cognitive processes and behavior of subjects in different consumption situations. This article aims to show how the level of education can determine the effect and the degree of emotional intelligence of Moroccan consumers when making a purchase decision.

Keywords: Educational Level · Emotions · Emotional Intelligence · Goleman · Baron

1 Introduction

The concept of emotional intelligence occupies a central place in scientific, academic and entrepreneurial debates. Nowadays, several scientific and non-scientific fields reserve a considerable interest for this concept; it is often at the bottom of the questions raised by the researches which are interested in the development of the performances of the human being and of the company. Beyond its linguistic meaning proposed by some dictionaries, this concept is the point of intersection of several disciplines, of debates and especially one of the keys to understand the factors that participate in the evolution of human skills. In addition to its permanent topicality, emotional intelligence is indispensable today in the debates on business, leadership and decision-making. We hypothesize that the level of education determines the degree of emotional intelligence of consumers and influences their purchasing decisions, and is therefore a determining factor in understanding their behavior before and during the purchase of a given product.

2 Literature Review

For a very long time, the concept of emotional intelligence has been neglected by scientific research. Emotional intelligence was first used by Salovey and Mayer in 1990 to describe the ability of individuals to manage their feelings and those of others, to

M. Lazaar et al. (Eds.): BDIoT 2022, LNNS 625, pp. 428–434, 2023.
https://doi.org/10.1007/978-3-031-28387-1_36

discriminate between different types of emotions, and to use this information to guide thought and action. In their article, Salovey and Mayer define emotional intelligence as a set of skills that are thought to contribute to the accurate assessment and expression of emotion in oneself and others, and the effective regulation of emotion in oneself and others. They clarify the adaptive and maladaptive qualities of emotions by exploring the literature on intelligence, and in particular social intelligence, to examine the place of emotion in traditional conceptions of intelligence. Emotional intelligence is thus a new construct. This paradigm was later developed by Goleman in 1995, who confirmed that it is the characteristics of the personality that constitute emotional intelligence: empathy, the ability to motivate oneself or to persevere in adversity, to control one's impulses and to achieve the satisfaction of one's desires with patience, the ability to maintain an even mood and not to let oneself be dominated by grief to the point of not being able to think anymore, the ability to hope. In the same perspective, Bar-On (1997) defines the concept of emotional intelligence as a set of emotional and social skills that determine the ways in which a person relates to himself and to others, and is able to cope with the pressures and demands of the environment. Emotional intelligence will become a key concept in the history of psychology, as it will be an important step in understanding the relationship between reason and passion. The Greek and Roman Stoics kept emotions away from rational thought, because they were often associated with women, considered therefore as weak and inferior aspects of man. This Stoic vision of the irrational character of the emotions persisted until the XXth century (Daisy Grewal and Peter Salovey, 1999). Emotional intelligence has thus aroused the interest of psychological research and has nevertheless attracted considerable criticism from scientists. For most, it represents any quality that cannot be measured by an IQ test, such as motivation, confidence, optimism, or "good character.

Emotional intelligence therefore plays a major role in the effective stimulation of our behaviors. In other words, it is the psychological skill that promotes success, optimism, hope and productivity. It should be remembered that with the emergence of the concept of emotional intelligence, it was linked to social intelligence (George, Legree, Sternberg and Smith, Wong, Day, Maxwell, and Meara). Emotional intelligence thus has roots in other constructs such as social intelligence which has a relatively long history. However, Mayer, Salovey, Caruso, and George suggest that emotional intelligence is richer than social intelligence because it is more intrinsically focused. For Goleman (1997), emotional competence is a learned intelligence-based ability that translates into outstanding performance at work. It involves emotional capabilities and critical elements such as effective communication and excellent influence, leading others to respond in desired ways. At the heart of this competency are two abilities: empathy, which involves reading the feelings of others, and social skills, which enable one to manage those feelings skillfully. Emotional intelligence determines the potential for learning practical skills based on its five domains: self-awareness, motivation, self-regulation, empathy and relationship management.

In light of these definitions, we believe that the concept of emotional intelligence is a primary way to understand and analyze individual behaviors and performance. It is considered today as an indicator of psychological maturity and as a condition for academic and professional success. In other words, emotional intelligence is a capacity that brings out learning, human relationships and is one of the foundations of personal development. In this spirit, personal sense and social sense are likely to be built together in an individual with a very high degree of emotional intelligence.

3 Procedure and Population

3.1 Sample

In Morocco, purchase decision making is one of the phenomena that still requires extensive sociological and psychological studies to understand the effect of emotions on consumer behavior. Since this study focuses on the analysis of individual variables, it is appropriate to circumscribe as precisely as possible the population from which we selected our sample. The population for this study is Moroccan consumers. We distributed 300 questionnaires in the North of Morocco and obtained a final total of 200 valid responses. The sample obtained has the following characteristics: out of 200 consumers, we have 69 young people, 63 adults and 68 consumers over 50 years old. Regarding the gender and the level of education, the majority of the consumers who answered our questionnaire are women (51.50%) while (48.50%) are men. The most frequent levels of education in the sample are (Bac + 3 and more) with a percentage of (57%) and (Bac + 5 and more) with a value of (33%), as well as 10% of the respondents in our sample say that they have a level of education of (Bac + 1 and more). Regarding the function of these respondents, our sample consists of (52.51%) civil servants, (10.55%) senior managers and (8.54%) accounting assistants. The rest of the population consists of controllers (6.03%), salesmen and planners (4.55%), accountants (3.52%), employees and contractors (3.02%), a digital marqueter and two unemployed people.

3.2 Methodology for Conducting the Research

In order to answer our research problem, the quantitative method seems to be the most appropriate. We then opted for a hypothetical-deductive approach. This involves testing the hypotheses deduced from the literature on a supposedly representative sample. This will allow us to confirm or refute the validity of the hypotheses developed. The present research is based on the articulation of two approaches: exploratory qualitative and confirmatory quantitative by questionnaire (Evrard et al. 1993). As a first step, we were inspired by recent works and literature to extract a number of items in order to carry out our questionnaire. Then, it was via Google forms that we divulged our work and also we tried in some cases to give personally the questionnaire to the consumers to be able to clarify with them some elements that could seem obscure to them.

3.3 The Questionnaire and Variable Measures

In this study, the self-administered questionnaire was our preferred data collection tool and measurement instrument. Respondents were asked to respond to a five-point Likert scale ranging from "strongly disagree" to "strongly agree". For the emotional intelligence variables, the measurement scales used are largely based on the work of Salovey et al. (1995), Schutte et al. (1998) and Wong and Low (2002).

4 Results and Discussion

4.1 Purification of Measures

To better understand the relationship between educational level and the degree of influence of emotional intelligence on the purchase decision making of the surveyed consumers, we analyzed the variance between these variables using the analysis of variance in parametric test (ANOVA) method which allowed us to highlight the existing significant differences. To confirm our hypotheses, we tested the significance of emotional intelligence using a Student's t-test analysis. We also used the Principal Axis Factorization method as a factor extraction method. We eliminated items with a very low contribution or quality of representation.

4.2 Results

We tried to show the influence of education level on the degree of emotional intelligence among the surveyed consumers, especially in decision making. The table below shows the difference in EI according to the consumers' education level (Table 1):

Table 1. Emotional intelligence by level of education.

	Bac + 1 and more		Bac + 3 and more		Bac + 5 and more		Average	
	M	E	M	E	M	E	M	E
Self-awareness	3.88	1.01	4.10	1.06	4.81	1.55	3.74	1.23
Self regulation	3.96	1.24	4.35	1.15	4.67	1.43	3.38	1.11
The motivation	4.08	1.05	4.27	1.33	4.29	1.08	3.91	1.43
Self-esteem	4.12	1.03	4.55	1.13	4.96	1.50	3.78	0.97
Empathy	3.87	0.99	3.91	1.03	4.13	0.98	2.71	1.52
Pulse control	3.98	1.06	4.11	1.18	4.77	1.56	2.93	1.62
Score	31.08	4.15	32.55	4.71	36.61	5. 39	34.10	5.79

It can be noticed that the three levels of education manifest a significant degree of emotional intelligence, their means are higher. Comparing the three scores, we can see that consumers with a level of education of (Bac + 5 and more) show the highest degree (m = 36.61), followed by consumers with a level of (Bac + 3 and more) with a score of (m = 32.55), and finally consumers who have a level of education of (Bac + 1 and more). These results mean that as the level of education increases, the degree of emotional intelligence becomes more important. To confirm this hypothesis, we tested the significance of intelligence by a T-test analysis as shown in the following Table 2:

Table 2. Comparison of emotional intelligence by level of education

Comparison of emotional intelligence by educational level	N	Average	Standard	T	Meaning
Bac + 1 and more	29	31.08	4.14		
Bac + 3 and more	67	32.55	4.71		
Bac + 5 and more	104	36.61	5.39		
Between (Bac + 1 and more) and (Bac + 3 and more)				4.780	0.000*
Between (Bac + 3 and more) and (Bac + 5 and more)				1.456	0.129
Between (Bac + 1 and more) and (Bac + 5 and more)				7.519	0.000*

* The difference is significant at the 0.05 level.

In fact, the difference in emotional intelligence is significant between consumers with (Bac + 1 and above) and those with (Bac + 5 and above) with a score of ($t = 7.519$; $p = 0.000 < 0.05$), and also significant between consumers with (Bac + 1 and above) and those with (Bac + 3 and above) ($t = 4.780$; $p = 0.000 < 0.05$). While it is not significant between consumers of (Bac + 3 and more) and those with (Bac + 5 and more) ($t = 1.456$; $p = 0.129 > 0.05$).

The reliability of the constructs was checked by evaluating the Cronbach's Alpha and composite reliability coefficients (see Table 3). These coefficients meet the thresholds of 0.7 recommended by Wong, K. (2013) respectively. They thus confirm the reliability of the constructs:

Table 3. Reliability and convergent validity

	Cronbach's Alpha	Jöreskog's Rhô	Composite reliability	AVE
Self-awareness	0.853	0.860	0.880	0.681
Self regulation	0.860	0.870	0.903	0.745
Motivation	0.871	0.888	0.912	0.708
Empathy	0.960	0.972	0.977	0.851
Self-esteem	0.865	0.890	0.927	0.760
Managing emotions	0.905	0.907	0.929	0.721

In fact, the Mean Extracted Variances (MEV) of the different constructs respect the minimum threshold of 0.5 foreseen by this approach, moreover the values of the Jöreskog's Rhô vary between 0.860 and 0.972, which confirms their convergent validity (see Table 3). The discriminant validity of the constructs is checked by comparing the square root of the AVE of each construct with the values of the correlations between the constructs. As shown in Table 4, the square roots of the AVEs are all greater than the correlations. This confirms the discriminant validity of the measures.

Table 4. Discriminant validity

	Estime de soi	Conscience de soi	Empathie	Autorégulation	Motivation	Gestion des émotions
Self-esteem	0.759					
Self-awareness	0.106	0.835				
Empathy	0.127	0.375	0.803			
Self regulation	0.242	0.582	0.434	0.909		
Motivation	0.253	0.269	0.352	0.482	0.929	
Managing emotions	0.462	0.241	0.535	0.363	0.452	0.845

Our study found the following result: the significant effect of educational level as a determinant of the impact of emotional intelligence on the purchasing decisions of Moroccan consumers. Our results confirm the link between the level of education and the degree of emotional intelligence, i.e., consumers with a higher level of education are more emotionally intelligent, and therefore less impacted by marketing strategies when making purchase decisions. Indeed, the higher the level of education, the higher the level of emotional intelligence. Our results are consistent with previous research, including studies by (Carrigan and Attala 2001) (De Pelsmacker et al. 2005); Dickson and Littrell 1997 and (Roberts (1995). Studies of consumer behavior have also shown that educational attainment develops strong emotional intelligence (Kinnear, Taylor 1974; Van Liere and Dunlap, 1980). Roozen (1997) also showed that consumers with a high level of education increase the probability of having a very high level of emotional intelligence, which allows them to rationalize their purchasing decisions. We found that the higher the level of education, the less impacted consumers are by the flattery and advertising of marketing strategies. Consumers with the highest level of education can therefore manage their emotions in a more rational way, which means that their decision making is based on the knowledge of the product to be purchased. We recall that studies that have addressed the issue of emotional intelligence by education level are rare.

Our results confirmed the importance of educational level in determining the degree of emotional intelligence of consumers, including self-awareness, self-regulation, motivation, self-esteem, emotion management and empathy. The majority of respondents demonstrated a moderately high level of emotional intelligence. This finding validates our hypothesis regarding the presence of a significant relationship between educational level and the effect of emotional intelligence as a moderating factor determining purchase decision making. Most studies compare the degree of EI with age, gender, job function, or income, thus neglecting educational attainment. In short, we observe that respondents with higher levels of education will tend to seek more information before making a purchase decision.

5 Conclusion

Given these results, education level proves to be an important factor that could influence consumer behavior. Our study showed that consumers with a very high level of education are the most emotionally aware, and thus they have the ability to rationalize their purchasing decisions.

References

Bar-on, R.: The emotional quotient inventory (EQ-i): Technical manuai, Toronto, Multi-Health Systems (1997)
Bar-On, R.: Emotional and social intelligence: Insights from the Emotional Intelligence Inventory (EQ-I). Handbook of emotional intelligence. JosseyBass, San Francisco (2000)
Carrigan, M., Attalla, A.: The myth of the ethical consumer—do ethics matter in purchase behaviour? J. Consum. Mark. **18**, 560–577 (2001)
De Pelsmacker, P., Driesen, L., Rayp, G.: Do consumers care about ethics? willingness to pay for fair-trade coffee. J. Consum. Aff. **39**, 363–385 (2005)
Dickson, M.A., Littrell, M.A.: Consumers of clothing from alternative trading organizations: societal attitudes and purchase evaluative criteria. Cloth. Text. Res. J. **15**, 20–33 (1997)
Goleman, D.: L'intelligence émotionnelle. Robert Laffont, Paris (1997)
Mayer, J.D.: What is emotional intelligence? In: Salovey, P., Sluyter, Y D. (Eds.). Emotional Development and Emotional Intelligence: Implications for Educators, pp. 3–31 (1997)
Salovey, P., Mayer, J.D.: Emotional intelligence. Imagin. Cogn. Pers. **9**(3), 185–211 (1990)

Improving Health Care Services via Personalized Medicine

Fatima Ezzahrae El rhatassi[1(✉)], Btihal El Ghali[1], and Najima Daoudi[1,2]

[1] ITQAN Team, LYRICA Lab, Information Sciences School (ESI), Rabat, Morocco
{fatima-ezzahrae.el-rhatassi,bel-ghali,ndaoudi}@esi.ac.ma
[2] SSLab, ENSIAS, Mohammed V University, Rabat, Morocco

Abstract. In addition to genes, people's habits and environment are crucial factors in deciding whether or not they will get an illness and how quickly it will progress. This review paper gives an overview of current research in the application of a variety of tools, including artificial intelligence, to customize medicine. Many applications are utilized to personalize medical care, including machine learning, natural language processing and chatbots for the disease detection, treatment, and prevention. The main goal of this survey is to consider the various diseases, techniques, features, and limitations that exist in order to improve the patient's well-being and to propose a conversational agent - a chatbot as a viable option for detecting the patient's diseases and traits, as well as gathering symptoms, treatments, and therapies that could help the patient be guided and better manage his disease.

Keywords: Personalized medicine · Machine learning · Natural Language Processing · Chatbot

1 Introduction

In general, personalization is about people, it is person-centered; it starts from detecting the person's needs to deliver a service that suites them. This concept interested many fields where it means different things to different people for different objectives bounded to the individual. These fields include cognitive science, sociology, architecture, computer science, e-commerce, e-learning, marketing and medicine [1].

For the domain of cognitive science, personalization is "a system that makes explicit assumptions about users' goals, interests, preferences and knowledge based on an observation of his or her behavior or a set of rules relating behavior to cognitive elements" [2]; For computer scientists, so many definitions exist but they turn around the tools and applications used to build an interface personalized with the user's needs. E-commerce and marketing are about combining the use of technology and the customer service by understanding these needs and trying to satisfy or convince him with a tailored service [1] while e-learning targets the learner by putting at his disposal a personalized education; taking

M. Lazaar et al. (Eds.): BDIoT 2022, LNNS 625, pp. 435–449, 2023.
https://doi.org/10.1007/978-3-031-28387-1_37

into consideration his pace, his learning preferences and his interests. This individualized learning will caption him to better use the educational system, get access to platforms and taking the courses [3,4].

In recent years, medicine has evolved to treat diseases in a way that is tailored to the individual and committed to treating the individual while considering his or her uniqueness (genes, medical history, lifestyle and environment). Several limits imposed by vast amounts of diverse data must be addressed and explored. Fortunately, computer technology allowed for a more efficient study of this massive amount of data. The application of artificial intelligence to medicine aided precision medical diagnostics and therapies, as well as progress in the health-care system [5].

Artificial intelligence applications should pay particular attention to psychological illnesses and developmental abnormalities. One of these diseases is autism. The World Health Organization estimates that one in 100 children has autism, while this number could be higher given that many low- and middle-income countries are unaware of this condition. While some autistic persons can live independently, others require ongoing care and assistance in order to develop their social and communication skills and integrate into society.

The review is divided into the sections below. The concept of customized medicine and the many ways used to customize medical services will be discussed in Sect. 2. Section 3 provides an overview of some applications of artificial intelligence in treating various diseases and achieving various goals. Section 4 examines the features, objectives, and limitations of each technique employed before proposing our solution for a chatbot to better assist the patient in managing his sickness and the next research's future work in the last section.

2 Context and Problem Statement

In the medical field, the concept of personalization is not recent because the interest of medicine was and is always centered on human by promoting his wellbeing and his health condition, but the concept is developing with time. First, it was based on scientific experimentation by using chemicals and specific doses to having now the capability to go further and to analyze our individuality by proposing treatments adapted to our gene, medical history, lifestyle and the environmental factors that impact our lives [6].

2.1 Personalized Medicine's Concept

Many words can refer to personalized medicine as: Individualize care, precision medicine, tailored medicine, targeted medicine, patient-centered healthcare and stratified medicine. The most frequent used term is personalized medicine followed by individualized medicine [7] and each of the precited terms are often used to describe an angle of the concept. As an example, precision medicine is described by Kevin B. Johnson and his colleagues [8] as the knowledge consisted by sequencing the human genome. In other words, precision medicine combines the genomic and nongenomic determinants to facilitate diagnosing a patient with a personalized diagnosis and therapy. However, concerning personalized medicine approach, it gathers environments, lifestyle, inheritance, genetics and medical history factors to precise the risk of having a disease, the response of the patient to a treatment and to tailor the treatment to its specific needs and constrains [9].

All these terms have as a goal to use the data to target a group or a patient with a specific treatment. However, we can notice a difference that may exist in the degree of personalizing to whom the treatment is delivered:

- Generic degree that goes from the concept of "one size fits all". This concept is addressed to a population not individuals and the treatment proposed is supposed and intended to suit everyone; not the patient but the group;
- Personalized degree which is addressed to a person by its name considering all its needs to provide a suitable treatment in the case of the response to a drug.
- Targeted degree gives certain treatments to a subgroup of patients having some characteristics in common. The treatment is supposed to be adapted and efficient to these patients.
- Tailored degree provides a specific treatment to the individual with some specific characteristics. This treatment is designed for this person taking into consideration his or her disease history, allergies, lifestyle and genes [10].

However, one size doesn't fit all. People are not the same in many things as their genes, medical history, ethnicity, environment and lifestyle. In order to improve treatment efficacy and reduce the risk of adverse, we will choose to consider personalized medicine as the main term of our research for:

- People can respond differently to the treatment if it is not adapted to its specific needs;
- Prevention, treatment and therapy strategies will be more efficient for patients if we take into account the large amounts of high dimensional, heterogeneous data ranging from different resources: clinical data, electronic health records (EHRs), mobile applications... Gathering this data and exploring it is feasible via artificial intelligence which will combine these factors into a model. By knowing the main factors responsible of a disease and modifying in the risk factors, we will be able of changing the progression of the disease.

Individualizing therapy and prevention are the most essential goals of customization medicine. Studying each case independently based on the needs and conditions of each patient is a revolution in medicine in general, especially with the application of artificial intelligence to aid in the advancement of medicine's services.

2.2 Techniques of Personalization in the Medical Field

With the aim of improving our understanding, prevention some diseases and proposing the best treatment at the right time for the right person, the huge amount of data extracted from the different resources needs to be developed and explored using many techniques:

The establishment of an ontology allows deducing the patient's profile through the constitution of a complete knowledge base. This database contains all the similar profile vectors for a population of patients as well as clinical outcome variables such as survival time or reaction to certain drugs. For each patient, an analysis is done in relation to other profiles which are considered as reference profiles to deduce the best match of the patient's profile to propose recommendations for the diagnosis, the prognosis and the therapy to the patient [11].

Artificial intelligence (AI) is a huge concept whose objective is to create intelligent machines that can perform tasks and act as human thinking does. AI includes many techniques such as machine learning, deep learning, natural language processing and chatbots. Machine learning improves its performance by generating an algorithm through experience based on data rather than defining rules and using traditional approaches and statistics to make predictions or decisions without being explicitly programmed to do so [30]. Deep learning uses a complex structure of algorithms referring to the human brain. It consists of using layers to learn representations of data, to deal with unstructured data such as documents, images, and text to understand and capture essential correlations among input data to improve by the end accuracy [11,12]. Natural language processing (NLP) makes the computers understand the language spoken by human. It converts an unstructured narrative text into a structured form [13]. NLP has as a domain of application sentiment analysis where while exchanging with the user, emotions and state of mind of the user can be extracted to better understand the user's needs and adapt the intervention maintenance programs on his state of emotion [13–15].

A conversational agent or a chatbot in the field of medical support is an agent that employs dialog systems to communicate with users using speech, text, or both. The personalization form shows when it comes to interpret their needs and deliver a suitable response after detecting the disease [15]. Chatbots often uses machine leaning, natural language processing and deep learning to expand their knowledge by conversing with the user to better understand what he suffers from and what he wants to respond appropriately, to better fit the assistance and the treatment provided to his needs and emotions.

2.3 Problem Statement

We are interested in personalized medicine and our goal is to identify diseases whose treatment is most influenced by the individual's parameters, the most relevant characteristics that can be taken into account and how to use artificial intelligence to provide personalized medical services that contribute to improve the patient's health conditions and life.

The works that have addressed this issue will be presented in the following section as related work. So we can offer a medical service that responds to the patient's needs, many techniques are represented and applied to different diseases as chronic, neurological, oncological diseases. Certainly, each method has its advantages and its limits and we will propose our own method for the disease chosen to treat.

3 Related Work and State of the Art

Artificial intelligence (AI) is being applied frequently in the field of medicine to better understand the disease and to prevent it in the future. It also allows to have a positive response to the treatment of the patient, establish assistance for him, and make a follow-up after the cure.

3.1 Machine Learning and Deep Learning in Personalized Medicine

The personalized approach via machine learning studies how to improve the detection, evaluation, treatment, and prevention of Hypertension by combining biological, lifestyle, and environmental factors. It is recommended to focus less on conventional factors (Blood Pressure alone) and to combine many other variables that are changeable and that may play a central role as dietary, environmental and psychological factors and where wearable technologies are potentially a real opportunity for these type of data as part of precision medicine in the hypertension disease [16].

Cardiovascular disease as an example may be preventable by changing and modifying some lifestyle behaviors as quitting smoking, increasing healthy nutrition and providing more physical activity. The availability of a function model may evaluate the modifications' utilities, reduces the risk of having CVD and identify the suitable recommendation approach for each patient under his preferences to change the disease progression uncertainties factor [17].

Diabetes is a chronic health condition that occurs when the blood glucose is elevated. So many studies and many AI applications have been done to decrease its existence and its progression. AI tools are used to detect and to predict threats related to diabetes and recommend certain changes to the everyday lifestyle, especially Diabetes Mellitus Type II because Mellitus Type I requires insulin. However, Diabetes Mellitus Type II may be prevented or well managed by self-management in terms of nutrition, sleep activity, physical activity, stress management and many more factors that affects the well being of the patient [18,19].

Another form of following up with patients is by using apps, devices to recognize food items on a plate and estimates their calorific and nutrition content. This way aims to adopt a healthier lifestyle and helps patients to be more informed about their food choice decisions [21].

Machine learning is also used in cancer research by improving the accuracy of predicting the susceptibility of having cancer, recurrence, survival and mortality. Researches can benefit from the large volumes of data by the help of new technologies and make some predictions for prostate or breast cancer patients such as suggesting personalized treatments to patients newly diagnosis, developing and validating performance metrics of surgeons [22,23].

In the aim to face mental illnesses, proposing a personalized treatment is the only way to reduce its symptoms and increase the patient's quality of life. Bipolar disorder as a serious disease needs to be monitored to detect the recurrent episodes of depression and mood elevation to prevent and avoid some weaknesses or suicidal moments. Many ways can be used to generate a meaning of the disturbance in mood and to monitor it to a better management of the mood disorder or to intervene in emergency situations by specialists [24,25].

3.2 NLP and Chatbots in Health Care Precision

Nowadays people are more interested in using chatbots for the fact that they need the information instantly without taking too much time to get an answer and the chatbot is always available to provide this information.

Also, many people prefer using chatbots because they are task-oriented. Instead of surfing the net and getting many incorrect information, conversing with a chatbot will be a time-saving and efficient search.

Chatbot technologies are being used to provide a medical assistance in many fields as improving some developmental disorders such as autism. Lissa (Learning Intelligent System for Student Assistance) is an online technology that is used in many cases to give a behavioral therapy by conversing with an autist, make him feel comfortable to share information in the aim to identify psychological distress indicators and to improve the user's nonverbal behavior in communication [26].

Chronic diseases are difficult to treat since they linger for a long time, if not forever. Kbot is a customized chatbot meant to assist asthmatic children in gaining better control of their condition. Mobile devices are used in conjunction to track a patient's health and to continuously monitor his medicine consumption as well as environmental data. The triggers of an asthma attack are ranked based on their occurrence score with the patient-reported symptom to generate a personalized alert whenever the triggers are in an unhealthy range and to help the patient self-manage his disease [27]. In addition to the medication history entered by the patient, the gathered data from different resources: answering questionnaires and day-to-day conversations, and environmental data, the triggers of an asthma attack are ranked based on their occurrence score with the patient-reported symptom to generate a personalized alert whenever the triggers are in an unhealthy range and to help the patient to self-manage his disease [28].

For the treatment of some mental illnesses, there are numerous chatbots available. The majority of them are devoted to improving communication skills of older adults who are isolated and depressed, to monitoring some mental health status of perinatal women during pregnancy or after childbirth, or to self-help psychological intervention and assistance for students who are depressed [25,28,29]. The technology engages users in conversation, makes them feel comfortable speaking, and provides valuable feedback based on analyzing: eye contact, smiling, speaking volume, and the valence of speech content to provide personalized guidance for developing each user's communication skills [28].

Cancer chatbots leads the patient in terms of diagnosis and treatment of the disease by providing an imaging diagnostic, symptom screening. People with cancer need help by giving treatment recommendations, ensuring the appropriate monitoring of the treatment. They also need an emotional support during their fight with cancer that's why providing discussions, emotional support by a conversational agent is a need especially in the aim to avoid waiting for real consultations for this kind of support [13,30].

4 Analyses and Discussion

Each disease requires personalization in treating each patient differently from the other, but the need to this personalization is seen in a very important way in some cancer or mental disorders. Clinicians will be able to observe the symptoms, the signs and the evolution of the patient's state not only during his appointment in hospital but while he's at home resting or doing his daily activities [26,31]. In that way, controlling the disease and ensuring its prevention may reduce the deaths or may avoid having severe cases by changing some individual's behaviors.

Many studies showed a deep relation between some diseases (oncology, chronic, neuro-diseases...) and environmental and lifestyle factors. When we talk about environmental and lifestyle factors, we refer to pollution, nutrition, stress, physical activity, the use of smoking and drinking alcohol and many other factors... All these factors are important to precision medicine, and they're applied to customize prevention and treatment strategies for an individual by identifying the factors that impact the disease that he suffers from or he can get later [26]. These features may be responsible for developing a disease and they can be categorized into changeable features and unchangeable features. Changeable features affect the disease in a direct or indirect way while unchangeable features don't affect the disease but it characterizes the patient by some characteristics as gender, weight, height... Precision medicine must be tailored to the individual by taking into account the different characteristics that define him, both in terms of features that are directly related to the disease and whose modification automatically impacts the disease, and in terms of the different features that characterize his lifestyle and the activities that he performs in his daily life and that allow to indirectly impact the disease.

For achieving the purpose of personalized medicine, it is recommended to develop a huge amount of data and combine biological, non-biological, and environmental factors to prevent diseases and treatment of complex diseases using artificial intelligence. As it is known, behavioral, socio-economical, physical, and psychological data account for 60% of influencing individual health, the impact of our genes on the health accounts for 30% and medical history accounts only for 10% [8]. Therefore, combining all the features and studying which factor is playing a central role in developing a risk of having a disease is a must, especially those relied on lifestyle, nutrition, and environmental factors.

The concept of personalized medicine is explored using different models with machine learning and deep learning. We gathered some diseases where the personalization is used, the objective of its use, techniques implemented to tailor

Table 1. Categorization of Artificial Intelligence in health care (Machine learning and deep learning)

Disease	Purpose of personalization	Techniques	Features
cardiovascular disease [16,33]	Analysis of CVD risk, Risk reduction based on recommended lifestyle modification	- Random forest - Support vector machine(SVM) - CNN classifiers - Deep neural network (DNN)	**Directly changeable features:** Alcohol intake (g) per day, smoking status, total activity hours per week, carbohydrate, protein, saturated fatty acid (g) per day, total fat (g) per day, body mass index, dietary cholesterol (mg), dietary fiber, total energy intake **Indirectly changeable features:** HDL cholesterol in mg/dl, LDL cholesterol in mg/dl, Total cholesterol in mmol/L, Total triglycerides in mmol/L, 2nd and 3rd Systolic blood pressure average, 2nd and 3rd Diastolic blood pressure, blood pressure average, Waist girth to nearest, Hip girth to nearest, Heart rate **Unchangeable features:** Cigarette years of smoking, Education level, Sex, Age, Race, Diabetes [0, 1], Hypertension [0, 1], Hypertension medication in the past 2 weeks, Cholesterol-lowering medication use [0, 1], Smoking status history
Neurological disorders (autism, Parkinson) [26,33]	Identifying the probability of developing Parkinson Disease, Identifying persons with an increased risk of mortality at 6 and 12 years, Behavior understanding from video streams	- Logistic regression model - Hidden Markov model	**Directly changeable features:** Heavy smoking, alcohol consumption, milk and carbohydrates intake and polyunsaturated fat intake, rural living (including years of rural residency and ground-water use), pesticide use, and male lifestyle head trauma, male-dominated occupations) **Indirectly changeable features:** Gender **Images features**
Mental illness [29]	Analyzing the mental health of perinatal women (degree of anxiety, hypomania and depression)	- Support Vector Machines (SVM) - Convolutional Neural Network (CNN)	**Directly changeable features:** Angry, tension, Mood swings Overeating, Anorexia, Physical discomfort, panic, Happiness quotient, Negative thinking, Positive Thinking, Suicidal tendency, Work Time, Stress, Anxiety, Depression, Intervention needs, Therapist Needs. **Indirectly changeable features:** Sleep, Irregular Wake-up, Gloomy, Irregular sleep, Jobs. **Unchangeable features:** Age, Gender, Nationality, Race, Sexual Orientation, Weight, Pregnancy stage
Chronic diseases [16,18,32]	improving the detection, evaluation, treatment, prevention of (hypertension, diabetes) and improving adherence	- Unsupervised learning - multivariate linear regression - Deep learning (DL) -Neural network model	**Directly changeable features:** Blood pressure, systolic blood pressure (SBP), diastolic blood pressure (DBP), body mass index (BMI), Low-density lipoprotein cholesterol (LDL), high-density lipoprotein cholesterol (HDL), Glucose concentration. **Indirectly changeable features:** Heart rate (HR), triglyceride, Diet, sleep activity, stress, physical activity. **Indirectly changeable features:** Age, gender, diabetes, smoking.

the treatment to the patient and features used to personalize the healthcare as described by Table 1. As we can notice, the published articles included in this study have different objectives in the health care domain from screening, detecting to monitoring and preventing many diseases. Data used is gathered from clinical sources, devices and apps, questionnaires and forms. The features used can be categorized into directly changeable features, indirectly changeable features and unchangeable features.

When we talk about personalization medicine, we mean identifying the features that lead to personalized insights where the patient will be able to be engaged, confident and will succeed in monitoring his disease and maintaining his health condition under a better control [21]. To achieve this purpose, patients should be adequately informed in the right time to promote awareness and develop their management of the disease especially with this huge information existing everywhere. Many patients use internet-based information to understand and manage their diseases and using incorrect or misguided information can make their health worse. One way to face this challenge is by putting at his disposal a chatbot; disease-oriented, capable of giving him a credible information and guide him at any moment and today, since everyone has cell phones, we can take advantage of their use to offer the patients (individuals) a chatbot that allows them to be in continuous assistance and that "understands" them through natural conversations, persuade them to change and build a healthier behavior by means of speech, text, or both. These conversational agents won't replace specialists, doctors, therapists but it will complete and generate the information that patients receive from traditional educational models and that may be false. Many chatbots exist already in the medical field and they are used for several objectives (as shown in Table 2).

The features used in chatbots to provide an individualized medicine generally bounded to the disease are captured from the text conversations exchanged with the patient and from other apps and devices and the symptoms that may change the individual's appearance as facial characteristics, voice and head pose.

There are several limitations and ethical problems that should be considered in a chatbot, including the privacy of information exchanged and the chatbots' limited capacities and emotional intelligence, which may limit their application in daily life.

5 Chatbot's Design

An efficient chatbot should gather image recognition, voice recognition and emotional intelligence to capture and understand the state of mind of the patient, the possibility of putting feedback by the patient to improve its services, and the ability to access and extract information from multiple platforms and especially to react to the unexpected response.

The chatbot that we will propose will be dedicated to autistic patients. This vulnerable population has difficulties in verbal, non-verbal communication and repetitive behavior. Because Autism Spectrum Disorder (ASD) patients suffer

Table 2. Categorization of Artificial Intelligence in health care (Chatbot and NLP)

Disease	Purpose of personalization	Techniques	Features	Limitations
Neurological disorders [26]	Identify psychological distress indicators, Improving users nonverbal behavior.	- Chatbot with Hidden Markov Model (HMM) - Natural language processing	Head pose, smile, facial action units, volume, and voice pitch.	- The chatbot is unable to discuss whatever the user wants - Interactions are insufficiently individualized, and the chatbot should avoid topics in which the user appears uninterested - The content is unrelated to nonverbal feedback
Asthma [27]	Tracking patients' health and their medication intake. Continuous monitoring of the patient's health, Alerting the patient about asthma triggers and help him self-manage their asthma.	- Chatbot using natural language processing	Weather, Cough and sneezing history days and moods, air quality, pollen level, Smoking, weight	- Each symptom should be assigned a severity level, and each patient should be treated individually based on his or her severity level - Training the language model on real-life patient-doctor dialogue data will be helpful in delivering a more human-like conversation
Mental illness [25,28]	self-help psychological intervention for depression	A chatbot with three machine learning models: - Natural language processing - Intention classification - Emotion recognition	Scale, mean (SD) Depression (PHQ-9), Anxiety (GAD-7), Positive affect, Negative affect, Eye contact, smiling, speaking volume and valence of speech content	- The chatbot's content was restricted - The conversations should be compared on a regular basis to keep track of the patient's symptoms and needs so that a personalized treatment plan may be developed - When a negative word is used in a positive context, the solution is unable to recognize it, and vice versa
Oncology [13,30]	Enquiring about cancer symptoms, therapies, and survival, as well as providing consolation to cancer patients	Chatbot using natural language processing	The user's mood (sentiment analysis), as well as other limited features such as simply supporting text-based messaging	- Voice recognition features should be added. - The chatbot should be trained to access data directly from the web

from interacting with other people, an interactive technology or a chatbot will help them to be engaged to improve their communicational skills and affective behavior.

There have been numerous efforts made to develop conversational agents for autistic people. Some of them focused on text-based inquiries with the autistic patient in a goal to make him feel at ease while disclosing information and to determine the severity of the illness using some predefined signs [35]. Others have attempted to enhance a user's nonverbal communication skills through conversation by providing real-time feedback through four flashing icons that indicate if the user is speaking normally or needs to modify his conduct [26]. The subjects that the user could chat with the chatbot in all of these studies were constrained. We suggest a system that will initiate the patient to converse, exchange and to share informations in many topics by talking and text chatting in order to identify the illness, afford an assistance to make him feel comfortable and improve his communicational skills. The architecture of the chatbot will be as shown in Fig. 1.

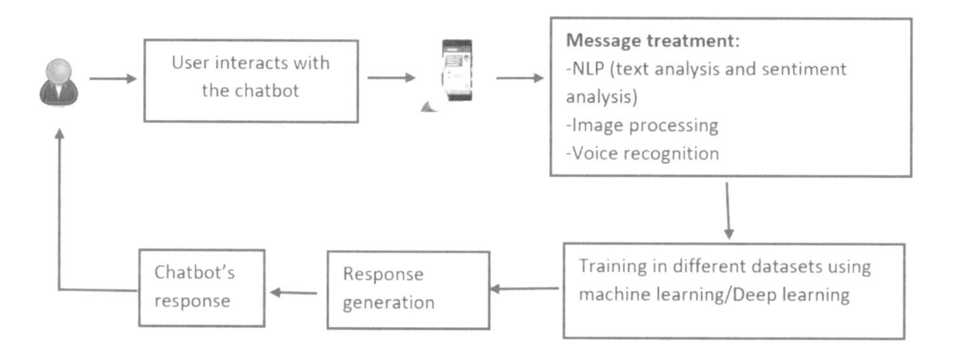

Fig. 1. Architecture of the chatbot

Through this architecture, we propose that the user interacts with the chatbot. The message addressed to the conversational agent is transformed and processed through natural language processing (NLP). If the message addressed to the chatbot is a speech one, then it will be transformed into text to be treated after that by the use of the automatic speech recognition system (ASR). Also, when the response will be generated, we'll keep the same type of receiving the message and converts the response back from text to speech if the message is a speech one in the aid of the text to speech system (TTS).

NLU (natural language understanding) as a part of NLP helps to better understand the received request, analyze it and to better provide an answer. NLU deals with three tasks in general:

-Dialogue act classification is crucial for figuring out the user's function since it helps us discover if the user is making a request, making a statement, or asking a question with his message which is important to choose the format of the chatbot's response [36].

-Intent classification restricts the user's goal and identifies his intent's domain [37].

-Slot filling has as a goal to gather any additional information that, when combined with the dialog and the intent, will help the system better understand the user's request. Tokenization, which divides a text into units by words, punctuation, or numbers, latent semantic analysis, which compares the meaning of the words rather than the actual words, text normalization, and other techniques are used to analyze the context of the input by breaking the text into smaller pieces [37].

Sentiment analysis employs natural language processing (NLP) to infer the user's attitude from the text's content; in other words, it aids in understanding human psychology by categorizing the text as having a positive, negative, or neutral sentiment. Sentiment analysis plays an important role in healthcare by determining the patient's level of wellbeing, happiness, or unhappiness, allowing medical personnel to assist in urgent situations, adapt the course of treatment to the patient's psychological state, and provide continuous patient monitoring to track his mental evolution as he manages the disease in his daily life [38].

These technologies will aid in the extraction of useful data (text, spoken language manipulation through video or voice recognition, and sentiment analysis) and the chatbot will provide real-time responses by accessing numerous platforms and using machine learning or deep learning algorithms to be more trained.

This is a general design of a chatbot. This chatbot's primary objectives are to examine, diagnose autistic disorder and help patients with ASD become more socially adept. In upcoming developments, the architecture will be clarified more and implemented with more features and techniques to accomplish the goals. The chatbot's field of application can be expanded and applied to a wide range of diseases that are similar to autism and treat mental disorders such as attention deficit hyperactivity disorder (ADHD) and adding more features specific to this ailment will be beneficial. These patterns should be improved in the future to discover which causes are responsible for the development of these diseases and the improvement of the patient's health conditions.

6 Conclusion and Futur Work

In this study, we have presented personalized medicine and its domain of applications. Many techniques are used in delivering a tailored medical service that takes into account the medical history of the patient, his genes, environmental and lifestyle factors. We conducted multiple studies of using conversational agents (chatbots) for detecting, treating and preventing some diseases as chronic, neurological, mental disorders and others for managing daily stress and improving physical activity. We then proposed a design of our chatbot. The chatbot aims to analyze the mental health and to assist the autist patient to improve his communicational skills. Many features will be used to characterize the patient as voice recognition features and video analysis to better collect user information and ensure a suitable intervention. These two types of analysis, in addition to text chatting, will assist us in detecting the patterns associated with autism and its severity, such as late responses to requests, difficulties maintaining a conversation, limited eye contact, and repetitive speech, in order to track the development of communication and improve the patient's wellbeing.

References

1. Fan, H., Pool, M.: Perspectives on personalization. In: Americas Conference on Information Systems (AMCIS), pp. 2123–2125 (2003)
2. Kobsa, A.: User modeling as a key factor in system personalization. In: CHI 2000 (2000)
3. Dolog, P., Henze, N., Nejdl, W., Sintek, M.: Personalization in distributed e-learning environments. In: WWW Conference (2004)
4. Assami, S., Daoudi, N., Ajhoun, R: Personalization criteria for enhancing learner engagement in MOOC platforms. In: IEEE Global Engineering Education Conference (EDUCON), Tenerife (2018)
5. MacEachern, S., Forkert, N.: Machine learning for precision medicine. Genome **64**(4), 416–425 (2021)
6. Gordon, E., Koslow, S.: Integrative Neuroscience and Personalized Medicine. Oxford University Press, Oxford (2011)
7. Pokorska-Bocci, A., Stewart, A., Sagoo, G., Hall, A., Kroese, M., Burton, H.: 'Personalized medicine': what's in a name? Pers. Med. **11**(2), 197–210 (2014). https://doi.org/10.2217/pme.13.107
8. Johnson, K., et al.: Precision medicine, AI, and the future of personalized health care. Clin. Transl. Sci. **14**(1), 86–93 (2021)
9. Zhang, S., Bamakan, M.H., Qu, Q., Li, S.: Learning for personalized medicine: a comprehensive review from a deep learning perspective. IEEE Rev. Biomed. Eng. **12**, 194–208 (2019). https://doi.org/10.1109/RBME.2018.2864254
10. Kocaballi, A.B., et al.: The personalization of conversational agents in health care: systematic review. J. Med. Internet Res. (2019). https://doi.org/10.2196/15360
11. Emmert-Streib, F., Dehmer, M.: A machine learning perspective on personalized medicine: an automized, comprehensive knowledge base with ontology for pattern recognition. Mach. Learn. Knowl. Extr. **1**, 149–156 (2019). https://doi.org/10.3390/make1010009
12. He, J., Baxter, S.L., Xu, J., Xu, J., Zhou, X., Zhang, K.: The practical implementation of artificial intelligence technologies in medicine. Nat. Med. **25**(1), 30–36 (2019). https://doi.org/10.1038/s41591-018-0307-0
13. Belfin, R.V., Shobana, A.J., Manilal, M., Mathew, A.A., Babu, B.: A graph based chatbot for cancer patients. In: 2019 5th International Conference on Advanced Computing & Communication Systems (ICACCS), pp. 717–721. IEEE (2019)
14. Zhang, J., Oh, Y.J., Lange, P., Yu, Z., Fukuoka, Y.: Artificial intelligence chatbot behavior change model for designing artificial intelligence chatbots to promote physical activity and a healthy diet: viewpoint. J. Med. Internet Res. **22**(9), e22845 (2020). https://doi.org/10.2196/22845
15. Prajapati, N., Mhaske, V., Dubey, S., Kumar Soni, P.: Chatbot for medical assistance: a review. Int. J. Recent Adv. Multidiscip. Topics **3**(3), 66–70 (2022). https://journals.resaim.com/ijramt/article/view/1853
16. Krittanawong, C., Bomback, A.S., Baber, U., Bangalore, S., Messerli, F.H., Wilson Tang, W.H.: Future direction for using artificial intelligence to predict and manage hypertension. Curr. Hypertens. Rep. **20**(9), 1–16 (2018). https://doi.org/10.1007/s11906-018-0875-x
17. Dogan, A., Li, Y., Odo, C.P., Sonawane, K., Lin, Y., Liu, C.: A utility-based machine learning-driven personalized lifestyle recommendation for cardiovascular disease prevention (2022)

18. Arab, K., Bouida, Z., Ibnkahla, M.: Artificial intelligence for diabetes mellitus type II: forecasting and anomaly detection. In: 2019 IEEE Wireless Communications and Networking Conference (WCNC), pp. 1–6 (2019). https://doi.org/10.1109/WCNC.2019.8885802
19. Ellahham, S.: Artificial intelligence: the future for diabetes care. Am. J. Med. **133**(8), 895–900 (2020). https://doi.org/10.1016/j.amjmed.2020.03.033
20. Yom-Tov, E., Feraru, G., Kozdoba, M., Mannor, S., Tennenholtz, M., Hochberg, I.: Encouraging physical activity in patients with diabetes: intervention using a reinforcement learning system. J. Med. Internet Res. **19**(10), e338 (2017). https://doi.org/10.2196/jmir.7994
21. Dankwa-Mullan, I., Rivo, M., Sepulveda, M., Park, Y., Snowdon, J., Rhee, K.: Transforming diabetes care through artificial intelligence: the future is here. Popul. Health Manag. **22**(3), 229–242 (2019). https://doi.org/10.1089/pop.2018.0129
22. Wong, N.C., Shayegan, B.: Patient centered care for prostate cancer-how can artificial intelligence and machine learning help make the right decision for the right patient? Ann. Transl. Med. **7**(Suppl 1), S1 (2019). https://doi.org/10.21037/atm.2019.01.13
23. Ozer, M.E., Sarica, P.O., Arga, K.Y.: New machine learning applications to accelerate personalized medicine in breast cancer: rise of the support vector machines. OMICS **24**(5), 241–246 (2020). https://doi.org/10.1089/omi.2020.0001
24. Zulueta, J., et al.: Predicting mood disturbance severity with mobile phone keystroke metadata: a BIaffect digital phenotyping study. J. Med. Internet Res. **20**(7), e241 (2018). https://doi.org/10.2196/jmir.9775
25. Liu, H., Peng, H., Song, X., Xu, C., Zhang, M.: Using AI chatbots to provide self-help depression interventions for university students: a randomized trial of effectiveness. Internet Interv. **27**, 100495 (2022). https://doi.org/10.1016/j.invent.2022.100495
26. Ali, M.R., et al.: A virtual conversational agent for teens with autism spectrum disorder: experimental results and design lessons. In: Proceedings of the 20th ACM International Conference on Intelligent Virtual Agents, pp. 1–8 (2020)
27. Kadariya, D., Venkataramanan, R., Yip, H.Y., Kalra, M., Thirunarayanan, K., Sheth, A.: kBot: knowledge-enabled personalized chatbot for asthma self-management. In: 2019 IEEE International Conference on Smart Computing (SMARTCOMP), pp. 138–143. IEEE (2019). https://doi.org/10.1109/smartcomp.2019.00043
28. Ali, M.R., et al.: Aging and engaging: a social conversational skills training program for older adults. In: 23rd International Conference on Intelligent User Interfaces (2018)
29. Wang, R., Wang, J., Liao, Y., Wang, J.: Supervised machine learning chatbots for perinatal mental healthcare. In: 2020 International Conference on Intelligent Computing and Human-Computer Interaction (ICHCI), pp. 378–383 (2020). https://doi.org/10.1109/ICHCI51889.2020.00086
30. Xu, L., Sanders, L., Li, K., Chow, J.: Chatbot for health care and oncology applications using artificial intelligence and machine learning: systematic review. JMIR Cancer **7**(4), e27850 (2021). https://doi.org/10.2196/27850
31. Seyhan, A.A., Carini, C.: Are innovation and new technologies in precision medicine paving a new era in patients centric care? J. Transl. Med. **17**, 114 (2019). https://doi.org/10.1186/s12967-019-1864-9
32. Subramanian, M., Wojtusciszyn, A., Favre, L., et al.: Precision medicine in the era of artificial intelligence: implications in chronic disease management. J. Transl. Med. **18**, 472 (2020). https://doi.org/10.1186/s12967-020-02658-5

33. Wang, H., et al.: Deep learning in systems medicine. Brief. Bioinform. **22**(2), 1543–1559 (2021). https://doi.org/10.1093/bib/bbaa237
34. Mendes-Soares, H., et al.: Assessment of a personalized approach to predicting postprandial glycemic responses to food among individuals without diabetes. JAMA Netw. Open **2**(2), e188102 (2019). https://doi.org/10.1001/jamanetworkopen.2018.8102
35. Mujeeb, S., Javed, M.H., Arshad, T.: Aquabot: a diagnostic chatbot for achluophobia and autism. Int. J. Adv. Comput. Sci. Appl. **8**(9), 39–46 (2017). https://doi.org/10.14569/IJACSA.2017.080930
36. Tur, G., Deng, L.: Intent determination and spoken utterance classification. In: Tur, G., de Mori, R. (eds.) Spoken Language Understanding: Systems for Extracting Semantic Information from Speech. Wiley, Chichester, pp. 93–118 (2011). https://doi.org/10.1002/9781119992691.ch4
37. McTear, M., Callejas, Z., Griol, D.: The Conversational Interface. Springer, Cham (2016). https://doi.org/10.1007/978-3-319-32967-3
38. Onyenwe, I., Nwagbo, S., Mbeledogu, N., Onyedinma, E.: The impact of political party/candidate on the election results from a sentiment analysis perspective using #AnambraDecides2017 tweets. Soc. Netw. Anal. Min. **10**(1), 1–17 (2020). https://doi.org/10.1007/s13278-020-00667-2

Network Technologies & IoT

Design, Development and Assessment of an Immersive Virtual Reality Serious Game

Yassine Tazouti[1(✉)], Siham Boulaknadel[2], and Fakhri Youssef[1]

[1] Department of Computer Science, Ibn Tofail University, Kénitra, Morocco
yassine.tazouti@uit.ac.ma
[2] IRCAM, Rabat, Morocco

Abstract. This paper presents ImALeG v2 "Immersive Amazigh Learning Game" an immersive serious game project to learn Amazigh language. This serious game is an extended version of our previous project which aims to enrich an old language practiced for centuries, not just in Morocco but by many African countries. Following the encouraging results obtained from the first version of ImALeG which is a 3D virtual environment for learning Tifinaghe alphabets. This extended version that support virtual reality aims to broaden students' knowledge by adding, two more levels to learn animal names as well as household objects vocabulary. We checked the educational and emotional impact of ImALeG for two student groups to compare ImALeG-based learning and the traditional language courses of Amazigh. Results presented in the evaluation section shows a significant improvement over traditional learning and a positive impact on ImALeG users.

Keywords: Serious Games · Game-based language Leaning · Virtual Reality · Language Learning · Educational Game

1 Introduction

Research in the field of teaching and learning languages, devotes a huge part of its work to the problem of motivation which is the most significant factor that determinates the efficacy of learning, therefore it can be easily achieved, by motivating learners who recognize their specific goals and interests. Serious game approach is not only designed for fun, but also for an educational purpose. Currently, it's attracting the interest of many stakeholders in the scientific research, professional field, education, health, vocational training, governance and defense. Serious games allow learners to acquire many skills effectively, as well as lots of disciplines [1]. Henceforth the integration of serious games in learning is a highly complex strategy that involve the expertise of a wide range of professionals from programmers, animators, designers, artists, and musicians [2]. Therefore, many methodologies for designing and developing serious games are dealing with a principal problematic which is the adaptability defined [3] as the possibility to monitor the ingame learning performance, and adapt the game to match this performance

M. Lazaar et al. (Eds.): BDIoT 2022, LNNS 625, pp. 453–463, 2023.
https://doi.org/10.1007/978-3-031-28387-1_38

and avoid serious games that suffer from system inflexibility, which is originated from stereotyped training scenarios and predictable game-play [4].

The presented work has been developed without any founding under the research case to apply serious game for learning Amazighe vocabulary in virtual worlds. This paper comes to complete the results obtained during the evaluation of the first version of ImALeG [5]. This paper presents ImALeG VR, a virtual reality serious game, which aims to engage the player interactively to learn Tifinaghe alphabet, Amazigh names of animals, household names, through the interaction mechanisms of the virtual reality system that improves the learner's sense of presence in a virtual world. Allows in an immersive way to learn by perceiving objects, interacting and communicating with other non-player characters (NPCs). The Player tracking system allows us to monitor learning progress in real time in order to compare it to the traditional teaching methods on scholars of the 4-year primary.

2 Related Works

Several serious game works have been noticed dealing with language learning. Silva et al. [6] presented a serious game called REAP.PT for learning Portuguese. The learning situation takes place in 3D rooms. To be able to open the door of each room, the player must complete the exercises presented in a desk. Additionally, a recent research presents a serious game called MFEG [7] which has features designed for people with hearing disabilities to improve English learning as a main objective to teach children with disabilities the English language. Another research project called VirtUAM [8, 9] composed of a 3D in class game with five levels, each one deals with a specific learning objective, tracks and gathers the players' data they used a several tests integrated in the beginning and at the end of the game session. Another example related to our field, is the project ARGuing [10], an alternative reality game developed by the European commission Comenius with a 6 project partners and used by 328 secondary school students from different European countries. They described, in their paper, the additional value that ARGuing came with, to supporting modern European language teaching and affecting positively the student skills, delivering them a motivational experience. Another example for language game called IFLEG which is a single player prototype game to learn French vocabulary language in a 3D environment and which supports virtual reality. The game contains only one level. The environment contains several rooms in a house. The player is represented by an avatar who can navigate freely in the environment. To win the game the player must search in the environment for hidden objects which constitute the universal language machine. The tasks to be accomplished are generated automatically by the system [11].

To verify the educational effectiveness of serious games, comparative studies have been carried out between different educational supports. They highlighted the attractiveness of serious games, which surpasses more austere media, based on the fixed image and printed text (manuals, books, maps, etc.), as hypertext systems or videos [12]. Specifically, the positive effects of animated and interactive 3D representation could be shown by a comparative study of 2D and 3D versions of the game Bilat dealing with the negotiation and intercultural communication [13]. However, motivation and intensity of engagement increases with the illusion of living a realistic situation, which is presumed to be better prepared.

From these projects we notice that serious games are not used only to learn a word or sentence, but more importantly to practice and communicate meaningfully with others, this is the ultimate goal of language learning games. More importantly, clarifying the rules and methods of the game must ensure fair participation [14].

In order for the language Amazigh game activities to be effective, the rules of the game must be clear. Before the game is launched, the game must provide instructions to introduce the rules, requirements, and rewards, and avoid the phenomenon of constantly adding rules while playing games during the game.

2.1 Increasing Realism Using FSM Methods

Artificial intelligence in game has been used in most video games development tasks to increase user's immersion. The artificial intelligence techniques used in games include FSM (Finite State Machine), NPC (Non playing character) group formation, to control the state of NPCs agents, (Neural Network), Genetic Algorithm and PathFinding, which determine the state and behavior of characters using evolution theory.

The game contains two type of animal NPCs a static animal NPCs [22] which the players can communicate with and NPCs implemented in FSM mode, contains a patrolling state to move along a specified path defined.These agent aims to improve realism of game, and a tutor humanoid NPC that have an idle state when the player stop, a state of approaching the player when a Tifinaghe character, animal or a home item founded, and a state of displaying information to the player. Also, since the state is unconditionally determined by the presence or absence of an item, the player is engaged in a searching mode with tutor NPC and gather useful information from static animals NPCs placed in various places. NPCs can bypass the obstacles and reach the destination through a pathfinding formed maps and calculate the shortest path to reach destination. This algorithm is called A Star algorithm or A* algorithm.

2.2 Players Tracking System

The start scene is mainly the UI layout (see Fig. 3) that contains the login and registration menu which help us to keep tracking the players evolution during a play session. The authentication system secured by an Md5 hash method connected to MySQL database using post and get methods to communicate over the local or internet network. Players must create an account indicating their first and last name, Age, gender and school grade. The tracking system collect information that will be transmitted to the database, such as the energy level reached by the player, acquired Tifinaghe alphabet, animals' names, and home objects acquired. Tracking system collects information about the player movement cliques, reached energy in each level, Tifinaghe alphabet, animals and house items names collected and whether they are approved by an evaluation. This information are required to carry out a general balance sheet on the learning evolution of each player and a general balance sheet to identify the players difficulties to adapt and improve our learning and the game experience. The player can also see their evolution in the game compared to the acquisition level of other players according to the score and their level reached (Fig. 1).

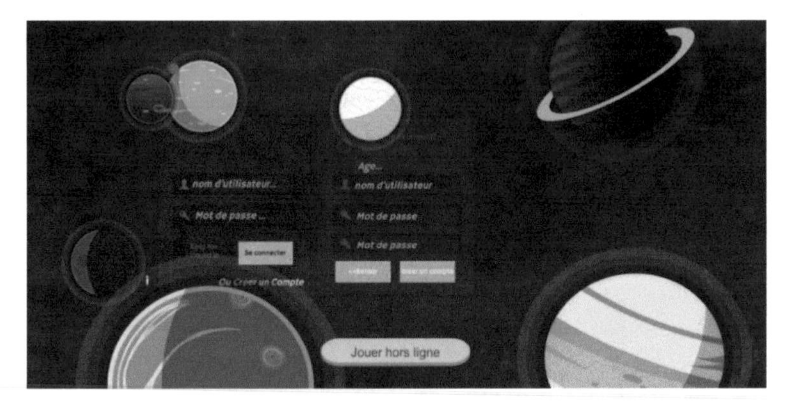

Fig. 1. UI login system

3 Game Play Description

ImALeG can be played on several platforms under windows Linux, android tablet, game consoles, or Oculus VR. Depending on the situation of the player, Tifinaghe alphabets, animals and items are presented visually in the virtual world. To be explored, the player must conduct conversations with NPCs and identify the location of each learning item through a mini-map.

The progression in ImALeG organized by 33 pedagogical units or situations in the first level matches 33 Tifinaghe alphabets to explore. A level takes between 15 to 20 min, according to the player speed and error rate. Progression is made so that each unit will include graphical and phonological aspects previously seen in Amazigh classic courses. To guide the player; a linear order of the game storyline and storytelling as a cinematic scenario is presented in advance.

The storyline of the first level, managed through a tutor agent (NPC) who gives the player rewards, tracks his movements all over the scene and presents reached Tifinaghe alphabet. The player can click on the discovered Tifinaghe alphabet to display a characteristics window that contains its type, equivalence in Latin and a sound button to listen to the corresponding pronunciation and eventually the collected alphabets appears in his inventory. For each discovered alphabet the player increases his energy with 3%. When the player reaches 30% of his energy which corresponding to 10 Tifinaghe alphabets collected, a selection menu appears asking the player for an assessment on these collected alphabets or to keep playing until the second 20 alphabets "see Fig. 4", if the player accepts, he will be redirected to a new scene to perform evaluation about the collected alphabets. In the evaluation scene the player must click on the alphabet that matches the sound, if the player selects the exact alphabet, another stimulus is proposed until the player completes the evaluation. If the player succeeds in the assessment, he will validate acquired alphabets, else the player's health bar will be reset to 0%, 30% or 60% corresponding to evaluated fragment that must be learned. The process is repetitive 3 times in each fragment of Tifinaghe alphabets. During the game, the system records the players' answers and store them in an external database (Fig. 2).

Fig. 2. Tutor NPC guidance and rewards in ImALeG

The second level "see Fig. 5" presented as a zoo contains serval animal NPCs divided in two types; statics NPCs with them the player can interact and receive advice to accomplish a quest and a dynamic NPCs who moves randomly to increase realism in the scene.

To facilitate detecting the position of theses statics NPCs, a navigation map presents the position of each static animal. The player must conduct conversation with these NPCs in order to learn their Amazigh names by reading and hearing conversations and by clicking on a static animal a window displays more information about it such as their category, name spelling in Amazigh and equivalence in Latin. For each discovered animal, the player receives energy and an image illustrating the item acquired appears in their inventory. This level contains 25 static animals that can be improved at each update to enhance the players experience "see Fig. 5". When the player reaches 100% of his energy, he will be teleported to an evaluation scene similar to the first level "see Fig. 4". During this assessment the player hears and reads the animals' name and must match the corresponding animal. If the player can identify 23 among the 25 animals, he will be taken to the next level. Above this number the player fails and replays the entire level.

Fig. 3. Screen shot of the perspective design environment in mobile version at Level 2.

In the 3rd level the player is put in a house containing 4 rooms "kitchen, bathroom, living room, bedroom". Each room contains several objects that the player must explore "see Fig. 6". The first learning situation is the living room. The player must seek and explore Amazigh names of each items in this room. After completing the exploration of all the items, an assessment is presented similar to situations seen in the 2nd level. If the player succeeds, he will win the key to open the next door, else, the player's inventory will be reset to none and he will began exploring the room items again.

Fig. 4. Home items Learning Level

This brief presentation indicates that ImALeG is a serious game that presents auditory stimuli at different unit proportions (phoneme, syllable, words). ImALeG reinforces auditory perception, and provides access to phonological representations of the Amazigh language. Thanks to the massive repetition, ImALeG aims to stabilize the orthographic and phonological coding. The game contains feedback that can be experienced as negative Feedbacks, are generally seen as negative feedback from the player's point of view, but it does not alter the player's sense of competence. While a safe progression without feedbacks gradually improves the competence feeling.

4 Evaluation Strategy and Results

Sixty-eight students took part in the event. Group A followed game-based learning at school using ImALeG, with 33 students. While Group B, with 35 students, took a normal Amazigh course (see Fig. 7). The calculated principal values were used to measure the changes before and after the event as well as to clarify the differences between the two teaching methods.

Both groups have gained knowledge, which emerges from the questions. However, the knowledge gain of group A is much greater. The determination of knowledge gain was checked using two pre and post-test methods.

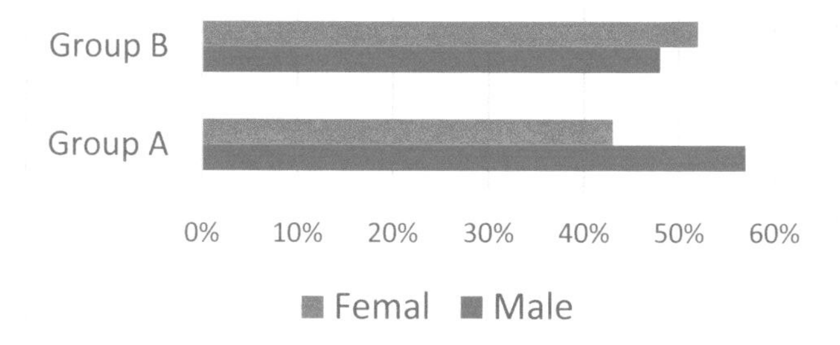

Fig. 5. Target population and gender variety in each group

The user experience was evaluated through a questionnaire that was offered to students in group A after using ImALeG. It consisted of closed questions whose answers were given on a four-point scale (0 = strongly disagree, 1 = somewhat disagree, 2; 3 = somewhat agree, 4; 5 = completely disagree).

The actual assessment of ImALeG carried out through activities in the primary school. We conducted a preliminary formative knowledge assessment of the both groups on pedagogical objectives (Pretest) "Tifinaghe alphabets, animal names, home objects", these are defined in the official manual of the Amazigh language. This formative evaluation helps us to identify skills of each student in both groups. Group A took 3 play sessions guided by a teacher for 3 days, while group B had a classical course with the same duration through Amazigh manual. After 3 days, both groups received a post-test divided into three parts, each part deals with specific pedagogical teaching skills that have been already dealt with in every course session.

Table 1. Pre-test and post-test results for both groups

		Min	*Max*	*Avg*	δ
Group A	*Before game session*	3	9	5.61	3.4
	After game sessions	5	10	7.17	1.41
Group B	*Before classic course*	3	9	5.69	3.1
	After classic courses	4	7	6.11	0.8

As we have mentioned, the user experience was evaluated through a pre and post-test survey that was offered to students before and after using ImALeG. We compared obtained results from group A with group B, who received a classic course. The pre-test and post-test are composed of a closed survey whose answers were given on a scale from 0 to 10. Table 1 shows students' level of knowledge, before and after the game sessions.

Table 2. Emotion felt by participants

	Min	*Max*	*Avg*	*δ*
Surprise	4	5	4,88	0,332
Pleasure	4	5	4,83	0,383
Excitement	4	5	4,89	0,323
Joy	4	5	4,89	0,323
Perplexity	3	5	4,83	0,514
Fear	0	3	0,78	1,003
Sadness	0	0	0,00	0,000
Disgust	0	0	0,00	0,000

We then asked the participants of group A in a scale from 0 to 5 to measure their feeling level of different emotions. Table 2 shows the average responses expressed.

Table 3. Students feedbacks

	Feedback number	*Frequency*	*Percentage*
Level 1	0	21	63,6%
	1	12	36,4%
Level 2	0	27	81.8%
	1	6	18.2%
Level 3	0	18	54.54%
	1	15	45.46%

Feedbacks in Table 3 refer to the number of attempts that took place by the players to move on to the next level (without any attempts; feedback = 0, with one attempt; feedback = 1). Table 3 presents the feedback frequency made by the students of group A (F = 12 for the first level, F = 6 for the second level and F = 15 for the 3rd level).

5 Discussion

The objective of this study was to assess the impact of virtual reality in the case of learning the Amazigh language. The pupils of group A practiced the game in class, for 3 days. In terms of learning, the first results presented for group A compared to group B here are encouraging (Table 1). Indeed, our study shows that learning is better if the tools and objects explored during the training are carried out in Virtual Reality rather than in a more symbolic way as is the case with the first version. The rationale for this result, in our opinion, is linked to the fact that the degree of realism and immersion made possible by VR allows participants to more simply make the connection between the elements with which they interact in the game and the reality of the game. Activity that is simulated.

It is deduced that the level of knowledge of the Amazigh language before the game experience is almost the same for the two groups (m = 5.61 for the group A and m = 5.69 for the group B) and after the sessions It's clearly remarkable that the group A has better overall skill scores than group B (m = 6.11 for group B vs. m = 7.17 for group A).

The results obtained during the ImaLeG VR emotional impact assessment questionnaire (Table 2) show that our concept is useful, relevant, fun and easy to use. They report positive emotions towards learning tools such as pleasure, excitement, joy and do not experience negative emotions such as embarrassment, disappointment, fear, sadness, disgust. They also do not express a sense of stress related to the pace of the activity or the errors that handling VR devices would induce. Thus, we can conclude that the design of ImaLeG does not create a usability problem. The most often expressed emotions are positive like pleasure (m = 4.83), excitement (m = 4.89) and joy (m = 4.89). The students did not express negative emotions. Some students expressed a fear feeling of choosing the wrong answers (m = 0.78) or misuse of the workstation.

The percentage rate of students who succeed at the first level without feedback was is important. In the second game session, we notice that the group A has become familiar with the game system, the feedback percentage rate is almost reduced P = 18.2%. However, this result remains important. The feedback decrease is explained by the increase of the knowledge level for the second level. Regarding the 3rd level of the game. The percentage of students who had only one feedback was high (F = 45.46%), due to the increased difficulty level. This result invites us to divide the 3rd level into 2 levels to reduce the complexity of the learning situations presented.

Our study shows that Amazighe learning through ImALeG VR is better understood than a traditional course. Students could play individually at first and then discuss the different group choices based on the effects they produce. These activities would have the effect of deepening the discussions and thus understanding the deployment methods of ImALeG. These results should be considered with caution, however, as the size of the sample of participants was relatively small (68 participants in total).

6 Conclusion

The VR version of ImALeG presented in this paper is an improved version of our previous project. The system allows collecting learners data through a tracking system, to analyse and identify all long-term positive and negative trends. The results obtained during the tests allow us to say that ImALeG VR is therefore functional. We attribute this success to our game's original approach of using virtual reality to learn the Amazigh language. The ImALeG scenario also contributed to this result in two ways. Firstly, it motivates students by making them actors in the unfolding of history. Secondly, it gradually introduces the educational content. Although ImALeG was developed with a clear vision, it has been systematically examined whether individual design decisions to achieve the desired goals present in the introduction section. One aspect that deserves special attention here are the gamification of learning elements and associated rules used. Also based on users' feedbacks, we suspect that the participant experience in ImaLeG can be further improved by integrating other levels as a future work.

References

1. Susi, T., Johannesson, M., Backlund, P.: Serious games: An overview (2007)
2. Westera, W., Nadolski, R.J., Hummel, H.G., Wopereis, I.G.: Serious games for higher education: a framework for reducing design complexity. J. Comput. Assist. Learn. **24**(5), 420–432 (2008)
3. Van Oostendorp, H., Van der Spek, E.D., Linssen, J.: Adapting the complexity level of a serious game to the proficiency of players. EAI Endorsed Trans. Serious Games **1**(2), e5 (2014)
4. Lopes, R.: Scenario adaptivity in serious games. In: Proceedings of the Fifth International Conference on the Foundations of Digital Games, pp. 268–270 (2010)
5. Tazouti, Y., Boulaknadel, S., Fakhri, Y.: ImALeG: a serious game for amazigh language learning. Int. J. Emerg. Technol. Learn. (IJET) **14**(18), 28–38 (2019)
6. Silva, A., Mamede, N., Ferreira, A., Baptista, J., Fernandes, J.: Towards a serious game for portuguese learning. In: Ma, M., Fradinho Oliveira, M., Madeiras Pereira, J. (eds.) SGDA 2011. LNCS, vol. 6944, pp. 83–94. Springer, Heidelberg (2011). https://doi.org/10.1007/978-3-642-23834-5_8
7. Paredes, M., Rocha, Á.: Information technology and systems, RISTI - Rev. Iber. Sist. e Tecnol. Inf., no. E17, pp. IX–X (2019)
8. Berns, A., Gonzàlez-Pardo, A., Camacho, D.: Implementing the use of virtual worlds in the teaching of foreign languages (level A1). In: Proceedings of Learning a Language in Virtual Worlds: A Review of Innovation and ICT in Language Teaching Methodology. Warsaw, pp. 33–40 (2011)
9. Berns, A., Gonzalez-Pardo, A., Camacho, D.: Game-like language learning in 3-D virtual environments. Comput. Educ. **60**(1), 210–220 (2013)
10. Connolly, T.M., Stansfield, M., Hainey, T.: An alternate reality game for language learning: ARGuing for multilingual motivation. Comput. Educ. **57**(1), 1389–1415 (2011)
11. Amoia, M., Brétaudière, T., Denis, A., Gardent, C., Perez-Beltrachini, L.: A serious game for second language acquisition in a virtual environment. J. Systemics Cybern. Inf. **10**(1), 24–34 (2012)
12. Wong, W.L., et al.: Serious video game effectiveness. In: Proceedings of the International Conference on Advances in Computer Entertainment Technology, pp. 49–55 (2007)
13. Lane, H.C., Hays, M.J., Auerbach, D., Core, M.G.: Investigating the relationship between presence and learning in a serious game. In: Aleven, V., Kay, J., Mostow, J. (eds.) ITS 2010. LNCS, vol. 6094, pp. 274–284. Springer, Heidelberg (2010). https://doi.org/10.1007/978-3-642-13388-6_32
14. Vidoni, C., Lee, C.H., Azevedo, L.B.: Fair play game: a group contingency strategy to increase students' active behaviours in physical education. Early Child Dev. Care **184**(8), 1127–1141 (2014)
15. Sadouk, L., Gadi, T., Essoufi, E.H.: Handwritten tifinagh character recognition using deep learning architectures, pp. 1–11 (2018)
16. Perlman, S.B., Camras, L.A., Pelphrey, K.A.: Physiology and functioning: parents' vagal tone, emotion socialization, and children's emotion knowledge. J. Exp. Child Psychol. **100**(4), 308–315 (2008)

17. Steiner, K.E., Moher, T.G.: Graphic StoryWriter: an interactive environment for emergent storytelling. In: Proceedings of SIGCHI Conference on Human Factors in Computing System, pp. 357–364 (1992)
18. Yang, H.J.: Factors affecting student burnout and academic achievement in multiple enrollment programs in Taiwan's technical-vocational colleges. Int. J. Educ. Dev. **24**(3), 283–301 (2004)
19. Bakkes, S.C.J., Spronck, P.H.M., van Lankveld, G.: Player behavioural modelling for video games. Entertain. Comput. **3**(3), 71–79 (2012)
20. Petrovic, V.M.: Artificial intelligence and virtual worlds-toward human-level AI agents. IEEE Access **6**, 39976–39988 (2018)
21. Hoang, H., Lee-urban, S., Muñoz-avila, H.: Hierarchical plan representations for encoding strategic game AI. In: Artificial Intelligence and Interactive Digital Entertainment Conference, pp. 63–68 (2005)
22. Tazouti, Y., Boulaknadel, S., Fakhri, Y.: Design and IMPLEMENTATION OF ImALeG serious game: behavior of non-playable characters (NPC). In: Saeed, F., Al-Hadhrami, T., Mohammed, E., Al-Sarem, M. (eds.) Advances on Smart and Soft Computing. AISC, vol. 1399, pp. 69–77. Springer, Singapore (2022). https://doi.org/10.1007/978-981-16-5559-3_7

Efficient Autonomous Pathfinding Method for Virtual Environments

Yassine Tazouti(✉), Yasser Lamalem, and Khalid Housni

Department of Computer Science, Ibn Tofail University, Kénitra, Morocco
yassine.tazouti@uit.ac.ma

Abstract. Pathfinding is a core component of most games today. Characters, animals and vehicles all move in a deliberate way. The NPCs must be able to avoid obstacles and identify the most efficient route from the starting point to the destination. A large number of games needs to solve similar multi-intelligent body movement. In this paper, we propose an improved flow field pathfinding algorithm that show its efficiency over other pathfinding methods during the experiments and allowing a large number of moving intelligences to reach the same destination in less time.

Keywords: Pathfinding · Path planning · Heuristic methods · A* · Video game

1 Introduction

AI has been introduced into the various elements that make up games to increase the challenge for players, the playability of the game, and the realism of the simulated world. Games as one of the earliest games with the largest number of AI applications, FPS games not only have AI applications for NPC AI, tactical AI, battle layout, hazard estimation, and terrain analysis, but also have important applications for pathfinding algorithms that include simple AI pathfinding and group AI pathfinding. Intelligent path search algorithm is the most basic and core problem in the game, and it is also an important part of video game. The path search algorithm is an algorithm to find a feasible path between the starting point and the end point, and has a rich theoretical basis and many applications in many fields, including network traffic, robot path planning, military simulation and computer games. In 1968, Hart et al. [1] proposed an A* algorithm based on heuristic search, which shortened the path finding time; Zhao et al. [2] applied the A* algorithm to video games to improve single object path finding; R Silveira et al. [3] applied the potential field method to interactive path finding in video games; Sartoretti et al. [4] combined reinforcement and imitation learning to propose a new framework of Multi-Agent Path Finding (MAPF), Fu et al. [5], uses a batch processing approach to achieve multi-intelligent path finding in UAV traffic management. However, these pathfinding algorithms do not meet the need of fast pathfinding for multiple intelligences, and most of them optimize the speed of single path planning to improve the overall movement rate,

M. Lazaar et al. (Eds.): BDIoT 2022, LNNS 625, pp. 464–473, 2023.
https://doi.org/10.1007/978-3-031-28387-1_39

but do not improve the group path planning at the global macro level, so new algorithms are needed to solve the concurrent pathfinding problem for a large number of moving intelligences in a short time. In 2013, the flow field pathfinding algorithm proposed by Pentheny [6] introduced the flow field dynamics in physics to game pathfinding and successfully applied it to the games Supreme Commander 2 and Hold the Line 2, which realized the simultaneous fast pathfinding of multiple objects and broke the limitation of the number of smart bodies, so that the smart bodies in the game can simulate the real human movement and can gather and disperse when moving. In this paper, we design a new path search algorithm in video games - flow field pathfinding algorithm, and improve the algorithm in three aspects: data structure storage, path cost calculation and flow field generation algorithm, to further improve the operation efficiency of flow field pathfinding algorithm and solve the collision blocking problem. Within our paper we present a problem description.

2 Problem Description

The map in the game is generally constructed by rasterization for path computation, and then different path finding algorithms are used for computation. Figure 1 shows a rectangular map of size 200×160, divided into small squares of length 4, the number of grids is 50×40, each grid is considered as a node, and the edges are the paths between nodes. After the initialization of the map, n smart bodies are randomly generated and different pathfinding algorithms are used to calculate the generated path for each smart body. The earliest pathfinding algorithm used is Dijkstra's algorithm, which is widely used in various situations, such as: Geographic Information System (GIS) path planning [7], urban road optimization [8], logistics distribution [9], disaster relief routes [10], and it can be seen that the above scenarios are all about finding an optimal path between two points. The problem of finding an optimal path between two points is only considered to find the shortest path between two points, ignoring the global group movement and the blocking collision problem of the exit bottleneck in the actual problem, and is not applicable to the game of instantaneous uncertainty of multiple intelligent body movement. Subsequently, the A* algorithm emerged to solve the intelligent pathfinding problem of a single Non-Player Character (NPC) in the game, such as reducing the search area, lowering the cost time of the search, and ensuring the relatively suitable optimal path in a short time, which has become the most widely used pathfinding algorithm in the game, and is also used in urban traffic, vehicle It also has important applications in urban traffic, vehicle scheduling, mosaic extraction, parking lot pathfinding, and soldier combat.

20	20	20	20	20	20	20	20	20	20
20	20	20	20	20	20	20			20
	20				20	20	20	20	20
	20		20		20	20	20	20	20
			20		20				
	20		20		20		20		20
	20		20	20	20		20		20
20	20		20	20	20		20		20
20	20	20	20	20	20	20	20	20	20
20	20	20	20	20	20	20	20	20	20

Fig. 1. Schematic diagram example of cost field

The artificial potential field method is also introduced into video games as a common local path planning method. The algorithm mainly simulates a virtual force field, where the target point generates gravitational force on the object, the obstacle generates repulsive force on the object, and the combined force of gravitational force and repulsive force forms the movement path of the object. Intelligent bionic algorithms such as genetic algorithms and ant colony algorithms are also often used in path planning, which mainly simulate various behaviors and phenomena in nature such as genetic variation and ant foraging, but such intelligent algorithms need several iterations to obtain more accurate paths. All of the above methods can achieve path planning, but the traditional path planning is slow, the artificial potential field method is suitable for local planning, and the intelligent path planning requires several iterations, which greatly affects the fluency and user experience of the game. The flow field pathfinding algorithm is a good solution to the requirements of video games for pathfinding time and quantity, and is suitable for large scale simulation of crowd movement. The main idea is to generate a "field" about the path based on the map and the destination, which marks the direction of movement from any location in the map to the destination node.

3 Flow Field Pathfinding Algorithms

3.1 Basic Data Structure

There are three data types for flow field pathfinding, which are used to store the cost field type when passing through a node, the complete cost field type that stores the minimum generation value of the node to reach the target node, and the flow field type that stores the direction to the target node when passing through the node.

The cost field type is a cost type with a pre-defined value for the cost of each node when initializing the map (see Fig. 1), usually with a maximum value for the impassable area, e.g., walls, mountains, etc., and a positive integer for the surrogate value of passing, with different values for different terrain features, e.g., swamps, oceans, deserts, and slopes with a surrogate value greater than the surrogate value of passing through plains, with a minimum surrogate value of 1. If there is no explicit surrogate value, a static global variable is assigned to the cost field to facilitate the calculation and representation of the flow field.

The flow field stores the index field type of an enumerated direction lookup table, the direction of movement of the smart body when it passes a node in the process of reaching the target node, which will generally be an enumerated type with eight enumerated values: east, west, south, north, southeast, southwest, northeast, and northwest. If the next node in the direction of arrival is detected as an obstacle, the node is skipped, as shown in Fig. 2.

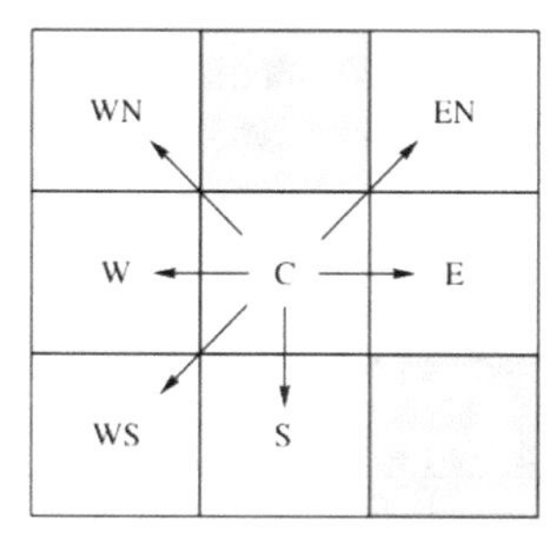

Fig. 2. Schematic diagram of flow field

3.2 The Basic Idea of the Algorithm

In virtual environment, the map is usually divided into a regular grid of squares, and the grid points are considered as nodes, and pathfinding is performed on the basis of the grid. The flow field pathing only needs to calculate the flow field direction of all the nodes in the map once, and then it can move regardless of the number of intelligent bodies. The basic steps of the algorithm are as follows.

<u>Algorithm 1: Flow field direction</u>

1: Initialize the smartbody and the map, define the set S, in which the grid, obstacles and target nodes are drawn, and the set S is used to store the nodes for which the flow field direction has been calculated but not determined.
2: Add the target node v to the set S and assign the complete generation value of v to 0.
3: Select the node u with the smallest complete generation value from the set S.
4: Record node u and remove it from the set S.
5: Calculate the complete generation value of the 8 neighboring nodes of the updated node u, and add the nodes that are not added to the set S to the set S.
6: Determine the 8-neighbor node flow field field values based on the complete generation values.
7: Repeat the above processes from 3) to 6) until the set S is empty.

The generation value of the neighboring nodes of the node in the set S is obtained through a loop, and the generation value is compared to determine which of the 8 neighboring nodes is the next node after passing the node, and is marked to generate the flow field value, which indicates the flow field direction of each moving intelligence passing the node. The above is the data structure and the whole algorithm idea of the flow field pathfinding. Although the flow field pathfinding algorithm solves the problem of fast pathfinding of multiple intelligences at the same time and makes all the moving intelligences in the video game look very smooth and intelligent, the flow field pathfinding algorithm still has shortcomings, such as: unnecessary double counting during the algorithm operation, the data does not have a good arrangement and calculation order, the method of calculating the generation value in the map and the method of obtaining the flow field direction. However, the flow field pathfinding algorithm still has some shortcomings, such as: unnecessary double counting in the algorithm, no good arrangement and order of data, complicated method of calculating the value of generation and obtaining the flow field in the map, and even the need to calculate nonlinear partial differential equations.

4 Improvement of Flow Field Pathfinding Algorithm

4.1 Optimal Path Cost Calculation Using Penalty Function

In the calculation method of cost value proposed by Pentheny [11], the eigenfunction equation is used for calculation. The eigenfunction equation is a kind of non-linear partial differential equation.

It is defined in the form of the Hamiltonian-Jacobi partial differential equation (HJ-PDE) [12], as shown in Eq. (1):

$$H(x, \nabla\phi(x)) = |\nabla\phi(x)|^2 - \frac{1}{f^2(x)} = 0; \forall x \in \Omega \tag{1}$$

where Ω is the domain in R^n, $\phi(x)$ is the distance or time to the source point, and $f(x)$ is the forward velocity function defined in Ω. Usually, nodes are defined in terms of n-tuples, and taking the two-dimensional node $x = (i, j)$ as an example, the algorithm in this paper defines the line segment directly connecting two nodes as an edge, and the length of the edge is along the corresponding axis $p(p \in \{x, y\})$ to define the length hp of the grid, and the adjacent neighboring nodes are defined as nodes connected by a single edge, such as there is node $y = (i + h_x, j)$, which is the node $x = (i, j)$ in the x-axis direction adjacent neighboring nodes. The Godunov upwind discretization [30] is used to discretize the equations of the equations of the equations, and the discretization result in the x-direction is $g(x)$, as shown in Eq. (2) .

$$g(x) = \left[\frac{(U(x) - U(x)^x_{min})^+}{h_x}\right]^2 + \left[\frac{(U(x) - U(x)^y_{min})^2}{h_y}\right] - \frac{1}{f^2(x)} \tag{2}$$

$U(x)$ is the discretized approximation of the node $x = (i, \phi)$ pointing in the direction of ϕ, and $U(x)^p_{min}$ is the minimum U value of two adjacent neighbors of $U(x)$ along the p direction, $(n)^+ = max(0, n)$. From the above analysis, it is clear that the constraints of the Michadon-Jacobi based equation are $U(x) - U(x)^p_{min}$, and it is obtained that each direction of the coordinate axis will affect the calculation of the integrated surrogate value. The value of $g(x)$ decreases as n takes on smaller values. Therefore, we try to avoid changes in direction during the calculation of the integrated generation value, and penalize $g(x)$ for changes in x and y directions. In the calculation of the complete generation value, it is necessary to repeat the calculation, because the calculation of the nonlinear partial differential equation is complicated, and there are two constraints, $U(x)^p_{min}$ and $(n)^+ = max(0, n)$, to convert the solution of the constrained nonlinear programming problem into the calculation of the unconstrained minimal value problem, according to the actual situation of the problem, the penalty function is used to calculate, and then the calculation of the integrated generation value.

The constrained problem min $g(x)$, satisfying the restriction $c_i(x) \geq 0, \forall x \in I$, is transformed into a sequence of unconstrained problems, as shown in Eq. (3):

$$\min h(x) = g(x) + dirCost; \tag{3}$$

$$dirCost = 10 \times \sqrt{|e.x - s.x| + |e.y - s.y|}$$

According to the experimental results, a penalty factor of 10 is chosen to penalize the direction, ensuring ensure that the generation value of diagonal direction is greater than that of horizontal or vertical direction.

4.2 Flow Field Calculation

The basic algorithm of flow field pathfinding uses the line of sight method to pre-process the data and perform the calculation of the whole but when the "wave" is pushed forward, it is necessary to compare all the calculated complete generation values to this node and ensure the minimum uniqueness of the values. This results in an increase in node information storage during the computation. Inspired by the literature [13] introduced Dijkstra's algorithm, which uses a breadth traversal approach to reverse the calculation. This method uses a generalized traversal to solve the multi-path problem based on the leading and trailing neighbors, which makes it possible to calculate multiple paths simultaneously. The computation process calculates multiple paths simultaneously, and then obtains the optimal solution. The target node is considered as the initial node, and in the process of "wave" forward Dijkstra's algorithm is used to calculate the predecessor neighbors of each node, which will have predecessor neighbors in the flow field. In the direction of the flow field, there will be predecessor neighbors pointing to this node, and the specific implementation process is as follows.

Algorithm 2: improved flow field calculation method

1: **While** (OPEN table is not empty)
    2: Take the node with the smallest value of the OPEN table generation.
    3: Deleting the node and inserting it into the CLOSE table.
    4: Update the nodes around the current point to the OPEN table.
// update the nodes in the 8 surrounding neighbors by traversal
    5: Determine whether the predecessor neighbor of the surrounding nodes is the current node.
    6: Determining the direction of the flow field of the surrounding nodes.
7: **EndWhile**

The node with the smallest value of the complete generation is taken for calculation each time, and the current node is deleted after calculation. After calculation, the previous node is deleted and the next integrated node with the smallest value is selected, and the operation is cyclic until the tree is empty.

Figure 3 shows the table of the complete cost values obtained from the interception section, and the target node (0, 0) is shown in the higher right corner. The result of the generated flow field is shown in Fig. 4. After completing the flow field calculation, there is no restriction on the number of smart body movements. The intelligent body at any position in the map can move in the direction of the flow field, and in the large map the intelligent body tends to move in the direction of the flow field, and it is more consistent with the group movement law when multiple intelligent bodies move. The intelligent body tends to move to the open area where the terrain is gentle and easier to reach the target point.

This experiment improves the flow field finding in three ways: data storage method, path cost calculation method and flow field calculation method. The improved flow field pathfinding algorithm is based on three aspects: data storage method, path cost calculation method and flow field calculation method. The improved flow field pathfinding algorithm has more advantages in data reading, and simplifies the calculation steps of the algorithm. The approach adopted in this paper simplifies the implementation process of the algorithm, improves the speed of path the algorithm is optimized from an overall perspective.

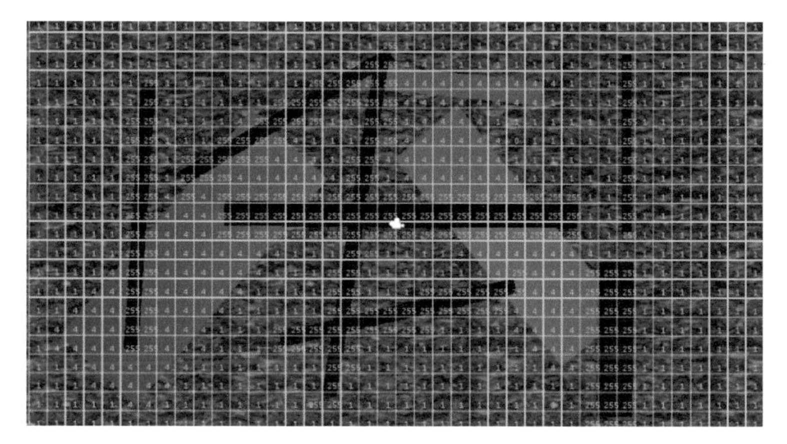

Fig. 3. Cost value of each node

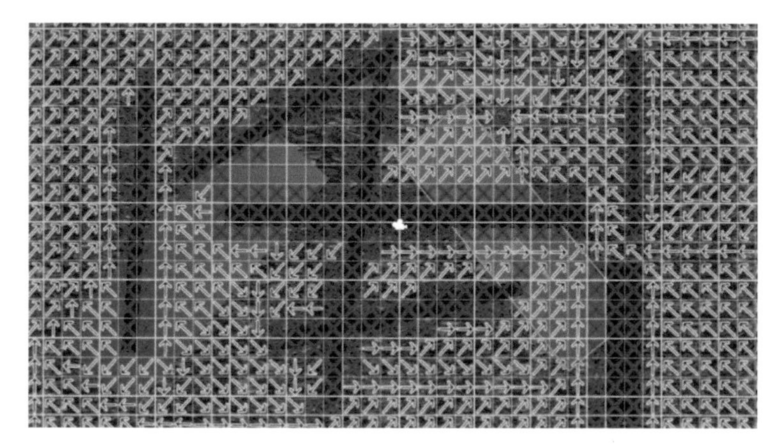

Fig. 4. Result diagram of flow field

5 Experimental Results and Analysis

Simulation experiments are implemented in Visual Studio 2017 platform using OpenGL the algorithms are visualized for different numbers of intelligent bodies, and the path generation and completion times of different algorithms are compared. The time to generate paths and complete the movement is compared for different number of intelligences. The experimental results show that flow field pathfinding is very useful in virtual environment games enabling a large number of moving intelligences to reach the same destination in a shorter time.

The experiments compare Dijkstra's algorithm, A* algorithm, artificial potential field method, flow field pathing and improved flow field pathing. The number of mobile intelligences in Dijkstra's algorithm, A* algorithm, artificial potential field method, flow field pathing and the improved flow field pathing algorithm are compared. The pathfinding and completion times are compared for the cases of 10, 20, 50, 100 and

200 for the Dijkstra algorithm, A* algorithm, artificial potential field method, flow field pathfinding and improved flow field pathfinding algorithm.

Experimental data the experimental data prove that the flow field pathfinding algorithm is more advantageous when the number of moves is important. It can perform pathfinding in the same time without any limitation. Table 1 shows the time required to calculate the paths for different algorithms with different number of intelligences, while Table 2 shows the time required for different algorithms to complete the movement for different number of intelligences.

Table 1. Algorithms' average pathfinding times.

	Completion times (Unit: s)					
	1	10	20	50	100	200
Dijkstra's algorithm	0.2694	0.7208	1.1788	2.8621	4.4377	8.4058
A* algorithm	0.2003	0.7277	1.0922	2.4401	4.1493	8.3825
Artificial potential field method	0.4075	0.6538	0.9237	1.9592	3.5376	6.8302
Flow field pathing	0.5636	0.5556	0.5828	0.4759	0.5254	0.4819
Improved flow field pathing	0.3534	0.4288	0.3272	0.4135	0.4584	0.3381

The data in Tables 1 and 2 show that the flow field pathfinding algorithm has an advantage in multiple pathfinding for large maps, and it can also be fast pathfinding in maps with many obstacles and complex terrain. The flow field pathfinding algorithm has the advantage of multiple pathfinding in large maps, and it can also fast pathfinding in maps with many obstacles and complex terrain, and has good stability.

Table 2. Algorithms' average movement times.

	Completion times (Unit: s)					
	1	10	20	50	100	200
Dijkstra's algorithm	1.88810	8.8658	22.189	53.759	98.207	213.280
A* algorithm	1.9880	9.3480	19.446	48.112	98.114	212.720
Artificial potential field method	1.7089	8.0545	17.263	43.970	73.748	147.570
Flow field pathing	7.5564	15.6490	18.786	19.901	20.582	21.341
Improved flow field pathing	7.6318	15.0910	18.307	19.596	19.431	19.961

6 Conclusion

A large number of 3D games need to solve similar multi-intelligent body movement.In this paper, we propose a flow field pathfinding algorithm and combine the data storage method, integrated generation value calculation method and flow field generation method. This paper proposes a flow field pathfinding algorithm, and improves the data storage method, integrated generation value calculation method and flow field generation method. The computational speed of flow field pathfinding and the efficiency of intelligent body movement are improved, which also improves the game intelligence level, making it possible for tens of thousands of intelligences to simultaneously This increases the playability and realism of this type of game and further optimizes the game experience. This paper mainly focuses on improving the algorithm itself and implementing it, which can also realize multi-targeted flow field calculation for the problem of avoiding obstacles, walls, and other smart objects while moving, the computation will be performed later.

References

1. Hart, P.E., Nilsson, N.J., Raphael, B.: A formal basis for the heuristic determination of minimum cost paths. IEEE Trans. Syst. Sci. Cybern. **4**(2), 100–107 (1968)
2. Zhou, R., Hansen, E.A.: Beam-stack search: integrating backtracking with beam search. In: ICAPS, pp. 90–98 (2005)
3. Silveira, R., Fischer, L., Ferreira, J. A. S., Prestes, E., Nedel, L.: Path-planning for RTS games based on potential fields. In: Boulic, R., Chrysanthou, Y., Komura, T. (eds.) MIG 2010. LNCS, vol. 6459, pp. 410–421. Springer, Heidelberg (2010). https://doi.org/10.1007/978-3-642-16958-8_38
4. Damani, M., Luo, Z., Wenzel, E., Sartoretti, G.: PRIMAL $ _2 $: pathfinding via reinforcement and imitation multi-agent learning-lifelong. IEEE Robot. Autom. Lett. **6**(2), 2666–2673 (2021)
5. Fu, Z., Mao, Y., He, D., Yu, J., Xie, G.: Secure multi-UAV collaborative task allocation. IEEE Access **7**, 35579–35587 (2019)
6. Pentheny, G.: Efficient Crowd Simulation for Mobile Games. Game AI Pro, pp.317–323 (2013)
7. Vorobieva, H., Glaser, S., Minoiu-Enache, N., Mammar, S.: Automatic parallel parking in tiny spots: path planning and control. IEEE Trans. Intell. Transp. Syst. **16**(1), 396–410 (2014)
8. Miandoabchi, E., Daneshzand, F., Szeto, W.Y., Farahani, R.Z.: Multi-objective discrete urban road network design. Comput. Oper. Res. **40**(10), 2429–2449 (2013)
9. Chang, F.S., Wu, J.S., Lee, C.N., Shen, H.C.: Greedy-search-based multi-objective genetic algorithm for emergency logistics scheduling. Expert Syst. Appl. **41**(6), 2947–2956 (2014)
10. Macone, D., Oddi, G., Pietrabissa, A.: MQ-routing: mobility-, GPS-and energy-aware routing protocol in MANETs for disaster relief scenarios. Ad Hoc Netw. **11**(3), 861–878 (2013)
11. Pentheny, G.: Efficient crowd simulation for mobile games. In: Game AI Pro 360: Guide to Movement and Pathfinding, vol. 8, p. 77 (2019)
12. Deng, J., Anton, C., Wong, Y.S.: High-order symplectic schemes for stochastic Hamiltonian systems. Commun. Comput. Phys. **16**(1), 169–200 (2014)
13. Bhadoria, A., Singh, R.K.: Optimized angular a star algorithm for global path search based on neighbor node evaluation. Int. J. Intell. Syst. Appl. **6**(8), 46 (2014)

A Distributed Architecture for Visual Data Processing in Visual Internet of Things (V-IoT)

Afaf Mosaif(✉) and Said Rakrak

Laboratory of Computer and Systems Engineering (L2IS), Faculty of Science and Technology, Cadi Ayyad University, Marrakesh, Morocco
afaf.mosaif@edu.uca.ac.ma

Abstract. As wireless sensors and computer vision have advanced, Visual Internet of Things (V-IoT) has emerged as a form of IoT that can offer different levels of intelligence based on its application and algorithms. It is important to note, however, that V-IoT nodes collect a substantial amount of visual data that must be transmitted to the network for further analysis. The purpose of our paper is to present a new architecture that will reduce the information flow from the V-IoT nodes to the base station and balance the traffic load on the network. Our architecture is based on the Message Queue Telemetry Transport (MQTT) protocol, and it distributes the data processing across the network levels taking into account the limited resources of the V-IoT nodes. To evaluate our architecture, we built a V-IoT node prototype using a Raspberry Pi board and then used it as part of an experiment to recognize and track faces. The results of the prototype implementation indicate that our proposal is effective.

Keywords: Face recognition · Message Queue Telemetry Transport · Object detection · Video Sensor · Visual data Processing · Visual Internet of Things

1 Introduction

Internet of things (IoT) is a well-known paradigm in which ubiquitous heterogeneous objects (e.g. sensors, actuators, mobile devices, radio frequency identification (RFID) tags, on-board computers, etc.) are connected to the Internet through a wired or wireless network to interact with each other to achieve a desirable goal. Generally, these objects are resource-constrained. IoT has been defined by different authors in many different ways. Vermesan et al. [1] define the IoT as simply an interaction between the real/physical and the digital/virtual worlds. While according to Nauman et al. [2], several researchers have defined IoT as "The integration of tiny devices known as Smart Objects (SO), usually, battery-operated equipped with a Microcontroller (MCU) and transceivers into the global Internet. The services offered by these smart objects are known as Smart Services (SS)".

Recently, most advanced IoT devices are equipped with visual sensors, forming the so-called Visual IoT (V-IoT). The V-IoT significantly relies on computer vision processing techniques to sense and process massive amounts of visual data [3], and it

M. Lazaar et al. (Eds.): BDIoT 2022, LNNS 625, pp. 474–485, 2023.
https://doi.org/10.1007/978-3-031-28387-1_40

can be used in various applications, as shown in Fig. 1, due to the miniaturized sizes and lightweight designs of its devices, for example, it can easily be loaded into unmanned drones or vehicles to reach terrains that are inaccessible to humans to enable intelligent surveillance, remote-sensing in these areas [4].

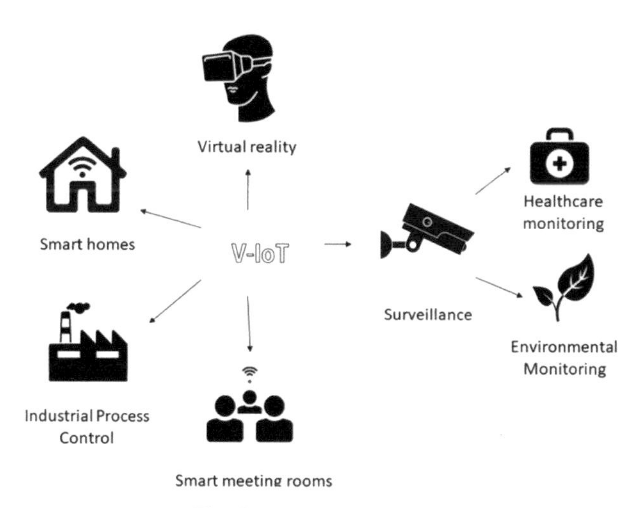

Fig. 1. V-IoT Applications

For a very long time, traditional surveillance systems based on cameras, such as Closed-circuit television (CCTV) systems, have been used for surveillance and monitoring to ensure public and wildlife safety. However, these systems are passive, which means they can only monitor and cannot take further action based on what is going on [5]. Therefore, nowadays, Wireless V-IoT (WV-IoT) has emerged as one of the most promising technologies for surveillance applications, with which the deployment of large-scale cameras is becoming easier and the expansion or the repair of V-IoT nodes (called camera nodes) on the network is becoming possible without the need for new cables installation [6]. In addition, the camera nodes can provide different levels of intelligence depending on the application and the processing algorithms used. R. Cucchiara [7] presented other advantages of surveillance systems based on WV-IoT over the traditional monitoring systems, which include: Enlarging the view of an event by providing a close-up view, enhancing the view of an event by using camera nodes with different capabilities (e.g. different field of view, different mobility, etc.) [8], and exploring multi-resolution views and multiple viewpoints for the same event. However, due to the nature and the amount of the visual data sensed by the V-IoT nodes, these later have to preprocess the collected data locally before sending it to the network. Therefore, they must employ different visual data processing algorithms depending on the type of information to be extracted (e.g. vehicle detection, face extraction, etc.). Nevertheless, these nodes are resource-constrained and oftentimes they cannot keep all the necessary processing algorithms in their constrained memory. Therefore, the objective of our paper is to propose a V-IoT-based architecture that aims to reduce the data flow in the network and distribute the visual data processing across the network levels. In this architecture, the Message Queuing Telemetry Transport protocol (MQTT) will be used as a messaging protocol and specific lightweight processing algorithms will be used by the V-IoT nodes due to their limited resources. To show the utility of our proposed architecture, a use case of face recognition and tracking will be simulated, in which the V-IoT node will be presented by a Raspberry Pi node equipped with a camera module.

The rest of the paper is organized as follows: Sect. 2 presents the materials and methods used in this work, followed by Sect. 3 that describes in detail our proposed system. Section 4 presents the simulation and results. Finally, the conclusions and future works are presented in Sect. 5.

2 Materials and Methods

2.1 MQTT Protocol

MQTT is a lightweight, open and simple event-driven protocol that was specially developed for Machine to Machine (M2M) communication. As MQTT is a publish/subscribe-based protocol, it decouples the client that sends a message (called the publisher) from the clients that receive the messages (called the subscribers). Therefore, the MQTT clients do not need to be aware of each other since the connection between them is handled by a third component called the Broker. This latter is mainly responsible for receiving and filtering all incoming messages and then distributing them correctly to all subscribed clients. All MQTT communication depends on the "Topic", which is a form of addressing that defines a unique connection between the publishers and subscribers.

The MQTT Broker may have one or more topics and can handle up to thousands of concurrently connected MQTT clients that publish/subscribe to a message related to the topic [9]. It is important to note that an MQTT client can be a publisher or subscriber or both and the subscriber gets a message if the topic published by the publisher matches the topic subscribed by this subscriber. Figure 2 illustrates publish and subscribe modal of the MQTT protocol.

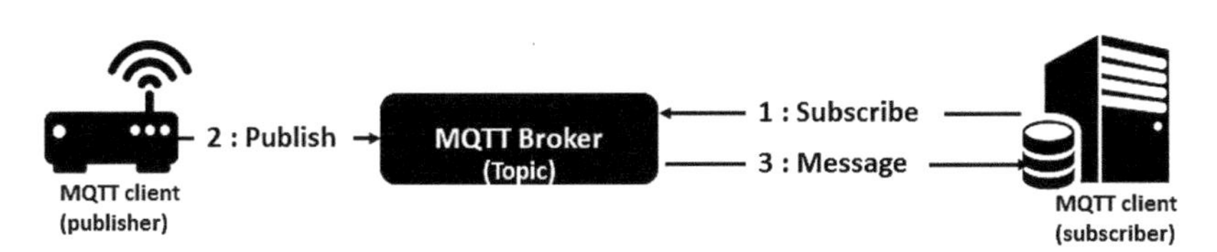

Fig. 2. MQTT publish and subscribe modal

MQTT have been used in many works in the literature as a messaging protocol such as in [10–14], to ensure exchanging data between heterogeneous IoT devices, avoid a direct communication between them, and guarantee the interoperability of the data.

MQTT has many other advantages such as it enables messages to be pushed efficiently to clients while minimizing required client resources and network bandwidth. It increases the scalability, enables broadcasting messages to groups of things, supports bidirectional communication between devices and servers, and supports persistent sessions over unreliable networks. In addition, it is very secure with permission-based security. However, MQTT has some limits such as the need for its clients to support TCP/IP protocol.

2.2 Raspberry Pi as a V-IoT Node

V-IoT consists of tiny, low-power nodes equipped with visual sensors and capable of collaboratively collecting, processing, and communicating visual data in the network. Generally, each node (called camera node) consists of a sensing unit composed of one or more visual sensors that can detect only a certain field of vision or a limited direction [15] called Field of view (FoV), a processing unit that performs visual data processing locally to reduce the amount and size of data that must subsequently be transmitted to other nodes in the network via the communication unit. The camera node receives its power supply from a power unit which is generally powered by a battery [8]. Figure 3 presents the hardware components of a typical camera node.

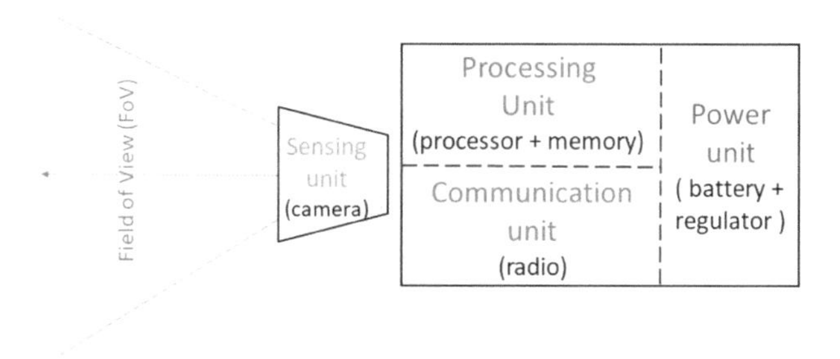

Fig. 3. Typical camera node

The camera nodes operate with respect to their available and limited resources. The processing unit should be small and should have high-speed calculation and low power consumption. In this work, we build a prototype of a camera node based on a Raspberry Pi 3 Model B as sensor board interfaced with Raspberry Pi Camera Module Rev 1.3 as a visual sensor.

Raspberry Pi have been used in the literature as camera node in many works such as in [5, 16–18]. With recent advances in Computer Vision (CV), the Raspberry Pi can provide different levels of intelligence depending on the application and the processing algorithms used. Hence our camera node prototype is programmed using the open-source language Python and the computer vision library OpenCV (Open-sourced Computer Vision library) to pre-process the collected data before sending it to the network and hence reduce the network data flow. We choose the OpenCV library as it is an open-source and free library that includes several hundreds of CV algorithms ranging from facial recognition to object identifying, classifications of human actions in videos, edge mapping, image transformations, detailed feature analysis and more. Hence, we can use these CV algorithms together to test our proposed approach.

3 Proposed Approach

The camera nodes are resource-constrained and oftentimes they cannot keep all the necessary processing algorithms in their limited memory. However, these devices generate a massive amount of image/video data that needs to be processed and transmitted to other nodes and the base station (BS) for further analysis.

Therefore, this section presents our proposed distributed architecture for visual data processing in a WV-IoT. Our architecture aims to distribute the visual data processing to the different levels of the network and hence decrease the information flow from the camera nodes to the BS and balance the traffic load in the network. Figure 4 presents our proposed architecture.

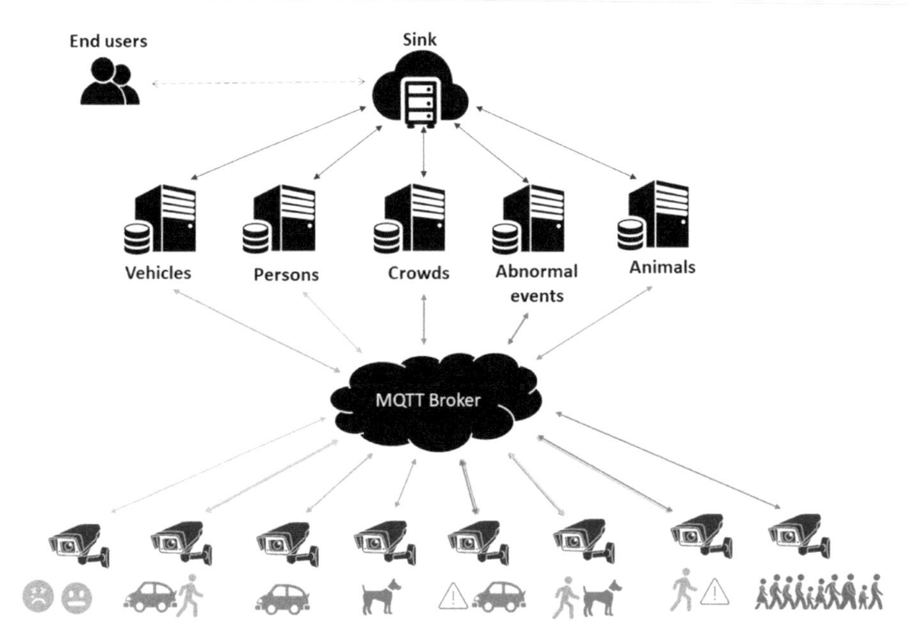

Fig. 4. Proposed architecture for visual data processing in WV-IoT

As shown in Fig. 4, our architecture is composed of camera nodes (V-IoT nodes), MQTT broker, local servers, the BS (called also sink) and the end-users.

In order to decrease the information flow from the camera nodes to the BS, the processing in our proposed architecture is dispatched across the network levels. As shown in Fig. 3, a typical camera node consists of a processing unit, which gives it the possibility to process locally the captured scene data (called image frame), extract useful information, and transmit it then to the local servers. These latter ones, collaborate, share data with each other and perform more complicated tasks (e.g. data fusion) to transmit just the processed data to the BS for further analysis. Table 1 summarizes the tasks of the proposed architecture elements.

Table 1. Proposed architecture elements tasks

Proposed architecture elements	Tasks
Camera nodes	Object detection Object classification
MQTT Broker(s)	Dispatching all messages between the senders and the rightful receivers Messages classification
Local servers	Regroups all the necessary image processing algorithms needed for a specific topic (persons, vehicles …) Contains a database that stores information about searched objects Can collaborate if needed
Base Station	Managing the cameras in the network Tracking across camera nodes Target position prediction Target recovery Video analyzing
End users	Add new searched objects to local servers' databases Can be alerted by the system if a wanted target is detected Can see the tracking of the target in video

To extract the necessary information from the collected data, different image processing algorithms for different types of images (the face detection algorithm, for example, is different from the vehicle detection algorithm) must be used and kept at the camera node, which is impossible because of its constrained memory [19]. As a solution to this problem, the camera nodes of the proposed architecture use only specific lightweight algorithms to detect and extract useful information from the collected data based on the specific tasks assigned to them and the area where they are installed. For example, a camera installed in a residence will be just responsible for face detection while another one installed on a highway will be responsible for both faces and vehicle plates detections.

After this pre-processed step, the camera nodes send the detected information back to the corresponding local servers for further processing. The camera nodes messages use a JSON data format that contains the fields shown in Fig. 5, where the message type represents a normal or urgent message, time represent the time of message generation, and other information represents the additional fields that can be added depending on the use case.

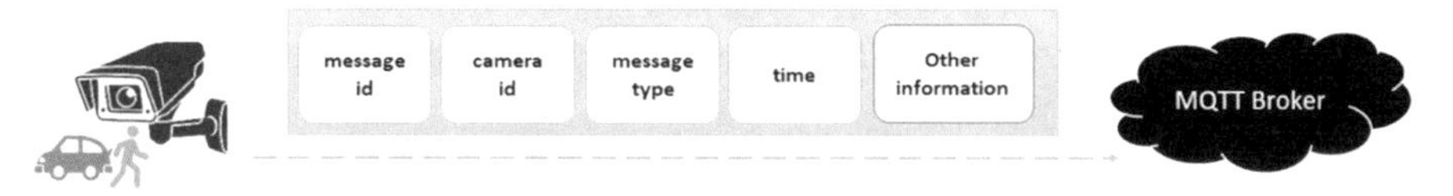

Fig. 5. Camera nodes messages

As described in Fig. 4, the protocol MQTT is used in the proposed architecture to exchange messages between the camera nodes and the local servers. Each local server is subscribed to a topic to receive the messages related to it. Table 2 presents some processing algorithms that can be used according to the specific topic.

Table 2. Some processing algorithms

Topic	Algorithms for
Vehicles	Vehicle detection Vehicles plate number detection License plate recognition Vehicle classification Vehicle matching Vehicle tracking
Persons	Face detection Pedestrian detection Facial Expression Recognition Face recognition Person Re-Identification Gesture recognition Hands tracking Eyes tracking Person tracking
Crowds	Crowd analyzing Crowd tracking Crowd counting People flow estimation
Abnormal events	Natural Disasters detection Fall detection Violence detection Abandoned object detection Abandoned object tracking
Animals	Animals detection Animals classification Animals tracking

Our proposed architecture presents the following advantages:

- The intelligence is present in all levels of the architecture.
- Lightweight algorithms are used in the camera nodes to save energy and to extract just useful information.
- The architecture is based on MQTT which enables highly scalable solutions without dependencies between the data producers and the data consumers.
- The data processing and storage are distributed across all levels of the architecture.

4 Simulation and Results

In this section, a simulation for face recognition and target tracking is carried out to test our proposed architecture in which the visual data processing is distributed across the network levels. Therefore, as shown in Fig. 6, the open-source Eclipse Mosquitto MQTT broker [20] is used in this simulation to enable the communication between publishers/senders and subscribers/receivers. In addition to two MQTT clients, where a local server is used as subscriber to the topic related to "Persons", and our camera node prototype presented in Subsect. 2.2 as publisher to same topic. The MQTT clients uses the Eclipse Paho MQTT Client API [21] to interact (e.g. publish message) with MQTT broker.

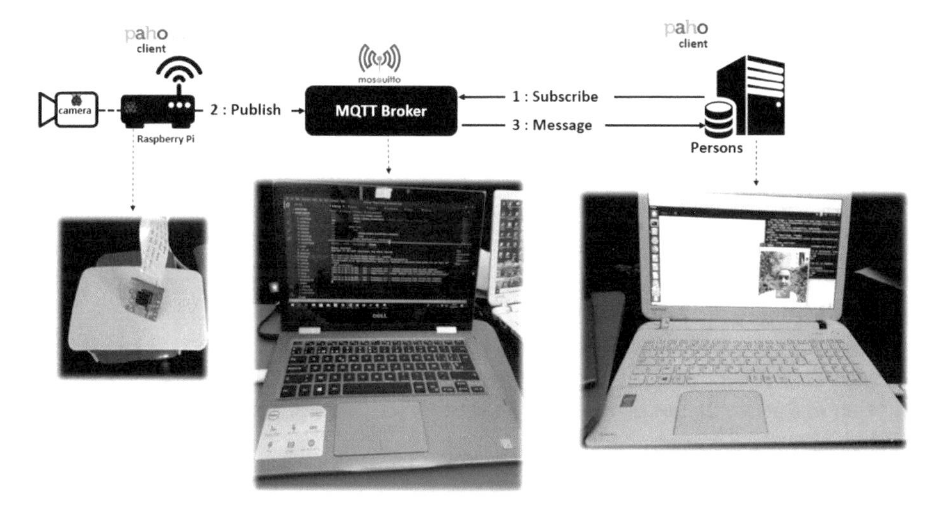

Fig. 6. Simulation environment

The processing tasks are distributed between the camera node and the local server as follows:

- The camera node is responsible for face detection, information extraction, and message generation and publishing.
- The local server is responsible for face recognition and target tracking.

The process diagram of this simulation is presented in Fig. 7.

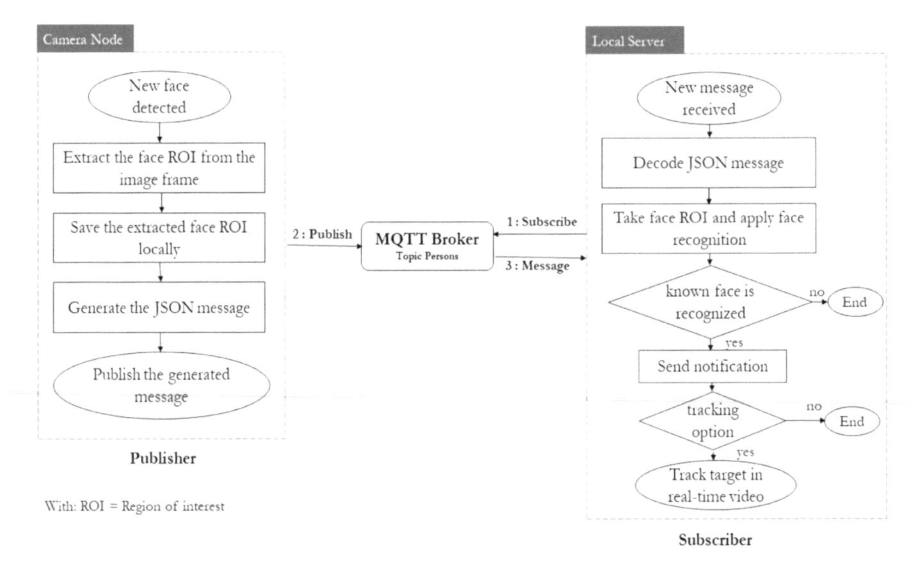

Fig. 7. Process diagram of the simulation

In this simulation, we used Adam Geitgey's face_recognition module [22], which is built using many built-in libraries such as Dlib and it uses machine learning to recognize the faces with an accuracy of 99.38%.

Before we can apply face recognition at the server-side, we first need to receive the faces we want to recognize and which are collected by the camera node. Therefore, the first step in this simulation is to detect and locate the faces present at the camera node Field of View.

To not overload the camera node's limited memory with a complex face detection method, the HOG (Histogram of Oriented Gradients) method [23] is used in this simulation for face detection, in which, the distribution of local intensity gradients or edge directions are used to describe the local appearance and shape of the object on a grayscale image. After detecting and locating faces on an image frame, the camera node crops the faces region of interest (ROI) and generates messages containing important information about the detected targets, and then transmits them, through the MQTT broker, to the local server for facial recognition and target tracking. The format of the messages generated by the camera node is described in Fig. 8.

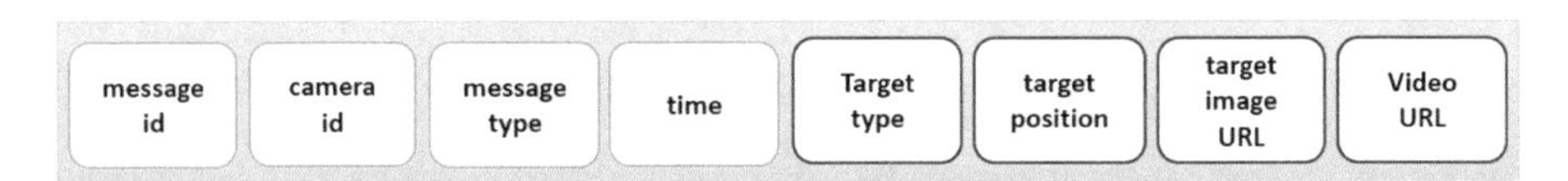

Fig. 8. Camera node message format

In addition to the fields presented in Fig. 5, the camera node transmits to the local server other information such as target type (in this case face) and position, the URL of the target image (person face) and the URL of the real-time video (streaming from the camera node). Once the local server receives a camera node's message containing the URL of the face ROI, it decodes the JSON message and performs face recognition to identify who is in the image. To do this, the face_recognition module in the local server uses an algorithm [24] for face landmark estimation to figure out the pose of the face and warp the image so that the eyes and mouth are centered. Then it passes this centered face image through a pre-trained neural network to compute 128 face measurements called embeddings. Finally, it uses a simple linear SVM (support-vector machines) classifier to find the person in our database of known people who has the closest measurements to the detected face. Once the detected face matches a known person in our database, the local server sends a notification to the end-user and asks him if he wants to track the person in real-time video. If so, the real-time video from the camera node is displayed with the detected face and the information about the recognized person. To be able to recognize faces at the local server, this latter is connected to a database containing face images and information about known persons such as their name, age, and gender. Figure 9 presents some face images used in our dataset, in addition to the results displayed at the local server level for face recognition and tracking from the video broadcast by the camera node.

Fig. 9. Faces dataset and results of face recognition at the local server side

As shown in Fig. 9, the faces and information of three known persons are stored at the local server. Once these faces are detected by the camera node, and the local server is requested to track the persons in real-time video, the recognized faces are displayed with their personal information and are tracked in real-time.

It is important to note that our proposed architecture is not limited to the topics described in Sect. 3. Other topics and related processing algorithms can be added depending on the application (e.g. monitoring in a greenhouse, medical images processing, etc.).

5 Conclusions and Future Works

This paper presented our contribution to visual data processing in V-IoT. Our main objective was to decrease the information flow from the V-IoT nodes to the base station by proposing a new architecture that can distribute the visual data processing to the different levels of the network. Since V-IoT nodes are resource-constrained, lightweight algorithms are used in our architecture to detect and extract useful information from collected visual data. Then, the preprocessed data is sent to the local servers for further processing using the MQTT protocol. To test the performance of our proposed architecture, we conducted a face recognition and tracking simulation, where our built Raspberry Pi board was used as a V-IoT node. The experimental results show that our objective was reached.

For future works, other complicated scenarios will be tested where different types of information and processing algorithms will be used (e.g. tracking vehicles and faces at the same time). In addition, we intend to improve our proposition by adding an algorithm for message scheduling at the broker level to process the V-IoT nodes messages according to their levels of importance and delay tolerance.

References

1. Vermesan, O., et al.: Internet of things strategic research roadmap. In: Internet of Things: Global Technological and Societal Trends. IEEE, pp 9–52 (2011)
2. Nauman, A., Qadri, Y.A., Amjad, M., et al.: Multimedia internet of things: a comprehensive survey. IEEE Access **8**, 8202–8250 (2020). https://doi.org/10.1109/ACCESS.2020.2964280
3. Du, B., Duan, Y., Zhang, H., et al.: Collaborative image compression and classification with multi-task learning for visual internet of things. Chin. J. Aeronaut. **35**(5), 390–399 (2021). https://doi.org/10.1016/j.cja.2021.10.003
4. Ji, W., Xu, J., Qiao, H., et al.: Visual IoT: Enabling internet of things visualization in smart cities. IEEE Netw. **33**, 102–110 (2019). https://doi.org/10.1109/MNET.2019.1800258
5. Farouk Khalifa, A., Badr, E., Elmahdy, H.N.: A survey on human detection surveillance systems for raspberry Pi. Image Vis. Comput. **85**, 1–13 (2019). https://doi.org/10.1016/j.imavis.2019.02.010
6. Peixoto, J.P.J., Costa, D.G.: Wireless visual sensor networks for smart city applications: a relevance-based approach for multiple sinks mobility. Futur. Gener. Comput. Syst. **76**, 51–62 (2017). https://doi.org/10.1016/j.future.2017.05.027
7. Cucchiara, R.: Multimedia surveillance systems. In: Proceedings of the 3rd ACM International Workshop on Video Surveillance & Sensor Networks. ACM, pp 3–10 (2005)

8. Chew, L.A., Seng, K.P., Wern, L.C., et al.: Wireless multimedia sensor networks on reconfigurable hardware (2013)
9. Malche, T., Maheshwary, P., Kumar, R.: Environmental monitoring system for smart city based on secure internet of things (IoT) architecture. Wireless Pers. Commun. **107**(4), 2143–2172 (2019). https://doi.org/10.1007/s11277-019-06376-0
10. Jang, S.Y., Lee, Y., Shin, B., Lee, D.: Application-aware IoT camera virtualization for video analytics edge computing. In: 2018 IEEE/ACM Symposium on Edge Computing (SEC). IEEE, pp 132–144 (2018)
11. Lachtar, A., Val, T., Kachouri, A.: Elderly monitoring system in a smart city environment using LoRa and MQTT. IET Wireless Sens. Syst. **10**, 70–77 (2020). https://doi.org/10.1049/iet-wss.2019.0121
12. Ahmed, A., Saha, S., Saha, S., et al.: Real-time face recognition based on IoT: a comparative study between IoT platforms and cloud infrastructures. J. High Speed Netw. **26**, 155–168 (2020). https://doi.org/10.3233/JHS-200636
13. Forestiero, A., Kurdi, H., Thayananthan, V.: A Multi-tier MQTT architecture with multiple brokers based on fog computing for securing industrial IoT (2022).https://doi.org/10.3390/app12147173
14. Tsao, Y.-C., Cheng, F.-J., Li, Y.-H., Liao, L.-D.: An IoT-based smart system with an MQTT broker for individual patient vital sign monitoring in potential emergency or prehospital applications (2022). https://doi.org/10.1155/2022/7245650
15. Enayet, A., Razzaque, M.A., Hassan, M.M., et al.: Moving target tracking through distributed clustering in directional sensor networks. Sens. (Basel, Switzerland) **14**, 24381–24407 (2014). https://doi.org/10.3390/s141224381
16. Kamath, R., Balachandra, M., Prabhu, S.: Raspberry Pi as visual sensor nodes in precision agriculture: a study. IEEE Access **7**, 45110–45122 (2019). https://doi.org/10.1109/ACCESS.2019.2908846
17. Kavalionak, H., et al.: Distributed video surveillance using smart cameras. J. Grid Comput. **17**(1), 59–77 (2018). https://doi.org/10.1007/s10723-018-9467-x
18. Idoudi, M., Bourennane, E.-B., Grayaa, K.: Wireless visual sensor network platform for indoor localization and tracking of a patient for rehabilitation task. IEEE Sens. J. **18**, 5915–5928 (2018). https://doi.org/10.1109/JSEN.2018.2838676
19. Soro, S., Heinzelman, W.: A survey of visual sensor networks. Adv. Multimedia **2009**, 21 (2009). https://doi.org/10.1155/2009/640386
20. Eclipse mosquitto broker. https://mosquitto.org/. Accessed 10 Jul 2021
21. Eclipse paho client. https://www.eclipse.org/paho/clients/java/. Accessed 10 Jul 2021
22. Geitgey a face recognition library. https://github.com/ageitgey/face_recognition/blob/master/face_recognition/face_recognition_cli.py. Accessed 20 Jun 2021
23. Dalal, N., Triggs, B.: Histograms of oriented gradients for human detection. In: 2005 IEEE Computer Society Conference on Computer Vision and Pattern Recognition (CVPR'05). IEEE, pp 886–893 (2005)
24. Kazemi, V., Sullivan, J.: One millisecond face alignment with an ensemble of regression trees. In: 2014 IEEE Conference on Computer Vision and Pattern Recognition. IEEE, pp 1867–1874 (2014)

An Overview on Machine Learning Approach to Secure the Blockchain

Abdellatif Bakar(✉), Abdelhamid Zouhair, and El Mokhtar En-Naimi

Faculty of Sciences and Technologies, Abdelmalek Essaâdi University (UAE), Tetouan, Morocco
bakar.abdellatif@etu.uae.ac.ma

Abstract. As a part of artificial intelligence, machine learning has been widely used in recent research to solve problems such as classification, clustering, and regression, using different approaches.

The fact that blockchain technology has made far-reaching changes in the world will be displayed in this article, which represents a benchmark for research on these technologies together. As a set of security issues that threaten blockchain technology (BT) have been listed, some of them have already been solved through machine learning with its different approaches, whether it is supervised learning, unsupervised learning, or reinforcement learning, and that gives us an idea of the different points covered by the enumerated technologies.

Keywords: Security · Blockchain · Machine Learning · Blockchain attacks

1 Introduction

Machine learning has become an important and prominent component of Artificial Intelligence (AI) in recent years. Not surprisingly, it's applications are increasingly being used across industries. The growth of open-source and production-ready frameworks, such as those used by machine learning, contributes to the ubiquity of the digital transformation process.

The ability to run multiple programs on the same machine is also one of the most important changes that the computer revolution has made in our lives. However, without the help of machine learning, many tasks still seem out of reach for computers [1].

Artificial intelligence (AI) has been used to transform various electronic tools into cognitive extension agents capable of interacting with humans in real time [2]. Some of them include speech recognition, spam filtering and natural language processing.

The use of BT will disrupt traditional business models by making them obsolete. Blockchain has become an integral part of modern society. It is used to store records from various applications. Originally created as a solution for the Bitcoin recording system as well as Ethereum [3], blockchains are now used to store all kinds of data.

Blockchains were first described in the 1990s by Stuart Haber and W. Scott Stornetta [4]. The blockchain is represented as a growing list of records, called blocks that are linked using cryptography and in which transactions are recorded and distributed on a number of computers called linked nodes in a peer-to-peer network.

M. Lazaar et al. (Eds.): BDIoT 2022, LNNS 625, pp. 486–500, 2023.
https://doi.org/10.1007/978-3-031-28387-1_41

BT can be applied to areas where central authority is needed to conduct their transactions. For example, when it comes to transferring money, people have to trust a central authority to carry out their transactions. With blockchain technology, a person can easily program it to replace the intermediary and therefore blockchain represents safety problems, use and efficiency. Beyond all the attention that blockchain technology has attracted, reasonable efforts are focused on solving important blockchain problems. These problems are the subject of much research aimed at improving blockchain security.

In the rest of this document, we will present in the first section the machine learning by focusing on the problems that it can solve, as well as it's different types to move on to a concrete example of the use of this technology. The second section is dedicated to blockchain technology, in which we will present the different properties on which the technology is based, many security problems will be set out later, in order to close the document with different solutions based on machine learning dedicated to the different problems related to blockchain.

2 Machine Learning

Machine learning (ML) is a science for discovering patterns and making predictions from data based on statistics, data drilling, pattern recognition and predictive analyzes.

It powers many popular modern services. Examples include recommendation engines used by Netflix [5], YouTube [6], Amazon [7], or Spotify [8]. The same is true for web search engines. Social media news feeds such as Facebook and Twitter are based on machine learning [9], as are voice assistants such as Siri and Alexa.

Machine learning is also used for automatic analysis and classification of medical x-ray images. AI is very good at this, sometimes even better than human experts at detecting anomalies or diseases.

2.1 Problems Solved by Machine Learning

Machine learning can solve many types of problems [10, 11]:

Regression: is a fundamental part of ML. This involves learning how a variable affects another variable by estimating its relationship to the input characteristics. This is done through supervised learning. Regression analysis is a type of mathematical method used in ML to predict a continuous result. It is commonly used by data scientists to predict the future. Linear regression is the most commonly used type of regression analysis.

Classification: is a process of identifying and understanding objects and ideas in a given set of predefined subpopulations. Using pre-categorized training data sets, ML programs can categorize future data sets MLinto relevant categories. ML programs use input data to predict the probability that a given data set falls into one of the predefined categories. For example, to filter e-mail, a service provider might use the classification to identify spam or non-spam content.

Classification is a process of identifying and understanding a set of models in a given data set. These models can be used to predict the probability that a given data set falls into one of the predefined categories.

Clustering is the process of identifying data points that are similar to each other and then grouping them into different clusters. The similarities are taken into account when determining the data points to be aggregated. The importance of clustering is that it helps to determine the intrinsic clustering of the various data points not labeled in the present.

The type of criteria a user should use when clustering depends on the situation. For example, they may want to find data types that are natural clusters, or they may need to find useful data objects suitable for particular applications.

2.2 Types of Machine Learning Algorithms

Depending on the nature of the problem, there are different approaches [12]. In this section, we talk through the categories of machine learning.

Supervised learning involves learning a prediction function from annotated examples, since it usually begins with a well-defined set of data with a certain well-defined classification.

The purpose of this technique is to unveil models within the data and then apply them to analytical processes.

This data consists of several characteristics associated with labels that define their meaning, for example, creating a machine learning application that can distinguish several million animals, based on images and written descriptions.

Unsupervised learning is learning without a supervisor. The aim is to extract classes or groups of individuals with common characteristics. The quality of a classification method is measured by it's ability to discover some or all of the hidden patterns.

In unsupervised learning, the responses that are being predicted are not available in the datasets. Here, the algorithm uses an unlabeled data set. So, we ask the machine to create it's own answers. It thus proposes answers based on analyzes and data aggregation.

Supervised and unsupervised learning is distinguished. In the first apprenticeship, it is to learn how to classify a new individual among a set of predefined classes: we know the a priori classes. While in unsupervised learning, the number and definition of classes are not given a priori.

An example of this is the spam detection technology sent by e-mail. Basic e-mail and spam have too many variables to allow an analyst to label bulk unwanted e-mail. By contrast, machine learning discriminants based on clustering and association are applied to identify unwanted emails [13].

Semi-supervised learning is a type of supervised learning that involves using unlabeled data for training. This usually involves collecting a large amount of data and a small amount of labeled data. One technique used in this type of learning is to detect abnormal behavior in a system. This can be done by comparing the parameters of a system to a basic data set.

The basic data set used in this method is generally composed of various parameters that represent the various possible variations in the operating state of a system. This method can be used to diagnose various health problems. In addition to in-depth learning networks, this method also involves the use of unsupervised components called Boltzmann machines [14, 15].

Reinforcement learning is a behavioral learning model. The algorithm receives feedback from the data analysis and guides the user to the best result. Reinforcement learning differs from other types of supervised learning, as the system is not formed with a sample data set. Instead, the system learns through trial and error.

Reinforcement learning involves learning from experience what actions to take for an autonomous agent (robot, etc.) in order to optimize a quantitative reward over time. The agent is immersed in an environment, and makes decisions based on his current condition. In turn, the environment provides the agent with a reward, which can be positive or negative. The agent seeks, through repeated experiments, an optimal decision-making behavior (called strategy or policy, and which is a function associating the action to be executed with the current state), in the sense that it maximizes the sum of the rewards over time.

Example: Detection of Fake Profiles in Social Networks: When people use social media platforms like Facebook, Twitter, Instagram and LinkedIn, they are constantly exposed to the various features and activities of these platforms. Unfortunately, there are also fake profiles designed to encourage users to click on their links or share their personal information.

Research has been conducted to find a solution to identify fake profiles on Twitter using machine learning [3]. These models are able to distinguish between real and fake profiles according to the different functionalities of the platform.

Different methods were used to detect fake Twitter profiles. Each model focuses on detecting a false profile using only visible features. The accuracy of these models is compared to the accuracy of other models. Models are driven using various loss and optimization functions.

Different ML approaches have been tested, such as Neural Networks, Random Forest and XG Boost. False profiles detected can be blocked/removed to avoid future cybersecurity threats [16].

3 Blockchain

The blockchain is a distributed ledger, visible to all members of the network that contains records. This technology allows all kinds of information of such value to be recorded and exchanged in a peer-to-peer network: transactions, tangible assets (a house, a car, a piece of land) or intangible assets (intellectual property, patents, copyright).

The transactions are recorded in the form of blocks, so that the blockchain is constructed by linking several blocks forming a chain, the link between its blocks is maintained by means of cryptography, so that each block contains a cryptographic hash of the previous block, a timestamp and transaction data.

BT is based on the following three properties:

Decentralization: In blockchain, decentralization refers to the transfer of supervision and decision-making from a centralized association (individual, company or group of people) to a dispersed network. This means that information is stored by all entities that are part of the network, which means that the blockchain does not need to rely on a centralized node, and that all nodes own the information.

Transparency: This concept consists of putting the records in appearance for all nodes of the network while maintaining anonymity, taking the example of cryptocurrency, the transactions are recorded in such a way that the identity of the persons concerned is hidden by a complex cryptography. The following figure shows an example of Ethereum transactions (Fig. 1).

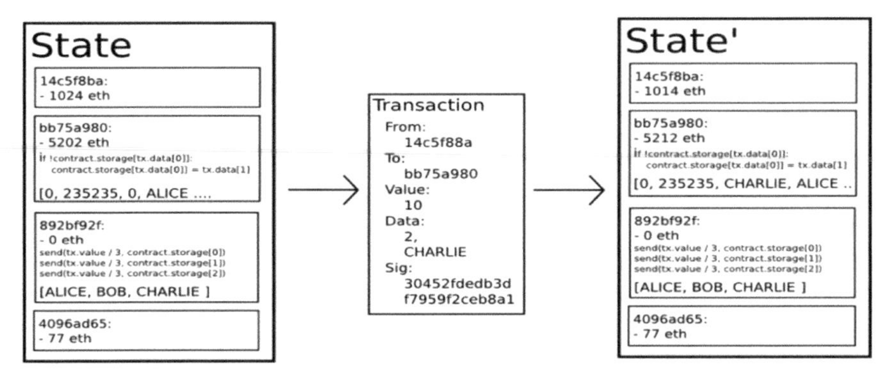

Fig. 1. Example of Ethereum transactions [17].

In the case of cryptocurrency, each node has two keys, the first is public, the other is private, the utility of this public key in the blockchain is to keep track of the transaction you have made. Your private key is never shared. It is linked to your public key to make the transaction valid.

This is how a person's identity is hidden, if you look at a person's transaction history you will see "1MF1bhsFLkBzzz9vpFYEmvwT2TbyCt7NZJ sent 1 Ether…", so the transparency of the blockchain lies in the fact that even if the person's real identity is protected, you will still see all the transactions that have been made by their public address. No other financial system has ever had this kind of transparency.

Immutability: The blockchain technology creates immutable records, this means that once the information is stored in the blockchain, one cannot go back, which means that the information introduced by this technology cannot be altered.

Blockchain immutability is ensured by using cryptographic hashing functionality, by transforming input strings of any length into an output string of fixed length.

3.1 The Structure of Blockchain

The block is usually made up of different types of data. These include master data, previous block, current block, and timestamp, as shown in the following figure (Fig. 2):

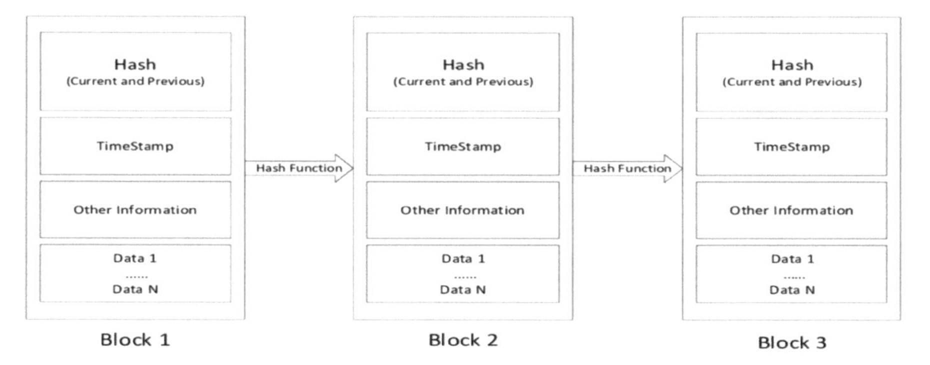

Fig. 2. Structure of the blockchain [18].

Main Data: This type of data depends on the service to which the blockchain applies, it may be either bank clearing records, IOT data records or transaction records etc.

Hash: When a transaction is executed, it is sent to each node on the network. The resulting transaction record will then be stored in code, and then distributed to the other nodes. In order to generate a final hash value, the blockchain use the Merkle tree function [18]. This method significantly reduces the amount of IT resources used to transmit and receive data. The final hash value will be stored in the block header of the current block.

Timestamp: Time of the generated block.

Other Information: Such as block signature, nonce, or other user-defined data.

3.2 Types of Blockchain

Blockchain technologies can be divided into three types:

Public Blockchain: Everyone can participate in the process of validating and verifying a transaction. Similar to how blockchains like Bitcoin and Ethereum work, public blockchains allow people to verify a transaction.

Private Blockchain: Not all nodes can participate in the blockchain and they have strict authority over their data. Whatever the type of blockchain, cryptocurrencies and public blockchains have advantages. Public blockchains are convenient, but sometimes we need private control over the type of blockchain we use, for example, a service that allows users to perform certain tasks.

Blockchain Consortium: A consortium blockchain is a hybrid type of blockchain that combines the characteristics of private and public chains. It should be noted that the level of unanimity between the two is higher in the consortium blockchain.

Unlike a traditional open system, a consortium blockchain uses a small number of powerful individuals as validators. This eliminates the need for a single party to select block producers.

A consortium blockchain can be useful in the case of a transfer of funds abroad, the funds in this case are transferred from the payer's account to the payee's account via the central bank and the messaging network that involves intermediaries and related charges. In contrast to national remittances, which are always immediate, more practical and available [19].

4 Blockchain - Security Issues

Although blockchain technology relies on a decentralized system, as well as on cryptography that makes it inviolable in the eyes of operators, it still presents many security problems. In this section we will present some attacks carried out on the blockchain. The following figure shows the distribution of security attacks [20] (Fig. 3).

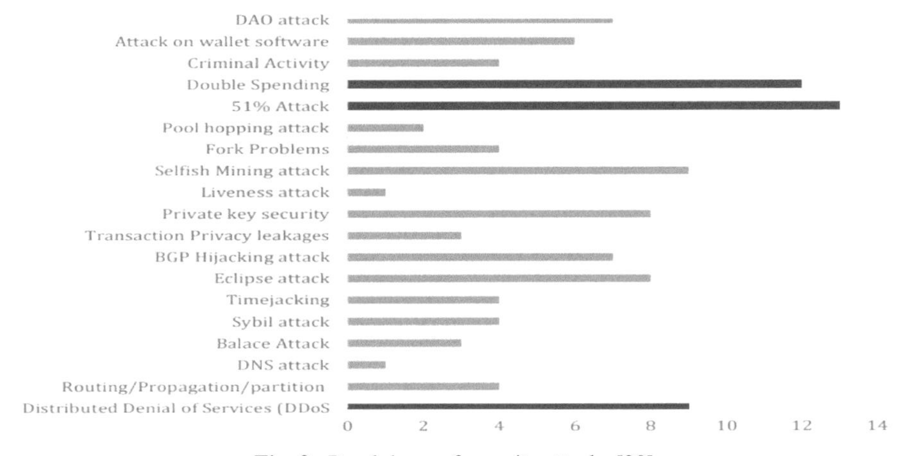

Fig. 3. Breakdown of security attacks [20]

Sybil attack is a type of network operation that allows an attacker to pose as several people at once. This can be very difficult to manage when connecting to a p2p network. In order to control the network, the attacker creates several false identities. These identities imitate those of regular users, but behind the scenes, an unknown attacker is able to control each one of them. These false identities are called Sybil nodes, their role is to surround a node to prevent it from connecting to the honest nodes of the network.

This way, an attempt could be made to prevent the sending or receiving of information on the network. The only way to mitigate Sybil attacks is to increase the cost of creating an identity. This cost must be balanced so that new participants are not prevented from joining the network and creating legit identities.

High cost is also a key factor that prevents the creation of a large number of identities in a short time. For example, in PoW blockchains, miners need enormous computing power to complete their verification and validation processes. This makes it difficult for them to carry out a successful attack [21].

Private key security attack is a security identifier used in the blockchain system. ECDSA is a digital signature algorithm used to generate this type of key. However, it has a deficiency that prevents it from generating enough random numbers [20].

If the key signature in your blockchain is poorly implemented, an attacker can easily access your private key. This may be due to the use of the same key for multiple signature operations instead of a Merkle tree. Having control over your private key is very important to ensure that all data associated with you is stored in a blockchain.

Although, it is very unlikely that all your parts will be stolen, the chances are very low unless you use a really buggy version of the blockchain. One of the biggest problems of blockchain is the mismanagement of private keys. For example, if you store it in a public dough, an attacker can have easy access to it.

Majority Attack/51% attack in most cases, the chain chosen in blockchain is the longest. Although the longest chain rule solves the problem of competing chains, it also makes these blockchains vulnerable to attack by the majority, also known as 51% attack [22].

This attack uses a string long enough to overwrite the transactions and blocks made by the honest nodes. It can be achieved by controlling the hash power of the network. This allows an attacker to generate blocks at a faster rate than the rest of the network.

Once an attacker has control over the blocks added to the blockchain, can change the priority of transactions in those blocks, and even bypass the old ones. This can lead to delays in confirming transactions and allows the company to carry out double-spend operations [28].

Double Spending: A group of researchers led by A. Begum explained that there are five steps in the double-spend problem [23].

Step 1: The process of adding blocks begins by disconnecting the user and requesting the transaction. Unconfirmed transactions are then entered into an unconfirmed transaction pool, which is then resolved by a process known as POW consensus. Through this process, a minor can achieve a single output by solving a complicated mathematical problem.
Step 2: When a good miner adds a block to the real blockchain, the miner block will be added to the blockchain. The other miners will then check the block and start their own chain. However, instead of starting his own chain, the corrupt miner spends all his money and transfers information to the real blockchain.
Step 3: At this point, the corrupted miner adds a block to its isolated string by performing computational power calculations on transactions. It is more powerful than miners who add blocks to the blockchain.
Step 4: When the size of the isolated blockchain exceeds that of the real blockchain, the real blockchain miner tries to add it's block to the isolated one.
Step 5: The rule states that blocks must be added to the largest by moving previous records. This means that the isolated does not know the transactions that occurred in the real blockchain.

When a block tries to add the isolated string, it deletes the previous transaction. This means that the corrupted miner will be able to spend all his previously spent currencies.

Forking when a software update is performed, a forking agreement is concluded between the various decentralized nodes. This is an important problem that involves several blockchains at the same time. Each time a new version of the software is released, a new consensus rule is also established between all nodes.

This process involves creating two types of nodes: old and new. The first is composed of nodes that have been previously established, and it may or may not be in agreement with the block sent by the elders.

The first is composed of nodes that have been previously established, and it may or may not be in agreement with the block sent by the elders. This problem is called the fork problem. There are two types of fork problems: soft fork and hard fork.

When a new version of the software is published, it is not compatible with the old version. This problem causes a hard fork. The new version of the software then creates it's own blockchain.

When a hard fork occurs, all existing nodes in the network must switch to a new chord. If the old ones are not upgraded, the network will continue to operate as a different string.

Soft fork occurs when the new version of the software conflicts with the old one. The new nodes don't agree with the mining of the old ones.

Unlike a hard fork, soft fork does not require existing nodes to immediately switch to the new chord. This process can be carried out gradually and should not affect the stability of the system.

A soft fork may also be caused by temporary divergence. For example, when a minor uses a non-updated software, it's clients on the other nodes do not understand the new consensus rule [24].

Selfish mining if the group blockchain is ahead of the honest state of the blockchain, it can introduce it's new block on the network. Miners could then steal other users cryptocurrencies. The network will then recognize the new block, and it will then invalidate the original blockchain [25].

The mining process is carried out by the nodes of the blockchain network, which validate and confirm the transactions. In exchange for their computational effort, miners earn new chips. With selfish mining, the group secretly extracts new blocks from the main chain, which later reveals them to the public.

In 2013, Emin Sirer and Ittay Eyal discovered that by hiding new blockchain blocks, miners could earn more bitcoins. They then claimed that this method could lead to a fork and create a new network. It is thought that minors could introduce the new block on the network at the right time [26].

Different cryptocurrencies such as Bitcoin rely on software developed by miners to solve the randomly generated hash number of their own block. When the number is resolved, a new block is created on the blockchain and the miner who resolved it receives a reward and transaction fees.

In their paper [26], the researchers noted that by hiding new blockchain blocks, miners can increase their share of revenue and improve the efficiency of their mining process. This method also eliminates the need for additional hardware and improves the discovery process.

5 Some Blockchain Solutions Based on Machine Learning a Subsection Sample

5.1 Reinforcement Learning Approaches

Cryptocurrency Trade: Due to the growing popularity of cryptocurrencies, it has become more common for retail investors and financial institutions to trade them. There are more than 300 cryptocurrency exchanges that are currently monitored by the CoinMarketCap website. Today, most stock-market robots are powered by integrated systems based on machine learning. This makes it an ideal candidate for developing systems that can exchange cryptocurrencies.

The solution introduces a method that uses reinforcement learning to create a framework for developing systems that can exchange cryptocurrencies. The method is known as direct reinforcement learning, which is a type of reinforcement learning that uses feedback to the officer instead of immediate feedback. This method eliminates the need for an agent to run a price forecast model. Through DRL Direct Reinforcement Learning, researchers were able to create a system that could adapt to a specific time interval [27].

Mining Strategy Optimization: This section explores the application of reinforcement learning to optimize blockchain mining strategy so that we can use it to improve mining efforts and prevent mining resource exploitation. In this study, the researchers discovered that it was possible to implement reinforcement learning techniques to dynamically update the mining strategies of blockchain without an initial model. This method was more efficient than traditional mining techniques.

Traditional reinforcement learning methods rely on representing agents in an environment to maximize rewards. However, in blockchains, the network is dynamic, making it difficult to create a representative model, hence the development of the multidimensional version of the RL algorithm known as Q Learning.

The study showed that machine learning techniques could be used to develop effective mining strategies. Due to the growing popularity of cryptocurrencies and the extraction of bitcoins [28].

5.2 Supervised Learning Approaches

Anti-money Laundering in Bitcoin: Bitcoin has been used as a money laundering den, where the user's identity is concealed using a pseudonym called address. Despite this, Bitcoin's public registry allows investigators to gather intelligence. This feature allows investigators to perform forensic analysis and gather circumstantial evidence.

Many supervised learning algorithms have been used to combat money laundering, such as: Random Forest, Extra Trees, Gradient Boosting, Bagging Classifier, AdaBoost, k-Nearest Neighbors [29], the objective of this experiment is to develop a machine learning framework capable of identifying criminal entities in a large data set. By this method, the goal is to reduce false positives while maintaining the quality of detection.

In order to perform the task, the Scikit learning package was used, which is a Python programming language, to perform the classification of illegal transactions in the Elliptic dataset used in the search.

The purpose of this study was to compare the different supervised learning methods used to predict legal and illegal transactions. The proposed comprehensive learning method, which is a general meta-approach to machine learning that seeks better predictive performance by combining predictions from multiple models, has a better performance than the other different methods.

Detection of Illegal Entities in Bitcoin: Because of its nature, many users of illicit activities remain hidden behind Bitcoin's anonymity for it's role in easing illegal activities. This question is important for forensic investigations as it allows them to identify these individuals. Currently, machine learning can be used to identify these individuals, but it focuses only on a limited set of users. Research has been done to propose, to introduce a framework that will allow supervised learning to perform better in identifying these individuals [30].

The study was conducted on a set of public data, which included the characteristics and addresses of illegal bitcoin entities. An empirical analysis of the different learning strategies used in the study was then carried out to classify Bitcoin users. In other words, the researchers implemented a supervised learning approach to identify the most discriminating traits of these users.

The results show that even though the tree classification offers a little more than 60% of Bitcoin users in the right category but it still needs additional features to improve it's accuracy in order to conceal illicit entities in order to combat illegal activities.

5.3 Unsupervised Learning Approaches

Anomaly Detection: This search uses two different algorithms, one of which is considered to be an unsupervised learning algorithm, the objective of this search is to create a machine learning algorithm capable of identifying and detecting anomalies in bitcoin, by combining the KMeans algorithm and the OCSVM algorithm.

The ocsvm algorithm is used to detect outliers, while the KMeans algorithm allows grouping similar outliers with the same type of anomalies. The result of this combination of algorithms is a new model, to detect anomalies in the Bitcoin transaction. The first step of the project is to develop a single-class SVM method to identify outliers, while the second step is to develop a Kmeans algorithm to identify the type of attack that can be made against them. As we talk about the solutions provided by unsupervised learning for blockchain problems, we will insist on the second step where the k-means is put in place to identify attacks that can be carried out on the blockchain.

In this part the k-mans algorithm comes into play to make the ing cluster, in order to minimize the number of outliers in the S clusters, using only the negative output points of the first step of the SVM to a class to generate the input data for the second step of the K-means algorithm [31].

Blockchain consensus protocol Improvement a consensus protocol is proposed that considers the ledger as an acyclic graph directed instead of a blockchain. In order to make the system robust to double-expense attacks, a two-step strategy was developed that involves the use of a graph grouping algorithm and a ranking algorithm. The first step is to develop a strategy that divides the blocks into two groups, one of which is composed of the non-operational miners and the other of which is composed of the reference blocks included in the header.

The first step is to learn the relationships between the vertices of a graph. In order to counter double-spend attacks, the proposed algorithm uses a clustering algorithm to divide blocks into two groups.

The longest bitcoin string rule protocol uses the blockchain to perform it's transactions. It could also be said, it suffers from low transaction throughput due to it's overestimation of the end-to-end network propagation delay. Despite the security provided by the longest chain rule, bitcoin protocol suffers from it's limited transaction scalability.

The proposed consensus protocol can improve block creation rate and bitcoin transaction throughput by avoiding double-spend attacks. It also ensures that the security of the blocks is maintained [32].

6 Comparative Studies of Machine Learning Algorithms Applied to Blockchain Solutions

Table 1. Summary of improvements to blockchain technology through machine learning

ML Category	BT issue/improvement	Algorithm or technique
Reinforcement learning	Enhancing cryptocurrency exchange	Direct reinforcement learning [27]
	Optimizing mining strategies	Q-learning [28]
Supervised learning	Optimizing mining strategies	Random Forest, Extra Trees, Gradient Boosting, Bagging Classifier, AdaBoost, k-Nearest Neighbors [29]
	Illegal entity detection - bitcoin	Ensemble Decision Trees [30]
	Anomaly detection	One class svm [31]
Unsupervised learning	*Anomaly detection*	K-means [31]
	Consensus protocol improvement	Graph clustering [32]

According to Table 1, the blockchain technology can perfectly adopt machine learning solutions with its three main categories, either to combat known attacks, or to make blockchain more efficient and reduce its vulnerabilities.

Table 2. Algorithms used to solve some security issues

Security issue	Algorithm
Forking	Naive bayes, KNN, Decision tree, Multi-layer perceptron [33]
Selfish-mining	Random forest [34]
	SquirRL [35]
	Fully connected neural network [36]
Sybil attack	SVM, Random forest, Logistic regression, Decision tree [37]
Majority attack	Game theory algorithmic [38]
Double spending	K-means [39]

We have listed in Table 2, a set of works that aim to combat an attack set that has already been seen in this article, using machine learning algorithms.

7 Conclusion and Future Work

This article focuses on two revolutionary technologies that have contributed to the resolution of highly complex problems. Blockchain represents many security shortcomings, hence the need for machine learning, which helps to improve the services of blockchain, or to strengthen its systems so that it is not vulnerable to attacks.

This work forms the basis of our future research, which will be channeled towards the set of security problems detected in the blockchain, to remain within the framework of the solutions that machine learning can provide to the blockchain, our next research will aim to determine all the problems that do not yet have solutions proposed by this technology then to treat them, as well as the upgrade of some proposed solutions.

References

1. Thanaki, J.: Machine Learning Solutions. Packt Publishing, Birmingham (2018)
2. Zhang, J., Ye, Z., Li, K.: Multi-sensor information fusion detection system for fire robot through back propagation neural network. PLoS ONE **15**(7), e0236482 (2020)
3. Vujičić, D., Jagodić, D., Ranđić, S.: Blockchain technology, bitcoin, and Ethereum: a brief overview. In: 2018 17th International Symposium Infoteh-Jahorina (Infoteh). IEEE (2018)
4. Haber, S., Stornetta, W.S.: How to time-stamp a digital document. In: Conference on the Theory and Application of Cryptography. Springer, Berlin (1990). https://doi.org/10.1007/BF00196791
5. Steck, H., et al.: Deep learning for recommender systems: a Netflix case study. AI Mag. **42**(3), 7–18 (2021)
6. Orsolic, I., Pevec, D., Suznjevic, M., Skorin-Kapov, L.: A machine learning approach to classifying YouTube QoE based on encrypted network traffic. Multimedia Tools Appl. **76**, 1–35 (2017). https://doi.org/10.1007/s11042-017-4728-4
7. Rastogi, R.: Machine learning@ amazon. In: The 41st International ACM SIGIR Conference on Research & Development in Information Retrieval (2018)

8. Pichl, M., Zangerle, E., Specht, G.: Understanding user-curated playlists on spotify: a machine learning approach. Int. J. Multimedia Data Eng. Manag. (IJMDEM) **8**(4), 44–59 (2017)
9. Belkacem, S.: Machine learning approaches to rank news feed updates on social media.Diss. Université des Sciences et de la Technologie Houari Boumediene Alger (2021).
10. Huang, J.-C., Ko, K.-M., Shu, M.-H., Hsu, B.-M.: Application and comparison of several machine learning algorithms and their integration models in regression problems. Neural Comput. Appl. **32**(10), 5461–5469 (2019). https://doi.org/10.1007/s00521-019-04644-5
11. Kotsiantis, S.B., Zaharakis, I.D., Pintelas, P.E.: Machine learning: a review of classification and combining techniques. Artif. Intell. Rev. **26**(3), 159–190 (2006)
12. Pecht, M.: Prognostics and health management of electronics. Encycl. Struct. Health Monit. (2009)
13. Beigy, H.: Dynamic classifier selection using clustering for spam detection. In: 2009 IEEE Symposium on Computational Intelligence and Data Mining. IEEE (2009)
14. Suk, H.-I., et al.: Hierarchical feature representation and multimodal fusion with deep learning for AD/MCI diagnosis. NeuroImage **101**, 569–582 (2014)
15. Cao, P., et al.: Restricted boltzmann machines based oversampling and semi-supervised learning for false positive reduction in breast CAD. Bio-med. Mater. Eng. **26**(s1), S1541–S1547 (2015)
16. Srinivas, M., Sucharitha, G., Matta, A. (eds.): Machine learning algorithms and applications. John Wiley & Sons, Hoboken (2021)
17. Jani, S.: An overview of ethereum & its comparison with bitcoin. Int. J. Sci. Eng. Res **10**(8), 1–6 (2017)
18. Joshi, A.P., Han, M., Wang, Y.: A survey on security and privacy issues of blockchain technology. Math. Found. Comput. **1**(2), 121 (2018)
19. Patil, P., Sangeetha, M., Bhaskar, V.: A consortium blockchain based overseas fund transfer system. Wireless Pers Commun. **122**, 1367–1389 (2022). https://doi.org/10.1007/s11277-021-08953-8
20. Tuyisenge, M.J.: Blockchain technology security concerns: literature review (2021)
21. Aggarwal, S., Kumar, N.: Attacks on blockchain. Adv. Comput. **121**, 399–410 (2021)
22. Cilloni, T., et al.: Understanding and detecting majority attacks. In: 2020 IEEE International Conference on Decentralized Applications and Infrastructures (DAPPS). IEEE, (2020)
23. Begum, A., et al.: Blockchain attacks analysis and a model to solve double spending attack. Int. J. Mach. Learn. Comput. **10**(2), 352–357 (2020)
24. Krishnan, S., Balas, V.E., Golden, J., Robinson, Y.H., Balaji, S., Kumar, R. (eds.): Handbook of Research on Blockchain Technology. Elsevier Science, Amsterdam (2020)
25. Bai, Q., et al.: A deep dive into blockchain selfish mining. In: ICC 2019-2019 IEEE International Conference on Communications (ICC). IEEE (2019)
26. Eyal, I., Sirer, E. G.: Majority is not enough: bitcoin mining is vulnerable. In: Christin, N., Safavi-Naini, R. (eds.) FC 2014. LNCS, vol. 8437, pp. 436–454. Springer, Heidelberg (2014). https://doi.org/10.1007/978-3-662-45472-5_28
27. Koker, T.E., Koutmos, D.: Cryptocurrency trading using machine learning. J. Risk Financ. Manag. **13**(8), 178 (2020)
28. Wang, T., Liew, S.C., Zhang, S.: When blockchain meets AI: optimal mining strategy achieved by machine learning. Int. J. Intell. Syst. **36**(5), 2183–2207 (2021)
29. Alarab, I., Prakoonwit, S., Nacer, M.I.: Comparative analysis using supervised learning methods for anti-money laundering in bitcoin. In: Proceedings of the 2020 5th International Conference on Machine Learning Technologies (2020)
30. Nerurkar, P., et al.: Detecting illicit entities in bitcoin using supervised learning of ensemble decision trees. In: Proceedings of the 2020 10th International Conference on Information Communication and Management (2020)

31. Sayadi, S., Rejeb, S.B., Choukair, Z.: Anomaly detection model over blockchain electronic transactions. In: 2019 15th International Wireless Communications & Mobile Computing Conference (IWCMC). IEEE (2019)
32. Reddy, S., Sharma, G.V.V.: Ul-blockdag: unsupervised learning based consensus protocol for blockchain. In: 2020 IEEE 40th International Conference on Distributed Computing Systems (ICDCS). IEEE (2020)
33. Mohammadi, S., Rabieinejad, E.: Prediction forks in the blockchain using machine learning
34. Peterson, M., Andel, T., Benton, R.: Towards detection of selfish mining using machine learning. In: International Conference on Cyber Warfare and Security, vol. 17, no. 1 (2022)
35. Hou, C., et al.: SquirRL: automating attack analysis on blockchain incentive mechanisms with deep reinforcement learning. arXiv preprint arXiv: 1912.01798 (2019)
36. Wang, Z., et al.: ForkDec: accurate detection for selfish mining attacks. Secur. Commun. Netw. **2021**, 1–8 (2021)
37. Mounica, M., Vijayasaraswathi, R., Vasavi, R.: Detecting sybil attack in wire-less sensor networks using machine learning algorithms. In: IOP Conference Series: Materials Science and Engineering, vol. 1042, no. 1. IOP Publishing (2021)
38. Dey, S.: Securing majority-attack in blockchain using machine learning and algorithmic game theory: a proof of work. In: 2018 10th Computer Science and Electronic Engineering (CEEC). IEEE (2018)
39. Kumari, R., Catherine, M.: Anomaly detection in block chain using clustering protocol (2018)

Virtual Reality Based on Machine Learning: State of the Art

Ghalia Mdaghri-Alaoui(✉), Abdelhamid Zouhair, El Mokhtar En-Naimi, Nihad Elghouch, and Aziz Mahboub

DSAI2S Research Team, Faculty of Sciences and Technologies, Abdelmalek Essaâdi University Tetouan, Tetouan, Morocco
ghaliaalaoui96@gmail.com

Abstract. The use of machine learning techniques has exploded in recent years as well as the use of virtual reality techniques it's can be used in many fields such as medicine, entertainment, education, military, etc. Machine learning visualization can be extended to virtual reality. For machine learning, data visualization is important. In addition, virtual reality can use machine learning to visualize digital educational simulations in three dimensions. Virtual reality transforms the learning process from passive to active, allowing users to engage with the material and apply what they have learned in a real-world setting. Using machine learning, we can design motion interaction systems with motion examples rather than coding, we design interactions by moving. This can make it possible to design much more natural interactions. In this article, we will discuss the contribution of machine learning to virtual reality, and compare the different uses of these two great technologies in some fields.

Keywords: Machine Learning · Virtual Reality · Supervised learning · unsupervised learning · reinforcement learning · Augmented Reality

1 Introduction

Artificial intelligence means that machines can perform "intelligent tasks" and make independent decisions. Machine Learning goes even further and allows machines to improve over time, as algorithms become more and more sophisticated [1].

Virtual reality and machine learning are currently the two most prominent actors in the technology world. Both are developing rapidly, trying to meet the technological needs of their respective fields, we are beginning to see the two combine to form a new dynamic combination, and many are looking at how Machine Learning can be used to enhance Virtual reality [2].

By integrating machine learning and virtual reality, we could add more touches that users are looking for; in addition, virtual reality adds a new level of immersion to the training environment, while the addition of machine learning allows for more in-depth training, not only in medicine but also in different domains such as education, entertainment, military, etc.

M. Lazaar et al. (Eds.): BDIoT 2022, LNNS 625, pp. 501–512, 2023.
https://doi.org/10.1007/978-3-031-28387-1_42

Virtual reality technology can place people in scenes, allowing them to immerse themselves and interact with the landscape. In this way, users conduct inquiry-based learning under the guidance of their interests and actively acquire knowledge [3].

Machine Learning is being used for the analysis of the importance of clinical parameters and their combinations for prognosis, e.g. prediction of disease progression, extraction of medical knowledge for outcome research, therapy planning and support, and overall patient management.

One of the most important uses of machine learning in healthcare is the detection diagnosis of diseases and conditions that are otherwise difficult to identify. This can range from tumors that are difficult to detect in their early stages to other hereditary illnesses. We summarize some major Machine Learning applications in healthcare.

Medical diagnostic reasoning may be a very important application area of intelligent systems. This framework, expert systems, and mod-el-based schemes provide mechanisms for the generation of hypotheses from patient data [4]. This approach is often extended to handle cases where there is no previous experience in the interpretation and understanding of medical data. For example, Hau and Coiera describe an intelligent system, which takes real-time patient data obtained during cardiac bypass surgery and creates models of normal and abnormal cardiac physiology to detect changes in the patient's condition [4].

Another field of application is biomedical signal processing Machine Learning methods use these sets of knowledge, which may be produced easier and can help to model the nonlinear relationships that exist between these data and extract parameters and features that can improve medical care [4].

In addition, ML is disrupting every industry and leading innovations such as assurance (Fraud detection), pharmaceuticals (Predict and prevent life-saving drug shortages), Farming (Assess harvest quality), Banking and Finance (Personalized services to customers), etc.

The rest of this paper is organized as follows: In Sect. 2 of this article, we defined the term machine learning. In the next section, we present a definition of virtual reality and augmented reality and a comparison between them, Sect. 4 presents some projects that use virtual reality based on machine learning in some fields, Sect. 5 aims to make a comparative study, and finally, in Sect. 6 we give a conclusion and our future work.

2 Machine Learning

Machine learning is a field of artificial intelligence that relies on mathematical and statistical approaches to give computers the ability to learn from data [5].

Machine learning used in several useful applications that simplify our daily life they are in all fields, one of the most popular applications is voice assistants like (Alexa, Siri, and Google Assistant) these assistants understand our language and respond to our requests, and recommendations systems like (Netflix, YouTube, Tiktok and Amazon) that recommends content close to our interests.

Another example in the field of health is the detection of cancerous cells. Microsoft's project (InnerEye-Democratizing Medical Imaging AI) uses machine-learning algorithms to differentiate tumor cells from patient radiographs better [6].

Machine translation and spelling correction there are also even more impressive applications like the automatic text generation application (Rytr. me), the list of Machine Learning applications is long but what exactly is Machine Learning?

Machine learning is an application of AI that enables systems to learn and improve from experience without being explicitly programmed.

In 1959 ARTHUR SAMUEL, a former researcher at IBM, introduced the term Machine Learning he defines machine learning as «Programming computers to learn from experience should eventually eliminate the need for much of this detailed Programming effort» [7].

In 1997 TOM MITCHELL, professor and head of the Machine Learning department at Carnegie Mellon University, defined machine learning as «Machine Learning is the study of computer algorithms that improve automatically through experience» [8].

What I like in these two definitions is the term experience because for the machine experience is nothing else than the mass of data that it can process.

Machine Learning encompasses three categories that are distinguished by the degree of supervision required of the user, analyst, or researcher, so these three categories are supervised learning, unsupervised learning, and reinforcement learning.

- First, supervised learning is a set of algorithms that attempts to identify the best mapping between input data and output, supervised learning is used to essentially meet either of the following two objectives it could be that we want to predict a continuous variable such as (profits, stock market shares, age…) its regression. As we can predict a class, a category, or an event like predicting if the employee is going to quit or not if the picture designates an animal or a human we are particularly interested here in a classification problem. It is the type of the response variable that determines the type of problem if regression (continuous target) or Classification (categorical target) [9].
- Secondly, in unsupervised learning, as its name indicates, the machine learns without any concrete supervision this means that the data does not contain any target values there is no response variable to predict or evaluate, here we have a set of input variables and we try to extract relevant information from these data, unsupervised learning is called exploratory analysis. For example, in unsupervised learning, observations can be grouped (grouping similar customers, identifying different types of tumors, segmenting machines, etc.) [9].
- Finally, The last category of machine learning is reinforcement learning. This type of learning is based on theories of behavioral psychology where the machine called an agent observes the environment, chooses an action, and gets a reward in return. It is more in the field of robotics or games [10].

Suppose we build a mini robot and present it to the outlier with an entrance door and an exit door, if the robot goes in the right direction it gets points, if he makes a mistake, he will be deducted points it is expected that the robot learns by itself to browse the aberrant without mistakes.

This figure summarizes the different machine learning approaches.

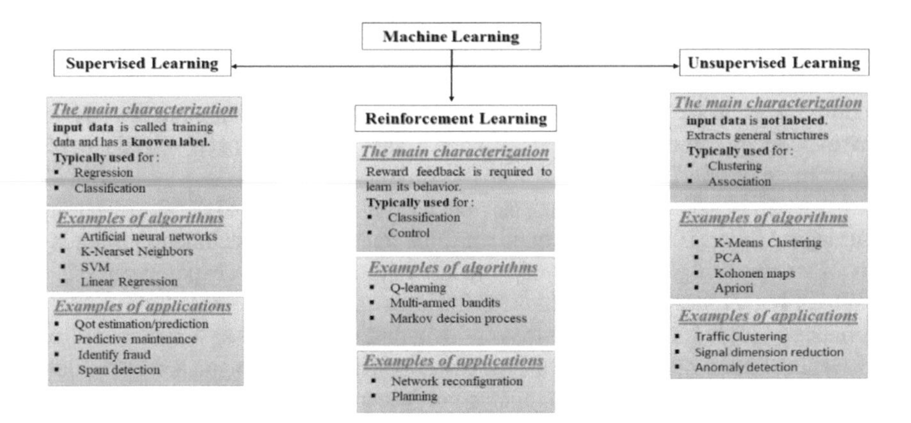

Fig. 1. The different approaches to machine learning.

Figure Fig. 1 presents the three categories of Machine Learning, including the algorithms used according to the types of problems, the domains of use, the types of data, and some examples of applications.

3 Virtual Reality

Virtual reality is a fully immersive user environment that modifies or alters sensory data. This environment allows users to interact with this sensory data as they interact with the virtual world. More specifically the term virtual reality encompasses a series of computer technologies that aim to immerse one or more people in a virtual environment created by the software. An environment that more or less faithfully reproduces a real setting [11].

Fuchs proposed two definitions of VR, one technical and one functional:

- Technical definition: VR is a scientific and technical field exploiting computer science and behavioral interfaces to simulate in a virtual world the behavior of 3D entities, which are interacting in real-time with each other and with one or more users in pseudo-natural immersion through sensor-motor channels [12].
- Functional definition: virtual reality will allow extracting from physical reality to virtually change time, place, and/or type of interaction with an environment simulating reality or interaction with an imaginary or symbolic world [12].

Fuchs proposed a definition that encompasses all definitions. For this, they base it on the purpose of virtual reality: "The purpose of VR is to enable a person or persons, sensory-motor and cognitive activity in an artificial, digitally created world that can be imaginary, symbolic or a simulation of some aspect of the real world".

Similarly, we can define augmented reality (AR) as a technique to display additional information about the real world. With this definition, there is no need to talk about specific hardware but we can specify techniques and applications and focus on technology development [13].

Virtual reality (VR) and augmented reality (AR) seem identical. Besides that, augmented reality (AR) is the display of data in the real environment, creating a visual experience. Virtual images appear in the real environment. Therefore, augmented reality (AR) is more prevalent because it is easier to use than traditional virtual reality (VR) (no need to wear a VR headset) [13].

In order to better understand the spectrum of immersive computing, let us take a look at this Fig. 2:

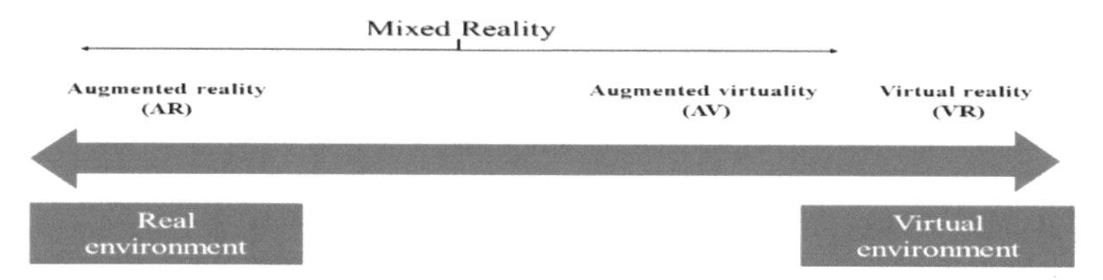

Fig. 2. Reality-virtuality continuum [13].

The preceding figure illustrates how the level of immersion affects the user experience, within the virtual continuum; we find the subset of mixed reality, which has been defined as everything between reality and a totally virtual environment.

4 Virtual Reality Based on Machine Learning

The main use of machine learning in virtual reality is data visualization [14]. For machine learning, data visualization is important, particularly in medicine, for knowledge extraction [14]. In addition, virtual reality can utilize machine learning to visualize numerical simulations in three dimensions. As a result, machine learning can facilitate the spatial understanding of complex data [14]. Machine learning and virtual reality user can work simultaneously to work on specific data-oriented tasks.

The current use of virtual reality in medicine is growing rapidly. Machine Learning has become in a few years an essential technology to automate many tasks and push the limits of traditional computing.

The objective is to demonstrate the contribution of machine learning for virtual reality in different fields specifically in medicine, several studies and several projects combine these two great technologies. These two technologies have a very important impact in the different fields and now we will see some researches that combine these two technologies.

The project by Raya et al., to diagnose autism spectrum disorder (ASD) is based primarily on behavioral symptoms in the sensory areas, On one side, cutting edge of technologies such as cameras, sensors, and virtual reality can accurately detect and classify the behavior of biomarkers such as body movement in real-world simulations. On the other hand, machine-learning techniques show the potential to identify and classify patient subgroups in addition; machine-learning methods improve the predictive value of motor behaviors [15].

In the same direction, a project by Siyar et al. describes the first investigation of the application of machine learning to differentiate between (skilled) and (novice) psychomotor performance in a virtual reality brain tumor resection task. Removal of a series of virtual brain tumors without damaging surrounding tissue, the results obtained highlight the potential of machine learning applied to virtual reality simulation data to help rebalance the educational paradigm based on proven performance criteria, shifting traditional learning to more objective models [16].

In the same context, Brouwera et al. with the addition of eye tracking to such simulations can provide very detailed outcome measures with great potential for neuropsychological assessment. In this study, 83 stroke patients and 103 healthy controls were asked to find 3 or 7 items from a shopping list in a virtual supermarket environment while recording eye movements. Using logistic regression and support vector machines we attempted to predict participants' tasks and whether they were in the stroke or control group [17].

While in the entertainment field, Hell and Argyriou described a new framework to obtain an automatic assessment of motion sickness using neural networks. An application that allows users to create roller coasters directly in virtual reality to share them with other users and to browse and evaluate them to collect in-game behavior with players, the tracks themselves, and the SSQ (Simulator Sickness Questionnaire) based on the application. A deep neural network-based machine learning architecture was trained using these data to predict the degree of motion sickness [18].

However, Li Ma focuses on the field of education. The goal is to improve students' ability to learn English, comparing the teaching experience of two classes at a university; the experimental class carries out immersive virtual situational teaching based on virtual reality technology from a constructivist perspective, while The Control Classroom adopts general multimedia equipment and traditional teaching methods. In the classroom, the teacher occupies most of the time; the students are only passively receiving a lot of information from the teacher, rarely have the opportunity to participate in the exchange of information and to express ideas in the target language, and most of the time is in a state of "immersion" in a Chinese environment. The overall level of English was also better than that of the control class [19].

In the same direction, Harbova et al. use the capabilities and tools of virtual reality and machine learning to build a learning platform for children with disabilities. The platform is designed to address the potential learning and social problems of students. The study concludes by emphasizing the need for further research and development [20].

In the same context, Wan's study makes it possible to integrate gesture recognition into classroom teaching and introduce a dynamic gesture recognition method. This study details the acquisition and preprocessing of the data to convert the data of the gesture action region into gray value images that are then classified using an improved algorithm.. In addition, this study designed a controlled experiment to analyze the performance of the algorithms in this study and compared the recognition accuracy of the algorithms from the perspective of simple background and complex background. Research results show that gesture recognition in distance education can effectively improve teaching effectiveness [21].

In addition, Yuanyuan and Tingting use different channels to automatically learn global and local features related to facial expression recognition tasks. They integrate the mechanism of soft attention in the proposed model to the model automatically learns the feature maps that are more important for facial expression recognition and salient regions in feature maps. They perform a weighted fusion on the features extracted from different branches and use the fused features to recognize student characteristics [22].

5 Comparative Study of Machine Learning and Virtual Reality Applied to Different Domains

Virtual Reality (VR) and Machine Learning (ML) are two big players in the technological world right now. Both of them are advancing at a rapid pace in an attempt to fuel the tech needs in their respective fields. So, just how are Virtual Reality and Machine Learning being used together? Several projects are combined between these two biggest technologies.

In the medical field the project by Moncada et al. who diagnoses patients with epileptic seizures caused by photosensitivity, this study proposes an alternative to the conventional IPS (Intrusion prevention system) procedure for PPR (photo paroxysmal response) detection using virtual reality and machine learning. The main contributions of this research are: introducing virtual reality in a close loop with machine learning can lead shortly to more advanced diagnostic tests and procedures, also to develop ML models to detect anomalies in the EEG (electroencephalogram) recordings when the patient is flashed using either VR-ML IPS or conventional IPS [23].

In the same context, Andrew et al. have a new proposition of a new method of machine learning for processing automated performance measures (APM) to assess surgical performance and predict clinical outcomes after robot-assisted radical prostatectomy (RARP). This study shows that APMs and ML algorithms may help assess surgical RARP performance and predict clinical outcomes. With further accrual of clinical data (oncologic and functional data), this process will become increasingly relevant and valuable in surgical assessment and training [25].

While in Neuroergonomics, Abujelala et al. have a new solution for firefighters that they trained in a virtual environment that includes virtual perturbations such as fires, alarms, and smoke. The objective is to use machine learning methods to discern encoding and retrieval states in firefighters during a visuospatial episodic memory task and explore which regions of the brain provide suitable signals to solve this classification problem [26].

In the same field, Alkadri et al. have demonstrated the benefits of artificial neural network algorithms in assessing and analyzing virtual surgical performances. This study applies the algorithm to a virtual reality simulated annulus incision task during an anterior cervical discectomy and fusion scenario. An artificial neural network model was trained on nine selected surgical metrics, spanning all three categories, and achieved 80% testing accuracy [27].

However, Cavedoni et al. have a proposition for the detection of Mild Cognitive Impairment (MCI) Classical neuropsychological measures underlying a categorical model of diagnosis could be integrated with a dimensional assessment approach involving Virtual Reality and Artificial Intelligence. VR can be used to create highly ecologically controlled simulations resembling daily life contexts. In addition, employs Machine Learning to analyze them in combination with clinical and neuropsychological data. This integrated computational approach would enable the creation of a predictive model to identify specific patterns of cognitive and motor impairment in MCI [29].

In the entertainment field, de Lima et al. have proposed a new method to identify what players fear in a virtual reality horror game. With the use of machine learning and player modeling techniques to create a model of players' fears, which can be used in real-time to adapt the content of horror games to intensify the fear evoked in players. They want to predict what future players fear in a virtual reality game by using the use of machine learning techniques [24].

As for education, Timothy et al. emphasize identifying machine learning predictors that could be utilized for cognitive performance classifiers, Participants were categorized into either high-performing or low-performing categories. The goal was to identify specific machine learning predictors that can be utilized in the development of an adaptive framework for the social cues and environmental distractors occurring in the virtual environment during the Stroop Task in a Virtual School Environment for potential attention enhancement or remediation purposes [28].

In industry, Carletti et al. propose an approach for defining a 'feature importance' in Anomaly Detection problems that has enormous applicability in industrial scenarios. Indeed, it is extremely relevant for quality monitoring. Moreover, it is often the first step towards the design of a Machine Learning-based smart monitoring solution because Anomaly Detection can be implemented without the need for labeled data [30].

In the table below, we will compare the different uses of machine learning and virtual reality in different fields based on the following criteria (Machine-learning category, the algorithm used, the application domain, etc.) (Table 1).

Table 1. Comparative studies of machine learning algorithms applied to different domains.

Virtual Reality based on Machine learning	BT issue/improvement	ML Category	Algorithm	Application domain
The Design of Adaptive Virtual Reality for Horror Games	Create a model of players' fears, which can be used to adapt in-game horror Elements to intensify the fear evoked in players [24].	Supervised Learning	Artificial Neural Network ANN	Entertainment
A Robot-Assisted Radical Prostatectomy Performance and Predict Outcomes	The processing of automated performance measures (APM) to assess surgical performance and predict clinical outcomes after robot-assisted radical prostatectomy (RARP) [25].	Supervised Learning	RandomForest-50	Medicine
Assessing Learning States in VR under Stress among Firefighters	Use machine learning methods to discern encoding and retrieval states in firefighters during a visuospatial episodic memory task and explore which regions of the brain provide suitable signals [26].	Supervised Learning	Random Forest	Medicine
A virtual reality surgical procedure	Applies the algorithm of machine learning to a virtual reality simulated annulus incision task during an anterior cervical [27].	Supervised Learning	Artificial Neural Network ANN	Medicine

(*continued*)

Table 1. (*continued*)

Virtual Reality based on Machine learning	BT issue/improvement	ML Category	Algorithm	Application domain
A Virtual School Environment: Virtual Reality Stroop Task	Identify predictors that can be utilized in the development of an adaptive framework for the social cues and environmental distractors occurring in the virtual environment for potential attention enhancement or remediation purposes [28].	Supervised Learning	Support Vector Machines SVM, Naïve Bayes NB, and k-Nearest Neighbors KNN.	Education
The automatic photo par response detection The Early Detection of Mild Cognitive Impairment	Diagnosis of patients with epileptic seizures caused by photosensitivity [23] Identify people with mild cognitive impairment (MCI). Using virtual reality to create highly environmentally controlled simulations that resemble everyday life contexts. In addition, machine learning (ML) is used to analyze them in conjunction with clinical and neuropsychological data [29]	Supervised and unsupervised learning Supervised and unsupervised learning	K-nearest neighbors KNN K-means clustering, PCA, Artificial Neural Networks (ANN)	Medicine Medicine
Anomaly Detection	Propose an approach for defining a 'feature importance' in Anomaly Detection problems [30].	Unsupervised Learning	Isolation Forest	Industry

Through the comparative study of some projects in which virtual reality and machine learning were used together, we noticed that the most field that combines these two biggest technologies is medicine also supervised machine-learning approach has been used extensively as well, KNN and ANN algorithms are the most used.

6 Conclusion and Perspectives

Machine learning and virtual reality are two completely different technologies. The first technology helps computers to reason as well as humans, and the second technology allows us to enrich the world around us by adding useful information, according to what we have already seen, these two technologies are used in different fields and applications, mainly in the medical field, education, and entertainment, etc. The combination of these two technologies will change the world in the future and facilitate daily life.

As we know, many projects have used virtual reality and machine learning this comparative study showed us some fields such as education, entertainment, industry, and especially medicine that combine these two technologies by using different algorithms in order to boost virtual reality.

In future work, we plan to develop our architecture of virtual reality based on machine learning with applications in different domains such as medicine and distance learning.

References

1. Claverie, B.: Pour une Histoire Naturelle de l'Intelligence Artificielle. Hal Open Science (2018)
2. Yeh, S.-C., et al.: Machine learning-based assessment tool for imbalance and vestibular dysfunction with virtual reality rehabilitation system. Comput. Methods Program. Biomed. **3**(116), 311–318 (2014)
3. Noël, F., Pinquié, R.: Formation à la réalité virtuelle pour l'industrie 4.0. Hal Open Science (2021)
4. Magoulas, G. D., Prentza, A.: Machine learning in medical applications, pp. 300–307 (1999)
5. Contardo, G.: Machine learning under budget constraints. In: Pierre and Marie Curie University (2017)
6. Garnier, Ch., Collet, N., Jaremko, S., Abtan, M.: L'intelligence artificielle dans les industries de santé. AEC Partners (2020)
7. Petrik, M.: Machine Learning Introduction to Machine Learning (2017)
8. Mitchell, T.: Machine Learning, 2nd edn. McGraw-Hill Science/Engineering/Math (1997)
9. Brownell, J.: Supervised and unsupervised machine learning. https://machinelearningmastery.com/supervised-and-unsupervised-machine-learning-algorithms/. Accessed 12 June 2022
10. DataScientest. https://datascientest.com/reinforcement-learning,2022/06/13
11. Bellazi, A., et al.: Virtual reality for assessing visual quality and lighting perception: a systematic review. Build. Environ. (209) (2022)
12. Leubou Ngu, R.: Impact de la réalité virtuelle sur la formation à distance. Hal Open Science, University of Limoges, France (2021)
13. Nincarean, D., Alia, M.B., Halim, N.D.A., Rahman, M.H.A.: Mobile augmented reality: the potential for education. Proc. – Soc. Behav. Sci. 657–664 (2013)
14. Hall, L., et al.: Using virtual reality and machine learning techniques to visualize the human spine. EPiC Ser. Comput. 123–132 (2021)

15. Raya, M.A., Marin-Morales, J., Minissi, M.E., Garcia, G.T., Abad, L., Giglioli I.A.C.: Machine learning and virtual reality on body movements' behaviors to classify children with autism spectrum disorder. Clin. Med. (2020)
16. Siyar, S., et al.: Machine learning distinguishes neurosurgical skill levels in a virtual reality tumor resection task. Amirkabir University of Technology, Iran (2018)
17. Brouwer, V.H.E.W., et al.: Applying machine learning to dissociate between stroke patients and healthy controls using eye movement features obtained from a virtual reality task. Elsevier Direct Science (2022)
18. Hell, S., Argyriou, V.: Machine learning architectures to predict motion sickness using a virtual reality rollercoaster simulation tool. In: International Conference on Artificial Intelligence and Virtual Reality (AIVR) (2018)
19. Ma, L.: An immersive context teaching method for college English based on artificial intelligence and machine learning in virtual reality technology. Hindawi (2021)
20. Harbova, M., Andrunky, V., Chyrun, L.: Virtual reality platform using ml for teaching children with special needs. In: Lviv Polytechnic National University, Ukraine (2020)
21. Wan, J.: Gesture recognition and information recommendation based on machine learning and virtual reality in distance education. In: School of Intelligence Science and Information Engineering, Xi'an Peihua University, Shaanxi, China (2021)
22. Yuanyuan, C., Tingting, Z.: Performance analysis of distance teaching classroom based on machine learning and virtual reality. In: North China University of Science and Technology, Tangshan, China (2021)
23. Moncada, F., et al.: Virtual reality and machine learning in the automatic photoparoxysmal response detection. In: Computational-Based Biomarkers For Mental And Emotional Health (CBMEH) (2021)
24. de Lima, E.S., Silva, B.M.C., Galam, G.T.: Towards the design of adaptive virtual reality horror games: a model of players' fears using machine learning and player modeling. In: 19th Brazilian Symposium on Computer Games and Digital Entertainment (SBGames) (2020)
25. Hung, A.J., et al.: Utilizing machine learning and automated performance metrics to evaluate robot-assisted radical prostatectomy performance and predict outcomes. J. Endourol. 438–444 (2018)
26. Abujelala, M., Karthikeyan, R., Tyagi, O., Du, J., Mehta, R. K.: Brain activity-based metrics for assessing learning states in VR under stress among firefighters: an explorative machine learning approach in neuroergonomics. Brain Sci. (2021)
27. Alkadri, S., et al.: Using a multilayer perceptron artificial neural network to assess a virtual reality surgical procedure. Comput. Biol. Med. (2021)
28. McMahan, T., Duffield, T., Parsons, Th.D.: Feasibility study to identify machine learning predictors for a virtual school environment: virtual reality stroop task. Front. Virtual Reality U.S. (2021)
29. Cavedoni, S., Chirico, A., Pedroli, E., Cipresso, P., Riva, G.: Digital biomarkers for the early detection of mild cognitive impairment: artificial intelligence meets virtual reality. Front. Hum. Neurosci. **14**, 245 (2020)
30. Carletti, M., Masiero, Ch., Beghi, A., Susto, G.A.: Explainable machine learning in industry 4.0: evaluating feature importance in anomaly detection to enable root cause analysis. In: International Conference on Systems, Man and Cybernetics (SMC), Bari, Italy. IEEE (2019)

QoS in IoMT: Towards Using TCP Header to Manage Priorities

Fathia Ouakasse[1,2](✉), Radia Belkeziz[1], and Said Rakrak[2]

[1] Private University of Marrakesh, Marrakesh, Morocco
fathia.ouakasse@gmail.com

[2] Faculty of Sciences and Techniques, Laboratory of Computer and Systems Engineering (L2IS), Cadi Ayyad University, Marrakesh, Morocco

Abstract. Recently, with the advances in technologies and the integration of Internet of Things (IoT), many fields face difficulties and challenges in implementing IoT based systems and applications. One of the most important domain where IoT has been widely used and proved its performance is healthcare. In the medical field, connected objects are called Internet of Medical Things (IoMT). Besides, in this field, a large amount of critical and sensitive data is generated and the network may undergo the problem of congestion or traffic overload. Due to the nature of the generated data, it is important to ensure getting data in real time and without delay. In this paper, we direct our attention to deal with the Quality of Services in IoMT. We propose an approach able to classify connected devices into smart devices and IoMT devices, and then classify IoMT gathered data according to data emergency by prioritizing urgent data. We found our solution based on the reserved field in the Transmission Control Protocol (TCP) header to manage priorities for data transmission within a medical system.

Keywords: IoMT · QoS · Medical System · TCP Header · Reserved Field · Priority

1 Introduction

The use of connected devices and smart gadgets has invaded all aspects of our life with the emergence of different types and applications. So far, IoT has been implemented in different fields such as agriculture [1, 2], transport [3], water and energy consumption [4, 5] and so on and so far. Besides, e-health is an important and a fertile field where smart connected devices are implemented and tested. Internet of Medical Things (IoMT) is the term related to smart gadgets used to connect medical objects. Nevertheless, regarding the nature of medical data like sensitive and critical, the challenge is to collect and transmit data without delay and get the information when an event of interest occurs in order to intervene in real-time.

The IoMT is a platform that tolerates everything for information processing, data communication, and collaborative context analysis to predict medical issues and decide the case of a patient. Hence, in order to perform IoMT systems, a massive amount of data having different content and formats must be processed.

M. Lazaar et al. (Eds.): BDIoT 2022, LNNS 625, pp. 513–523, 2023.
https://doi.org/10.1007/978-3-031-28387-1_43

Furthermore, according to a report published in January 2022 by TechForge Media [6], the number of remotely-monitored patients hit 45.6M in 2020. Besides, the leading research firm's analysts expect 7.4 million IoMT devices to be deployed globally, with over 3,850 per smart hospital. The report estimates, likewise, that there will be 120 billion connected devices by 2030, with an average of 15 per person.

Whereas these great achievements, there are many phenomena that might slow the transmission of data such as traffic congestion and interference. The reason why it is important to manage the Quality of Service (QoS) in data transmission within IoMT networks.

As mentioned above, the widespread use of IoMT engenders a large amount of circulated data and notifications. Thus, in this paper, we direct our attention to QoS management in order to ensure that critical medical data is transmitted within an IoMT network without latency. We propose to use the field reserved in TCP header in the transport layer to specify the nature of the transmitter node and the priorities. Firstly, we classify the transmitter node, the source of a generated data, if it is about an IoMT device or another connected device like tablet, smartphone....Then, secondly, we depict the priorities based on the emergency of gathered data. Moreover, the field reserved in the TCP header includes 6 bits; 3 bits are consacred for congestion management and the remaining 3 bits are reserved for future use (these bits will be used in our proposition).

The remainder of this paper is organized as follows: in the second section, the architecture of IoMT network and the QoS management are presented. The third section is devoted to related works. Then, in the fourth section, the proposed mechanism to ensure priorities in an IoMT network is presented. Finally, we conclude with a conclusion and future works in the fifth section.

2 IoMT Networks and QoS Management

2.1 IoMT Network Architecture

The IoMT is the blend of medical devices with the IoT. IoMTs represent the base of the future healthcare systems. So far, every medical device will be connected via an infrastructure over the Internet and monitored through applications thanks to doctors and healthcare professionals [7].

Therefore, using medical sensors, medical data can be gathered and the connected patient can be tracked. After the collection, data is stored and processed via smart devices and gateways. Then, the information is stored and analyzed in the cloud. Finally, it is transmitted either to the healthcare professionals to process for a decision making, or to the family to get the tracking [8]. Figure 1 describes the aforementioned steps.

Besides, the architecture of IoMT consists of several technology suites to support the healthcare system. It consists of the integration of a set of technologies related to each other to develop the IoMT system modularity and scalability. The architecture of IoMT system includes several layers and functionalities namely [9]:

- Medical sensor layer: the first layer includes different medical sensors used to connect the real world and the physical measurements for real-time information processes.

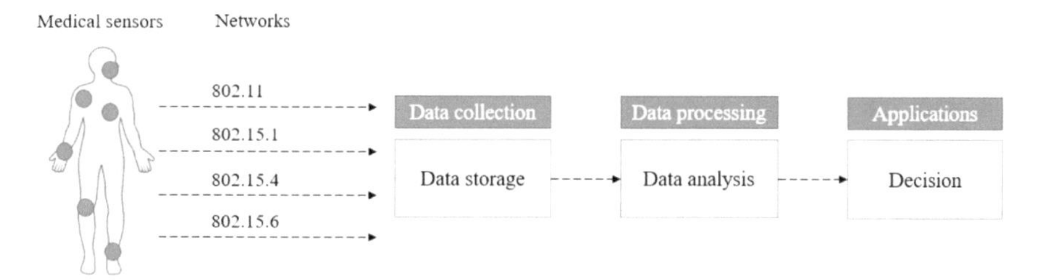

Fig. 1. Architecture of data processing in IoMT

Data collected by medical sensors through gateways in the form of personal area networks (PAN), such as Bluetooth, ZigBee, Ultra-Wideband (UWB), local area network (LAN), or WLAN.

- Network layer: is a layer operating between the cloud and the physical layer. This layer includes servers that serve to manage and administer the security and integrity of the system, and gateways that serve to redirect the data from these servers to the cloud for processing.
- Application layer: this layer is interested in the storage and the processing of data. Here, based on data analysis, healthcare professionals drive decision-making systems.

The architecture of IoMT as presented in [10] is shown in Fig. 2.

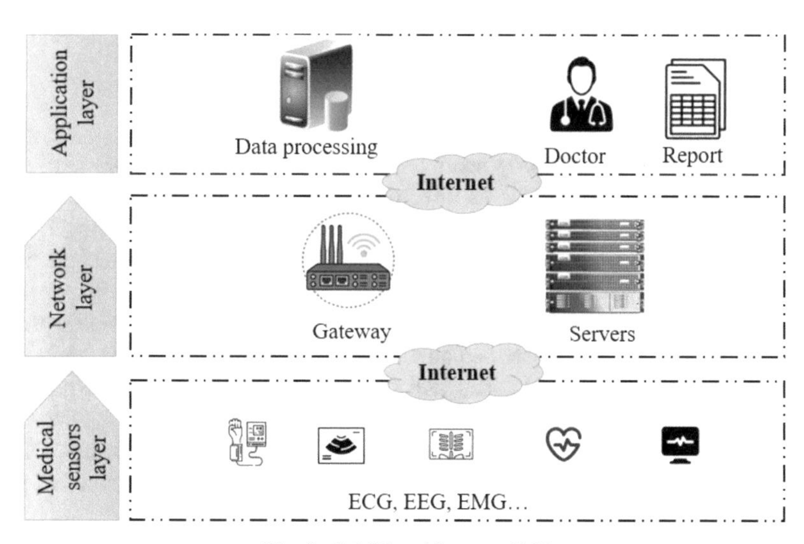

Fig. 2. IoMT architecture [11]

2.2 QoS Management

In the IoT network, the use of QoS is important to define priorities for devices to have a fluent network access and avoid delays. QoS is defined as a set of mechanisms and technologies used to control traffic and ensure the performance of critical applications. It enables users to adjust the overall network traffic by prioritizing specific high-performance applications [12].

Indeed, in traditional computer networking protocols such as IP and TCP/IP, the traffic is served on a first-come first-serve basis, which is termed as best-effort service. This means that all resources of the network are shared equally between connected devices [13]. However, the data emanating from critical sources like medical devices are more sensitive to packet delay, jitter and loss. Whereas, data emanating from voice or video streaming applications are less sensitive to packet loss.

Today, the Internet and many other TCP/IP networks only support the so-called best effort characteristic for traffic. This technique proves its insufficiency to support the requirements of real-time applications, critical reliability applications, and applications requiring any form of guarantees [14].

So, there is no doubt that the QoS management is an important key to successful communication [15]. In Fig. 3, priority order of traffic according to the type of data is shown.

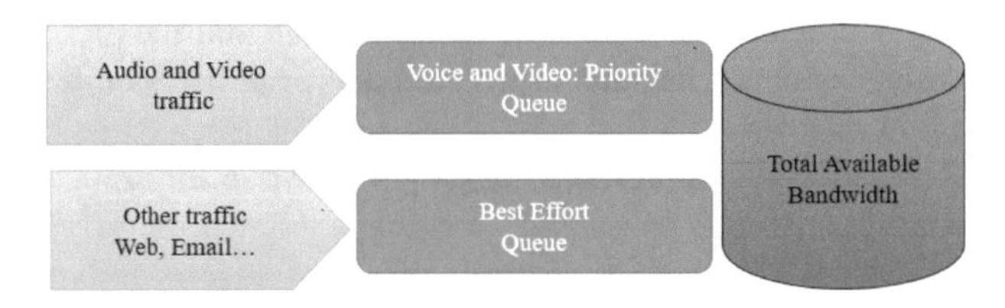

Fig. 3. Traffic Priority according to data types [16]

Consequently, the Internet Engineering Task Force (IETF) defines two QoS control architectures commonly presented as Integrated Service (IntServ) [17] and Differentiated Service (DiffServ) [18]. IntServ is based on resource allocation using Resource ReSerVation Protocol (RSVP). IntServ uses the QoS requirements of the application to compute the required bandwidth between the corresponding nodes, then allocate the network resource adequately [19]. On the other hand, DiffServ control architecture proposes to separate the traffic by classes. Based on a classifier, the mechanism differentiates traffic flows into different classes of traffic, and assigns network resources to certain classes of traffic rather than allocating resources for traffic flows, each time. Then, according to traffic class QoS requirements, network resources are allocated [20]. Besides, DiffServ uses Differentiated Services Code Point (DSCP) values in the header of an IPV4 packet to distinguish packets coming from different traffic classes.

Furthermore, to ensure how QoS network traffic works, it is important to take on consideration the characteristics and measurements it is defined by. Firstly, the bandwidth that is the speed on the link. Secondly, the delay that is the time taken by a packet to get from the source node to the destination node. Thirdly, packet loss, that is the amount of data, lost during the transmission due to network congestion and queuing. Fourthly, jitter that is the variation in the inter-arrival time of packet sent [12].

Therefore, QoS is crucial for sensitive applications to guarantee the best performance of their most critical data. It is vital to ensure that high-bandwidth service is accorded to medical sensors in order to transmit data without latency or lag.

3 Related Works

QoS is the ability of a set of network technologies to guarantee suitable network services to the users. Based on various technologies and network parameters, many QoS control mechanisms were proposed. Here we list some works used to guarantee QoS in IoMT networks.

The IoMT services are represented as dynamical resources organized at edge level in a resource-oriented infrastructure as described by authors in [21]. The model presented consists of a multilevel architecture characterized jointly with an analytical methodology. This methodology is composed of several interoperable modules that can evaluate different aspects of QoS. The proposed model takes into consideration the evaluation of Acceptability, Usability and User Experience. Each level of the model is set as the base of the next level which makes the evaluation rest on a solid building block, driven by the context. Authors conducted the model as to suit the forthcoming 6G and considering its enabling technologies: more use of edge intelligent capabilities such as machine learning.

The artificial intelligence technology of preference logic is applied in [22], in order to come up with the Preference Logic Model. This model deals with both qualitative and quantitative cases by using preferences to describe QoS requirements of flows. Then according to preferences, the flows are aggregated into different QoS classes after eliminating the potential conflicts in the preferences. This model helps to provide differentiated services (Diffserv) and improvement of QoE (Quality of Experience).

A window-based Rate Control Algorithm (w-RCA) was proposed in [23] to optimize the medical Quality of Service (m-QoS) for remote healthcare, especially, Tele-surgery over 5G based mobile edge computing healthcare. The w-RCA is periodically applied by considering the network parameters: peak-to-mean ratio (PMR), standard deviation (Std.dev), delay and jitter during 8 min medical video stream. A video source creates one data frame per time unit and is transferred through work ahead transmission by delaying that video W time units. Medical server or transmitter gets information about future frames and recomputes transmission schedule. The goal is to control transmission to reduce the high peak data rate and Std.dev which are the key components in m-QoS optimization.

In [24], an energy-efficient routing protocol for wireless sensor network based IoT applications was designed. It implements OF based routing methodology to convey data from source to destination by selecting the best (optimal) path. This results in improving QoS and enhancing network performance. Three parameters are considered to optimal path selection: the reliability, the lifetime of a node and the possible traffic intensity. The protocol evaluation using NS-2 network simulation resulted in a better performance concerning end-to-end delay, packet delivery ratio, residual energy compared to other contemporary protocols.

An integrated energy and QoS-Based protocol were proposed in [25]. This protocol takes into consideration energy, end-to-end latency, and reliability requirements of BAN communication. It consists of modules divided into two categories: i) MAC layer modules and network layer modules. The Mac layer includes four modules: MAC receiver that receives data or hello packets from other nodes, then based on the packets' MAC address, forwards them to the network layer, ii) the reliability module calculates the numbers of packets sent to neighbor node *j* and the number of acknowledgements received from j, iii) the delay module monitors the time required to capture the channel. Network layer introduces four modules: the Packet Classifier that differentiates and forwards the data packets and Hello packets, the Hello Protocol Module, the Routing services. Modules that contain four submodules: QoS classifier, routing table constructor algorithm, routing table, and path selector algorithm The ZEQoS protocol also introduces three algorithms: neighbor table constructor, routing table constructor, and path selector which are submodules to Hello Protocol and Routing services Modules. Simulations also show that the ZEQoS also offers better performance in terms of higher throughput, less packets dropped on MAC and network layers, and lower network traffic compared to other protocols.

In LOCALMOR [26], a localized multi-objective routing protocol was proposed. It consists of four modules: a power-efficiency module, a reliability-sensitive module, a delay-sensitive module, and a neighbor manager. The protocol relies on the traffic diversity of biomedical applications and guarantees differentiated routing, based on using QoS metrics. Energy efficiency, reliability, and latency are considered in this protocol, and the data traffic is divided into four classes: regular, delay-sensitive, reliability-sensitive, and critical. Simulation showed that LOCALMOR brings improvements compared to other protocols according to QoS metrics.

In [27], authors talked about the integrated Clinical Environment, which counts 3 primary components: Supervisor, ICE interface description language, and network controller. The network controller is the primary channel of device control messages and signal datastreams. It also provides security services for device authentication and data encryption and ensures proper QoS in terms of data partitioning of datastreams and time.

4 TCP Header Based QoS in IoMT

In the IoMT system, data transmission may be delayed due to many phenomena such as traffic congestion, link delay or interferences, whereas in IoMT data must be transmitted with a high throughput and a minimal delay. The reason why it is important to manage the QoS in data transmission within IoMT networks. It is in this perspective that our work fits, we propose a mechanism to manage the data transmission priorities within a hospital. Hence, we propose to use the field reserved in TCP header in the transport layer to specify the nature of the transmitter node and the priorities.

Transmission Control Protocol (TCP), is a reliable, connected-mode transport protocol documented in RFC 7931 [28] of IETF. TCP sits above IP. In the OSI model, it corresponds to the transport layer, intermediate to the network layer and the session layer. Applications transmit data streams over a network connection. TCP cuts the stream of bytes into segments whose size depends on the MTU of the underlying network (data link layer).

<table>
<tr><td colspan="5">Source Port</td><td colspan="4">Destination Port</td></tr>
<tr><td colspan="9">Sequence Number</td></tr>
<tr><td colspan="9">Acknowledgement Number</td></tr>
<tr><td>Data Offset</td><td>Reserverd</td><td>U
R
G</td><td>A
C
K</td><td>P
S
H</td><td>R
S
T</td><td>S
Y
N</td><td>F
I
N</td><td>Window Size</td></tr>
<tr><td colspan="8">Checksum Bits</td><td>Urgent Pointer</td></tr>
<tr><td colspan="9">Options and Paddings</td></tr>
<tr><td colspan="9">Data</td></tr>
</table>

Fig. 4. TCP header [29]

In Fig. 4, we draw the TCP header including its different fields.

The field reserved in the TCP header is coded in 6 bits; 3 bits are used as follow:

- ECN/NS: signals the presence of congestion as described in RFC 31683 [30]; or Nonce Signaling, RFC 35404 [31].
- CWR: Congestion Window Reduced: indicates that a packet with ECE has been received and the congestion has been resolved.
- ECE: ECN-Echo: SYN = 1 indicates ECN management capability, and SYN = 0 indicates IP signaled congestion as described in RFC 3168 [28].

The three remaining bits in the reserved field are reserved for future use. Therefore, in this paper we propose the use of these three bits to manage priorities using suitable and specific applications supporting this mechanism within a hospital. To conduct this research, we adopt two classifications.

a. **First classification**

The mechanism firstly classifies connected devices into two categories as follows:

- Smart devices such as tablets, smartphones, computers…
- And IoMT such as remote patient monitoring, glucose monitoring, heart-rate monitoring, connected inhalers…

Indeed, in a hospital, several devices are connected to the network. Medical machines that can be equipped with devices used to monitor patients or to carry out the various checks and examinations of patients.

Each department has its own machines. For example, in radiology, there are MRIs (Fig. 5) (Magnetic Resonance Imaging), echocardiography (Fig. 6), ultrasounds (Fig. 7) etc.… and each of these machines can communicate the results to a computer or to another terminal of doctors to make decisions.

On the other hand, patients can get visits from family or friends who may use their terminals by connecting them to the network. There can be smartphones (Fig. 8), tablets (Fig. 9), smart watches (Fig. 10), etc.…

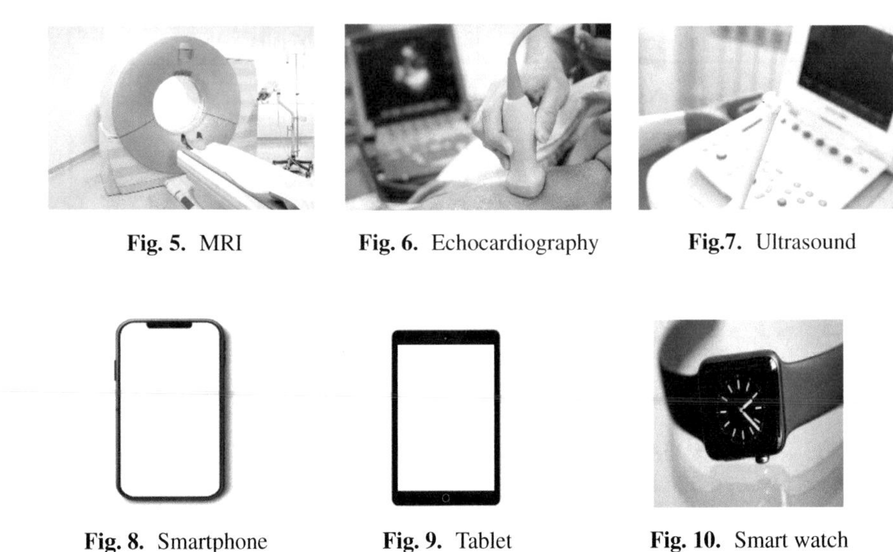

Fig. 5. MRI **Fig. 6.** Echocardiography **Fig.7.** Ultrasound

Fig. 8. Smartphone **Fig. 9.** Tablet **Fig. 10.** Smart watch

The purpose of the first classification is to recognize medical equipment by differentiating them from personal devices of users (patients and visitors).

Therefore, for smart devices, the remaining bits of reserved field are set to 000 and for IoMT in 001. This classification will allow us to have a vision concerning the hospital network, to create an IoMT database, and to determine the need in terms of QoS. Figure 11 illustrates this first classification.

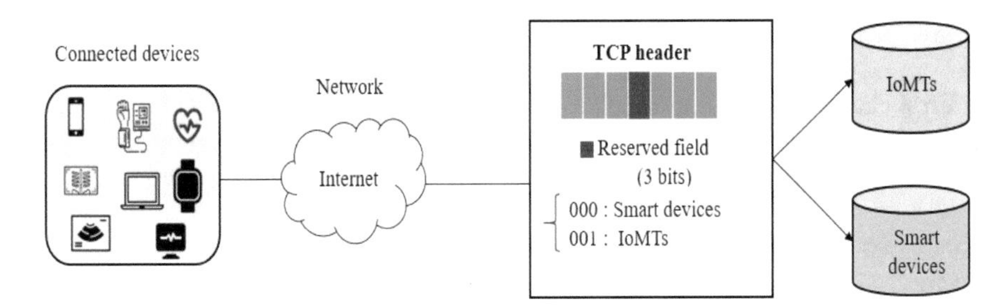

Fig. 11. First classification of connected devices

b. **Second classification**

According to gathered data emergencies, the priorities are managed in the TCP header. Furthermore, TCP reserved fields can take six values depending on data emergencies as shown in Fig. 12.

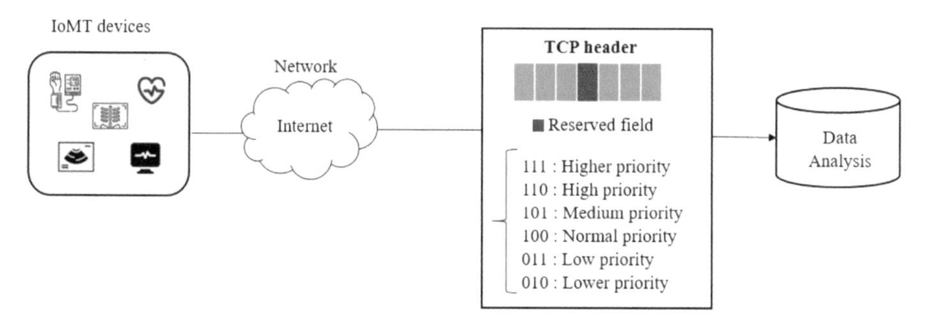

Fig. 12. Second classification for priority management

Apart from the general policies and unified procedures of hospital departments, each of the latter privileges certain data in decision-making. These data are qualified as critical and thus the reserved field in the TCP header is granted the highest value of priority e.g. 111.

Indeed, in each department, vital signs are critical data; such as body temperature, pulse rate, respiration rate and blood pressure. Therefore, in each hospital department, critical threshold or abnormal rates are defined and based on these thresholds and rates, a data can be declared as urgent. For example; a pulse rate less than 60/min or greater than 100/min, a temperature less than 35°C or greater than 38°C, a blood oxygen level less than 85%, an hypertension defined by a systolic blood pressure (SBP) measurement greater than 130 mm Hg or a diastolic blood pressure measurement (DBP) greater than 80 mm Hg, an hypotension was defined by an SBP less than 90 mm Hg or a DBP less than 60 mm Hg. In these cases, the reserved field takes the higher priority value.

Moreover, in certain circumstances, doctors might decide that some patients be given priorities according to their condition. To this end, all the data related to these patients are given priority and should follow the specification presented.

We should note here that the application used within the hospital must support the QoS as described in our proposal. So, it should be programmed in order to respect the specification of the reserved field used in this proposal.

5 Conclusion

In medical centers and hospitals, data generated by IoMT devices should be transmitted with emergency and in real time in order to promote health assessment such as physical checkup, counseling, curative treatment… So far, regarding the nature of medical data like sensitive and critical, the challenge is to collect and transmit data without delay or latency. Nevertheless, IoMT generates a large amount of data that can lead to network congestion. In this paper, we propose a mechanism to manage the QoS in the IoMT network. The approach proposes to classify connected devices into smart devices e.g. personal use devices (patients, doctors and visitors) and IoMT devices e.g. medical machines equipped with smart devices. The three bits of the reserved field are set to 000 in case of a personal use device and to 001 in case of a medical device. Then, we classify IoMT gathered data according to data emergency by prioritizing urgent data.

The priority is illustrated with the use of the reserved field in the TCP header. The bits are set to 111 for the highest priority and to 001 for the lowest priority.

A part of the research perspective associated with this work is to proceed to the implementation and the performance evaluation.

References

1. Quy, V.K., et al.: IoT-enabled smart agriculture: architecture, applications, and challenges. Appl. Sci. **12**(7), 3396 (2022)
2. Ramachandran, V., et al.: Exploiting IoT and its enabled technologies for irrigation needs in agriculture. Water **14**(5), 719 (2022)
3. Ganapathy, J.: design of algorithm for IoT-based application: case study on intelligent transport systems. In: García Márquez, F.P., Lev, B. (eds.) Internet of Things. ISORMS, vol. 305, pp. 227–249. Springer, Cham (2021). https://doi.org/10.1007/978-3-030-70478-0_11
4. Philip, M.S., Singh, P.: An energy efficient algorithm for sustainable monitoring of water quality in smart cities. Sustain. Comput.: Inform. Syst. **35**, 100768 (2022)
5. Razmjoo, A., et al.: An investigation of the policies and crucial sectors of smart cities based on IoT application. Appl. Sci. **12**(5), 2672 (2022)
6. Daws. R.: Vodafone Connected Consumer 2030: How smart tech will transform our world. TechForge Media (2022). https://www.iottechnews.com/news/2022/jan/26/vodafone-connected-consumer-2030-how-smart-tech-transform-world/. Accessed 27 June 2022
7. Balestrieri, E., De Vito, L., Picariello, F., Tudosa, I.: A novel method for compressed sensing based sampling of ECG signals in medical-IoT era. In: IEEE International Symposium on Medical Measurements and Applications (MeMeA), pp. 1–6. Istanbul, Turkey (2019)
8. Josephin Arockia Dhiyya, A.: Architecture of IoMT in healthcare. The Internet of Medical Things (IoMT): Healthcare Transformation, Wiley Online Library, Chapter 8 (2022)
9. Razdan, S., Sharma, S.: Internet of Medical Things (IoMT): overview, emerging technologies, and case studies. IETE Tech. Rev. **39**, 775–788. Taylor & Francis Online (2021). https://doi.org/10.1080/02564602.2021.1927863
10. Sun, L., Jiang, X., Ren, H., Guo, Y.: Edge-cloud computing and artificial intelligence in internet of medical things: architecture technology and application. IEEE Access **8**, 101079–101092 (2020). https://doi.org/10.1109/ACCESS.2020.2997831
11. De Vito, L., Picariello, F., Tudosa, I., Balestrieri, E.: A novel method for compressed sensing based sampling of ECG signals in medical-IoT era. In: Conference: MeMeA (2019). https://doi.org/10.1109/MeMeA.2019.8802184
12. What is Quality of Service (QoS) in Networking? https://www.fortinet.com/resources/cyberglossary/qos-quality-of-service#:~:text=Quality%20of%20service%20(QoS)%20is,prioritizing%20specific%20high%2Dperformance%20applications. Accessed 27 June 2022
13. Arindam, P.: QoS in Data Networks: Protocols and Standards. https://www.cse.wustl.edu/~jain/cis788-99/ftp/qos_protocols/index.html. Accessed 27 June 2022
14. Eckert, T., Bryant, S.: Quality of service (QoS). In: Toy, M. (ed.) Future Networks, Services and Management, pp. 309–344. Springer, Cham (2021). https://doi.org/10.1007/978-3-030-81961-3_11
15. Eswarappa, S.M., Rettore, P.: Adaptive QoS in SDN-enabled heterogeneous mobile tactical networks. Thesis for: Master of Science in Computer Science, Institute of Computer Science University of Bonn, Germany (2021). https://doi.org/10.13140/RG.2.2.21733.32483
16. Fowler, S.: Optimizing for Zoom conferencing, IQrouter (2022). https://evenroute.zendesk.com/hc/en-us/articles/360045162413-Optimizing-for-Zoom-conferencing. Accessed 29 Aug 2022

17. Braden, R., Clark, D., Shenker, S.: Integrated Services in the Internet Architecture: an Overview. Network Working Group (1994). https://www.ietf.org/rfc/rfc1633.txt. Accessed 27 June 2022
18. Blake, S., Black, D., Carlson, M., Davies, E., Wang, Z., Weiss, W.: An Architecture for Differentiated Services. Network Working Group (1998). https://www.ietf.org/rfc/rfc2475.txt. Accessed 27 June 2022
19. Jianting, L.: Design and implementation of Vo IPQoS model combining IntServ and DiffServ based on network processor IXP2400. In: 7th Annual International Conference on Network and Information Systems for Computers (ICNISC), pp. 60–64. Guiyang, China (2021)
20. Joung, J., Kwon, J., Ryoo, J.D., Cheung, T.: Asynchronous Deterministic Network Based on the DiffServ Architecture. IEEE Access **10**, 15068–15083 (2022). https://doi.org/10.1109/ACCESS.2022.3146398
21. Aiosa, G.V., Attanasio, B., La Corte, A., Scatá, M.: CoKnowEMe: an edge evaluation scheme for QoS of IoMT Microservices in 6G scenario. Future Internet **13**(7), 177 (2021). Academic Editor: Matthew Pediaditis. https://doi.org/10.3390/fi13070177
22. Tang, P., Dong, Y., Chen, Y., Mao, S., Halgamuge, S.: QoE-aware traffic aggregation using preference logic for edge intelligence. IEEE Trans. Wirel. Commun. **20**(9), 6093–6106 (2021). https://doi.org/10.1109/twc.2021.3071745
23. Sodhro, A.H., Luo, Z., Sangaiah, A.K., Baik, S.W.: Mobile edge computing based QoS optimization in medical healthcare applications. Int. J. Inf. Manage. **45**, 308–318 (2019). https://doi.org/10.1016/j.ijinfomgt.2018.08.004
24. Jaiswal, K., Anand, V.: EOMR: an energy-efficient optimal multi-path routing protocol to improve QoS in wireless sensor network for IoT applications. Wirel. Pers. Commun. **111**(4), 2493–2515 (2019). https://doi.org/10.1007/s11277-019-07000-x
25. Khan, Z.A., Sivakumar, S.C., Phillips, W., Robertson, B.: ZEQoS: a new energy and QoS-aware routing protocol for communication of sensor devices in healthcare system. Int. J. Distrib. Sens. Netw. (1–2), 1–18. Hindawi (2014) https://doi.org/10.1155/2014/627689
26. Djenouri, D., Balasingham, I.: New QoS and geographical routing in wireless biomedical sensor networks. In: Sixth International Conference on Broadband Communications, Networks, and Systems, pp. 1–8. IEEE, Madrid (2009)
27. Hatcliff, J., et al.: Rationale and architecture principles for medical application platforms. In: IEEE/ACM Third International Conference on Cyber-Physical Systems (2012). https://doi.org/10.1109/iccps.2012.9
28. Transmission Control Protocol, RFC: 793. Defense Advanced Research Projects Agency Information Processing Techniques Office (1981). https://datatracker.ietf.org/doc/html/rfc793. Accessed 27 June 2022
29. Masri, A.S.: Towards the distribution control of manufacturing systems: a component-based approach for taking into account the communication architecture in modeling. Thesis, HAL Id: tel-00578841, version 1 (2009)
30. Ramakrishnan, K., Floyd, S., Black, D.: The Addition of Explicit Congestion Notification (ECN) to IP. Network Working Group (2001). https://datatracker.ietf.org/doc/html/rfc3168. Accessed 27 June 2022
31. Spring, N., Wetherall, D., Ely. D.: Robust Explicit Congestion Notification (ECN) Signaling with Nonces. Network Working Group (2003). https://datatracker.ietf.org/doc/html/rfc3540. Accessed 27 June 2022

Evaluations of Some Routing Protocols Metrics in VANET

Oussama Sbayti(✉) and Khalid Housni

L@RI Laboratory, MISC Team, Faculty of Sciences, Ibn Tofail University, Kenitra, Morocco
{oussama.sbayti,housni.khalid}@uit.ac.ma

Abstract. Vehicular Ad hoc Network (VANET) is a type of Ad-hoc Network that attracts the attention of many researchers due to their different characteristics, such as high mobility of vehicles, topology changes, network overhead, and computing power. The choice of an optimal route is one of the main challenges in VANET. In this sense, this paper aims to evaluate some metrics of two types of routing protocols (proactive and reactive). We opted for Optimized Link State Routing (OLSR) for the proactive protocol and Ad hoc On-demand Distance Vector (AODV) for the reactive protocol. This study is based on a real scenario using Open Street Map (OSM), and simulations are performed using Simulation of Urban Mobility (SUMO). The trace files generated by SUMO are processed in the NS-3 network simulator.

Keywords: VANET · Performance analysis · OLSR · AODV

1 Introduction

VANET is a special case of Mobile Ad-hoc network (MANET) [1], with a dynamic topology, and various communication modes such as Vehicle to Vehicle (V2V), Vehicle to Infrastructure (V2I), and Vehicle to X (V2X). In VANET, vehicles communicate with each other using on-board units (OBUs) that are integrated on the vehicles, and with access points using roadside units (RSUs) within a distance limit of 100 m to 1000 m [2]. RSUs are fixed infrastructures placed in the street or at intersections [3]. Figure 1 illustrates the types of communication in VANET.

Packet routing presents a real problem in VANET. The choice of routing protocol is very interesting for the reliability of the routing process. In VANET we can differentiate between two types of routing protocols: proactive and reactive. For the proactive ones there is OLSR (Optimized Link State Protocol) [4], followed by its successor OLSRv2 [5], and DSDV (Destination-Sequenced Distance Vector) [6]. In proactive routing the nodes already know the network topology, and when the topology changes within a specified time the routing tables are updated automatically.

M. Lazaar et al. (Eds.): BDIoT 2022, LNNS 625, pp. 524–536, 2023.
https://doi.org/10.1007/978-3-031-28387-1_44

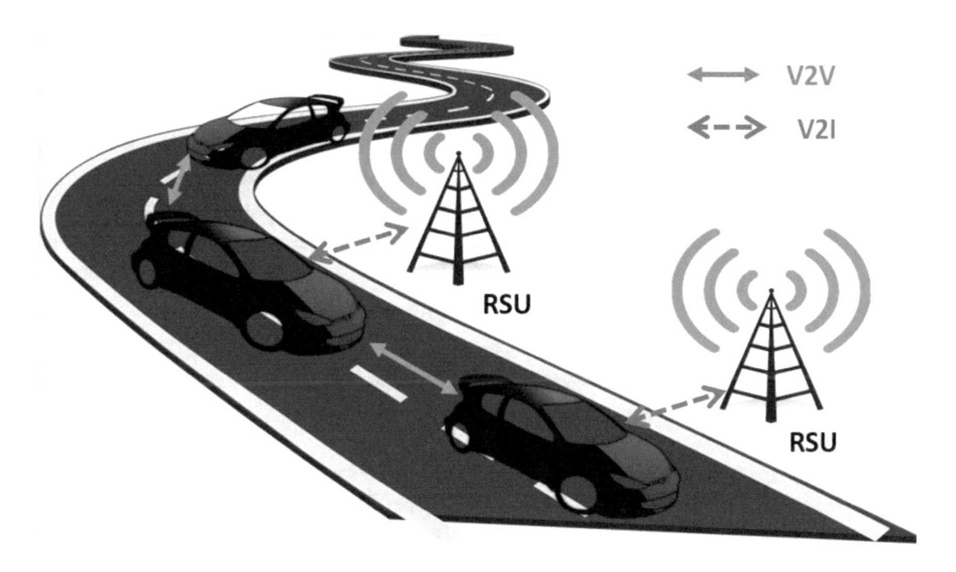

Fig. 1. The types of communication in VANET.

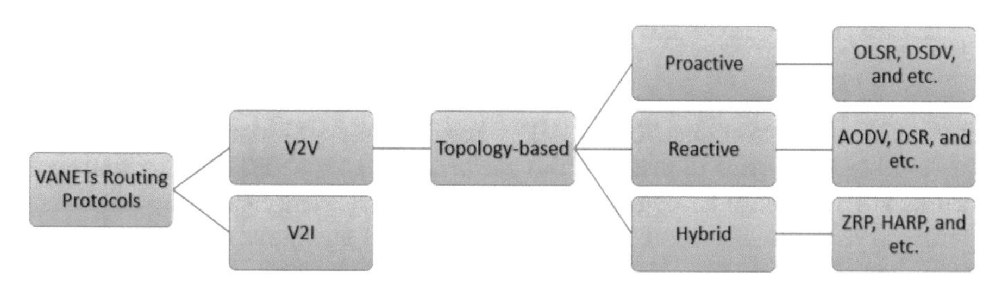

Fig. 2. The main types of routing protocols.

Contrariwise, reactive protocols such as AODV [7] and Dynamic Source Routing (DSR) [8], creates routes on demand and the transmission process starts with a transmission request. Figure 2 illustrates the main types of routing protocols.

The choice of OLSR from other proactive routing protocols is because OLSR is very intelligent, it used Multipoint Relay (MPR) mechanism to minimize the overhead and used a Dijkstra algorithm to calculate the shortest path to the destination and it is one of the most popular protocols in the proactive family. Also, for networks where communication is between random nodes rather than between the same sets of nodes OLSR is better suited [24]. On the other hand, the choice of AODV reactive protocol is because the majority of articles favor AODV in the reactive protocol family [9].

The routing problem in VANET is a current research topic. This paper aimed to evaluate some metrics of OLSR routing protocol and AODV routing protocol in terms of: throughput, packet delivery ratio (PDR), end-to-end delay (E2ED), and MAC/PHY overhead. The simulation uses a constant speed of 10 m/s, and varying the number of vehicles from 20, 40, 60, 80, 100, and 120. Specifically, the simulation was conducted by taking a real-time traffic scenario of Kenitra, Morocco, using SUMO [10]. The network simulation tool NS-3 [11] was used to run the simulation scenarios. This work analyzes the impact of simulated routing protocols by varying the number of nodes (vehicles) on network performance.

The scientific contributions of the paper is to find a routing protocol that performs well in terms of throughput, E2ED, PDR, overhead, and can minimized routing problems in VANET. In this paper we have discussed the most important points:

- Study and analysis of OLSR and AODV on VANET performance evaluation.
- The performance evaluation is based on throughput, PDR, E2ED, and MAC/PHY overhead observed for routing protocols.

The paper is organized as follows. Section 2 presents the existing studies. Section 3 introduces the proposed work. The performance metrics and simulation setup are described in Sect. 4. The results obtained are discussed in Sect. 5. Finally, the paper is concluded in Sect. 6.

2 Literature Review

Several studies have focused on the performance analysis of routing protocols in ad hoc network, especially MANET [12], and VANET [13]. Authors in [14] compare a range of metrics to measure the performance of VANET routing protocols such as PDR, average throughput, delay, and overall energy consumption. They confirmed that the high mobility of vehicles, and the change of data transmission speed, has an important role on the performance of VANET routing protocols.

Another study is presented in [15] with the objective of comparing and evaluating different routing protocols based on different performance metrics such as throughput, PDR, E2ED, and network stability. The authors stated that it is difficult to select a routing protocol that meets all requirements. On the other hand, they proposed that the AODV protocol is better than the DSR and DSDV protocols.

The paper [16] analyzed the throughput, PDR, and normalized routing load (NRL) of AODV, DSR, and DSDV. This paper favors AODV in terms of throughput and PDR. Therefore, the simulation under different number of vehicles shows that the NRL values for DSDV is lower than that of AODV and DSR.

Narayan et al. [17] compared a collection of VANET routing protocols: DSDV, AODV, AOMDV, and GPSR, using various network metrics for urban mobility. The results obtained favorites AODV in terms of throughput, delay, PDR, and AOMDV for values below the network overhead level. For congestion control, DSDV is more suitable.

From the state of the art on the performance of routing protocols in Ad-hoc network, there are some studies that have classified these protocols into two subcategories: one based on topology, and the other based on position [18].

Other works use less metrics to analyze the performance of routing protocols. Therefore, there are studies that process a set of routing protocols and do not make a protocol that performs 100% in terms of the metrics processed [19].

Therefore, the absence of a direct performance analysis between the OLSR and the AODV, we oblige to discuss our paper which is a brief interesting analysis of the performances of these two important families of VANET routing protocols. The choice to make this study, only between AODV and OLSR, is that in the literature a set of works prefer AODV in Ad-hoc networks [20]. However, for us it is very interesting to integrate OLSR in our paper; because it uses an intelligent mechanism to choose the shortest path.

3 Proposed Work

3.1 Study Objectives

This paper evaluates some OLSR and AODV metrics in VANET. In this evaluation we used two simulators in parallel. The first one is SUMO to trace a real environment with vehicles and base stations, while the second one is a NS-3 network simulator to test the performance of these routing metrics. SUMO also supports the import formats, such as OpenStreetMap.

After importing the file generated by SUMO, the process continued with the implementation of the routing scenarios. Finally, the simulation was run, and produced a trace file used to analyze the throughput, PDR, E2ED, and MAC/PHY Overhead values. The simulation process is illustrated in Fig. 3.

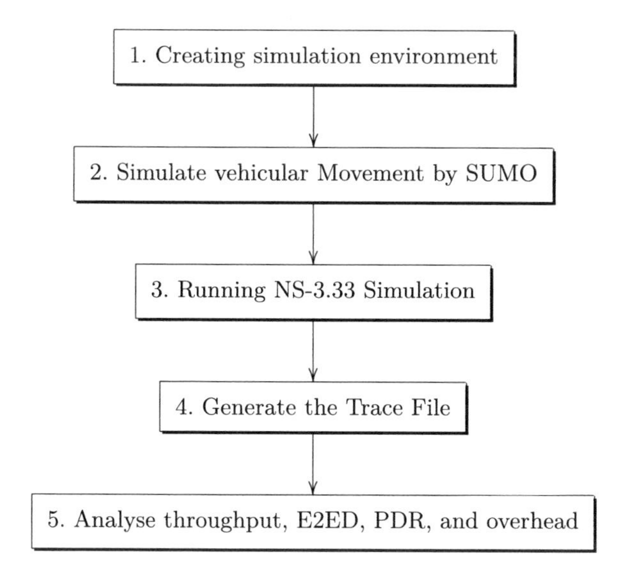

Fig. 3. The simulation process.

3.2 Routing Protocols

V2V, V2I, and V2X are types of communication used in VANET, these types directly influence the topology of the network which is going to be highly dynamic because of the high mobility of the nodes. That's why routing in VANET is a challenge for the researchers. The continuous movement of vehicles requires an efficient routing protocol that provides better traffic and network management. Here, we consider topology-based protocols: proactive protocol OLSR, and reactive protocol AODV.

AODV is a reactive routing protocol. Their working mechanism is started by sending a route request (RREQ) between all the neighbors of the network. Within a well-defined time TTL (Time to live) A particular node is established a route table and stores this information in its routing table. In order to establish the communication between the neighbors A HELLO message is transmitted. We say that the connection is established.

OLSR is a proactive routing protocol, uses the hop count as the default metric to calculate the shortest path according to the Dijkstra algorithm. In OLSR the selection of Multipoint Relay (MPR) is very important to minimize the overhead in the network. From a set of neighbors OLSR chooses MPRs using well-defined algorithms in this sense. The packet routing process in OLSR uses 2 main types of messages:

- HELLO messages: Their role is to allow each node of the network to know its one-hop and two-hops neighbors.
- TC messages: The MPRs transmit this type of message to get a global view of the network topology.

4 Performance Metric and Simulation Setup

This section presents the performance metrics and simulation parameters chosen for the evaluation and generation of the simulation.

4.1 Performance Evaluation Metrics

Deepak and Rajkumar confirmed that to evaluate the performance of the different routing protocols, the most important metrics to consider are: Throughput, PDR, and overhead [21]. And for this paper we added E2ED just to introduce another metric that will help us to confirm the results obtained.

The metrics used to evaluate the performance of the routing protocols (OLSR and AODV) in VANET are presented below. The summary of the notations used is presented in Table 1.

Table 1. Summary of notations.

Notation	Signification
P_R	Packet Received
P_S	Packet Sent
PDR	Packet Delivery Ratio
T_L	Time of last received packet
T_F	Time of first transmit packet
$E2ED$	End-to-End Delay
T_{Delays}	Total Delays
OvH	MAC-PHYsical Overhead
PHY_S	PHYsical layer packets send in bytes
APP_S	Application layer packets send in bytes
PHY_D	PHYsical layer packets delivered in bytes

Throughput. The sum of the number of successfully received packets in a routing process. In this paper, it is measured in kilobytes per second (Kbps). High throughput states better network performance, it is calculated using Eq. (1):

$$Throughput(Kbps) = TotalP_R/(T_L - T_F) \tag{1}$$

Packet Delivery Ratio. The ratio of packets received by receiving nodes to packets sent by sending nodes. When PDR values are higher, the network is considered to be performing better. The PDR is calculated using Eq. (2):

$$PDR(\%) = (TotalP_R/TotalP_S) * 100 \tag{2}$$

End-to-End Delay is the total transmission time of packets from the source to its destination [22], it is measured in milliseconds (ms). The network is performant when the value of E2ED is low. The E2ED is calculated using Eq. (3):

$$E2ED(ms) = T_{Delays}/TotalP_R \tag{3}$$

where T_{Delays} represents the summation of all delays of received packets.

MAC/PHY Overhead. In VANET, vehicles send and receive packets, which causes overhead in the routing process. The Overhead is the ratio of the sum of packets generated to the sum of packets delivered. Less overhead infers better network performance. The Overhead is calculated using Eq. (4):

$$OvH = (TotalPHY_S - TotalAPP_S)/TotalPHY_D \tag{4}$$

4.2 Simulation Parameters Settings

We used these simulation parameters to evaluate the performance of OLSR and AODV in VANET. The Table 2 presents these parameters.

Table 2. Simulation parameters setting.

Network Simulator	NS-3.33
Traffic Simulators	SUMO
Map Model	KENITRA-Morocco
Routing Layer	IEEE 802.11p
Number of Vehicles	20,40,60,80,100,120
Channel Type	Wireless
vehicles Speed	10 m/s
vehicles Pause	0
Base address	"10.1.0.0", "255.255.0.0"
Routing Protocol	OLSR, AODV
Transport Protocol	UDP
IEEE Scenario	VANET (802.11p)
Mobility Model	Urban Mobility
Simulation time	100 s

4.3 Generation of Simulation Environment

The simulation was realized in urban environment. To do this we downloaded the map of the city of KENITRA-Morocco using Open Street Map (it is the free wiki world map). SUMO used the .osm file to generate the mobility.tcl file which contains the information about each vehicle. The rest of the simulation is performed in NS-3 with the mobility.tcl file. A snapshot of the realistic scenario with the simulation SUMO and NetAnim generated is depicted in Fig. 4 and Fig. 5.

The literature does not contain any studies that perform simulations in Morocco. Our idea is to choose a Moroccan city for the simulation. The city chosen is Kenitra - Morocco -, however, we can then perform the simulation in other cities.

We have chosen NS-3.33 because it is the recent version, and it is the most popular simulator for academic research, it is usually criticized for its complex architecture. But its large community of users compensates, because many people help each other to solve their problems using mailing lists and forums.

We have chosen SUMO, because SUMO generates high mobility traffic to simulate the vehicle network due to its unique characteristics, also SUMO can simulate a single part and entire cities in a single simulation [23].

For the number of vehicles, we started with roads that do not contain many vehicles and then each time we added 20 vehicles to compare the impact of the number of vehicles on the network performance. We finished the simulation at 120 vehicles because we deduced motivating results.

For the duration of the simulation, we have chosen 100 s because if we choose a minimum duration of 100 s we cannot observe the effect of the routing protocols chosen in the VANET.

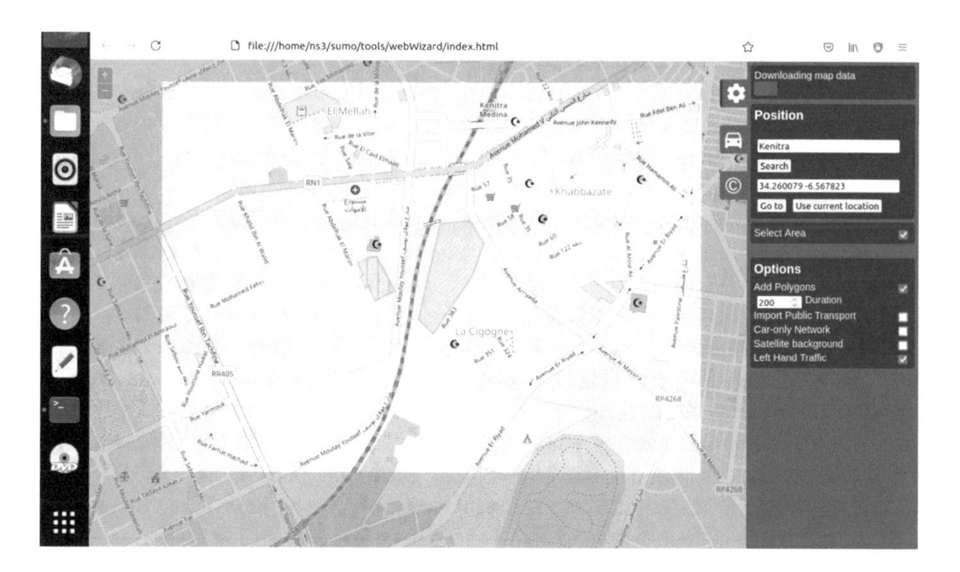

Fig. 4. Simulation Environment.

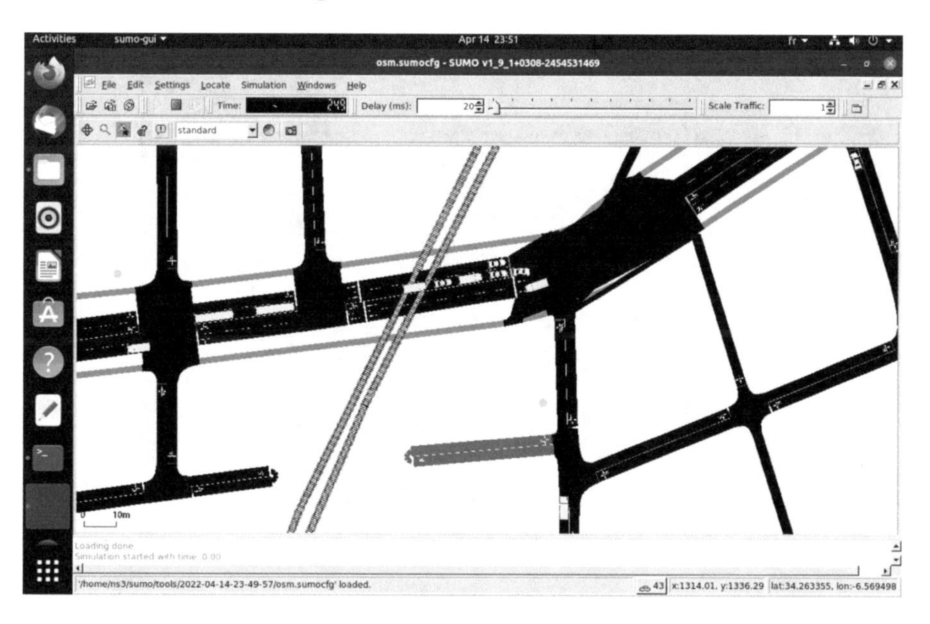

Fig. 5. Map of KENITRA-Morocco by OSM.

5 Results Discussion

The objective of this contribution is to evaluate the performance of OLSR and AODV in a VANET scenario, using a constant speed of 10 m/s and varying the number of vehicles by 20, 40, 90, and 150. The metrics used are: throughput, PDR, E2ED, and MAC/PHY Overhead.

Figure 6 illustrates the throughput analysis of OLSR and AODV by varying the number of vehicles. With higher numbers of vehicles in a V2V communication, the throughput of the AODV protocol becomes important. Our simulation shows that AODV is more performant in terms of throughput.

Figure 7 illustrates the PDR analysis of OLSR and AODV by varying the number of vehicles. With increasing numbers of vehicles in the network, the PDR of the AODV protocol increases, but the PDR of OLSR decreases. OLSR is more efficient in terms of PDR for VANET that do not exceed 100 vehicles. Therefore, if the network exceeds 100 vehicles AODV becomes more efficient.

Figure 8 illustrates the E2ED analysis of OLSR and AODV by varying the number of vehicles. According to the E2ED analysis, the results obtained show that the E2ED values change with the number of vehicles in the network. The simulation indicates that for the two routing protocols (OLSR and AODV), the E2ED values sometimes increase and sometimes decrease. Therefore, we can say that OLSR is efficient in terms of E2ED in a network containing between 60 and 90 vehicles. On the other hand, when a network contains less than 60 or more than 100 vehicles, AODV becomes more efficient.

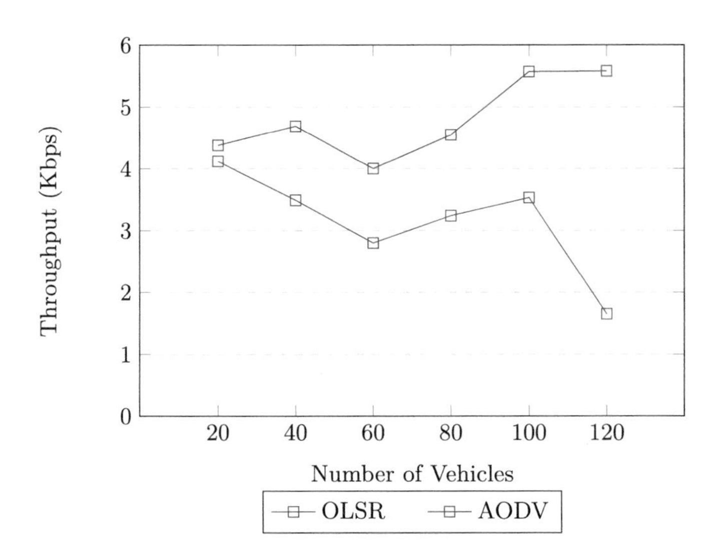

Fig. 6. Throughput analysis of Routing Protocols varying vehicles Numbers for VANET scenario.

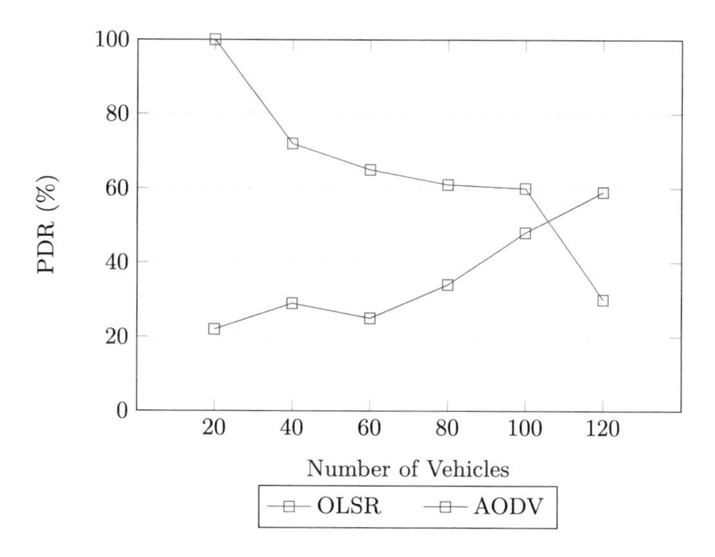

Fig. 7. PDR analysis of Routing Protocols varying vehicles Numbers for VANET scenario.

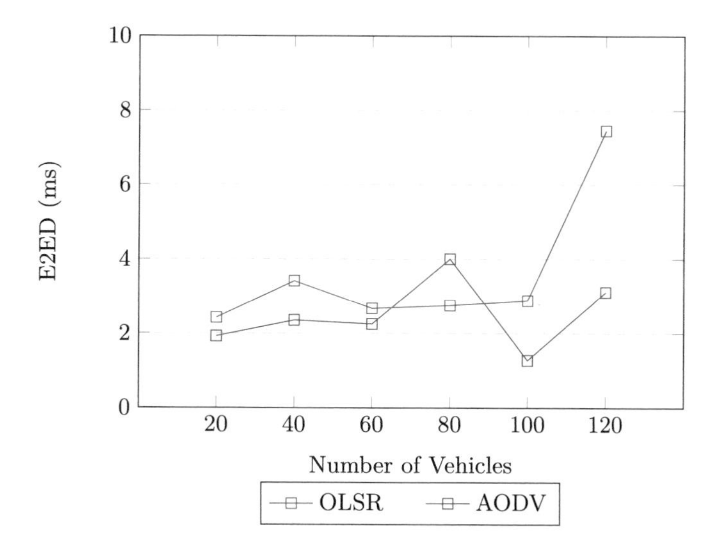

Fig. 8. E2ED analysis of Routing Protocols varying vehicles Numbers for VANET scenario.

Figure 9 illustrates the MAC/PHY Overhead analysis of OLSR and AODV by varying the number of vehicles. The MAC/PHY overhead value of two protocols increases as the number of vehicles in the network increases. The simulation shows that OLSR is performing well when the network exceeds 40 vehicles.

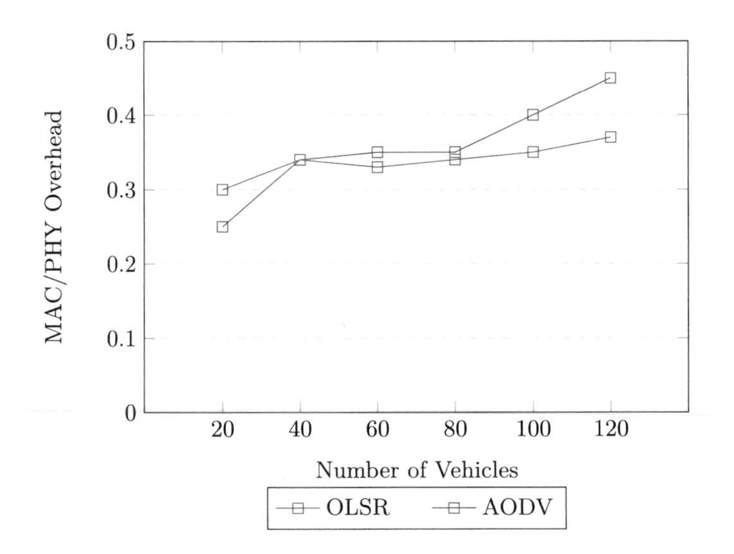

Fig. 9. MAC/PHY Overhead analysis of Routing Protocols varying vehicles Numbers for VANET scenario.

6 Conclusion

This paper discusses the performance evaluation of OLSR and AODV in VANET by varying the number of vehicles. The simulators used are SUMO and NS 3.33. The results indicate that OLSR is performant in terms of MAC/PHY overhead by using the MPR selection technique, and also in terms of PDR in a VANET that does not exceed 100 vehicles. On the contrary, AODV is more efficient in terms of throughput, and when the VANET contains more than 100 vehicles, AODV becomes more performant in terms of PDR and E2ED.

Future work focuses on improving the performance of routing protocols using an intelligent solution using AI (artificial intelligence) algorithms. We can do simulations in other cities, and also on the Moroccan highways.

References

1. Kumari, K., Sah, B., Maakar, S.: A survey: different mobility model for FANET. Int. J. Adv. Res. Comput. Sci. Softw. Eng. **5**, 1170–1173 (2015)
2. Sultan, A., Al-Doori, M., Al-Bayatti, H., Zedan, H.: A comprehensive survey on vehicular ad hoc network. J. Netw. Comput. **37**, 380–392 (2014)
3. Lopamudra, H., Biraja, N., Arun, K., Bibhudatta, S., Nawaz, A.: A performance analysis of VANETs propagation models and routing protocols. J. Sustain. **14** (2022)
4. Jacquet, P., Muhlethaler, P., Clausen, T., Laouiti, A., Qayyum, A., Viennot, L.: Optimized link state routing protocol for ad hoc networks. In: Proceedings IEEE International Multi Topic Conference, IEEE INMIC 2001. Technology for the 21st Century (2001). https://doi.org/10.1109/inmic.2001.995315

5. Clausen, T., Dearlove, C., Jacquet, P., Herberg, U.: The optimized link state routing protocol version 2. Journal Technical report (2014)
6. Perkins, C., Bhagwat, P.: Highly dynamic destination sequenced distance-vector routing (DSDV) for mobile computers. In: Proceedings of the Conference on Communications Architectures, Protocols and Applications - SIGCOMM 1994, vol. 24, pp. 234–244 (1994). https://doi.org/10.1145/190314.190336
7. Perkins, C., Belding-Royer, E., Das, S.: Ad hoc on-demand distance vector (AODV) routing. In: Proceedings WMCSA 1999, Second IEEE Workshop on Mobile Computing Systems and Applications (1999). https://doi.org/10.1109/mcsa.1999.749281
8. Johnson, D.B., Maltz, D.A.: Dynamic source routing in ad hoc wireless networks. In: Imielinski, T., Korth, H.F. (eds.) Mobile Computing. SECS, vol. 353, pp. 153–181. Springer, Boston (1996). https://doi.org/10.1007/978-0-585-29603-6_5
9. Sindhwani, M., Singh, R., Sachdeva, A., Singh, C.: Improvisation of optimization technique and AODV routing protocol in VANET. Int. Proc. Mater. Today **49**, 3457–3461 (2022)
10. Lopez, A., Behrisch, M., Bieker-Walz, L., Erdmann, J., Pang, Y.: Microscopic traffic simulation using SUMO. In: IEEE Intelligent Transportation Systems Conference (ITSC), Maui, HI, USA (2018)
11. Campanile, L., Gribaudo, M., Iacono, M., Marulli, F., Mastroianni, M.: Computer network simulation with NS-3: a systematic literature review. Electronics **9**, 1–25 (2020)
12. Kumar Nigam, G.: Performance analysis and evaluation of routing protocols for mobile adhoc networks. In: 13th International Conference on Contemporary Computing, IC3 2021, Code 173502, pp. 196–2025 (2021)
13. Kumareshan, N., Prakash, N., Arun Vignesh, N., Kumaran, G.: Performance analysis of various routing protocols for VANET environments. Int. J. Innov. Technol. Exploring Eng. (IJITEE) **8**, 4381–4384 (2019)
14. Khoza, E., Tu, C., Owolawi, P.: Comparative study on routing protocols for vehicular ad-hoc networks (VANETs). In: International Conference on Advances in Big Data, Computing and Data Communication Systems (icABCD), Durban, South Africa, pp. 1–6 (2018)
15. Rehman, M., Ahmed, S., Khan, S., Begum, S., Ahmed, S.: Performance and execution evaluation of VANETs routing protocols in different scenarios. In: EAI Endorsed Transactions on Energy Web and Information Technologies, vol. 5 (2018)
16. Kandali, K., Bennis, H.: Performance assessment of AODV, DSR and DSDV in an urban VANET scenario. In: Ezziyyani, M. (ed.) AI2SD 2018. AISC, vol. 915, pp. 98–109. Springer, Cham (2019). https://doi.org/10.1007/978-3-030-11928-7_8
17. Arvind Narayan, S., Rajashekar Reddy, R., Femilda Josephin, J.S.: Secured congestion control in VANET using greedy perimeter stateless routing (GPSR). In: Dash, S.S., Lakshmi, C., Das, S., Panigrahi, B.K. (eds.) Artificial Intelligence and Evolutionary Computations in Engineering Systems. AISC, vol. 1056, pp. 683–700. Springer, Singapore (2020). https://doi.org/10.1007/978-981-15-0199-9_59
18. Thangakumar, J., Rajeswari, M.: Simulation of vehicular ad-hoc network routing protocols with a performance analysis. J. Commun. Softw. Syst. **11**, 86–93 (2015)
19. Radwan, A., Mahmoud, T., Houssein, E.: Evaluation comparison of some ad hoc networks routing protocols. Egypt. Inform. J. **12**, 95–106 (2011)
20. Keshtgary, M., Rikhtegar, N.: Performance evaluation of routing protocols for wireless sensor networks in forest fire detection application. In: 5th Conference on Information and Knowledge Technology, pp. 248–251. IEEE (2013)

21. Deepak, R.: Performance comparison of routing protocols in VANETs using network simulator-NS3. Int. J. Res. Electron. Comput. Eng. **6**, 2097–2104 (2018)
22. Sisodia, D., Singhal, R., Khandal, V.: A performance review of intra and intergroup MANET routing protocols under varying speed of nodes. IJECE **7**, 2721–2730 (2017)
23. Krajzewicz, D., Erdmann, J., Behrisch, M., Bieker, L.: Recent development and applications of SUMO-simulation of urban mobility. Int. J. Adv. Syst. Measur. **5**(3–4), 128–138 (2012)
24. Rango, F., Cano, J., Fotino, M., Calafate, C., Manzoni, P., Marano, S.: OLSR vs DSR: a comparative analysis of proactive and reactive mechanisms from an energetic point of view in wireless ad hoc networks. Comput. Commun. **31**, 3843–3854 (2008)

Contribution to Smart Irrigation Based on Internet of Things and Artificial Intelligence

Ali Mhaned(✉), Mouatassim Salma, El Haji Mounia, and Benhra Jamal

Laboratory of Advanced Research in Industrial and Logistic Engineering (LARILE), Team OSIL, Department GIL, National High School of Electrical and Mechanical Engineering (ENSEM), University Hassan II, Casablanca, Morocco
ali.mhaned.doc21@ensem.ac.ma

Abstract. Agriculture is Morocco's primary sector and depends mainly on rainfall. Therefore, water is an essential resource that must be managed with care. Traditional irrigation methods waste a lot of water, and crops are often under or over irrigated. To guarantee the right quantities of water for plants, automatic irrigation systems are available. They make it possible to ensure the quantities of water necessary for the plant. In this paper we will present an intelligent irrigation system based on Internet of things (IoT) and artificial intelligence (AI). Node-MCU 32S boards were used to monitor physical parameters such as air temperature and humidity, soil temperature and moisture, rain and light. Collected data is routed to the raspberry pi 4 via MQTT protocol, then the program running inside the raspberry determinates how much water is needed to irrigate the plants. Based on that calculated amount, an instruction is sent to Node-MCU 32S boards to operate the pumps connected to the relay module. Weather data history is used to forecast reference crop evapotranspiration (ET_0) to predict the amount of water needed during each growing stage using neural networks especially long short-term memory (LSTM) techniques of recurrent neural networks (RNN).

Keywords: Smart Irrigation systems · Internet of Things (IoT) · Reference crop evapotranspiration · Long Short-Term Memory (LSTM) · Reccurent Neural Networks (RNN)

1 Introduction

Water is an essential resource for human being, nowadays, an average of 70% of the world's freshwater volume is attributed to agriculture [1]. Morocco is ranked among the top 25 most water-stressed countries in the world, and agriculture industry is the primary user of surface water accounting 87% of withdrawals [2, 3]. That makes the management of water resources one of the most important challenges in modern agriculture.

The importance of the agricultural sector is evidenced by its significant contribution to the formation of the national gross domestic product (GDP) (15% to 20%) and job creation (40%), particularly in rural areas where agriculture is the main employer (80%) and source of income (15 million farmers) [4]. Thirty-eight percent of cultivated crops are using irrigation, irrigated agriculture contributes to 45% of agricultural added value by intervening for 75% of agricultural exports, and 35% of agricultural employment [5].

M. Lazaar et al. (Eds.): BDIoT 2022, LNNS 625, pp. 537–549, 2023.
https://doi.org/10.1007/978-3-031-28387-1_45

During the last decade, agriculture has been evolving due to the main progresses in technologies. The technological advances have allowed to create new applications in telecommunications, applying low power and reduced costs in their equipment, thus achieving the evolution of new wireless networks or also denominated Wireless Sensor Network. These technologies allow the generation of measurements and analyses of environmental parameters of data and soil. Precision agriculture requires parameters for the improvement of production, obtained through WSN technologies [6]. The IoT is perfect match for precision agriculture due to its highly interoperable, scalable, pervasive and open nature. There are lots of IoT derived technologies and all of them bring various benefits including reducing the risk of vendor lock-in, adopting machinery and better sensing/automation systems [7].

The system in [8] targets on sensing soil quality, moisture and providing adequate moisture needed using motor pumps with limited human interaction. The soil moisture sensor is connected to an Arduino Uno which analyzes the data and decides whether the pump should start working or not. ESP8266 is used to connect the Arduino Uno card to a nearby WIFI, and so, data is sent to ThingSpeak servers and allows the user to visualize data captured by sensor and the pump status to help keeping a track from an isolated place.

In [9], Smart&Green framework is proposed to offer services for smart irrigation, such as data monitoring, preprocessing, fusion, synchronization, storage, and irrigation management enriched by the prediction of soil moisture. Outlier removal techniques allow for more precise irrigation management. For fields without soil moisture sensors, the prediction model estimates the matric potential using weather, crop, and irrigation information. The predicted matric potential approach was applied to the Van Genutchen model to determine the moisture used in an irrigation management scheme. We can save, on average, between 56.4% and 90% of the irrigation water needed by applying the Zscore, MZscore and Chauvenet outlier removal techniques to the predicted data.

A design of a generic IoT framework was presented in [10] to improve agriculture yield by effectively scheduling irrigation and fertilization based on the crop's current requirements, environmental conditions and weather forecasts. This work proposes the design of an affordable irrigation and fertilization system. The proposed fertilization system spreads fertilizers to the root directly. This reduces the amount of fertilizers required and thus reduces the cost and improves the soil health. A user-friendly mobile application delivers this information to the farmers in their regional language.

A Long short-term memory neural networks (LSTM), Gradient Boosting Regressor (GBR), Random Forest (RF) and Support Vector Regression (SVR) were developed in [11] to estimate ET_0 with climatic variables as input parameters. These models were evaluated in two different climatic regions in India. The results indicated that 99% accuracy could be achieved with all climatic input, whereas accuracy drops to 86% with fewer data. LSTM model performed better than other ML models with all input combinations at both the stations, followed by SVR and RF.

In [12], a low-cost system was designed to reduce both human intervention and water consumption. This system acquires data about soil moisture, ambient temperature and humidity as well as rain presence parameters. MQTT protocol is used to transmit data between nodes and the Raspberry Pi device, data is then sent to ThingSpeak in order to be checked by the user in real time. A Fuzzy Logic system was implemented to define the water needs of the plant based on acquired data.

In this paper, we are focusing on proposing a low-cost smart irrigation system architecture to acquire physical parameters from sensors' readings, store data acquired and allow the user to check in real time sensed samples.

The remaining paper is organized as: a presentation of materials and methods used in this research, the proposed system, obtained results and a conclusion.

2 Methodology

Evapotranspiration is one of the major components of the hydrologic cycle and its accurate estimation is of paramount importance for many studies such as hydrologic water balance, irrigation system design and management, crop yield simulation, and water resources planning and management [13]. In this section we will present the crop water requirement, reference evapotranspiration, crop coefficient and crop evapotranspiration:

2.1 Crop Water Requirement

The crop water requirement determines the amount of water needed to compensate the evapotranspiration loss from the crop field termed as *CWR*. The value of *CWR* is identical to evapotranspiration [14]. *CWR* represents the evapotranspiration *ET* under ideal crop growth condition, it can be calculated from the climate and crop data. *CWR* for a given crop, i, for the whole growing season can be calculated as mentioned in (Eq. 1), [14]:

$$CWR_i = ET_i = \sum_{t=0}^{m} (ET_{0t} \times K_{ct}) \tag{1}$$

where $CWRi$ is the crop water requirement for the growing period, in mm, ET_i is the crop evapotranspiration for the growing period, in mm, t is the time interval in days, m is the days to physiological maturity from sowing or transplanting (total effective crop growth period), in numbers, ET_{0t} is the reference crop evapotranspiration of the location concern for the day t, in mm, and K_{ct} is the crop coefficient for the time t day.

2.2 Reference Evapotranspiration

ET_0 Represents the evapotranspiration rate from a reference surface, not short of water. A large uniform grass field is considered worldwide as the reference surface. The reference crop completely covers the soil, is kept short, well-watered and is actively growing under optimal agronomic conditions. The input data required to compute ET_0 are radiation, air temperature, air humidity and wind speed [15]. FAO Penman-Monteith method (Eq. 2) to estimate ET_0 can be derived, [16]:

$$ET_0 = \frac{0.408\Delta(R_n - G) + \gamma \frac{900}{T+273} u_2(e_s - e_a)}{\Delta + \gamma(1 + 0.34u_2)} \tag{2}$$

where ET_0 is the reference evapotranspiration (mm day^{-1}), R_n is the net radiation at the crop surface (MJ $m^{-2}day^{-1}$), G is the soil heat flux density (MJ $m^{-2}day^{-1}$), T is the mean of daily air temperature at 2 m height (°C), u_2 is the wind speed at 2 m height (m s^{-1}), e_s is the saturation vapor pressure (kPa), e_a is the actual vapor pressure (kPa), (e_s − e_a) is the saturation vapor pressure deficit (kPa), Δ is the slope vapor pressure curve (kPa °C^{-1}), and γ is the psychometric constant (kPa °C^{-1}).

2.3 Crop Potential Evapotranspiration

Crop coefficient (Kc) is defined as the ratio of the actual evapotranspiration of a disease-free crop grown in a large field adequately supplied with water to the reference evapotranspiration [17]. Essentially, the crop coefficient is a coefficient expressing the difference in evapotranspiration between the cropped and reference grass surface. The crop evapotranspiration under standard conditions, denoted as ET_c, is the evapotranspiration from disease-free, well-fertilized crops, grown in large fields, under optimum soil water conditions, and achieving full production under the given climatic conditions. To calculate ET_c we multiply ET_0 to the crop coefficient Kc as shown in the equation (Eq. 3) bellow, [17]:

$$ET_c = ET_0 \times K_c \tag{3}$$

2.4 Soil Water Balance

The soil water balance equation (Eq. 4) is used to define the right irrigation amount based principally on precipitation P, crop evapotranspiration ET_c, and root zone depletion $D_{r,i}$ at the end of the day, and root zone depletion at the end of the previous day $D_{r,i-1}$, , [18]

$$D_{r,i} = D_{r,i-1} - (P - RO)_i - I_i - CR_i + ET_{c,i} + DP_i \tag{4}$$

where $D_{r,i}$ is the root zone depletion at the end of the day I [mm], $D_{r,i-1}$ is the water content in the root zone at the end of the previous day, i-1 [mm], P_i is the precipitation on day i [mm], RO_i is the runoff from the soil surface on day i [mm], I_i is the net irrigation depth on day i that infiltrates the soil [mm], CR_i is the capillary rise from the groundwater table on day i [mm], $ET_{c,i}$ is the crop evapotranspiration on day i [mm] and DP_i water loss out of the root zone by deep percolation on day i [mm].

3 Proposed System

To manage the irrigation process, we propose the system below (Fig. 1.). Two sensing nodes collect physical parameters from the crop field then transmit data collected to the base station. The base station is connected to the cloud to permit to the user supervising and controlling the system; From the cloud the system will obtain the weather current and meteorological data. According to the instructions, the base station will turn on/off the water pumps.

The proposed system can be divided to six units. The acquisition unit collect physical parameters from the field, ambient temperature and humidity, soil moisture and temperature, luminosity and rain detection. Collected date is sent through the communication unit to the treatment unit. The treatment unit receive real-time data from acquisition unit and meteorological data from the cloud. Based on data received, the treatment unit supply irrigation parameters (water needs, irrigation schedule…) to the command unit to irrigate the field. The cloud allows the connection between user, the irrigation system and the weather provider. The user can visualize real-time data, historical and forecasted weather, irrigation requirement and water use through mobile phones or web page. The storage unit is a database where real-time data and data from the weather API provider [19] are saved. Data stored on a MicroSD card can be lately uploaded to the database in case of missing data due to connectivity problems. A power supply unit feeds the other units with necessary energy.

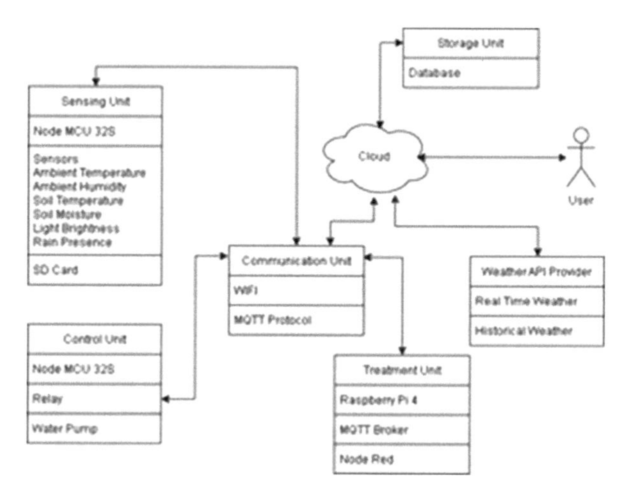

Fig. 1. Smart Irrigation system units' architecture

In this research, we designed two sensing nodes capable of reading physical parameters, send them to the base station and store a backup in the SD Card connected on an SD Card module. To ensure irrigation, two actuating nodes are charged to receive instructions from the base station to open or close the water pump.

Each sensing node (Fig. 2.) is composed of many sensors to collect ambient humidity and temperature, soil moisture, soil temperature, luminosity and rain presence. Sensors are connected to a Lua board. The board collects data from sensors and sends it to the base station above the MQTT protocol. To avoid data loss, due to communication failures, a MicroSD card module is connected to save sensed parameters. Actuator nodes (Fig. 3.) are a Lua Node-MCU32S based nodes, a relay module is connected to the node to actuate the water pump when receiving order from the treatment unit. The MicroSD Card module save irrigation data to ensure a backup history and check the system liability.

In MQTT protocol, publisher publishing messages and users subscribing to topics that are commonly considered as a Publish/Subscribe model [20]. Subscriber subscribes

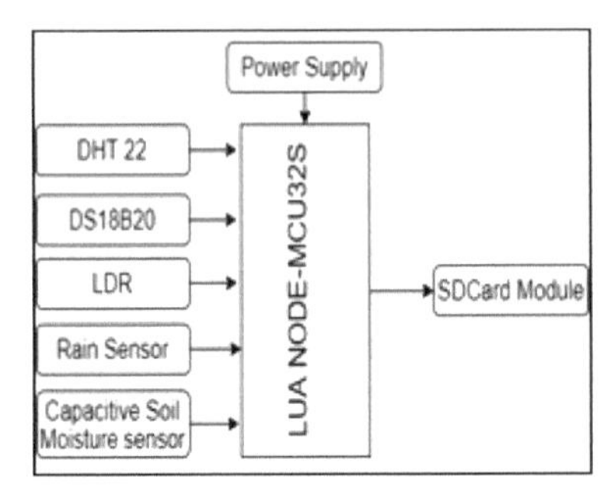

Fig. 2. Sensing node architecture

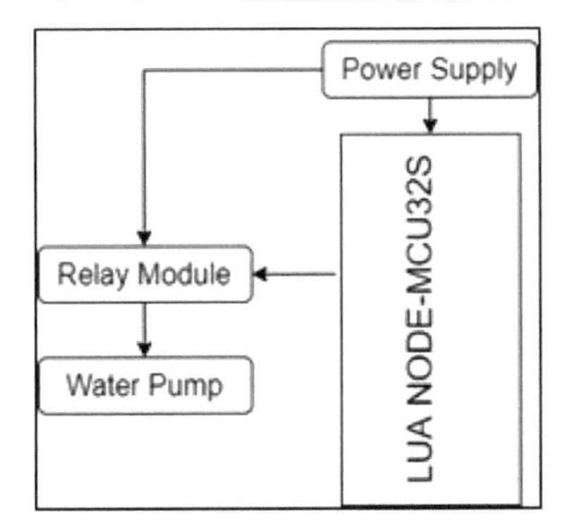

Fig. 3. Sensing node architecture

to particular topics which are related to them, and by that receive messages published to those topics [21]. On the other hand, clients can publish messages to topics, in such a way that allows all subscribers to access messages of those topics.

A smart irrigation system data flowchart is presented in (Fig. 4.), the system is composed of API data provider which deliver historical and current weather data, sensing nodes publishes collected values of physical parameters via MQTT protocol, the MQTT broker manage the communication, Node-RED is used to monitor received parameters.

The MQTT broker is the heart of each MQTT arrangement. It provides connecting link between applications or physical devices and enterprise systems. Brokers are in charge of the subscription, determined sessions, missed messages and general security, including authentication and authorization [22]. In this research, we will use Mosquitto Broker Mosquitto [23], an open source MQTT broker has been implemented on Raspberry Pi. It uses two services mosquittopub and mosquittosub to publish and subscribe to the application messages.

In MQTT, publishers publish messages to topics that can be considered as a message subject. Subscriber, thus, subscribes to topics to get specific messages. The subscriptions of topics can be expressed, which then restricts the data that is collected to the particular topic [20]. Topics contain two wildcard levels, to get data for a range of related topics. The MQTT architecture of the system publishers and subscribers is presented in Fig. 5.

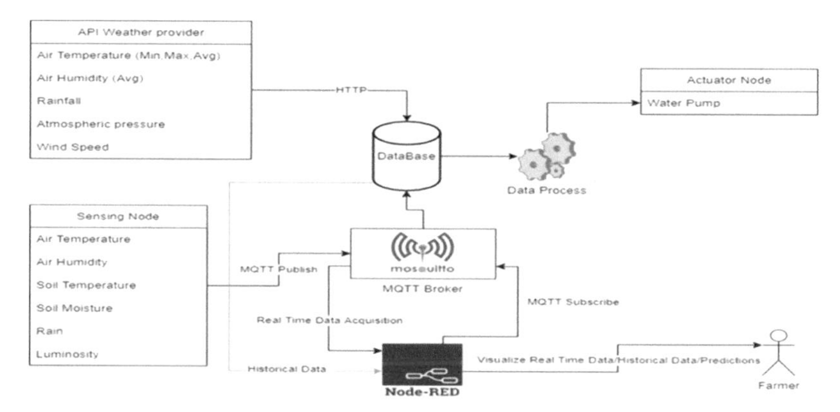

Fig. 4. Smart irrigation system data flowchart

In our case, each sensing node will publish data on the corresponding topic: node 1 will publish on "sensors/node1" and sensing node 2 will publish on "sensors/node2." To receive data, the raspberry pi needs to subscribe to the correspondent topics. Actuating node 1 & 2 will subscribe to "actuating/node1" and "actuating/node2" respectively. Instructions will be sent by the raspberry pi by publishing on the two topics.

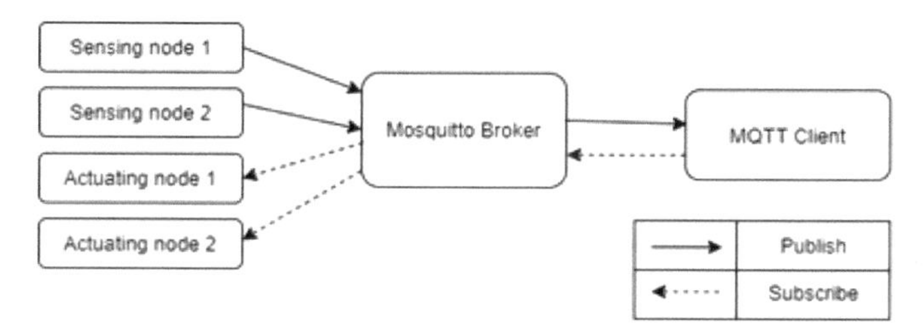

Fig. 5. MQTT communication flow

Node-RED is a programming tool for wiring together hardware devices, APIs and online services in new and interesting ways. It provides a browser-based editor that makes it easy to wire together flows using the wide range of nodes in the palette that can be deployed to its runtime in a single click [24]. In Fig. 6, the MQTT nodes were used to subscribe specified topics. Dashboard nodes are used to display data on charts and graphs.

Long short-term memory (LSTM) is a special type of recurrent neural network (RNN), with advantages over traditional RNN, used to handle sequential data [26]. When processing input data from time t, , information from previous time steps (t_{-1}, t_{-2}, …) is considered. An LSTM network is composed of LSTM blocks. The LSTM block as shown in (Fig. 7.) contains forget gate, input gate, output gate, hidden state and cell state. The forget gate decides what information should be removed from the cell state, generating a f_t value. The input gate decides what information from cell state should be updated, generating an i_t value. The output gate is responsible to produce o_t,

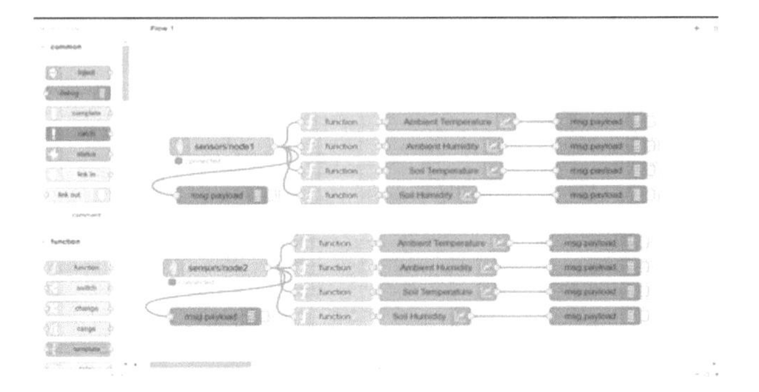

Fig. 6. Node-Red flow

which is used to compute the hidden state h_t based on a filtered version of the cell state [27].

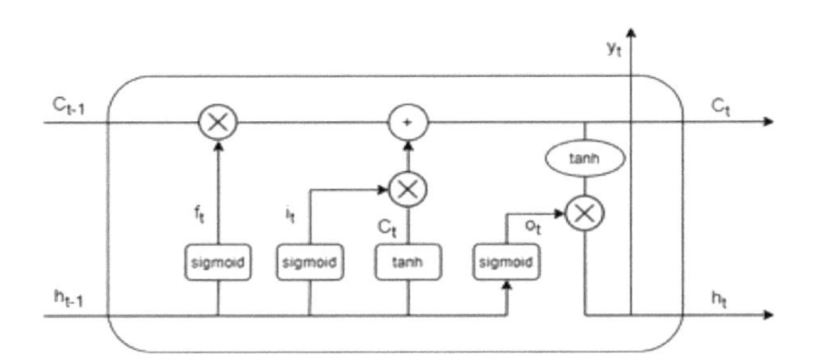

Fig. 7. LSTM schematic diagram of cell structure

The algorithm tuning is configured as in [28], the algorithms were configured according to several parameters of which we quote, the number of hidden layers, the number of neurons in each layer and the training methods, then the best result of root-mean-square error have been retained.

4 Results and Discussion

The sensing unit of each node (Node1 & Node2) are collecting physical parameters and send it thought the MQTT protocol. Data sensed is stored on the database as presented in Table 1. The interval between each sampling is defined in one minute, it can be varied depending on the case of its use.

Node-RED has been used to create a dashboard (Fig. 8.) that allows users to check current weather information provided by the API weather station, but can also easily get the recent sensing data from node installed in the field.

Historical weather dataset for Casablanca is used to compute reference evapotranspiration. The dataset contains 375063 measurements representing an hourly sample for

Table 1. Sample of data received over the sensing unit: (AM: Ambient Humidity; AT: Ambient Temperature; ST: Soil Temperature; SH: Soil Humidity; LB: Light Brightness; RP: Rain Presence)

Date and Time	Node ID	AT (°C)	AH (%)	ST (°C)	SH (%)	LB (%)	RP
03/04/2022 13:06:55	1	15.22	68	14.37	72	68	None
03/04/2022 13:06:52	2	14.93	67	14.56	69	67	None
03/04/2022 13:07:56	1	15.23	68	14.38	72	68	None
03/04/2022 13:07:53	2	14.90	67	14.56	68	66	None

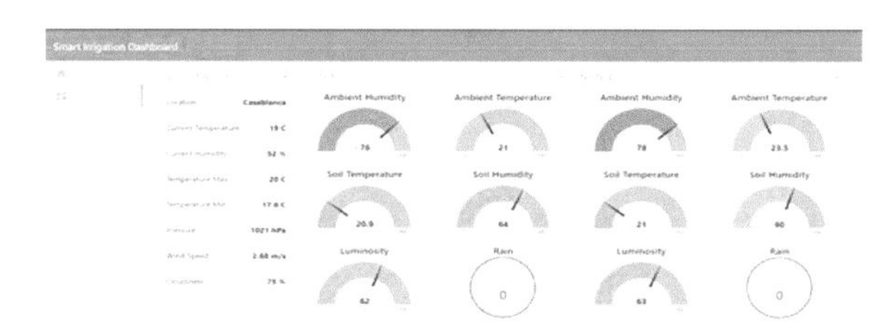

Fig. 8. Node-Red Dashboard

minimum, maximum and average temperature, relative humidity, wind speed, pressure and rainfall. The historical dataset represents 42 years of data collected from Casablanca meteorological station. Average temperature, relative humidity, wind speed and pressure parameters are used as inputs to calculate average daily reference evapotranspiration. Radiation values were not available so they are derived from air temperature differences [25].

An LSTM model was trained using calculated ET_0 over past 42 years, the Fig. 9 represents both calculated and predicted values of ET_0. Data was splitted into 80% train data and 20% test data. Evaluation criteria were used to study our model performance. Results presented in Table 2 represents average values of ET_0 by month of computed FAO-PM and forecasted values, the difference between the two values is about 0.051 mm which is an acceptable difference.

The Mean Squared Error (*MSE*) of an estimator (of a procedure for estimating an unobserved quantity) measures the average of the squares of the errors or deviations—that is, the difference between the estimator and what is estimated [29]. The root-mean-square error (*RMSE*) is a frequently used measure of the differences between values (sample and population values) predicted by a model or an estimator and the values actually observed [30]. The *RMSE* represents the sample standard deviation of the differences between predicted values and observed values. *MSE* and *RMSE* formulas are shown in Eq. 5 and Eq. 6:

$$MSE = \frac{1}{n}\sum\nolimits_{i=1}^{n}\left(Y_i^{\wedge} - Y_i\right)^2 \tag{5}$$

$$RMSE = \sqrt{\frac{\sum_{t=1}^{n}\left(Y_i^{\wedge} - y_i\right)}{n}} = \sqrt{MSE} \tag{6}$$

Table 2. Computed FAO-PM ET_0, Forecasted ET_0 and calculated difference between computed FAO-PM ET_0 and forecasted ET_0

Month	Computed FAO-PM ET_0	Forecasted ET_0	Computed FAO-PM ET_0 – Forecasted ET_0
Jan	1.523	1.574	– 0.051
Feb	1.717	1.780	– 0.051
Mar	2.193	2.258	– 0.051
Apr	2.517	2.572	– 0.051
May	3.336	3.414	– 0.051
Jun	3.726	3.823	– 0.051
Jul	3.710	3.817	– 0.051
Aug	3.611	3.710	– 0.051
Sep	3.248	3.331	– 0.051
Oct	2.958	3.081	– 0.051
Nov	2.158	2.221	– 0.051
Dec	1.521	1.555	– 0.051

Standard Error (STD error) and Mean Error are a standard deviation of train and test error [31]. Train and test errors are subtraction of targets from output in dataset. To evaluate our LSTM model accuracy MSE, RMSE, Error Mean and STD Error were used. According to Fig. 10, the MSE and the RMSE of the LSTM are respectively 0.93294 and 0.96589, the Error Mean is 0.05105 and the STD Error is 0.96454. An inferior value of the RMSE and MSE to 1 is considered as a good result. In [32] 31 combinations of inputs were tested: the RMSE varied between 0.996 and 0.209.

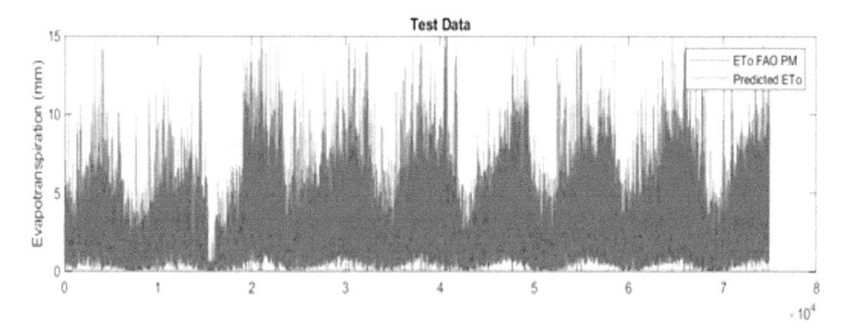

Fig. 9. Reference evapotranspiration FAO PM vs LSTM Predicted reference evapotranspiration

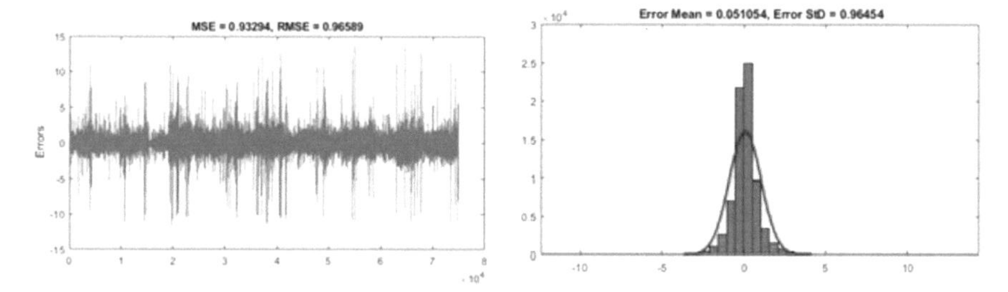

Fig. 10. Mean Squared Error (MSE), Root Mean Square Error (RMSE), Error Mean and Standard Error (STD Error)

5 Conclusion

Agriculture is one of the most water consuming sectors. Irrigation must therefore be well controlled in order to enable an efficient utilization of water resources and guarantee an optimal productivity in order to face up to agri-food and water scarcity challenges. In this project, a supervision system was designed by implementing IoT technologies, acquisition and control nodes NodeMCU_32S-based were inspired by WSN wireless sensor networks. Then, the data is transmitted to the Raspberry Pi base station via the MQTT protocol. The data is stored in a memory card and in the cloud. Data is monitored in real time via the Node-Red dashboard, as well as the weather data acquired via the weather data provider. To calculate the necessary quantity of water the FAO56 P-M method is used, the historical meteorological data of Casablanca was used to calculate the reference evapotranspiration over a period of 42 years. This data was used to rain an LSTM model to predict ET_0. To evaluate our model many errors were calculated, $MSE = 0.93294$, $RMSE = 0.996589$, $MeanError = 0.05105$, $STDError = 0.96454$. Our system performance can be ameliorated in future works. As long as fertilization is an important pillar for agriculture and the irrigation channels are the same as for dispersing fertilizers, incorporating fertilization management would be a plus for our proposed system. Solar Energy can be used to power wireless nodes in order to reduce energy consumption, as well as it can be used for water pumping. The integration of solar panels into our system will be an expansion on one more aspect of global challenges, that of reducing energy consumption and opening up to green energies.

References

1. [UNESCO] United Nations Educational Scientific and Cultural Organization. Securing the Food Supply. UNESCO, Paris (2001a)
2. García-Ruiz, J.M., López-Moreno, J.I., Vicente-Serrano, S.M., Lasanta-Martínez, T., Beguería, S.: Mediterranean water resources in a global change scenario. Earth Sci. Rev. **105**, 121–139 (2011)
3. Ait Kadi, M.: From Water Scarcity to Water Security in the Maghreb Region: The Moroccan Case. Environmental Challenges in the Mediterranean 2000–2050, pp. 175–185 (2004).https://doi.org/10.1007/978-94-007-0973-7_11

4. United Nations, 2014. Examen des performances environnementales–Maroc, synopsis. Commission Economique des Nations Unies pour l'Afrique, Bureau pour l'Afrique du Nord (2014)
5. World Bank. Poverty and social impacts analysis of the Moroccan green growth policy. Energy Axis, a general equilibrium. Departement du Developpement durable, p. 41 (2013)
6. Guaña-Moya, J., Sánchez-Almeida, T., Salgado-Reyes, N.: Measurement of agricultural parameters using wireless sensor network (WSN). In: AIP Conference Proceedings, vol. 1952, p. 020009 (2018)
7. Davcev, D., Mitreski, K., Trajkovic, S., Nikolovski V., Koteli, N.: IoT agriculture system based on LoRaWAN. In: 2018 14th IEEE International Workshop on Factory Communication Systems (WFCS), pp. 1–4 (2018)
8. Karpagam, J., Merlin, I.I., Bavithra, P., Kousalya, J.: Smart irrigation system using IoT. In: 2020 6th International Conference on Advanced Computing and Communication Systems (ICACCS), pp. 1292–1295 (2020)
9. Koduru S., Padala V.G.D.P.R., Padala P.: Smart irrigation system using cloud and internet of things. In: Krishna C., Dutta M., Kumar R. (eds.) Proceedings of 2nd International Conference on Communication, Computing and Networking. Lecture Notes in Networks and Systems, vol. 46. Springer, Singapore (2019).https://doi.org/10.1007/978-981-13-1217-5
10. Saini, A.K., Banerjee, S., Nigam, H.: An IoT instrumented smart agricultural monitoring and irrigation system. In: 2020 International Conference on Artificial Intelligence and Signal Processing (AISP), pp. 1–4 (2020)
11. Prabha, R., Sinitambirivoutin, E., Passelaigue, F., Ramesh, M.V.: Design and development of an IoT based smart irrigation and fertilization system for chili farming. In: 2018 International Conference on Wireless Communications, Signal Processing and Networking (WiSPNET), pp. 1–7 (2018)
12. Mhaned, A., Mouatassim, S., Benhra, J., Elhaji, M.: Low-cost smart irrigation system based on internet of things and fuzzy logic. In: The International Conference on Smart Applications and Data Analysis for Smart Cyber-Physical Systems, Marrakech
13. Kumar, M., Raghuwanshi, N.S., Singh, R., Wallender, W.W., Pruitt, W.O.: Estimating evapotranspiration using artificial neural network **128**(4), 224 (2002)
14. Allen, R.G., Luis, S., Pereira, D.R., Smith, M.: FAO irrigation and drainage paper No. 56. Rome: Food Agric. Organ. United Nations **56**(97), e156, p. 9 (1998)
15. Ali, M.H.: Fundamentals of Irrigation and On-farm Water Management: Volume 1. Crop Water Requirement and Irrigation Scheduling, Chapter 9, pp. 399–452 (2010)
16. Allen, R.G., Luis, S., Pereira, D.R., Smith, M.: FAO irrigation and drainage paper No. 56. Rome: Food Agric. Organ. United Nat. **56**(97), e156, pp. 24–25 (1998)
17. Allen, R.G., Luis, S., Pereira, D.R., Smith, M.: FAO irrigation and drainage paper No. 56. Rome: Food Agric. Organ. United Nat. **56**(97), e156, p. 103 (1998)
18. Allen, R.G., Luis, S., Pereira, D.R., Smith, M.: FAO irrigation and drainage paper No. 56. Rome: Food Agric. Organ. United Nat. **56**(97), e156, p. 170 (1998)
19. OpenWeatherMap Weather API. http://openweathermap.org/api
20. Lampkin, V., et al.: Building Smarter Planet Solutions with MQTT and IBM Websphere MQ Telemetry. IBM Redbooks, Durham (2012)
21. Hunkeler, U., Truong, H. L., Stanford-Clark, A.: MQTT-S—a publish/subscribe protocol for wireless sensor networks. In: 2008. 3rd International Comsware Conference on Communication Systems Software and Middleware and Workshops, 2008, pp. 791–798. IEEE, January 2008
22. Luzuriaga, J.E., Perez, M., Boronat, P., Cano, J.C., Calafate, C., Manzoni, P.A.: Comparative evaluation of AMQP and MQTT protocols over unstable and mobile networks. In: 2015 12th Annual Consumer Communications and Networking Conference (CCNC), IEEE, pp. 931–936. IEEE, January 2015

23. Eclipse. Mosquitto an open source mqtt broker. [Online]. Available: http//mosquitto.org/
24. OpenJS Foundation & Contributors, Node-Red. https://nodered.org
25. Allen, R.G., Luis, S., Pereira, D.R., Smith, M.: FAO irrigation and drainage paper No. 56. Rome: Food Agric. Organ. United Nat. **56**(97), e156, p. 60 (1998)
26. Hochreiter, S., Schmidhuber, J.: Long short-term memory. Neural Comput. **9**, 1735–1780 (1997)
27. Lee, T., Shin, J.Y., Kim, J.S., Singh, V.P.: Stochastic simulation on reproducing long- term memory of hydroclimatological variables using deep learning model. J. Hydrol. **582** (2020)
28. Rguiga, G., Mouttaki, N., Benhra, J.: CONTRIBUTION TO SALES FORECASTING BASED ON RECURRENT NEURAL NETWORK IN THE CONTEXT OF A MOROCCAN COMPANY. In: 13ème CONFERENCE INTERNATIONALE DE MODELISATION, OPTIMISATION ET SIMULATION (MOSIM2020), 12–14 Nov 2020, AGADIR, Maroc, Nov 2020, AGADIR (virtual), Morocco
29. Lehmann, E.L., Casella, G.: Theory of Point Estimation, 2nd edn. Springer, New York (1998)
30. Hyndman, R.J., Koehler, A.B.: Another look at measures of forecast accuracy. Int. J. Forecast. **22**(4), 679–688 (2006)
31. Everitt, B.S.: The Cambridge Dictionary of Statistics, CUP (2003)
32. Roy, D.K., et al.: Daily prediction and multi-step forward forecasting of reference evapotranspiration using LSTM and Bi-LSTM models. Agronomy **12**, 594 (2022). https://doi.org/10.3390/agronomy12030594

NSoV: An Efficient SDVC Assisted Forwarding Model in NDN Based IoV

Asadullah Tariq[1], Farag Salabi[1], Mohamed Adel Serhani[1](✉), and Irfan Ud Din[2]

[1] College of Information Technology, United Arab Emirates University, Al Ain, UAE
700039114@uaeu.ac.ae

[2] The Superior University, Lahore 54000, Pakistan

Abstract. Internet of Vehicles (IoV) is a fast-growing mechanism, identified as a good solution to connectivity and communication issues. The conventional Vehicular Ad Hoc Networks (VANETs) trend is shifting towards IoV. Most of the expectations of the IoV are achieved by a thriving internet paradigm called Named Data Networking (NDN). The main inspiration behind NDN is the limitations of current IP Internet architecture. Another emerging networking paradigm is Software-Defined Networking (SDN), which is highly capable of transforming complex networking architectures into simple architectures along with well-planned management of overall networks. In NDN-based IoV, broadcast storm is a major issue due to NDN's broadcasting nature. In IoV scenarios, unnecessary transmission delays and disconnected link problems are created by the fast-changing topology and high speed of vehicles. In this paper, we propose an intelligent forwarding strategy in NDN-enabled IoV using Software Defined Vehicular Controller (SDVC) and Edge Controller (EC) along with a reliable and durable path estimation model. We use NS3 and SUMO to simulate the model and run experiments. The delay is reduced by the In-network caching and naming strategy of NDN nodes. In the network, path failure minimization is mainly contributed by the properties of EC.

Keywords: Forwarding · Named Data Networking · IoV · SDN · EC

1 Introduction

From the past two decades, smart devices using network applications and services has been seen to modernize at a great level having differently new technologies. Withal, the demand of such devices, using networks and people is now pushing the boundaries. The trends are shifting from traditional VANET to IoV in fields like computing, sensation, networking technologies for vehicles, automation, and technological advancements. Researchers, Government agencies along with industries have already been made great endeavors because of the rapid growth of IoV for efficient vehicular communication, that would hand out significantly to the development and deployment of the Intelligent transport System (ITS). In ITS, IoV is representing a protruding instantiation of the IoTs. If we compare IoV with Mobile Ad Hoc Network (MANET), it has been seen that IoV covers some exclusive attributes. For instance, predictable mobility, variable density of

M. Lazaar et al. (Eds.): BDIoT 2022, LNNS 625, pp. 550–563, 2023.
https://doi.org/10.1007/978-3-031-28387-1_46

the network, high computation ability, and high-speed internet connectivity are notable. Limited battery power, random motion and computation are some of the limitations in MANETs. For ITS safety applications, Internet of vehicles (IoV) is more suitable, and its centralized management makes it different from VANET [1]. In VANET, we have quite a different scenario, vehicular nodes are the main access points to offer connectivity to the other vehicular nodes for communication. If we only talk about USA, by 2030 it is estimated to have 20.8 million vehicles. By looking at this estimation for VANET, it would be a difficult task to provide support for safety applications and traffic management on a large scale. Despite having more powerful attributes, implementation of IoV is more challenging, but it may be regarded as evolution in MANETs and VANETs.

To design and manage the network, SDN and NFV endows significant and new processes. New innovation ideas to get implemented and tested, SDN proposes a platform while seeking programmable behaviors and centralized control plane. SDN splits the control plane from the data plane to gives a centralized view of distributed network. Because of several problems with node density in internet of Vehicles, it is known as a huge challenging task to design routing protocols, heterogeneous communication technologies, efficient data transmission, intermittent connectivity, and varying mobility. For current vehicular network architecture, the more challenging task is to satisfy the basic demands of intelligent and advanced transportation besides flexibility and scalability. SDN modernizes the IoV to accomplish a perfect and systematic routing methodology. SDN-enabled IoV has already been discussed in many research papers. In big cities, increase in the number of vehicles is causing the increase in number of accidents, which is basically the outcome of bad traffic management. SDN provides a logically centralized environment in heterogeneous networks for IoVs. IoV along with SDN in access networks and data centers is currently taken as a traffic booster in smart cities, which boosts the traffic communication [2]. Practical implementation is a challenging and most highlighted task in mobile and vehicular networks. If we look at the recent research works, there is a lack of SDN-IoV implementation in fast changing, dynamic topology, and high-speed connectivity area. Our given plan of action gives the way for the realization of centralized node management to implement SDN-IoV architecture. We believe that using the benefits of SDN over Internet of vehicles (IoV), we can bridge the gap between road safety applications and Internet of vehicles (IoV). From the past years, we have seen a rapid shift of research towards SDN-enabled vehicular networking problems with some limitations like communication delay, broadcast storm and mobility of high-density nodes.

Named data networking (NDN) is an emanating content-based paradigm and internet architecture based on naming scheme, developed on the limitation of tradition IP architecture [3]. In NDN hourglass model, content names are added in place of IP address layer as shown in Fig. 1. NDN for internet of vehicles in terms of content retrieval, is very suitable. NDN shifts the current internet's host-based communication model by using "Named Data" instead of the "IP address". NDN is meeting an increased demand of the market for content fetching and data distribution by doing so [4]. NDN preserves the hourglass model of TCP/IP which is basically a network architecture model with seven layers but introduces its own modifications.

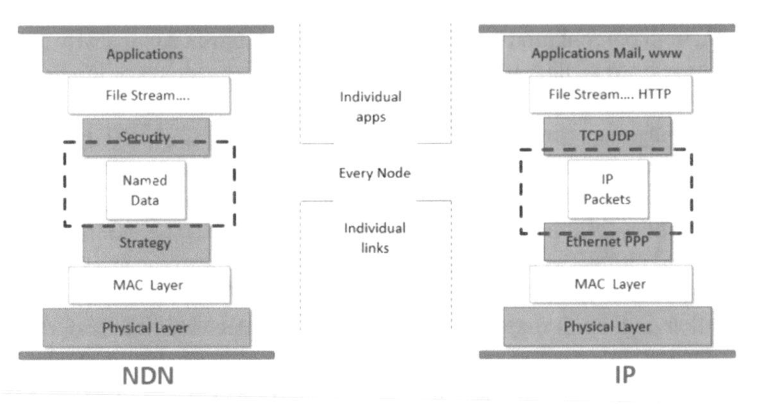

Fig. 1. Hourglass Models of NDN and IP

The lower layer is designed to adapt to the underline physical chain links and communications whereas the upper layer is designed to correspond to related applications. The most significant difference between NDN communication model and TCP/IP sits at the middle layer which basically replaces the IP address with named data. This fundamentally shifts the notion from pushing data packets based on the IP location to obtaining data based on name. In other words, instead of doing a server-side push of data, NDN does a demand pull of data.

There are two types of transmission packets in the NDN network: Interest packet and Data packet. In NDN model, the consumer requests data by sending interest packets to the network to all available nodes on the network and any of those nodes can receive an interest packet. If a particular node has a data content that corresponds to the data packet, then that data packet will be send back to the requester in the same way it was requested. The data packet contains the required data. The data producer binds the name of the content to the content information and puts the signature on top [5]. This is an important statement to make that the data packet would only be transmitted to the interest packet based on that interest and it will not be involving any third parties along the way. From what we have just described we see that NDN has two types of packets neither of which contain the IP address only the name of the data is used along with the signature that contains the information about the data. The forwarding model of NDN is shown in Fig. 2.

In our proposed work, the main strategy is to use a road aware technique that will look after the road-segments created by the gateway nodes and is completely suitable for IoV. For path selection, our strategy of using segment ID over vehicle ID makes it more enduring and reliable. In forwarding the data packets different technologies has deployed, since the packets from the vehicles to SDN and road will be received and forwarded through the cellular networks. In our strategy which will reduce overall response time and network overhead, in which after a defined time, real time data will be explored by the EC. To find the durable yet shortest communication path, we present a path estimation model, along with efficient forwarding scheme. EC in our proposed work is a key factor and having significance for real time information collection from the vehicular nodes. Seamless vehicular node's information without any delay is a major key in our strategy.

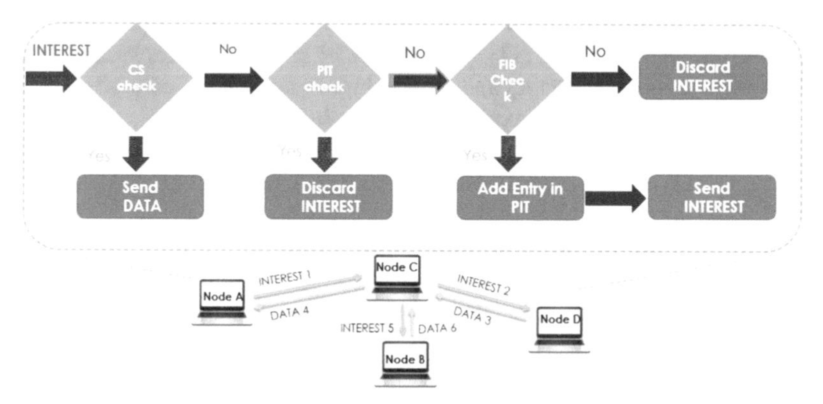

Fig. 2. Forwarding Model of NDN

The remaining organization of the paper is as following: Sect. 2 will present related works. Section 3 will elaborate the problem formulation along with propose methodology in Sect. 4, results and discussions for experiments in Sect. 5 and conclusion in Sect. 6.

2 Related Works

In this section, we present related research works in NDN based SDN-IoV. In V-NDN, BSMS in [6] is a forwarding mechanism know as broadcast storm mitigation strategy which reduces the broadcast storm issues. High speed of vehicles on the road causes vehicle unavailability and disconnection of links which results in data packets not following in the same path as interest packets. The distance between the receiver node and the sender node is calculated by a forwarding timer in broadcast storm mitigation strategy (BSMS). As a potential forwarder the selected node will be the farthest node. The timer which is used for rebroadcast and re-transmit a same request is referred to as a forwarder timer counter.

BSAM and vehicular network architecture (NSDVN) in [7] is defined by NDN based SDN. In this research work, broadcast mitigation routing mechanism has put forward. To calculate the distance between vehicles, a counter value is used in this work. Farthest node is used as relay node in forwarding mechanism. An SDN enabled Information Centric Networking (ICN) based consumer mobility supportive novel protocol is proposed in [8]. The model to support the consumer mobility is edge computing model. A directional forwarding mechanism is used in which author deployed directional antennas in vehicles, for the nodes to forward the data in certain direction. A contention-based forwarding scheme is proposed in [9] to support the vehicular mobility. To identify the failure messages, timer-based strategy is used. It is known that there is very less research work on the mobility of the Internet of vehicles when it comes to SDN enabled NDN environment. Our focus is on removing the issues related to the mobility of the vehicles on Internet of vehicles (IoV) scenario and removing the broadcast storm issues.

In [10], a controller based Selective forwarding (CSF) is another SDN assisted forwarding strategy in NDN. CSF is actually mitigating broadcast storm issue. CSF forwarding flows requested by the consumer are calculated by the SDN. Only selected nodes forward the packet in network. The focus of this paper is to handle average delay and Retransmissions. SRSC in [11] is another SDN based NDN implementation having two phases like bootstrapping and forwarding. In first phase, controller initiates a control message to all the other nodes to get their information. Nodes will then share their information will all the neighbors and controller. The sharing of information is creating an extra overhead here. In NDN enabled MANETs, NAIF in [12] is another neighborhood aware forwarding. NAIF is using relay node's content retrieval rate and distance between data node and relay node. This mechanism will filter ineligible nodes. A lot of computation has been done at nodes level here. Moreover, there is a lack of central controller here. In [13], a multipath forwarding strategy has been introduced. In case of unavailable requested data, this scheme will enable the nodes to forward the message at router. Controller will use updated messages in future and send back if the requester node is available.

3 Problem Formulation

In this section, we will formulate the problem statement that motivated our proposed solution. Handover of vehicles and mobility of migration state have been studied in various research papers [14, 15]. The maximum speed on which one can attain reliability is referred as mobility. In an SDN-based Internet of vehicles (IoV) comparing to the traditional cellular network, handover mechanism is very challenging. Reconsideration of radio resources is essential with a new SDN controller. There is a need for random changing topology to update SDN controller flow table. By using trajectory prediction component using position, destination of vehicle and velocity, the position of vehicles shortly in the future can be estimated. Which will help to update the service migration and flow table in SDN.

In NDN based IoV, there is a proper demand to have mobility support. To re-issue the packets in NDN, NDN's broadcast nature can cause broadcast storm issues. In NDN, broadcast storm is considered as a critical issue which ends up having extra transmission delays and to be a big waste of network resources. Disconnected link issue due to unstable communication path and disconnection of producer node, intermediate node, and consumer node in the network caused unnecessary delay and transmissions failure. Not only the updating of global tables is important but also the management of in-time information in IoV-NDN is of a great importance. To solve all these problems, we can get benefits from the combination of NDN, SDN and IoV. Broadcast storm and delay problems due to disconnected link and broadcast storm is presented in Fig. 3.

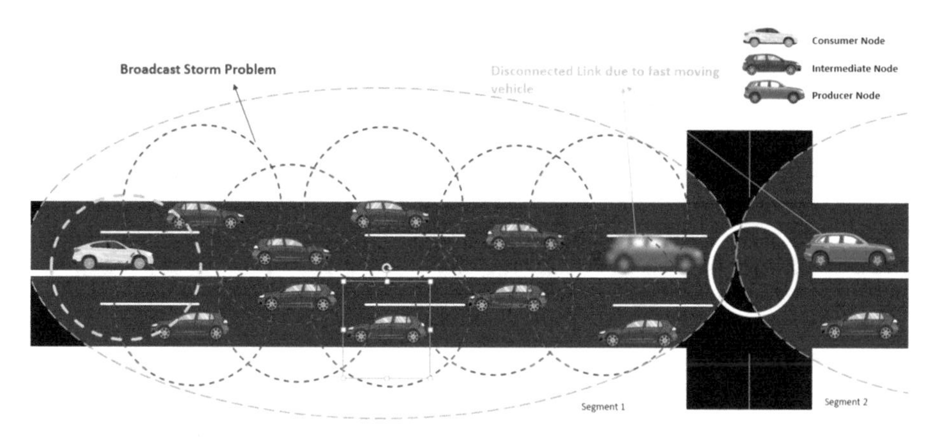

Fig. 3. Broadcast Storm and Delay Problem scenario in NDN-IoV

4 Proposed Methodology

This section is the main important part of our research article and contains proposed solution for above discussed problems. Two data structures, Software Defined Vehicular Controller (SDVC) and Edge Controller (EC) are used to manage the overall network in our proposed work. Edge Controller is a centralized controller over every road segment and having information of nodes of that specific road segment in Local Information Table (LIT). SDVC is a Global Controller having information of all the segments. It will use its computation power when we need to initiate intra segment communication. There is also a routing table that maintain the shortest path from one node to another node in SDVC. There is an entry against each vehicle in SDVC and EC with the information of segment ID, speed, vehicle ID, content names, direction and the position. Global Information Table (GIT) in SDVC holds all the information received from EC and LIT holds the information against each vehicle for a specific direction and segment. Another benefit of SDVC is the best path calculation from consumer node to the producer node after taking all the parameters in account.

Data_Pkt, Int_Pkt, EC_Info_Pkt, and L_Info_Pkt are the four types of packets that will be used in our methodology. In our two leveled routing strategy, each vehicle will send its information to EC by sending L_Info_Pkt. RSU's and the gateway nodes are responsible for the connectivity of road segments. EC_Info_Pkt then used to send the information from Edge controller to SDVC. Minimum hop, direction and relative velocity are the parameters for the computation of shortest path. Shortest path can compute only on the roads with vehicle density of 15%–75%.

At start all the nodes will send their information to EC. EC then updates its LIT against each node. All the ECs will send the information to SDVC, which will update its GIT. There would be a routing table in both EC and SDVC having the shortest and best paths from one node to another. Then nodes will update the EC after every fixed interval of time (100 ms) with updated information. On arrival of an Interest Packet at one hop neighbor, a vehicular node will first check its PIT to find the same request already received before. If it finds the same request in PIT, then it will discard the Interest Packet. In other case, node will check its CS and return Data_Pkt back to the Consumer node if it finds the data against the received Int_Pkt. If it did not find the data at one hop, then requesting node will send packet to Edge controller. A counter value is used to help nodes to analyze the Int_Pkt for one hop neighbors. EC will then check it's LIT and on successful finding the data, it will compute the shortest path between requesting node and the node having the data information. EC then return a packet back to requesting node with the path to follow for data. Requesting node will then again broadcast the packet with best and shortest path embedded in the packet. Any node having information in requesting path will accept the packet and all other nodes will drop the request. This way we can protect our model from excessive number of requests. After reaching to the final destination, the data will be send using the same path used by the request packet. If previous path does not exit, then it will follow the same procedure as for Int_pkt. Nodes can send packets directly to EC and EC then send it to SDVC to find the forwarding port and SDVC will then reply to them with the path estimation information that leads the packets towards the destination. Least speed is the reason behind more durable connectivity between two vehicles.

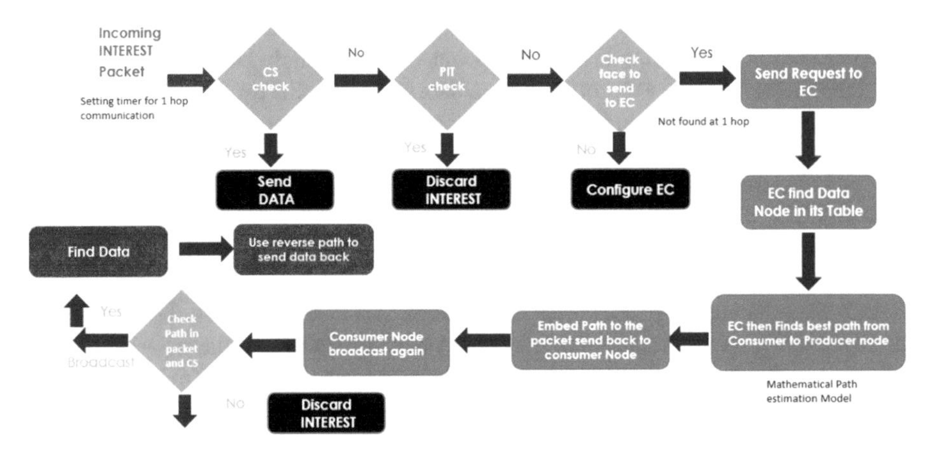

Fig. 4. Forwarding Model of Proposed NSoV

If the destination node is on another segment, then EC will send the request to SDVC and it will check its GIT and compute a routing path from requester node to destination node. Routing path is based on receiver timer counter, where transmission range, distance between Consumer and current vehicle and relative speed will be considered. Using this receiver timer counter, the farthest node in transmission range would be select as the potential forwarder. Forwarding timer counter is a lightweight formula that will compute the time taken by the transmission from Consumer to Producer node. In case of failure in transmission in forwarding timer counter value, node will again request to the EC. If a node is going out of the segment, then Edge Controller will wait for its acknowledgement. Because a node needs to send information to Edge Controller after every 100 ms. After waiting for 200 ms and no response then Edge controller will delete that entry.

Each node will update its flow table after receiving and forwarding the packets. EC will take the charge in case of failure with SDVC. Whenever a vehicle is going out of the range it will remove the saved entries and for this purpose, we used a hard time out. In hierarchical manners first node will check for the possible content and destination, otherwise it will send the request to EC and If EC dot have the information the request will send to the SDVC controller. A failure message is used to re-compute the shortest and best path in case of unsuccessful flow table investigation. An updated topology will be selected by the SDVC if there is a case in which a vehicle is going to leave the road segment with ongoing transmission. If the failure message is within the segment, then EC will handle it and in-case of outside the segment, SDVC controller will handle the update. On demand path estimation is used in our proposed work for which we introduced probabilistic path estimation model. SDVC will estimate the duration of a path to get a reliable path. In our strategy, nodes will broadcast the packets and on successful communication path will remain open till end of communication. The connection duration for reliable path depends upon the vehicle's velocity and position of its neighbors determined by the SDVC. The incorporated parameters in our proposed methodology are link connectivity, direction of vehicle, number of hops, and velocity (speed). Data forwarding in vehicular environment is based on traditional traffic flow principle in our estimation model. Forwarding flow of our proposed work is presented in Fig. 4.

Mathematical relation like duration of the link and number of hops are used to deduce the path duration between two vehicles. The path estimation model used in our paper along with simulation model is presented in Fig. 5. A path with minimum number of hops is used to find the more durable and stable path between requester vehicle and vehicle having data information. Closet node from the data vehicle and the farthest one from the requester vehicle will be selected as potential forwarder in communication. Area of the entire region can be computed by using Eq. (1)

$$AREA_{Total} = AREA_{Int1} + AREA_{int2} \tag{1}$$

$$AREA_{Total} = AREA_{int1} \approx \left[\frac{(\alpha - \sin(\alpha)).Range_C^2}{2}\right] + AREA_{int2} \approx \left[\frac{(\alpha - \sin(\alpha)) \cdot Range_C^2}{2}\right] \tag{2}$$

Direction and velocity of the vehicle are the most important parameters for path estimation. There will be a direct effect on link duration due to the direction of vehicle. We have four scenarios for this case in which velocity of and direction of both the vehicle will take in account. Vel_1 and vel_2 are the velocities of both the vehicles. Vel_R is the relative velocity, d is the distance, θ is the angle between both the vehicles and R is the range between the communication links. Table 1 will describe all the traffic scenarios.

$$\left|\overrightarrow{vel_R}\right| = \sqrt{vel_1^2 + vel_2^2 - 2vel_1vel_2\cos\theta} \tag{3}$$

To find the number of hops, count of average number of vehicles is essential. Equation (4) can be used to find the hop count.

$$hop_{count} = \frac{Dist_{C_P}}{Dist_{C_R}} \tag{4}$$

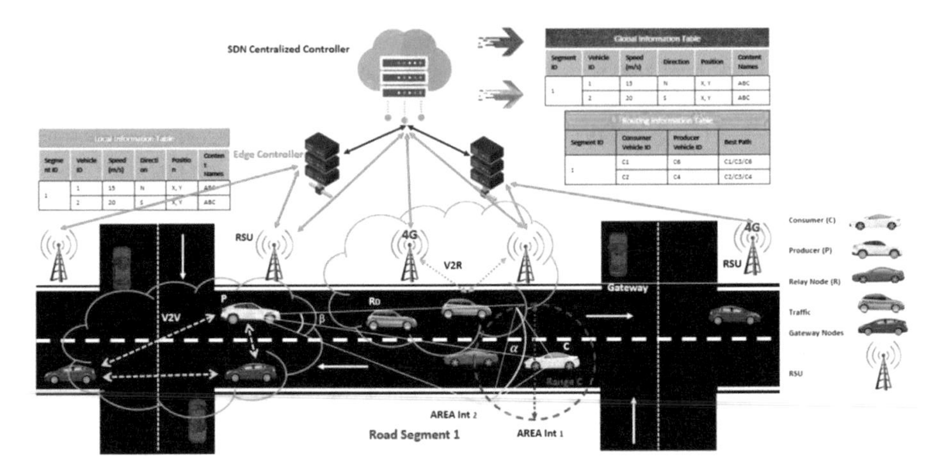

Fig. 5. Network Setting and Path estimation Model in NSoV

Table 1. Traffic Scenarios in proposed model

Scenario	Velocity	Angle θ	Relative Velocity
Same direction and same Velocity	$vel_1 = vel_2 = vel$	$\theta = o$	$\left\lvert \overrightarrow{vel_R} \right\rvert = o$
Opposite direction and same Velocity	$vel_1 = vel_2 = vel$	$\theta = \pi$	$\left\lvert \overrightarrow{vel_R} \right\rvert = 2vel$
Same direction and different Velocity	$\alpha vel_1 = vel_2$	$\theta = \pi$	$\left\lvert \overrightarrow{vel_R} \right\rvert = vel_1(\alpha - 1)$
Opposite direction and different Velocity	$\alpha vel_1 = vel_2$	$\theta = 2\pi$	$\left\lvert \overrightarrow{vel_R} \right\rvert = vel_1(\alpha - 1)$

Link duration is the most important estimation and we can find it using Eq. (5).

$$Time_{link} = \frac{Range_c - Dist_{C_R}}{vel_C - vel_R} \tag{5}$$

The distance after which a node can move out of the communication range can be computed as:

$$Dist_R = Range_C - Dist_{C_R} \tag{6}$$

The average path duration can be calculated as:

$$Time_{link}(avg) = \int_0^{\alpha} Time_{link} f(Time_{link}) dTime_{link} \tag{7}$$

The exact knowledge of the real time topology after collecting the data from each vehicle is the actual purpose of EC in our proposed work through which we can overcome the mobility and link breakage issue. L_I_Pkt and EC_I_Pkt can be send to the SDN controller using two different scenarios in which a packet can follow fixed time interval and after a particular position change of the vehicle as shown in Eq. (8) and (9).

$$Timeinterval_{hello} = \frac{Distance}{Speed} \tag{8}$$

$$Timeinterval_{hello} = MIN\left(C, \frac{Distance}{Speed}\right) \tag{9}$$

Table 2. Simulation Parameters

Parameter	values
Simulation time (s)	1200 s
Field size (m * m)	1000 * 1000
Number of nodes	20, 40, 60, 80, 100
Communication range (m)	250
Data rate	250 Kbps
Interest size (byte)	70
Data size (byte)	120
Simulations run	10
Packet size	1040 bytes
Technology	IEEE802.11OCB
Rx sensitivity	−86 dBm
Tx Power	18 dBm
Speed (m/sec)	0–24
Road Type	Two ways
Traffic model	SUMO

5 Simulation Results

To evaluate the performance of proposed strategy NSoV against other schemes, we used different simulation matrices and parameters. We used Network Simulator (NS-3) and ndnSIM [16] along with sumo for the possible simulations and experiments. IEEE 802. 11OCB face and 3G/LTE technologies are used, that are efficient in high-speed data connectivity and data mulling. SUMO is used as traffic generator with the vehicle range 20–100 in environment. The network range and transmission range of each node are 1000 m^2 and 250 m respectively. Different number of sender and receiver nodes in IoV are used to experience the diversity and density of the topology. The percentage of sender nodes is 70\% but it varies in producer percentage from 5%, 10%, 15%, 20% and 25% to get more clear results. Different simulation time from 200 s–1200 s are used to evaluate the results. 10 independent runs were used to evaluate NSoV. Point-to-point delay, total number of re-transmissions and Interest packet satisfaction ration are the performance matrices. We evaluated the performance of our model against simulation time, speed of the vehicle and number of consumer producer pairs. We compared our scheme with traditional NDN forwarding and SRSC [11]. Table 2 presents simulation parameters.

In Fig. 6, end-to-end delay is evaluated for NSoV and other comparative schemes. The value of Y-Axis is Average Waiting Time, and the value of the X-axis is Simulation Time. Distributed strategy of NSoV and efficient use of SDVC and EC results in better output because of the efficient maintenance and estimation of durable path. Broadcast mitigation process in NSoV results in decreased unnecessary network overhead and decreased congestion. With increase in consumer producer vehicles, end-to-end delay increased.

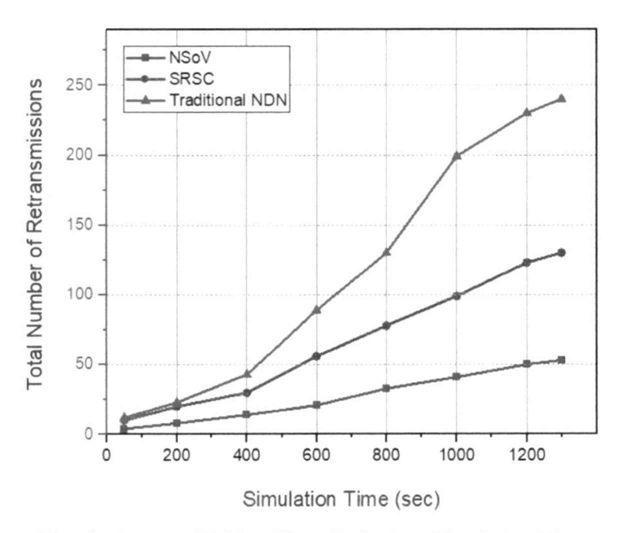

Fig. 6. Average Waiting Time (Delay) vs Simulation Time

In Fig. 7, we evaluated number of total retransmissions against simulation time (sec). The value on x-axis is simulation time (sec) and the value on y-axis is the total number of retransmissions. NSoV performed better in this case of total number of retransmissions Due to the efficient use of EC and SDVC, retransmissions are very less as compared to SRSC and traditional NDN models.

In Fig. 8, we compare Total number of packets with simulation time (sec). NSoV is performing efficiently against comparative studies and there is less overhead in network. Less retransmission is also benefiting these results. SDVC and EC are benefiting NSoV with efficient forwarding management that results in mitigation of broadcast storm.

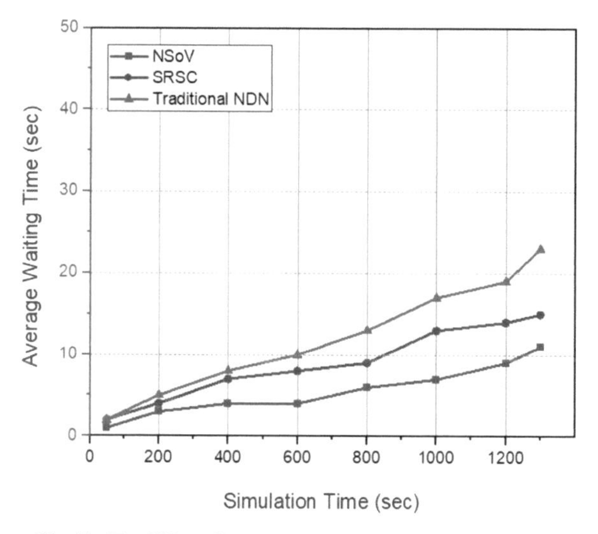

Fig. 7. Total No. of Retransmissions vs Simulation Time

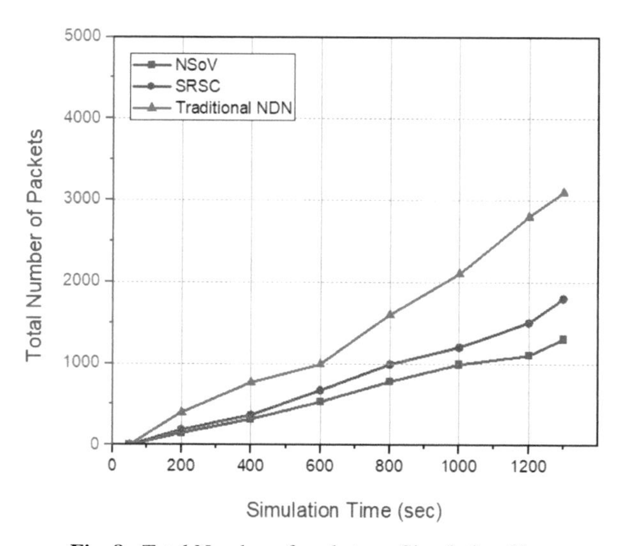

Fig. 8. Total Number of packets vs Simulation Time

In Fig. 9, we compared packet success ratio against total number of nodes. NSoV is performing better in packet success rate as well as compared to other techniques. Minimized broadcast storm issue and an excellent path estimation model using SDVC and Ec in IoV results in better packet success ratio.

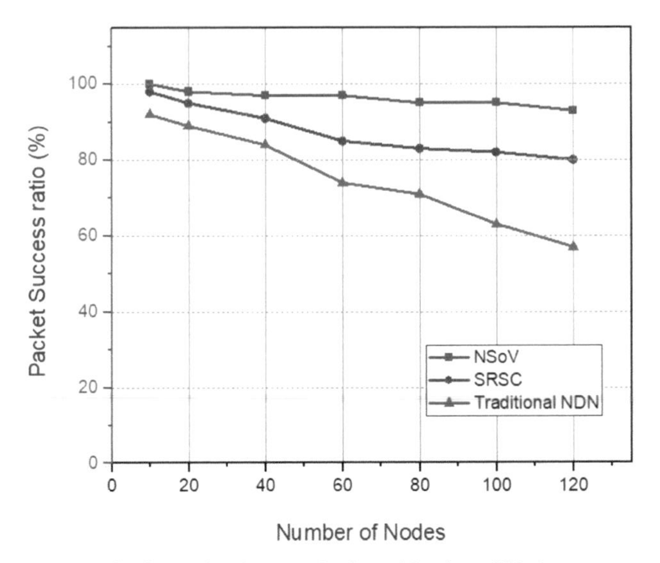

Fig. 9. Packet Success Ratio vs Number of Nodes

6 Conclusion

In this paper, we proposed NSoV, an efficient forwarding mechanism that utilizes the functionalities and benefits of SDVC, an SDN enabled vehicular controller and edge controller in NDN based IoV environment. The main focus of the paper is to design a forwarding strategy to mitigate the broadcast storm issue in NDN-IoV and reduced end-to-end delay in communication with support of mobility. This hierarchical forwarding strategy takes benefit from both SDVC and EC for efficient forwarding and provide mobility support as well in case of fast-moving topologies and intra-segment communication. We formulated a road aware strategy that take various parameters e.g., location, speed, direction, content, and controller information of a vehicular node into account. To manage and maintain the real-time and in-time vehicular topology the introduced idea is of EC in IoV. To find not only the reliable and sustainable path but also the best and the shortest path in our work a mathematical model has also been introduced. Our work outperforms comparative schemes in terms of total number of Interest, total number of retransmissions, average waiting time and packet success rate.

References

1. Grassi, G., Pesavento, D., Pau, G., Vuyyuru, R., Wakikawa, R.: VANET via named nata networking. In: IEEE INFOCOM WKSHPS, Toronto, Canada, pp. 410–415 (2014)
2. Bannour, F., Souihi, S., Mellouk, A.: Distributed SDN control: survey, taxonomy, and challenges. IEEE Commun. Surv. Tutor. **20**(1), 333–354 (2017)
3. Zhang, L., Afanasyev, A., Burke, J., Jacobson, V., Claffy, C.: Named data networking. ACM SIGCOMM Comput. Commun. Rev. **44**(3), 66–73 (2014)
4. Tariq, A., Rehman, R.A., Kim, B.S.: Forwarding strategies in NDN-based wireless networks: a survey. IEEE Commun. Surv. Tutor. **22**(1), 68–95 (2019)

5. Tariq, A., Ud din, I., Rehman, R.A., Kim, B.S.: An intelligent forwarding strategy in SDN-enabled named-data IoV. Comput. Mater. Continua **69**(3), 2949–2966 (2021)
6. Burhan, M., Rehman, R.A.: BSMS: a reliable interest forwarding protocol for NDN based VANETS. In: 3rd International Conference on Advancements in Computational Sciences (ICACS), Lahore, pp. 1–6 (2020)
7. Arsalan, A., Rehman, R.A.: Distance-based scheme for broadcast storm mitigation in named software defined vehicular networks (NSDVN). In: 16th IEEE Annual Consumer Communications & Networking Conference (CCNC), Las Vegas, USA, pp. 1–4 (2019)
8. Ahmed, S.H., Bouk, S.H., Kim, D., Rawat, D.B., Song, H.: Named data networking for software defined vehicular networks. IEEE Commun. Mag. **55**(8), 60–66 (2017)
9. Wang, J., Luo, J., Zhou, J., Ran, Y.: A mobility-predict-based forwarding strategy in vehicular named data networks. In: GLOBECOM, IEEE Global Communications Conference, Taipei, Taiwan, pp. 01–06 (2020)
10. Ansari, F., Rehman, R.A., Kim, B.S.: CSF: controller based selective forwarding in software defined named data based MANETs. In: IEEE Globecom Workshops (GC Wkshps), Hawaii, USA, pp. 1–6 (2019)
11. Aubry, E., Silverston, T., Chrismen, I.: Implementation and evaluation of a controller-based forwarding scheme for NDN. In: 31st International Conference on Advanced Information Networking and Applications (AINA), Taipei, Taiwan, pp. 144–151 (2017)
12. Yu, Y.T., Dilmaghani, R.B., Calo, S., Sanadidi, M.Y., Gerla, M.: Interest propagation in named data manets. In: International Conference on Computing, Networking and Communications (ICNC), San Diego, pp. 1118–1122 (2013)
13. Alhowaidi, M., Nadig, D., Ramamurthy, B., Bockelman, B., Swanson, D.: Multipath forwarding strategies and SDN control for named data networking. In: IEEE International Conference on Advanced Networks and Telecommunications Systems (ANTS), Indore, India, pp. 1–6 (2018)
14. Labriji, I., et al.: Mobility aware and dynamic migration of MEC services for the Internet of Vehicles. IEEE Trans. Netw. Serv. Manage. **18**(1), 570–584 (2021)
15. Lin, Y., Zhang, Z., Huang, Y., Li, J., Shu, F., Hanzo, L.: Heterogeneous user-centric cluster migration design for vehicular networks: the connectivity versus handover rate trade-off. IEEE Trans. Veh. Technol. **69**(12), 16027–16043 (2020)
16. Mastorakis, S., Afanasyev, A., Zhang, L.: On the evolution of ndnSIM: an open-source simulator for NDN experimentation. ACM SIGCOMM Comput. Commun. Rev. **47**(3), 19–33 (2017)

Coil Design and Misalignment Effects on Wireless Charging Parameters for Electric Vehicle

Naima Oumidou(✉), Sara Khalil, Mouad Bahij, Ali Elkhatiri, and Mohamed Cherkaoui

Engineering for Smart and Sustainable Systems Research Center Mohammadia School of Engineers, Mohammed V University, Rabat, Morocco
naimaoumidou@research.emi.ac.ma

Abstract. The range of electric vehicles is one of the main obstacles to their widespread adoption. Various charging methods are being used to address this issue and wireless charging technology is emerging as a promising solution. However, the misalignment between the transmitting and receiving coils affects the transfer efficiency. The geometry of the coils can also have an impact on the efficiency. This manuscript investigates the influence of misalignment on the main parameters of the WPT system and evaluates the Ansys Maxwell magnetic flux model under perfect alignment and in the case of misalignment along the X-axis for both the circular and the square design. The simulation results show that the circular design has a large coupling coefficient and a large mutual inductance compared to the square design, but the latter is less sensitive to misalignment. Under perfect alignment the magnetic field density is maximum at the center and decreases moving outwards, so the magnetic field becomes weak along the vertical Z direction.

Keywords: Coil design · misalignment · wireless charging · Coupling coefficient

Nomenclature

WPT: Wireless power transfer
PHEV: Plug-in hybrid electric vehicles
SCMR: Strongly Coupled Magnetic Resonance
SPS: Series-parallel-series
ICPT: Inductively coupled power transfer
QDQ: D-quadrature

1 Introduction

In response to the growing demand for energy and in order to preserve depleting fossil fuels and also to reduce the increasing levels of pollution, the world is moving towards

M. Lazaar et al. (Eds.): BDIoT 2022, LNNS 625, pp. 564–574, 2023.
https://doi.org/10.1007/978-3-031-28387-1_47

the use of electric vehicles as an alternative that will help reduce environmental pollution and CO_2 emissions.

However, this technology faces two major problems that still limit its widespread adoption: charging infrastructure and autonomy. The current conductive charging method involves plugging in the vehicle to charge it, which is inconvenient especially when it is raining or snowing, in addition to the risks of tripping and contact wear associated with the wiring, which must be taken into account. So to overcome these problems wireless charging comes as a promising solution that will ensure convenient and safe charging [1], a solution that requires no contact between the vehicle and the charger, there is no connector and no need for a cable.

This type of charging can be applied in two ways: static wireless charging where the vehicle is simply parked and charged, and dynamic wireless charging where the vehicle can be charged in motion when driving on roads equipped with wireless charging systems, which will reduce the size of the battery and increase its life [2–4]. Inductive power transfer technology is the most mature wireless transfer technology, using the concept of mutual induction by connecting the transmitter and receiver through a variable magnetic field. It has various applications in different fields such as the medical field for wireless powering of biomedical devices [5], powering of some electronic devices [6, 7], automotive systems [8, 9], and industrial manufacturing [10].

In WPT systems, the coupling coefficient K is an important parameter that represents the percentage of coupling between the two coils and has an impact on the system efficiency. This parameter is influenced by the misalignment between the transmitter and the receiver; the value of k decreases with increasing misalignment [1, 11]. Figure 1 shows the different types of misalignment positions possible [12]. In order to improve the efficiency of the system, many researchers are trying to address the challenge of misalignments that significantly reduce the efficiency of power transfer. The authors of [4] present a ferrite core design and optimization method for the wireless charging system of plug-in hybrid electric vehicles (PHEV).

While Daerhan and Liu proposed in [14] a wireless charging system with Strongly Coupled Magnetic Resonance (SCMR) which is less sensitive to misalignment and they tested the different possible misalignment positions). The researchers in [15] present a solution for misalignment that has been investigated in a series-parallel-series (SPS) topology based on a new approach to improve the misalignment behaviour for ICPT systems. While in [4] another method to deal with the misalignment problem is cited; this is the use of a D-quadrature coil (QDQ) design for the electric vehicle application.

In this work, an analysis of the magnetic field and a study of the misalignment effect for both circular and square designs will be performed. The aim of this study is to prove or disprove the idea that the X misalignment simulation results for these two designs are identical due to their symmetrical geometries, and to determine which geometry is less sensitive to misalignment.

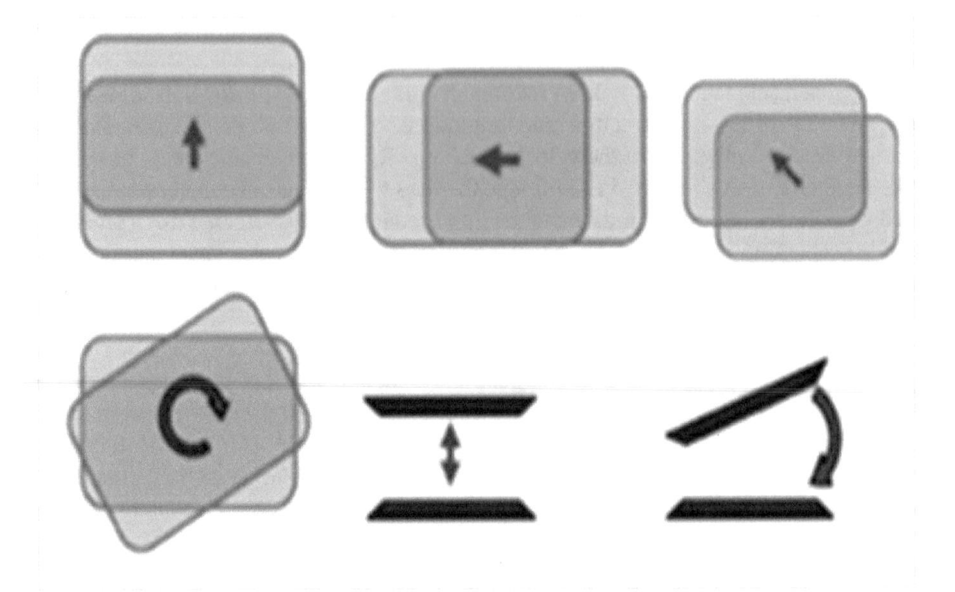

Fig. 1. Possible misalignments positions between the transmitter and receiver.

The remainder of this paper is structured as follows: After the introduction in Sect. 1, Sect. 2 presents the basic principle pf a wpt system for electric vehicle, the simulation model and lists the design parameters of the coils with which the simulation has been realized. The simulation results of the proposed conceptions are provided in Sect. 3. A discussion of results for both circular and square design is given in Sect. 4. Finally, conclusion and future works are given in Sect. 5.

2 Modeling and Simulation

2.1 Basic Principle

The WPT system consists of two main components, the first is the transmitter (Tx) which is usually fixed or embedded in the ground and the second is the receiver (Rx) which is located under the vehicle. When the transmitter is powered by an energy source, the transmitter and receiver are said to be magnetically coupled and the charging process begins when they are aligned.

An AC source provides a DC power supply which is rectified and then sent to a high frequency inverter which provides a time-varying voltage/current which induces a magnetic field. This field is transferred in a vacuum to an air gap to power the receiver coil. The magnetic flux coupled to the receiver coil is then converted into electrical energy to charge the electric vehicle battery [15, 16]. Figure 2 presents a typical elctric vehicle wireless charging system [2].

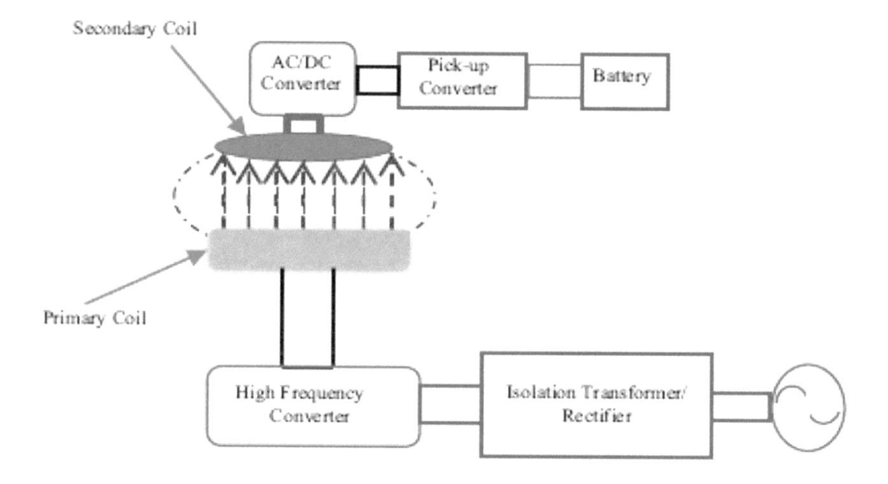

Fig. 2. EV wireless charging system components.

2.2 Design Parameters

In this section, the magnetic field strength has been analysed for two different coil designs; the circular and the square design. The simulation is performed using Ansys Maxwell software for the WPT between the transmitting and receiving coils separated by a distance d. The simulation model is shown in Fig. 3.

The simulation model for each case assumes that the transmitter and receiver coils considered have the same shape and size. Table 1 lists the design parameters of the coils with which the simulation was carried out.

Table 1. Design parameters.

Parameters	Values
Air gap distance, d	200 mm
Coil internal radius, r_i	8 mm
Number of turns in transmitter, N_T	25
Number of turns in receiver, N_R	25
Transmitter current, I_T	10 A
Receiver current, I_R	10 A

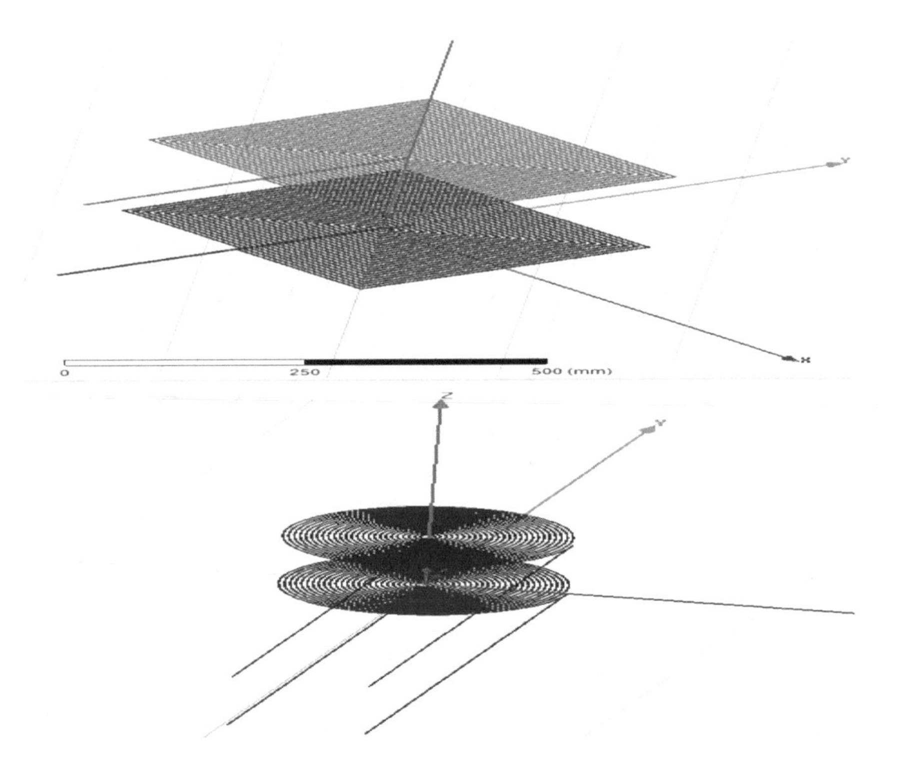

Fig. 3. Simulation model for WPT.

3 Simulation Results of Magnetic Field Plots

The alignment and distance between the two coils are among the factors that affect the coupling coefficient; the latter is maximum when the two coils are perfectly aligned and the air gap is small. Figure 4 shows the effect of increasing the air gap on the magnetic field strength at the center of the transmitting coil along the Z axis.

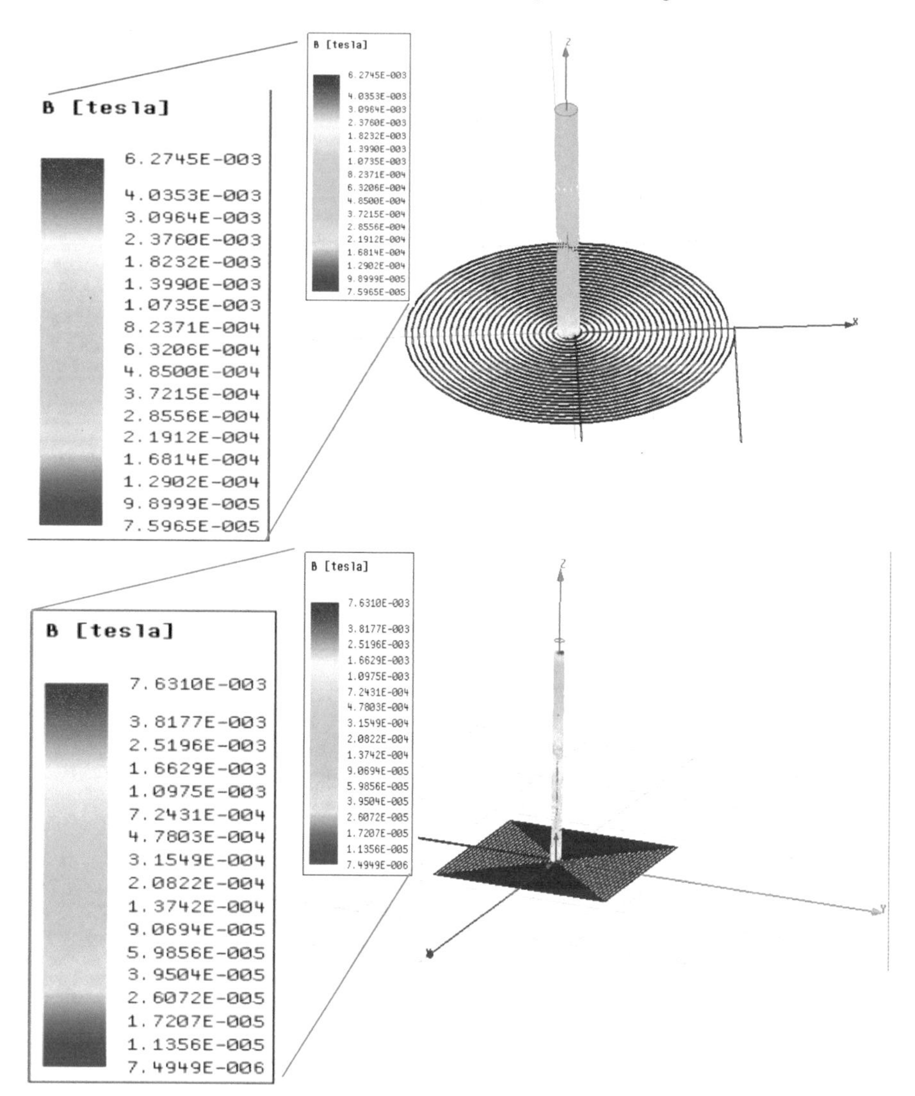

Fig. 4. Magnetic field strength at the center of the transmitter coil in the vertical direction.

The Fig. 5 presents the power transfer between the transmitting and receiving coils in the case of perfect alignment. As can be seen in both figures, the field strength is highest in the center and starts to decrease as it moves outwards. The plot in Fig. 6 represent the flow density in the case of misalignment along the X axis.

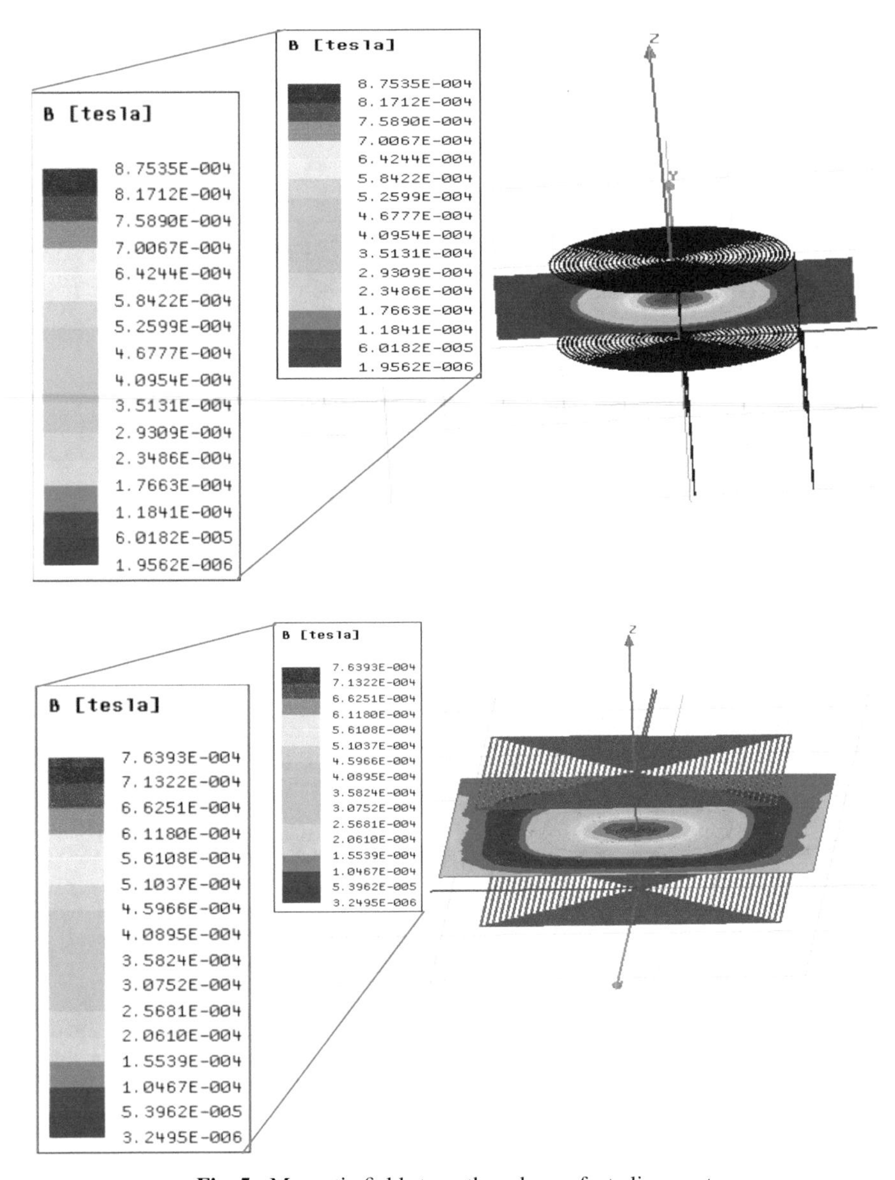

Fig. 5. Magnetic field strength under perfect alignment.

The values of self-inductance, mutual inductance, coupling coefficient between the two coils and magnetic flux produced by the two coils obtained from Ansys Maxwell for the two designs in the case of perfect alignment and in the case of misalignment are presented below in Table 2 for the circular design and in Table 3 for the square design.

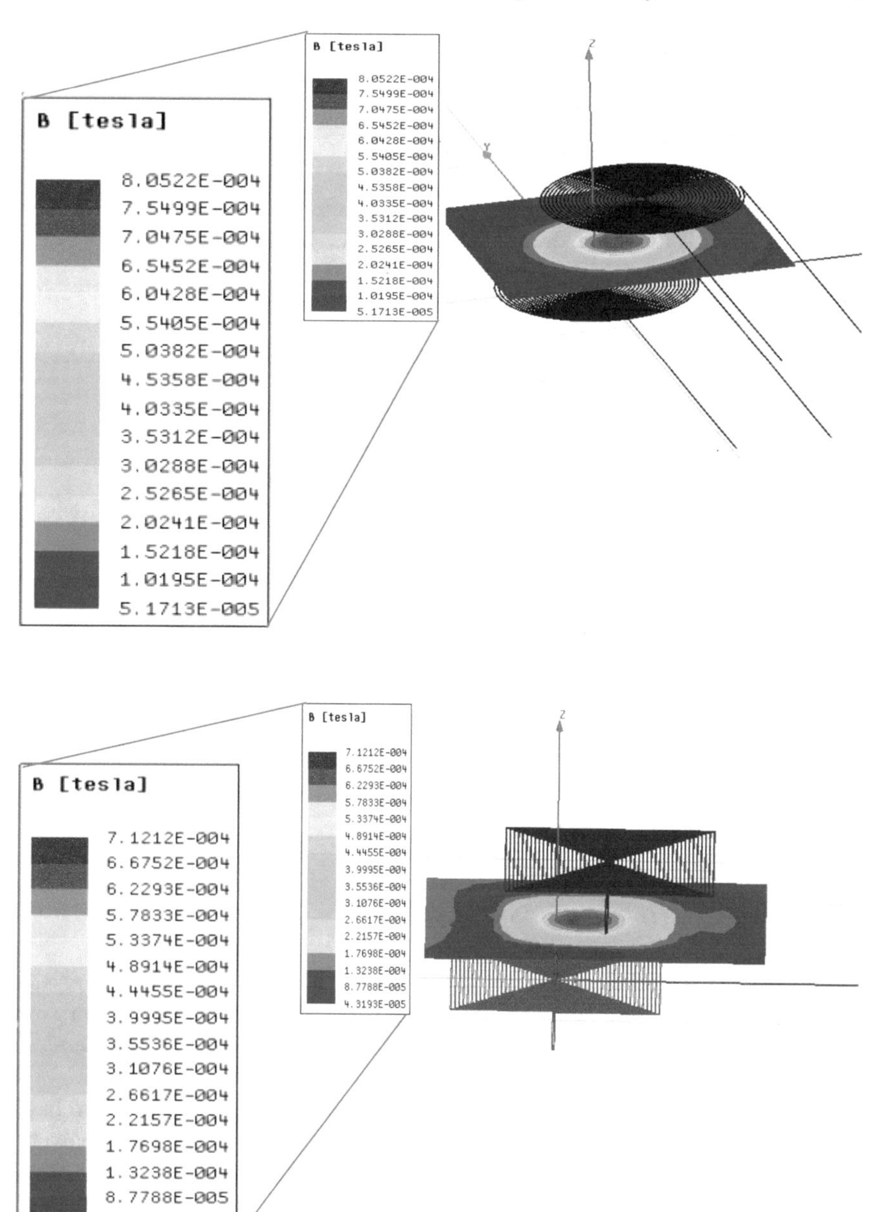

Fig. 6. Magnetic field strength at X-axis misalignment.

Table 2. Calculated parameters for circular design.

Parameters	Values	
	Perfect Alignment	Misalignment
Coupling Coefficient, k	0.099681	0.076313
Mutual Inductance, M	9.941827 μH	7.597928 μH
Self-Inductance, L_T	98.872508 μH	98.846053 μH
Self-Inductance, L_R	100.608627 μH	100.285429 μH
Magnetic Flux, φ_T	0.001088 Wb	0.001064 Wb
Magnetic Flux, φ_R	0.001106 Wb	0.001079 Wb

Table 3. Calculated parameters for square design.

Parameters	Values	
	Perfect Alignment	Misalignment
Coupling Coefficient, k	0.078163	0.068295
Mutual Inductance, M	7.825581 μH	7.043635 μH
Self-Inductance, L_T	98.166182 μH	100.066525 μH
Self-Inductance, L_R	102.110374 μH	106.298794 μH
Magnetic Flux, φ_T	0.001060 Wb	0.001071 Wb
Magnetic Flux, φ_R	0.001099 Wb	0.001133 Wb

4 Discussion

The coupling coefficient between coils and their mutual inductances are two main parameters that largely affect the power transfer efficiency in a WPT system. These two parameters depend on the positioning of the transmitting and receiving coil and therefore a misalignment between the two coils implies a reduction in the value of k and M.

From the results it can be seen that the use of the circular coil allows to have a significant coupling factor and a mutual inductance compared to the square geometry.

The results of the simulation show the influence of misalignment on these parameters, and it can be seen from the table of calculated parameters that following a misalignment along the X axis the value of k and M decreases. Also it can be seen that the square design is less sensitive to misalignment than the circular design.

The simulation results shows also that the magnetic field is maximum at the center of the transmitting coil and decreases with vertical movement away from it along the Z axis. This shows the importance of the air gap or the distance between the primary and the secondary coil; this means that the closer the two coils are to each other, the higher the magnetic field and the lower the flux losses and therefore the higher the transfer efficiency. Also, it is clear in the case of perfect alignment and in the case of misalignment that the magnetic field strength is maximum at the center and decreases moving outwards.

When transferring power between the transmitting and receiving coils and in case of perfect alignment, the flux losses are low and the magnetic field strength is maximum and consequently the transfer efficiency is maximum. Whereas a weak misalignment between the two coils leads to a decrease in the transfer efficiency.

5 Conclusion

In this work, the simulation of magnetic field densities under perfect coil alignment and in the case of misalignment along the X axis is carried out by Ansys Maxwell for two coil designs; square and circular.

The simulation results are not the same for both designs; the square design is less sensitive to misalignment compared to the circular design which has a large coupling coefficient k and mutual inductance M in the case of perfect alignment.

The results show also that the magnetic field is maximal at the center of the transmitter coil and decreases along the Z axis, also the field density decreases as it moves from the center to the outside of the coil. This reduction leads to a reduction in the value of the coupling coefficient and the mutual induction and consequently a reduction in the efficiency of the WPT system.

In future works, other designs will be compared in order to determine which design ensures efficient power transfer and at the same time is less affected by the misalignment problem.

References

1. Vatsala, A., Ahmad Aqueel, M.S.A., Chaban, R.C.: Efficiency enhancement of wireless charging for electric vehicles through reduction of coil misalignement. In: IEEE Transportation Electrification Conference and Expo (ITEC) (2017). https://doi.org/10.1109/ITEC.2017.7993241
2. Ahmad, A., Alam, M.S.: Magnectic analysis of copper coil power pad with ferrite core for wireless charging appplication. Trans. Electr. Electron. Mater. (2018). https://doi.org/10.1007/s42341-018-00091-6
3. Niculae, D., Iordache, M., Stanculescu, M., Bobaru, M.L., Deleanu, S.: A review of electric vehicles charging technologies stationary and dynamique. In: 11th International Symposium on Advanced Topics in Electrical Engineering (ATEE) (2019).https://doi.org/10.1109/ATTE.2019.8724943
4. Mostak, M., Choi, S., Islam, M.Z., Kwak, S,, Baek, J.: Core design and optimization for better misalignment tolerance and higher range of wireless charging of PHEV. IEEE Trans. Transp. Electrif. (2017). https://doi.org/10.1109/TTE.2017.2663662
5. Kopaei, M.K., Mehdizadeh, A., Ranasinghe, D.C., Al-Sarawi, S.: A novel hybrid approach for wireless powering of biomedical implants. In: IEEE Eighth International Conference on Intelligent Sensors, Sensor Networks and Information Processing (2013). https://doi.org/10.1109/ISSNIP.2013.529833
6. Olvitz, L., Vinko, D., Svedek, T.: Wireless power transfer for mobile phone charging device. In: IEEE Proceeding of the 35th International Convention MTPRO, 21–25 May 2012 (2012)

7. Boscanio, V., Pellitteri, F., Capponi, G., Rosa, R.L.: A wireless battery charger architecture for consumer electronics. In: IEEE Second International Conference on Consumer Electronics-Berlin (ICCE-berlin) (2012)
8. Jiang, H., Brazis, P., Tabaddor, M., Bablo, J.: Safety Considerations of wireless charger for electric vehicles-a review paper. In: IEEE Symposium on Product Compliance Engineering Proceedings (2012). https://doi.org/10.1109/ISPCE.2012.6398288
9. Lee, S., et al.: The optimal design of high-powered power supply modules for wireless power transferred train. In: IEEE Electrical Systems for Aircraft, Railway and Ship Propulsion (2012). https://doi.org/10.1109/ESARS.2012.6387396
10. Melki, R., Moslem, B.: Optimizing the design parameters of a wireless power transfer system for maximizing power transfer efficiency-a simulation study. In: Third International Conference on Technological Advances in Electrical, Electronics and Computer Engineering (TAEECE) (2015). https://doi.org/10.1109/TAEECE.2015.7113640
11. Yang, Y., Cui, J., Cui, X..: Design and analysis of magnetic coils for optimizing the coupling coefficient in an electric vehicle wireless power transfer system. Energies (2020). https://doi.org/10.3390/en13164143
12. Ahmad, A., Alam, M.S., Rafat, Y., Shariff, S.: Designing and demonstration of misalignment reduction for wireless charging of autonomous electric vehicle. eTransportation **4** (2020). https://doi.org/10.1006/j.etran.2020.100052
13. Liu, D., Hu, H., Georgakopulos, S.V.: Misalignment sensitivity of strongly coupled wireless power transfer system. IEEE Trans. Power Electron. (2016)
14. Villa, J.L., Sallan, J., Osorio, J.F.S., Llombart, A.: High-misalignment tolerant compensation topology for ICPT systems. IEEE Trans. Industr. Electron. (2012). https://doi.org/10.1109/TIE.2011.2161055
15. Uddin, M.K., Mekhilef, S., Ramasamy, G.: Compact wireless IPT system using a modified voltage-fed multi-resonant class EF2 inverter. J. Power Electron. **18**(1), 277–288 (2018)
16. Mohamed, N., et al.: A new wireless charging system for electric vehicles using two receiver coils. Eng. J. (2022). https://doi.org/10.1016/j.asej202108012

Optimal Resource Allocation in Mobile Edge Computing Based on Virtual Machine Migration

Sara Maftah[1(✉)], Mohamed EL Ghmary[2], Hamid El Bouabidi[1], Mohamed Amnai[1], and Ali Ouacha[3]

[1] Department of Computer Science, Faculty of Science, Ibn Tofaïl University, Kenitra, Morocco
{sara.maftah,hamid.elbouabidi,mohamed.amnai}@uit.ac.ma
[2] Department of Computer Science, FSDM, Sidi Mohamed Ben Abdellah University, Fez, Morocco
mohamed.elghmary@usmba.ac.ma
[3] Department of Computer Science, Mohammed V University, Rabat, Morocco
ali.ouacha@um5.ac.ma

Abstract. Computation offloading in Mobile Edge Computing offers limited computing resources compared to cloud infrastructures. However, the heterogeneity of Internet of Things devices and the emergence of the 5G imposed an extra overload on these resources to efficiently exploit them for a better offloading performance. Mobile Edge Computing servers host virtual machines and are deployed in the Radio Access Network. In this paper, we focus on mobility constraints, the availability of computing resources and its impact on successfully processing a load of tasks from multiple mobile users, we also identified a potential solution, which is virtual machine migration, a theme that is getting attention by scholars and researchers.

Keywords: resource allocation · virtual machine migration · mobile edge computing · computation offloading · internet of things · MANET

1 Introduction

The need for computing resources is growing due to the evolution in both computing and electronics fields, a thing that led to the rise of Internet of Things (IoT) which refers to a global intelligent network of wireless devices, objects and sensors. There are many aspects of connected living such as smart homes, smart vehicles and smart healthcare, the latter is one of the main fields that are considered to be important. Smart healthcare according too [1] includes several components such as IoT, edge computing and artificial intelligence. Latency-sensitive IoT applications benefit from Edge Computing since it enables data

M. Lazaar et al. (Eds.): BDIoT 2022, LNNS 625, pp. 575–585, 2023.
https://doi.org/10.1007/978-3-031-28387-1_48

processing and analyzing on premises and uses Artificial Intelligence to improve the ability of IoT devices in perceiving, learning, reasoning and behaving [2].

Due to the heterogeneity of IoT nowadays, the network interconnecting these devices is called Mobile Ad hoc Network (MANET) which supports the mobility of the connected devices and the ability of their self-organization. Moreover, mobile phones and wireless devices have exceeded the number of computers, its use has become intensive and urged the evolution of mobile network architectures to reach the deployment of 5G. Even though the components of IoT devices are in a constant evolution, their computing resources are limited as well as their battery life. Therefore, the main issues related to a MANET environment are power consumption and availability of resources. The IoT device's mobility can greatly impact on the number of failed tasks in either a MEC or cloud environment. One of the suggested solutions to such constraints is Mobile Edge Computing (MEC) which is the deployment of Edge Servers in the Radio Access Network (RAN), to make resources available within the edge of the network and close to the end user. Opting for this paradigm extends the services offered by Cloud Computing since MEC infrastructure alone has fewer processing capabilities [3].

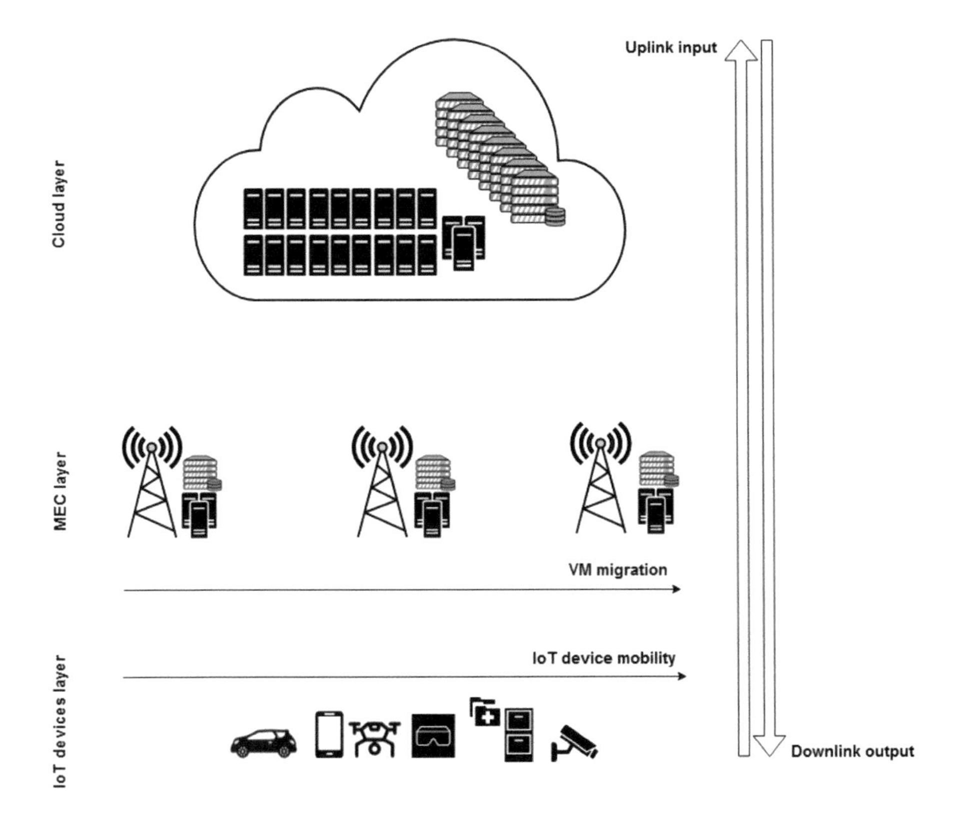

Fig. 1. IoT-MEC-CLOUD architecture

Figure 1 is an illustration of both MEC and Cloud Computing architecture. The IoT devices layer represents the various wireless devices and sensors, their mobility and need for computing resources urged the deployment of Mobile Edge Computing servers in the RAN, in this layer, the workload of tasks gets offloaded and processed, it is also the layer where many optimization services can be deployed, such as virtual machine migration. The MEC layer is an extension of the Cloud Layer which was the main infrastructure on which end users relied on to get their tasks processed and executed.

Task processing requires a certain environment which is represented by virtual machines, since the appearance of Cloud Computing, the deployment of computing paradigms, whether on a Cloud or on the Edge, relies mainly on virtualization. Therefore, in this paper, we will cover virtualization in MEC environment and study the migration of virtual machines from a congested node in the MEC to other more resourceful nodes within a set of hosts in a data-center.

After introducing the theme in Sect. 1, the rest of the paper is organized as follows. In Sect. 2, we introduced virtualization on Mobile Edge Computing and virtual machine migration. Section 3 is a literature review and brief virtual machine migration strategy is elaborated in Sect. 4. The simulation and results are discussed in Sect. 5. Finally, the paper is concluded in Sect. 6.

2 Virtualization in Mobile Edge Computing and Virtual Machine Migration

Before getting into virtualization, it is known that heterogeneous access networks have become more dominant and are expected to support vertical handover mechanism, in which users can maintain the connections with service continuity even when they switch from a network to a different one. For this purpose, IEEE 802.21 standard facilitates Media Independent Handovers (MIH) which is an add-on module that was provided by the NIST (National Institute of Standards and Technology) by providing higher layer mobility management functions with common service primitives for all technologies. The concept of a handover is keeping the connectivity of the mobile device and its service continuity even with the constant movement. Virtual machine migration represents also the idea of keeping an application running even with the mobility of the IoT device and the need to transfer the workload to another resourceful host.

Looking for ways to enhance the efficiency of resource utilization, virtualization is the backbone of Cloud and Edge environments. Unlike Cloud servers, Edge Computing impose many constraints to deploy a virtualized environment, such as resource-limited edge devices and mobility of IoT devices. Therefore, decision factors on virtualization selection in Edge Computing is based on virtualization features, richness of devices and IoT application requirements. Virtualization in MEC is divided into three distinct techniques, hypervisor-based, container-based and Uni-kernel virtualization. The main difference between these techniques is operating system sharing, and are compared according to their image size, where Uni-kernels are smaller than containers, followed by hypervisor-based virtualization.

- Virtual machines are useful when there is enough power and storage resources, every virtual machine runs its own operating system and is fully isolated, hence more secure. It also supports running multiple applications at a time. However, their performance are very limited, the booting time takes several minutes, hypervisor interactions cause a high overhead. Moreover, virtual machines are considered a heavyweight solution.
- Containers on the other side are lightweight with a superior performance where booting time is in milliseconds and the overhead is low, they are useful when power and storage resources are needed. Moreover, they operate on a shared kernel operating system, which makes the infrastructure less secure.
- Uni-kernels are also useful when there's a high demand on power and storage, they are even more lightweight than containers with a superior performance using a single specialized machine image which minimizes the overhead. The booting time is in milliseconds and relies on a fully isolated infrastructure. However, it only supports a single application at a time, unlike containers and virtual machines.

Virtual Machine placement is one of the techniques that ensure an optimized performance in Edge computing, it depends on four main objective functions: Execution time, energy consumption, resource utilization and cost [4].

Moreover, virtualization techniques in Mobile Edge Computing deployments enable the use of services and network intelligence in the RAN by providing computing capabilities. However, these resources may vary from a virtual machine to another, and the service on demand feature enables the end user to upgrade his resources if they were no longer enough to process his workload.

Moving a virtual machine from a physical machine to another one is referred to by virtual machine migration, and is categorized into two techniques: Live and offline (non-live) migration [5].

In offline migration, the VM is shutdown on the source host, transferred to its destination host to be restarted. This technique results in a dramatic downtime. Meanwhile, live migration allows the VM to be transferred from a host to another while running and with the least disruption of services, its process is divided into a pre-copy phase in which an appropriate target server is selected, its resources are reserved to begin transferring data from the virtual machine's memory to the selected target server and a post-copy phase that consists of sending a captured state of the virtual machine's resources to its destination in the target server followed by suspending the virtual machine on the source node and resuming it on the destination [5,6].

Therefore an efficient migration strategy does not only seek moving a VM from a physical host to another as fast as possible, but it also needs to minimize the consequences of the existing constraints. These constraints depend majorly on performance metrics such as total migration time, the VM downtime, total network traffic and computation overhead [7].

3 Related Work

The main feature that categorizes MEC is the ability to offload computation near the mobile device based on the idea of enabling the workload's migration from centralized clouds to Mobile Edge nodes deployed within the RAN, which enhances majorly the response time for delay-sensitive and heavy intense applications by relying [8], it also plays a major role in optimizing energy consumption [9,10].

However, the mobility of mobile devices can cause a significant performance degradation and migrating a certain service to another edge, when a mobile user moves from an area to another, can resolve the interruption. The authors in [11] presented live migration as a concept of service migration as well as hosting methods in MEC architectures such as virtual machines and containers. The service migration can be implemented either in virtual machines or containers, and the difference between the two methods according to [12] is that virtual machines emulate the hardware and kernel of the operating system, while in containers, the operating system kernel and hardware are shared with its host machine.

The authors of [3] improved the Quality of Service (QoS) of applications by studying the TCP throughput estimation while using a VM migration method that enables migrating a VM from a congested node to another resourceful one in the MEC within the RAN, either to a cellular base station or to a Wireless Local Area Network (WLAN) Access Point which is used as an Edge Sever.

Similarly, a better Quality of Experience (QoE) and lower latency are guaranteed when a VM can be migrated between nodes within the RAN according to device mobility [13]. Knowing that VM's data size is large and the process of migrating a VM from an edge node to another may burden the network traffic, the authors in [14] proposed an approach based on artificial intelligence to transfer a VM by subdividing it into smaller pieces, this technique is referred to as an ant colony algorithm.

The purpose of VM migration resides on providing edge services with the lowest latency as the device user moves across different places. Moreover, for a better service delivery, VM live migration is used to keep the services running normally during the migration, [15] aimed to maximize the average QoS during the VM migration while considering time constraint. However, VM migration can result a downtime period before the migration process is completed, [16] compared VM migration methods with container migration, which is considered lightweight but not mature enough to support service migration in MEC.

4 Virtual Machine Migration Strategy

Virtual machine migration can be deployed by optimizing an initial VM allocation to hosts using certain migration strategies. The first step resides on allocating a virtual machine to the simulated host where the configuration parameters of the virtual machine and tasks are defined and submitted to request the allocation of a VM to a desired host. After that, the initiation of VM migration, where the data-centers are responsible for the execution of the workloads.

The VM's processing metrics are updated throughout the simulation to track the progress of each running cycle. There's also the use of a method that accepts a list of active virtual machines and executes a sequence of instructions as shown below:

- Retrieve the list of the over-utilized hosts according to the deployed VM allocation policy.
- Backup the current VM-to-host mapping and identify the list of virtual machines to be migrated from the congested host.
- The process of finding a new resourceful host excludes the overloaded ones and sets in a temporary migration map a new vm placement and allocation mapping.
- For energy efficiency purposes, the underutilized hosts are also required to mark the virtual machines they host for migration, and these hosts, which will be turned off are also added to the excluded hosts list.
- The last step is the preparation of the migration map, aand the current allocation is restored to continue the simulation.

5 Simulation and Results

A number of experiments have been conducted using an Edge Computing simulating tool. In a multi-user environment, starting from 2 end users to reach 50 end users, the tasks are offloaded either to the MEC or Cloud servers, accordingly with the required resources to process these tasks.

A simulation starts with an initialization phase by loading the configuration files needed to run the scenario and by generating a random set of tasks that will be processed sequentially according to their start time. For each task, there will be a decision on whether to process it on the MEC or Cloud servers. This decision will be based on multiple factors.

Each generated task has certain stored information used while deciding task generation pattern via a task list, the properties of these tasks are:

- startTime: represents the beginning of a certain task's processing.
- length: is the parameter that defines the length of a certain task, in other words the task's data size.
- inputFileSize: is the task's input data size.
- outputFileSize: is the task's output data size.
- taskType: in our simulation, we have four different task types, each type refers to an example of IoT applications that requires computation offloading (Augmented reality, Health application, Heavy computation application and Infotainment application).
- pesNumber: refers to the responsible Processing Elements (PEs) of the task.
- mobileDeviceId: is the Mobile device identification that requested the task processing.

The simulation parameters are presented in the following Table 1:

Table 1. Simulation parameters

Parameter	Value
Task's usage percentage (%)	30, 20, 20, 30
Number of cores per host	16
Number of cores per VM	2
Number of mobile users	From 0 to 50
Number of virtual machines per server	4
Processing scenario	Cloud and Edge server
Offloading policy	Based on the WAN bandwidth and the MEC server's utilization
Processing speed per host (MIPS)	80 000
Processing speed per VM (MIPS)	10 000

We experiment at first using low computing resources and then we upgraded to high computing resources. We then compared the network delay in the two infrastructures, to find that even with the availability of computing and storage resources, network delay can be significant. Figure 2 is a graphical representation of the obtained results.

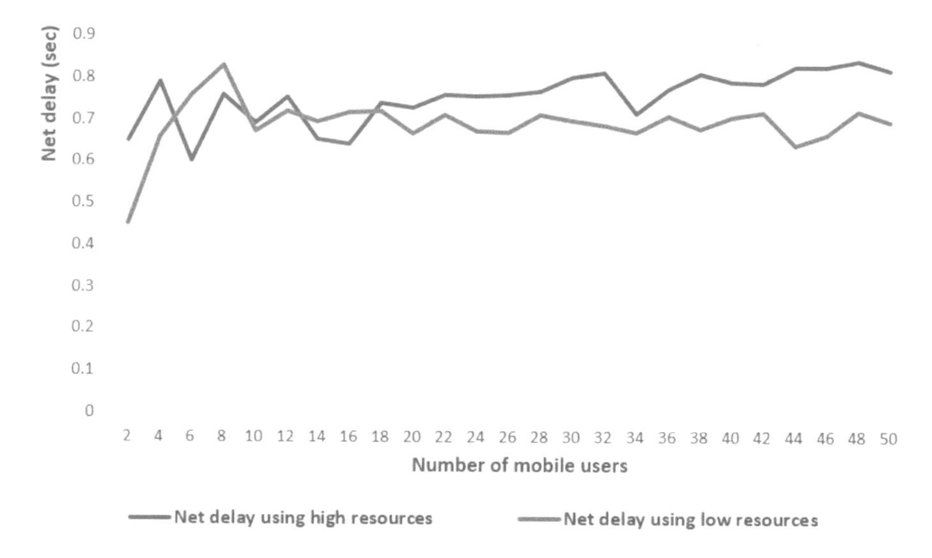

Fig. 2. Network delay in both high and low computing resources infrastructures

In the case of having low computing resources, the rate of success and failure fluctuates when only 2 end users offload their workload and when we reach 50 end users. Figure 3 shows the evolution of both the success to process the tasks and the failure to do so. The success shifts from 98.36% to 25.06% as the lowest rate per 50 end users, and failure varies between 1.63% and 74.93%.

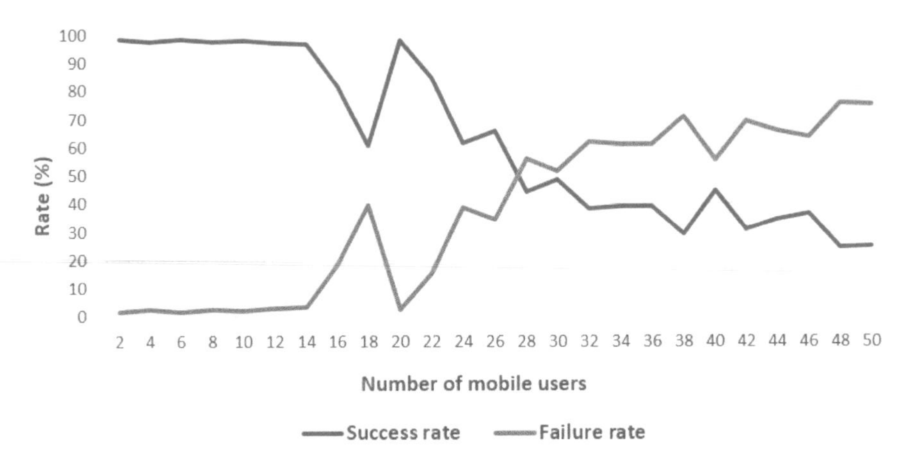

Fig. 3. Success and failure rate in a low computing resources infrastructure

It is also important to note that at the beginning of the simulation, the failure rate was 100% due to mobility and it reached the lowest rate of 8.54% at the end as shown in Fig. 4. Moreover, with the limited computing resources we had, the failure rate was at its highest by 91.45% at the end of the simulation.

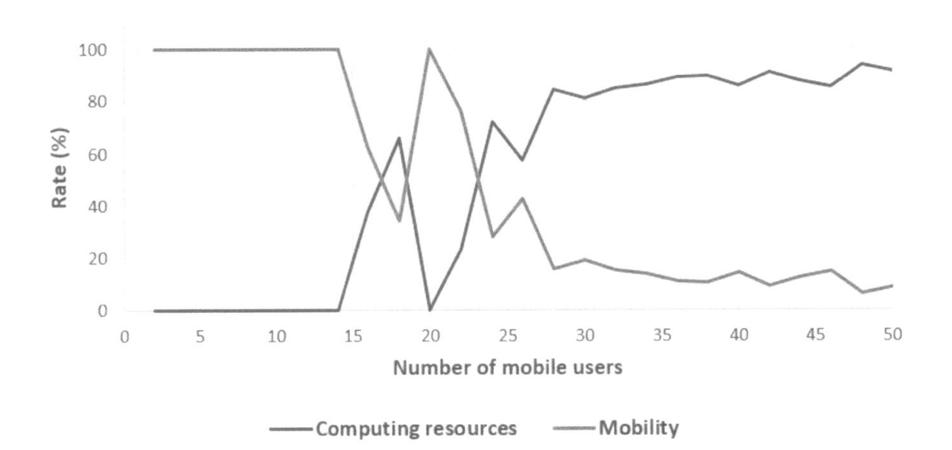

Fig. 4. Failure rate due to low computing resources and mobility

Meanwhile, when upgrading to more powerful resources, the success rate of offloading and processing the tasks varies between 98% and 99%, while the failure percentage did not exceed 1.67%.

Failure rate was successfully reduced to below 1.11% after providing the needed computing resources, to be left with a low rate of failure due to mobility.

The previous Figs. 2, 3 and 4 represent some of the obstacles that can get in the way of an optimal task processing on a Mobile Edge Computing infrastructure, such as the network delay due to the availability of computing resources and its impact on the success or failure of processing the set of tasks. We also identified the end user's mobility as another cause for task processing failure.

Another set of experiments were conducted to determine the use of virtual machine migration and its impact on energy consumption by the data-centers. We first adopted a scenario where data-centers relied on Dynamic Voltage and Frequency Scaling (DVFS) only, which is an energy saving technique of various applications by controlling voltage and frequency in computers.

Afterwards we ran a simulation where data-centers that allocates virtual machines based a static CPU utilization Threshold and adopts a VM selection policy based on the minimum migration time by following the virtual machine migration strategy we described in Sect. 4.

We used a data-center containing 50 hosts and started the simulation by having 20 running virtual machines until reaching 100 running virtual machines.

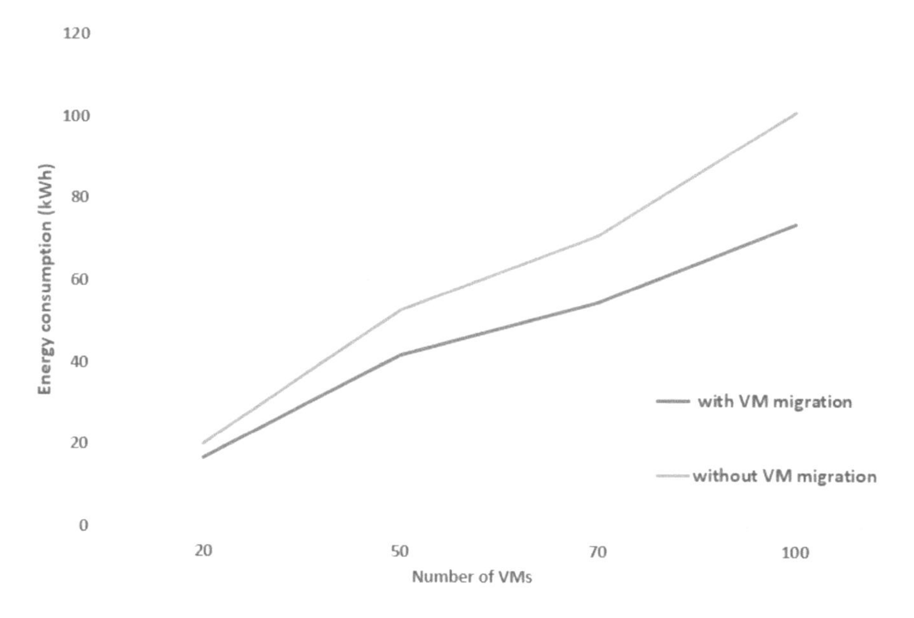

Fig. 5. Energy consumption by a data-center containing 50 hosts

Figure 5 is a graphical representation of energy consumption by a data-center that has 50 hosts ready to be used. The energy consumed when virtual machines are allocated accordingly to a static threshold and a minimum migration time VM selection policy is lower than the energy consumed when there was no optimization strategy of the VM allocation and selection.

6 Conclusion and Future Work

The purpose of this paper and the conducted simulations, is to examine in a first place the multiple existing constraints in an environment where multiple mobile users have a workload to be offloaded and processed, we analyzed the importance of high computing resources availability for a significant processing success rate. We also examined the impact of the end user's mobility, which have an impact on the offloading process, to which we found, according to a literature review that virtual machine migration is one of the solutions that can be deployed to optimize the migration time, thus the computation offloading time.

In order to study the impact brought by adopting a virtual machine migration strategy, the experiments we conducted have shown that the energy consumed my the data-center containing the machines that host the VMs is reduced compared to an experiment where there's no VM allocation nor selection optimization policy.

The results have shown that the delay caused by mobility and the inability to process the workload can be avoided by providing enough resources. Therefore, the study will be extended in a future work to deploy a virtual machine migration strategy that will optimize processing intensive heavy tasks and lift the burden from resource-limited IoT devices.

References

1. Alshehri, F., Muhammad, G.: A comprehensive survey of the Internet of Things (IoT) and AI-based smart healthcare. IEEE Access **9**, 3660–3678 (2020)
2. Zhang, J., Tao, D.: Empowering things with intelligence: a survey of the progress, challenges, and opportunities in artificial intelligence of things. IEEE Internet Things J. **8**, 7789–7817 (2020)
3. Kikuchi, J., Wu, C., Ji, Y., Murase, T.: VM migration in mobile edge computing for QoS improvement with wireless multi-hop access networks. In: Proceedings of The 12th International Conference on Ubiquitous Information Management And Communication, pp. 1–8 (2018)
4. Mansouri, Y., Babar, M.: A review of edge computing: features and resource virtualization. J. Parallel Distrib. Comput. **150**, 155–183 (2021)
5. Singh, G., Gupta, P.: A review on migration techniques and challenges in live virtual machine migration. In: 2016 5th International Conference on Reliability, Infocom Technologies and Optimization (Trends And Future Directions)(ICRITO), pp. 542–546 (2016)
6. Ahmad, R., Gani, A., Hamid, S., Shiraz, M., Yousafzai, A., Xia, F.: A survey on virtual machine migration and server consolidation frameworks for cloud data centers. J. Netw. Comput. Appl. **52**, 11–25 (2015)
7. Zhang, F., Liu, G., Fu, X., Yahyapour, R.: A survey on virtual machine migration: challenges, techniques, and open issues. IEEE Commun. Surv. Tutorials **20**, 1206–1243 (2018)
8. Lu, W., Meng, X., Guo, G.: Fast service migration method based on virtual machine technology for MEC. IEEE Internet Things J. **6**, 4344–4354 (2018)

9. El Ghmary, M., Malki, M., Hmimz, Y., Chanyour, T.: Energy and computational resources optimization in a mobile edge computing node. In: 2018 9th International Symposium on Signal, Image, Video and Communications (ISIVC), pp. 323–328 (2018)
10. Hmimz, Y., Chanyour, T., El Ghmary, M., Malik, M.: Energy efficient and devices priority aware computation offloading to a mobile edge computing server. In: 2019 5th International Conference on Optimization and Applications (ICOA), pp. 1–6 (2019)
11. Wang, S., Xu, J., Zhang, N., Liu, Y.: A survey on service migration in mobile edge computing. IEEE Access **6**, 23511–23528 (2018)
12. Gillani, K., Lee, J.: Comparison of linux virtual machines and containers for a service migration in 5G multi-access edge computing. ICT Express **6**, 1–2 (2020)
13. Lopes, M., Higashino, W., Capretz, M., Bittencourt, L.: MyiFogSim: a simulator for virtual machine migration in fog computing. In: Companion Proceedings Of The10th International Conference on Utility and Cloud Computing, pp. 47–52 (2017)
14. Ouacha, A., El Ghmary, M.: Virtual machine migration in MEC based artificial intelligence technique. IAES Int. J. Artif. Intell. **10**, 244 (2021)
15. Yang, L., Yang, D., Cao, J., Sahni, Y., Xu, X.: QoS guaranteed resource allocation for live virtual machine migration in edge clouds. IEEE Access **8**, 78441–78451 (2020)
16. Doan, T., et al.: Containers vs virtual machines: Choosing the right virtualization technology for mobile edge cloud. In: 2019 IEEE 2nd 5G World Forum (5GWF), pp. 46–52 (2019)

Natural Convection Heat Transfer of a Nanofluid Inside a Square Cavity with an Irregular Zig-Zag Wall Using the Global RBFs Collocation Method

Youssef Es-Sabry[1(✉)], Elmiloud Chaabelasri[1], and Najim Salhi[2]

[1] LPTPME, Faculty of Sciences, University Mohammed I, 60000 Oujda, Morocco
essabryoussef@gmail.com

[2] Faculty of Sciences, University Mohammed I, 60000 Oujda, Morocco

Abstract. The natural convection heat transfer in a square cavity with a zig-zag wall filled by the Al_2O_3-water nanofluid is investigated numerically in this study. We examined the situation for Rayleigh numbers ranging between 10^3–10^6. The governing equations are presented in dimensionless form. The spatial derivatives of the governing equations are approximated by the global RBFs approximation method. An explicit first-order Euler scheme is used to reach the steady-state solution. The obtained results were validated by comparing them with those of a previously published work. Streamlines, isotherms, and average Nusselt number are plotted to represent the effects of velocity, temperature and heat transfer in the cavity.

Keywords: Natural convection · heat transfer · Global RBFs collocation method · Nanofluid

1 Introduction

The heat transfer by convection is a very important phenomena in many industries fields; it appears in a range of engineering applications [1–3], such as electronic cooling devices, building heating and cooling systems and heat exchangers. However, the improvement of heat transfer is limited by the low thermal conductivity of pure fluids. In order to overcome this limitation, the nanoparticles are submerged in the pure fluid, which tends to change the fluid's thermophysical characteristics and improves the rate of heat transfer. The used of nanofluid can be observed in many engendering application, for example, heat exchangers, cooling of electronics, engine cooling, nuclear reactor safety, hyperthermia, biomedicine, vehicle thermal management, and many others [4]. The determination of the thermophysical properties of nanofluids has been the subject of several research studies. Bianco et al. [5]. discussed the fundamentals of this field, such as techniques for synthesizing nanoparticles, as well as the analysis and measurement of thermophysical properties of nanofluids.

M. Lazaar et al. (Eds.): BDIoT 2022, LNNS 625, pp. 586–596, 2023.
https://doi.org/10.1007/978-3-031-28387-1_49

Many researchers have studied heat transfer by natural convection in rectangular enclosures filled with nanofluid. Among them, Khanafer et al. [7] studied the natural heat transfer enhancement in a two-dimensional cavity using nanofluids for a limited range of Grashof numbers and nanoparticle volume fraction. Different models were compared with each other based on the physical properties of the nanofluids. Studied the natural convection heat transfer of a water-based nanofluid in a square tilted cavity. The results confirm that the average heat transfer decreases with increasing longitude of the heating element. For a smaller inclination angle, the average heat transfer rate starts to decrease as the heater longitude becomes larger. The natural convection heat transfer of a nanofluid in a square enclosure was investigated by Mahmoodi and Sebdani [8]. They considered the effect of volume fraction of the Cu nanoparticles and Rayleigh number. The results demonstrated that with increasing the nanoparticles concentration, the average Nusselt number enhances.

Wavy geometries are used in many engineering systems as a means of enhancing the transport performance. For example, The mixed convection and heat transfer in a square cavity with sinusoidal walls containing a heated rotating cylinder was numerically investigated by Jassim et al. [6]. They noticed that the heat transfer is significantly improved by the use of sinusoidal walls. A numerical analysis of the natural convection inside a two-dimensional cavity with a wavy right vertical wall has been investigated by Dalal and Das [9]. The results showed that the presence of undulation in the right wall affected the flow and heat transfer characteristics.

The present work investigates the natural convection heat transfer inside a square cavity with a zig-zag left vertical wall filled with the Al_2O_3-water nanofluid. Numerical treatment of the proposed model is being made via the global RBFs collocation method. The main objective of this study is to investigate the natural convection and heat transfer of the Al_2O_3-water nanofluid in a zig-zag wall cavity.

2 Problem Statement

Figure 1 depicts the geometrical configuration of the two-dimensional square cavity that we will study in this work. The left vertical wall of the cavity is zigzagged, while all other walls are plane surfaces. The cavity is filled by the suspension of Al_2O_3 nanoparticles in water. The hot temperature, T_h, is applied to the left zigzag wall, while the cold temperature, T_c, is applied to right wall. The other horizontal walls are adiabatic.

3 Mathematical Model

In this study, the Al_2O_3-water nanofluid is considered to be steady, Newtonian, incompressible, laminar, and it is simulated as a single-phase homogeneous fluid with no radiation effects and no viscous dissipation. The gravitational force acts in the vertical direction, and the thermophysical properties of the nanofluid are

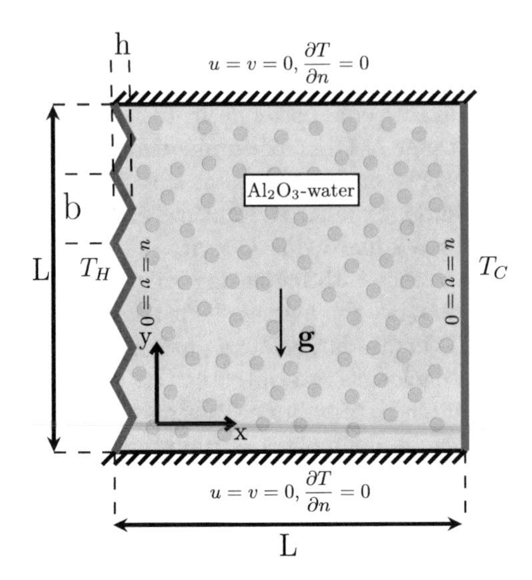

Fig. 1. The physical model and coordinate system

considered constant except for the density, which varies according to the Boussinesq model. Using these assumptions, the continuity, momentum and energy equations in dimensionless form for the nanofluid can be written as follows:

Continuity equation:

$$\nabla.\mathrm{V} = 0 \tag{1}$$

Momentum equation:

$$\mathrm{V}.\nabla\mathrm{V} = -\frac{\rho_\mathrm{f}}{\rho_\mathrm{nf}}\nabla\mathrm{P} + \mathrm{Pr}\frac{\rho_\mathrm{f}}{\rho_\mathrm{nf}}\frac{\mu_\mathrm{nf}}{\mu_\mathrm{f}}\nabla^2\mathrm{V} + \frac{(\rho\beta)_\mathrm{nf}}{\rho_\mathrm{nf}\beta_\mathrm{f}}\mathrm{RaPr}\theta\mathbf{g} \tag{2}$$

Energy equation:

$$\mathrm{V}.\nabla\theta = \frac{(\rho \mathrm{c_p})_\mathrm{nf}}{(\rho \mathrm{c_p})_\mathrm{f}}\frac{\mathrm{k_{nf}}}{\mathrm{k_f}}\nabla^2\theta \tag{3}$$

where **g** signifies the gravitational acceleration vector, V is the dimensionless velocity vector, ρ_{nf} is the density of the nanofluid, p is the dimensionless pressure, μ_nf is the dynamic viscosity of the nanofluid, β_{nf} is the thermal expansion coefficient of the nanofluid, $\mathrm{k_{nf}}$ is the nanofluid thermal conductivity, $(\rho \mathrm{c_p})_\mathrm{nf}$ is the nanofluid heat capacitance, and θ is the dimensionless temperature.

The dimensionless initial and boundary conditions of Eqs. (1)–(3) are presented as follows:

$$\mathrm{U} = \mathrm{V} = \theta = 0 \quad \text{at} \quad \mathrm{t} = 0 \quad \text{every where.} \tag{4}$$

$$\begin{cases} & \text{for} \quad \mathrm{t} > 0 \\ \mathrm{U} = \mathrm{V} = \frac{\partial\theta}{\partial \mathrm{Y}} = 0, & \text{On the adiabatic horizontal walls} \\ \mathrm{U} = \mathrm{V} = \theta = 0, & \text{On the right vertical wall} \\ \mathrm{U} = \mathrm{V} = 0, \theta = 1, & \text{On the left vertical wall} \end{cases} \tag{5}$$

The thermophysical properties of the nanofluid are given in this study as follows [11]:

$$\begin{aligned}
\rho_{\mathrm{nf}} &= (1-\phi)\rho_{\mathrm{f}} + \phi\rho_{\mathrm{p}} \\
(\rho\beta)_{\mathrm{nf}} &= (1-\phi)(\rho\beta)_{\mathrm{f}} + \phi(\rho\beta)_{\mathrm{p}} \\
\alpha_{\mathrm{nf}} &= \frac{\mathrm{k_{nf}}}{(\rho \mathrm{c_p})_{\mathrm{nf}}} \\
(\rho \mathrm{c_p})_{\mathrm{nf}} &= (1-\phi)(\rho \mathrm{c_p})_{\mathrm{f}} + \phi(\rho \mathrm{c_p})_{\mathrm{p}} \\
\frac{\mu_{\mathrm{nf}}}{\mu_{\mathrm{f}}} &= \frac{1}{(1-\phi)^{2.5}} \\
\frac{\mathrm{k_{nf}}}{\mathrm{k_f}} &= \frac{\mathrm{k_p} + 2\mathrm{k_f} - 2\phi(\mathrm{k_f} - \mathrm{k_p})}{\mathrm{k_p} + 2\mathrm{k_f} + \phi(\mathrm{k_f} - \mathrm{k_p})}
\end{aligned} \tag{6}$$

where ϕ is the volume fraction of the nanofluid and subscripts f, nf, and p indicating the base fluid, nanofluid, and solid, respectively.

The local Nusselt number $\mathrm{Nu_l}$ along the hot wall is defined as:

$$\mathrm{Nu_l} = -\frac{\mathrm{k_{nf}}}{\mathrm{k_f}} \left(\frac{\partial \theta}{\partial X}\right)_{X=0} \tag{7}$$

and the average Nusselt number is obtained by integrating the local Nusselt number along with the hot wall as follows:

$$\mathrm{Nu_{avg}} = \int_0^1 \mathrm{Nu_l dY} \tag{8}$$

4 RBF Based Approximation

Considered a spatial domain $\Omega \subset \mathbb{R}^2$. Let u(x) be an unknown function in $\mathbb{R}^2$, and we wish to approximate it on a set of distinct points x= $[\mathrm{x}_1, \mathrm{x}_2, ..., \mathrm{x}_N]$ in Ω. In the RBFs global approximation method, the function u(x) on the set x of N scattered points can be approximated as follows:

$$\mathrm{u(x)} = \sum_{\mathrm{j}=1}^{\mathrm{N}} \lambda_{\mathrm{j}} \phi(||\mathrm{x} - \mathrm{x_j}||) \tag{9}$$

where x denote a point in $\mathbb{R}^2$, $\mathrm{x_j}$ are the center points of the RBFs, ϕ is a radial basic function, $||.||$ is the Euclidean norm, and λ_{j} are a coefficients to be determined. Applying Eq. (9) to each point in Ω allows us to write the system of linear equations.

$$\mathrm{u} = \Phi\lambda \tag{10}$$

The expansion coefficients are then determined as follows:

$$\lambda = \Phi^{-1}\mathrm{u} \tag{11}$$

where the elements of Φ are $\Phi_{ij} = \phi(||x_i - x_j||)$, $\lambda = [\lambda_1, \lambda_2, ..., \lambda_N]^T$, and $u = [u_1, u_2, ..., u_N]^T$ is the vector of approximate solutions.

To approximate the space derivatives of the interpolant (9) the derivation from (9)–(11) can be used. Then

$$\frac{\partial^k u}{\partial x^k} = \sum_{j=1}^{N} \lambda_j \frac{\partial^k \phi}{\partial x^k} (||x_i - x_j||) \tag{12}$$

where index k represents the k_{th} derivative of u(x). By applying similar procedure for all points in Ω, we will get a system of equations

$$\frac{\partial^k u}{\partial x^k} = \Phi_x^k \Phi^{-1} u \tag{13}$$

where $\frac{\partial^k u}{\partial x^k}$ is a vector containing the k_{th} derivatives of u, and Φ_x^k is a matrix of (NxN) elements contain the k_{th} derivatives of RBFs.

5 Discretized Form of Navier-Stokes Equations Using RBFs Approximation

Let us consider that $F_c^{(n)}(U(x, y))$, $G_c^{(n)}(U(x, y))$, $F_v^{(n)}(U(x, y))$, and $G_v^{(n)}(U(x, y))$ are the convective, and viscous fluxes along the x and y direction at time $t = n\Delta t$, where $U(x, y) = [P, u, v, \theta]^T$ represents the vector of conservative variables. Using the same procedure as in the previous section, the space derivatives with respect to x and y of the fluxes can be approximated as:

$$\begin{aligned} \frac{\partial F_c^{(n)}}{\partial x}(U(x, y)) &= \Phi_x \Phi^{-1} F_c^{(n)} \\ \frac{\partial G_c^{(n)}}{\partial y}(U(x, y)) &= \Phi_y \Phi^{-1} G_c^{(n)} \\ \frac{\partial^2 F_v^{(n)}}{\partial x^2} U(x, y)) &= \Phi_{xx} \Phi^{-1} F_v^{(n)} \\ \frac{\partial^2 G_v^{(n)}}{\partial y^2}(U(x, y)) &= \Phi_{yy} \Phi^{-1} G_v^{(n)} \end{aligned} \tag{14}$$

In this study, an explicit first-order Euler scheme is used to reach the steady-state solution. The infinitely smooth multiquadric radial basis function is used to approximate the differentiation operators, this function is defined as:

$$\phi(||(x - x_j)||) = \sqrt{1 + \epsilon^2 r^2} \tag{15}$$

where ϵ represents the shape parameter, and r is the Euclidean norm.

6 Grid Testing and Validation

To ensure that the numerical results are independent of the number of grid points, an extensive grid testing procedure on the average Nusselt number was studied for Ra $= 10^4$ and 1% of the Al_2O_3-water nanofluid. The calculated results for each number of grid points are shown in Table 3. The table shows that the average Nusselt number is similar, with a difference of less than 1% for the number of grid points 2000, 3000, and 4000. We therefore choose the number of grid points 3000 for its better ratio between simulation time and accuracy of results.

In order to verify the validity and accuracy of the numerical method, the present model for Al_2O_3-water nanofluid inside of the square cavity with regular walls was compared with the results obtained by Kuznik et al. [10] in Table 2. The present results are in good agreement with those presented in the reference [10] (Table 1).

Table 1. Grid dependency of the local Nusselt number at the heated wall with Ra $= 10^4$, and $\phi = 0.1$

Number of points	$(Nu)_{avg}$
1000	2.3034
2000	2.3723
3000	2.3776
4000	2.3779

Table 2. Comparison between the present results and other works for the average Nusselt number

Ra	Present work	Kuznik et all
10^3	1.3446	1.117
10^4	2.3779	2.246
10^5	5.0012	4.518
10^6	9.7344	8.792

7 Results and Discussion

In this section, we present the results obtained for the natural convection heat transfer problem inside a square cavity with and without the zig-zag wall for a Rayleigh number in the rang 10^3–10^6. The volume fraction of the nanoparticles is fixed at $\phi = 0.1$. To show the effect of the zig-zag wall on the heat transfer, the values of the average Nusselt number are calculated for different values of Ra. The thermophysical properties of the pure fluid and solid Al_2O_3 phases shown in Table 3.

Table 3. Thermophysical properties of different nanoparticles and the base fluid

Property	Fluid phase (Water)	Al_2O_3(nanoparticles)
$c_p(J/kgK)$	4179	765
$\rho(kg/m^3)$	997.1	3970
$k(W/mK)$	0.6	40
$\beta(K^{-1})$	21×10^{-5}	0.85×10^{-5}
$\mu(kg/ms)$	10518×10^{-6}	–

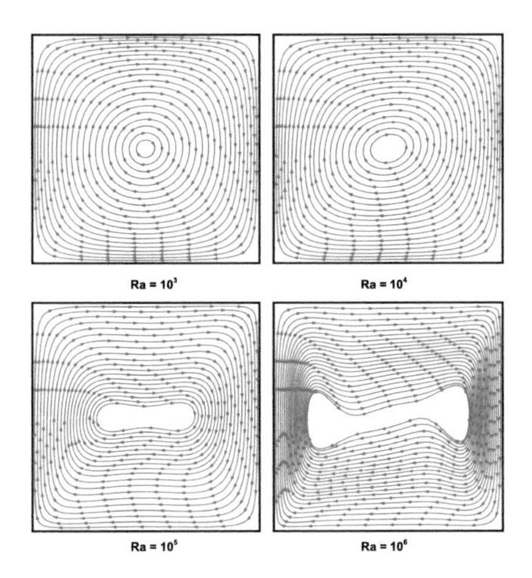

Fig. 2. Streamlines for different Rayleigh values

Figure 2 shows the effect of the Rayleigh number on the streamlines inside the cavity with regular walls. For a low Rayleigh number $Ra = 10^3$, we observe that there is a globally symmetrical vortex inside the cavity, and we also observe there is good contact between the streamlines and the heated wall, which makes cooling easy. A small counterclockwise deformation of the vortex is observed when $Ra = 10^4$. In the case where the Rayleigh number becomes very important ($Ra = 10^5$ or $Ra = 10^6$), we notice that the streamlines extend toward the corner of the cavity, due to the higher buoyancy forces. The symmetrical vortex disappears and the main vortex moves toward the source wall. This causes dense streamlines near the heated wall, resulting in improved cooling. Similarly, compressed streamlines can be observed near the right wall. The temperature lines inside the regular cavity for a Rayleigh number in the rang 10^3–10^6 are shown in Fig. (3). We notice that for low Ra $= 10^3$, the isotherms of the clod wall are nearly straight, parallel to the wall, and close to the left wall. Therefore, the isotherms are weakly deformed so that heat is transferred by conduction between the hot and cold walls. At horizontal walls, all isotherms are vertical. From Ra $=$ 10^4 we observe a deformation of the isotherms at the top of the cavity so that the transfer changes its mechanism from conduction to convection. We also notice that the deformation of the isotherms increases with the increase of the number of Rayleigh number and that the isotherms become parallel to the horizontal walls especially in the meddle of the cavity. On the left side of the cavity, we observe thermal stratification, and the thermal boundary layers become thinner with increasing Rayleigh number. The natural convection becomes more important, and the heat transfer by thermal convection is goes from the hot wall to the cold wall through the upper wall.

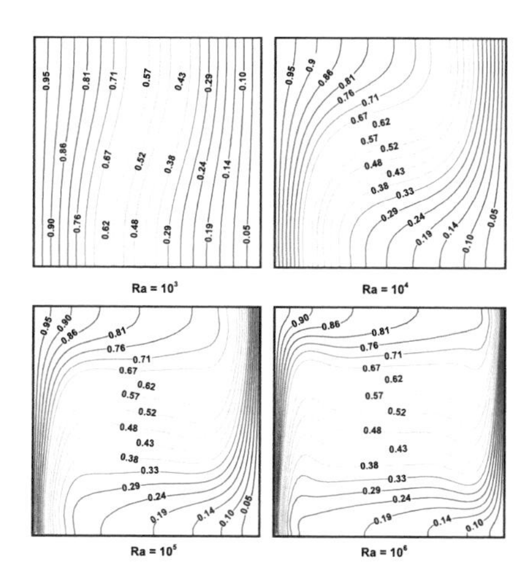

Fig. 3. Isotherms for different Rayleigh values

The effects of different values of the Rayleigh number on the Streamlines in the zigzag-wall cavity are shown in the Fig. 4. At low Rayleigh number, the flow is characterized by a symmetric vortex inside the cavity. We also observe a good contact between the fluid and the cavity walls, except in the corners of the zigzag wall. The increasing in Rayleigh number tends to increase the intensity of the streamlines and the number of vertexes.

Figure 5 shows the effects of different values of the Rayleigh number on the isotherms inside the cavity with a zig-zag wall. As can be seen, for the low Ra $= 10^3$, near the heated zig-zag wall the isotherm lines appear with a very low density, while in the middle of the cavity the isotherm lines takes an almost vertical line. When the Rayleigh number is increased (Ra $> 10^3$) the intensity of the isotherm lines near the heated zig-zag wall increases, while in the middle of the cavity the isotherm lines becomes parallel to the horizontal walls.

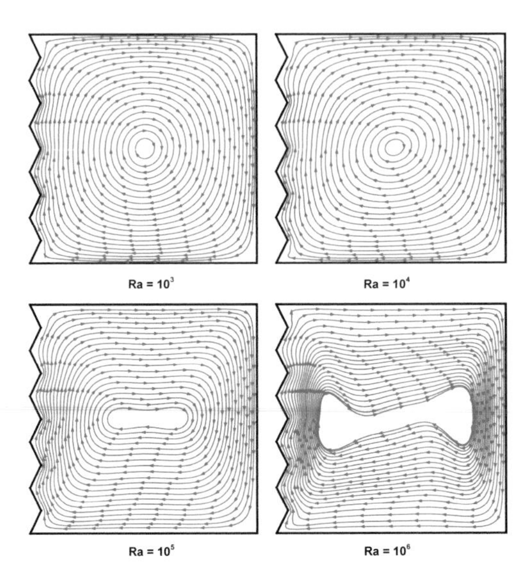

Fig. 4. Streamlines for different Rayleigh values

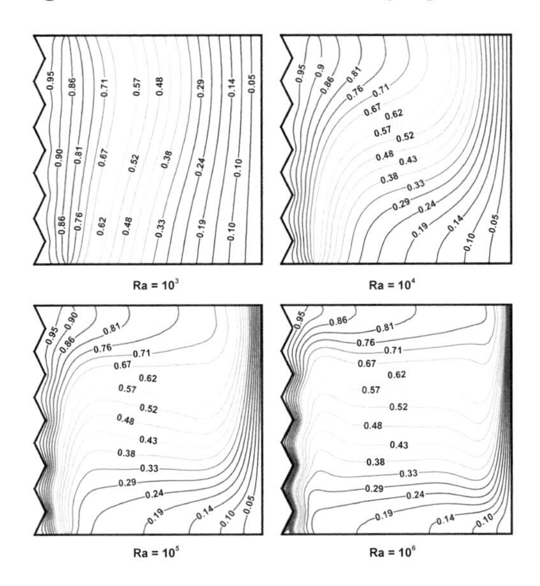

Fig. 5. Streamlines for different Rayleigh values

the changes in average Nusselt number respect to Rayleigh number are shown in Fig. 6. The figure shows a reduction in the average value of Nusselt number Nu for the case of the cavity with irregular zig-zag wall compared to the regular cavity especially at the high value of Ra. At low Rayleigh number ($Ra = 10^3$) the irregular zig-zag wall has no significant influence in the average Nusselt number.

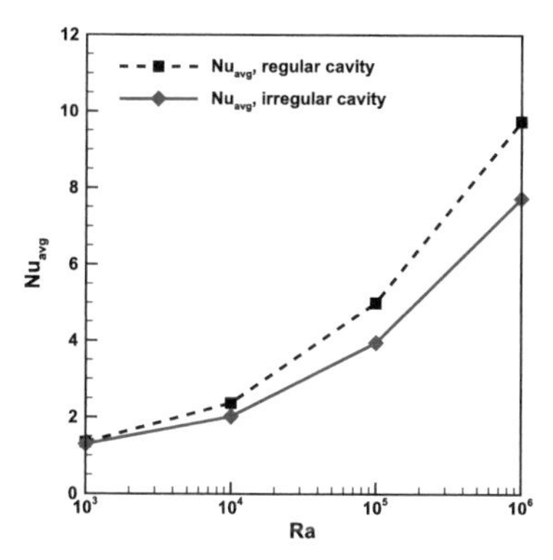

Fig. 6. The effect of Rayleigh number on average Nusselt number at $\phi = 0.1$

8 Conclusion

In this study, the natural convection heat transfer inside a square zigzag-wall cavity filled with the nanofluid has considered in this work. The Al_2O_3-water nanofluid was considered as a working fluid. The governing equations are solved numerically using the global RBFs collocation method. The structure of the flows and their thermal behavior are studied at a concentration of 1% of nanoparticles for a wide range of Rayleigh number. A comparative study with previously published work is being made to validate the proposed model. As a conclusion, We can see that the heat transfer rate decreases when we change the regular left vertical wall with the irregular zig-zag wall.

References

1. Yang, A.S., Wen, C.Y., Juan, Y.H., Su, Y.M., Wu, J.H.: Using the central ventilation shaft design within public buildings for natural aeration enhancement. Appl. Therm. Eng. **70**(1), 219–230 (2014). https://doi.org/10.1016/j.applthermaleng.2014.05.017
2. Fouladi, K., Wemhoff, A., Silva-Llanca, L., Abbasi, K., Ortega, A.: Optimization of data center cooling efficiency using reduced order flow modeling within a flow network modeling approach. Appl. Therm. Eng. **124**, 929–939 (2017). https://doi.org/10.1016/j.applthermaleng.2017.06.057
3. Pourhoseini, S.H., Naghizadeh, N., Hoseinzadeh, H.: Effect of silver-water nanofluid on heat transfer performance of a plate heat exchanger: an experimental and theoretical study. Powder Technol. **332**, 279–286 (2018)
4. Shenoy, A., Sheremet, M., Pop, I.: Convective flow and heat transfer from wavy surfaces viscous fluids, porous media, and nanofluids (2016). I
5. Bianco, V., Manca, O., Nardini, S., Vafai, K.: Heat Transfer Enhancement with Nanofluids. Taylor and Francis, pp. 1–458 (2015)
6. Jassim, H.M., Ali, F.H., Mahdi, Q.A., Hadi, N.J.: Effect of parallel and orthogonal sinusoidal walls on mixed convection inside square enclosure containing rotating cylinder. In: International Conference on Mechanical and Aerospace Engineering, pp. 365–370 (2017). https://doi.org/10.1109/ICMAE.2017.8038673.

7. Khanafer, K., Vafai, K., Lightstone, M.: Buoyancy-driven heat transfer enhancement in a two-dimensional enclosure utilizing nanofluids. Int. J. Heat Mass Transf. **46**(19), 3639–3653 (2003)
8. Mahmoodi, M., Sebdani, S.M.: Natural convection in a square cavity containing a nanofluid and an adiabatic square block at the center. Superlattices Microstruct. **52**(2), 261–275 (2012)
9. Dalal, A., Das, M.K.: Natural convection in a cavity with a wavy wall heated from below and uniformly cooled from the top and both sides. J. Heat Transfer **128**(7), 717–725 (2006). https://doi.org/10.1115/1.2194044
10. Kuznik, F., Vareilles, J., Rusaouen, G., Krauss, G.: Lattice Boltzmann method, double-population, non-uniform mesh, natural convection, square cavity, laminar and transitional flows. Int. J. Heat Mass Transfer 49(3), 727–739 (2006). (0017-9310). https://doi.org/10.1016/j.ijheatmasstransfer.2005.07.046.
11. Hashemi, H., Namazian, Z., Mehryan, S.: Cu-water micropolar nanofluid natural convection within a porous enclosure with heat generation. J. Mol. Liq. **236**, 48–60 (2017). https://doi.org/10.1016/j.molliq.2017.04.001

Author Index

M. Lazaar et al. (Eds.): BDIoT 2022, LNNS 625, pp. 597–599, 2023.
https://doi.org/10.1007/978-3-031-28387-1

Printed in the United States
by Baker & Taylor Publisher Services